Low Energy Electron Collisions in Gases

WILEY SERIES IN PLASMA PHYSICS

SANBORN C. BROWN   ADVISORY EDITOR
RESEARCH LABORATORY OF ELECTRONICS
MASSACHUSETTS INSTITUTE OF TECHNOLOGY

GILARDINI · LOW ENERGY ELECTRON COLLISIONS IN GASES
MCDANIEL AND MASON · THE MOBILITY AND DIFFUSION OF IONS IN GASES
MITCHNER AND KRUGER · PARTIALLY IONIZED GASES IN PLASMA PHYSICS
TIDMAN AND KRALL · SHOCK WAVES IN COLLISIONLESS PLASMAS
NASSER · FUNDAMENTALS OF GASEOUS IONIZATION IN PLASMA ELECTRONICS
LICHTENBERG · PHASE-SPACE DYNAMICS OF PARTICLES
BROWN · INTRODUCTION TO ELECTRICAL DISCHARGES IN GASES
BEKEFI · RADIATION PROCESSES IN PLASMA
MACDONALD · MICROWAVE BREAKDOWN IN GASES
HEALD AND WHARTON · PLASMA DIAGNOSTICS WITH MICROWAVES
MCDANIEL · COLLISION PHENOMENA IN IONIZED GASES

LOW ENERGY ELECTRON COLLISIONS IN GASES:

SWARM AND PLASMA METHODS APPLIED TO THEIR STUDY

ALDO GILARDINI

Head of the Research and Development Division,
Selenia, S.p.A., Rome, Italy,
Professor of Aerospace Electronics,
University of Rome, Italy.

A WILEY-INTERSCIENCE PUBLICATION

JOHN WILEY & SONS
New York · London · Sydney · Toronto

Library of Congress Cataloging in Publication Data:

Gilardini, Aldo, 1926–
Low energy electron collisions in gases.

(Wiley series in plasma physics)
"A Wiley-Interscience publication."
Bibliography: p.
1. Electrons—Scattering. 2. Gases.
3. Plasma dynamics. I. Title.
QC721.G44   539.7'2112   72-10291
ISBN 0-471-29889-1

Printed in the United States of America

10–9 8 7 6 5 4 3 2 1

TO MY PARENTS,
MY WIFE, FRANCA,
AND OUR CHILDREN,
ALESSANDRO, RICCARDO,
AND MARIA SAVERIA

Preface

Many scientific books, but not as many as one would expect, have titles that fully describe the subject the author plans to discuss. For other books this ideal seems impossible to realize if the title has to be short and comprehensive, as practice requires. I believe that the title of this monographic book, in spite of the addition of a clarifying subtitle, falls into the second class, and that only after a detailed explanation of the title and the addition of a few basic restrictions will the covered field and the limits become clear to the reader.

By rephrasing the title with more words, I may say that in this book I intend to consider the behavior of low-energy electron swarms in gases and their properties, which depend on the collisions of electrons with the gas molecules, ionized or not, the purpose being to derive from these properties the values of the parameters that characterize collisions.

Electron swarms are defined here as ensembles of electrons with a rather wide distribution of energy; so defined, their physical significance is quite broad. In fact, when the gas in which the swarm is formed contributes positive ions in such a number as to ensure an approximate electrical neutrality of the resulting medium, it may be called a plasma and swarm properties become plasma properties.

Only swarms of low-energy electrons, frequently referred to in the literature as slow electrons, are considered in this monograph. More precisely, this means that swarm electrons have energies well below those required for excitation of electronic states and for dissociation and ionization of the gas molecules. In a slightly different but equivalent way, one can say that in the swarm the average electron energy is sufficiently low so that the swarm properties to be investigated can be determined neglecting the few electrons in the high-energy tail of the energy distribution, which excite and ionize molecules. Consistently with this restriction, it has been also assumed that in general the gas is only slightly ionized.

According to the definition given above of the subject to be treated in this book, only those swarm and plasma properties that depend on electron-neutral and electron-ion collisions will be considered. "Collision" can be a difficult term to define, because we are accustomed to regard it as a discrete binary process, which almost instantaneously changes the straight-line, unperturbed motions of the incoming particles into new straight-line motions of the separating particles. A collision of this kind is thus the limiting case of an interaction, which can be represented by a finite potential of finite range, much smaller than the average distance between the particles. Actually, interactions between charged particles take place also at much longer ranges, and in this case the simultaneous effect of a great number of weak interactions must be considered. Formally the resulting scattering can still be resolved into the frame of a collision theory by regarding it as a sum of equivalent discrete binary interactions; this approach is followed also in this book, so that the same family of collision parameters can be used to describe all interaction processes.

On the other hand, in this monograph only conservative, that is, transformation-less, collisions will be considered. They can be defined as those in which no colliding particle disappears, as is the case on the contrary in such processes as recombination, attachment, and detachment.

The numerous restrictions that have been placed on the subject still leave a very large area to be covered. In fact, the properties to be treated fall under various, well-known names: electron mobility, electron diffusion, electrical conductivity, radiation temperature, thermal conductivity, and energy relaxation time. The possible presence of a uniform static magnetic field is considered throughout. Applied electric fields, however, are supposed to be weak enough to ensure that the swarm always remains a low-energy one.

The swarm and plasma properties are derived and written as functions of fundamental collision parameters, like the electron collision cross section for momentum transfer, the fractional energy loss per collision of the electron with the various existing particles, and the electron collision cross sections for the excitation of molecular rotational and vibrational states. Comparison of the theoretical expressions with the corresponding experimental data, taken under convenient, well-defined conditions, yields much information concerning the values of these parameters and their dependence on the electron velocity.

This topic is discussed at length in the book. Actually the main purpose for which this monograph was conceived was to collect the great quantity of recent important work in the field of low-energy electron collision processes, and thus to provide the specialist with a unified source of modern physical and technical data. For this reason ideas have been freely drawn from the many excellent articles that have appeared in the literature during the last twenty years on individual topics regarding electron collision phenomena in gases.

The significance of the content of the book is, however, larger. In fact, having succeeded in determining from experiment the basic collision parameters for the most common gases, one can use the general formulas again in order to estimate swarm or plasma properties under many conditions of practical interest. In addition to diagnostics, typical applications could be the evaluation of the radio-frequency conductivity of the ionosphere and of the transmission losses in microwave plasma devices.

In the discussion of the collision parameters and in the solution of the basic transport equations, quite general and synthetic mathematical expressions and procedures could have been adopted. However, I have rejected such a highly synthetic approach in favor of using expressions and developments that are formally simpler and easier to follow, even if they require more space. I hope that as a result the reader can follow the mathematical passages and can understand the physical significance of the formulas more easily and quickly than would be possible with a more compact presentation.

Reference to microwave techniques will be made frequently. No account of the standard microwave techniques is, however, included. For this the reader is referred to the many excellent books on microwaves that have been published in the last few years. Some knowledge of plasma physics, of vacuum techniques, and of electromagnetic theory is also required, but this background is certainly familiar to the interested readers.

Finally I want to emphasize that discussions on the theoretical evaluation of collision parameters using quantum physics have been deliberately avoided to any significant extent. Quantum theory predicts a few basic features of low-energy electron collisions, but for most gases it cannot yet be used to calculate precisely all parameters and their dependence on electron energy. Therefore, one must always resort to the experimental data. It is hoped that the collection of these data, of their derivation methods, and of a large number of expressions of the swarm and plasma properties, where they appear, will be very useful both to the plasma physicist and to the gaseous electronics engineer.

I am very grateful to Professors W. P. Allis and S. C. Brown, who inspired my interest in the field of plasma physics, and to the numerous physicists and friends with whom I had penetrating and helpful discussions on the topics of this book during the many years of its preparation. Professor Brown's review of my manuscript and his comments are also greatly appreciated.

Teaching courses of Aerospace Electronics at the University of Rome for many years has constantly renewed my interest on several of the major topics, which I treat in the book; my sincere thanks for this valuable opportunity are due to the Dean of the Aerospace Engineering School, Professor L. Broglis.

I express my sincere gratitude to Professor C. Calosi and to the present management of Selenia, for creating in the company a very stimulating atmosphere and for securing all the necessary facilities for documentation, drafting, and typing. My heartfelt thanks are also due to Miss Elena Salzano Gonnella, who always took good care of the manuscript, carefully and patiently typing its several revisions.

ALDO GILARDINI

Rome, Italy
June 1972

Contents

Introduction

Observation of the scattering of electrons from a monoenergetic beam in a field-free, low-pressure (below $\sim 10^{-2}$ torr) chamber is the classical and most natural method for determining collision parameters of electrons with atoms and molecules. However, this method fails to provide consistent and accurate data when the electron beam energies are below a few tenths of an electron volt. The reasons are well known [Crompton (1969)]: gas pressures below 10^{-2} torr are difficult to measure accurately, and, furthermore, unknown pressure gradients can exist within the chamber; truly monoenergetic beams are difficult to generate and to measure below ~ 1 eV, when momentum selection techniques and retarding potential measurements are used; contact potential differences, as well as stray electric and magnetic fields, may easily influence the outcome of the experiment.

For these reasons the determination of electron collision parameters below ~ 1 eV has become the realm of electron swarm methods, which do not suffer from the drawbacks of electron beam methods. In fact, a number of transport parameters of the electron swarm in a gas can be measured, and this is conveniently done at gas pressures above ~ 1 torr, so to ensure many collision processes per electron in the chamber. Furthermore, it is no longer necessary to measure single electron energies, since these methods allow for an electron energy distribution, to be inferred or determined from the experiment itself. Under these conditions measurements of the electron transport properties, either of pure electron swarms or of mixed quasi-neutral electron-ion swarms (plasmas), are generally easy and reliable, and, in spite of the complexity of the analytical inversion procedures, quantitative accurate information on the electron collision parameters can be derived. A loss of resolution, typical of all methods based on measurements of average quantities, is expected, but this limitation is much less serious than the systematic and unknown errors that may influence beam experiments. Furthermore, measurements of plasma properties at high electron and ion concentrations

1

offer unique possibilities for determining electron-ion and electron-electron collision parameters.

Therefore this book, which is concerned with the application of swarm techniques to the study of electron collision processes in gases, will treat elastic collision processes of low-energy electrons with neutral atoms or molecules, ions, and other electrons, and inelastic collisions of low-energy electrons involving the excitation of molecular rotational and vibrational states. Higher-energy processes, like dissociation, electronic excitation, and ionization of atoms and molecules by electron collisions, will be omitted from the discussion. Although swarm methods can be applied also to derive collision parameters at energies above 1 eV, here their reliability drops because of the usually large number of parameters involved at the same time; on the other hand, at these energies beam methods do not show the difficulties prevailing at the lower energies and therefore provide generally satisfactory results.

Furthermore, we shall omit from our discussion the determination of recombination, attachment, and detachment coefficients, namely, those coefficients that characterize collision processes in which the number of interacting particles changes. Actually swarm methods are used also for these determinations; but the emphasis is different, and in the writer's opinion this subject is sufficiently autonomous to be worth an independent and adequately extensive treatment. The interested reader is referred to the reviews of the subject by Prasad and Craggs (1962), Bates and Dalgarno (1962), Biondi (1963), McDaniel (1964), Hasted (1964), and Phelps (1969). The presence of recombination and attachment processes will be included in the treatment here when necessary, and mention will be made of the determination of the corresponding coefficients when this possibility follows from the application of the methods discussed.

It should also be noted that, in addition to the use of electron beam and swarm methods, another possibility exists for determining the very-low-energy characteristics of elastic electron collisions. This is the spectroscopic method of Fermi, based on measuring the frequency shift that the highest terms of the principal series of the spectra of alkali-metal atoms display when the gas under investigation is added in small but sufficient concentrations. The method is indirect, however, being based on the theoretical assumptions of the quantum-mechanical description of the elastic collision process, and is of limited value, insofar as it provides only the low-energy limit of the elastic collision cross section. A discussion of the method and of the results thereby obtained, which can be compared with those reported in this book, can be found in Massey and Burhop (1952) and in O'Malley (1963); significant, more recent work in the field has been performed by the Russian school [see, e.g., Mazing and Vrublevskaya (1966)].

To the writer's knowledge there is no other book with the same field coverage as the present one; however, most of the subjects that will be discussed here have been properly and authoritatively treated in earlier books and reviews. As the most significant of these surveys, which were published during the last ten years, the following should be mentioned: the books by McDaniel (1964), Hasted (1964), Shkarofsky, Johnston, and Bachynski (1966), and Delcroix (1968); the data collection by Laborie, Rocard, and Rees (1968); and the reviews by Huxley and Crompton (1962), Biondi (1963, 1968), Varney and Fisher (1968), Phelps (1968), and Crompton (1969).

Chapter 1

General Characteristics of Fundamental Electron Collision Processes

The properties of low-energy swarms of electrons in gases depend, among many various physical factors, on the values and functional relations of a few basic electron collision parameters. Inversely, it is possible to determine these parameters by measuring a suitable set of swarm properties. This first chapter is devoted to defining these fundamental collision quantities, which will be used throughout the book, and whose experimental determination is one of the main purposes of the studies reported in later chapters.

Most of the definitions and concepts to be introduced in this chapter have been in common use for a very long time, and appropriate discussions of their significance are found in all standard texts on collision phenomena in gases [e.g., Massey and Burhop (1952), McDaniel (1964), Hasted (1964)]. It is intended that the classical presentation given in these books be used to complement the present exposition.

For our purpose we consider first an elementary, ideal experiment of electron scattering, particularly appropriate for simple definitions of the electron collision parameters. Among the quantities that will be introduced in this way are various collision probabilities and frequencies (total, for momentum transfer, for energy exchange), gas particle cross sections, and energy loss factors. Some fundamental formulas related to these quantities will also be derived in this chapter, insofar as they follow from the discussion of the general characteristics of the collision experiment described. Reference to a few theoretical developments is included at the end, to provide formulas useful later on for analysis of the experimental results.

As explained in the Introduction, throughout this book we are interested in determining the collision parameters of electrons with energies below

4

about 1 eV, that is, well below the threshold energies required for dissociation, electronic state excitation, and ionization of gas atoms and molecules in their normal, thermal states. Therefore, in the scattering experiment to be discussed in this chapter, we shall always assume that electron energies are below these thresholds.

1.1 A BASIC ELECTRON SCATTERING EXPERIMENT

We consider here an elementary ideal experiment, which is particularly convenient for the definition of all fundamental electron collision parameters in gases. The experiment we shall now describe does not concern the overall behavior of an electron swarm, but makes use of scattering measurements on individual low-energy electrons for the later determination of the time-average properties of their motion in gases.

In discussing this experiment, we shall state that measurements indicate, under properly defined conditions, certain behaviors (e.g., a linear dependence on the gas pressure). This means, not that an actual experiment, exactly in the described form, has been performed with the mentioned result, but only that this would have been the outcome, so far as can be predicted on the basis of the results of other experiments actually performed, which involved derived quantities of the same physical processes.

The elementary ideal experiment consists of the following. An electron of known velocity **v**, whose energy is below a few electron volts, is shut (Fig. 1.1.1) at right angles into a slab of gas with given uniform density,

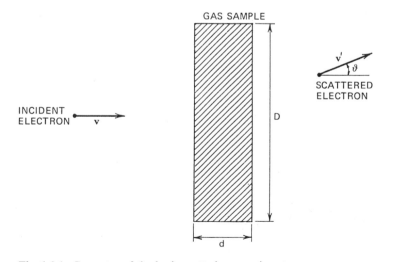

Fig. 1.1.1. Geometry of the basic scattering experiment.

composition, and temperature (no material walls and no external fields are present in this ideal situation). The slab is transversally bounded and, for convenience of discussion, is assumed to have the shape of a circular disk. The gas particles, which in the most general case are molecules, either in the normal state or excited or ionized, and free electrons, may perturb the motion of the incoming electron. This perturbation is interpreted as the effect of collisions of the electron with the gas particles.

The electron will finally leave the gas sample with a velocity $\mathbf{v'}$. The experiment is repeated a very large number of times with the same initial conditions. The value of $\mathbf{v'}$ is recorded for each case, so that the probability distribution of the final velocity $\mathbf{v'}$ can be determined according to the usual statistical definitions. The cases in which the electron is trapped in the gas by attachment or recombination with positive ions are excluded from the analysis.

To use these experimental results for the definition of the basic electron collision parameters, in which we are interested, the size of the slab and the composition and density of the gas must be properly chosen. In this respect it is clear that the gas density and composition in our experiment have simply to be chosen consistently with the range of values of these same quantities in gas discharge and plasma problems of usual interest. The size of the disk-shaped gas sample, on the other hand, cannot be decided so easily, and this problem must be more thoroughly discussed.

First we note that, because of the uncertainty principle of quantum mechanics, an intrinsic uncertainty will affect all $\mathbf{v}$ and $\mathbf{v'}$ measurements; therefore the effect of collisions on the electron velocity is determined only within this quantum-mechanical limit. It is necessary to discuss this point further, since in our cases it may actually impose significant limitations on the gas sample size.

The uncertainty principle, in one of its most familiar formulations, states that, for simultaneous measurements, the following limitation exists on the achievable minimum errors:

$$\delta v \cdot \delta l \simeq \frac{\hbar}{m} = 1.16 \times 10^{-4} \, \text{m}^2 \, \text{sec}^{-1} \qquad (1.1.1)$$

Here δv is the minimum error in the measurement of any component of velocity, δl is the minimum error in the measurement of its canonically conjugate spatial coordinate, m is the electron mass, and $2\pi\hbar$ is Planck's constant. We shall choose δv on the basis of the order of magnitude of the smallest significant electron velocity variations that are expected to take place as the result of collisions. For this, reference to average conditions is necessarily made, since the velocity or energy variations of the electron for each individual collision may cover a large range, including zero, and only their average values can be of interest for a statistical analysis of electron swarm properties.

To determine the smallest velocity variations, we consider the case of elastic collisions, since in inelastic collisions larger variations necessarily occur. Here the term "elastic collisions" means, as is customary in the physics of collisions, those types of collisions that involve only the exchange of kinetic energy, whereas "inelastic collisions" are those in which some kinetic energy is converted into internal energy of one or both of the colliding particles.

For an elastic collision between an incoming electron and a gas particle of molecular mass M ($\gg m$) having a much lower kinetic energy, the classical theory of collisions predicts that the electron velocity (modulus) variation is, on the average, of the order of the fraction m/M of the velocity itself (see also Section 1.3). It seems reasonable to assume that the experimental data are significant, from the point of view of deriving collision parameters useful for a theory of electron swarm behavior, only when the uncertainty δv is sufficiently smaller than this velocity change; a figure like 1% may be adequate. Hence our requirement for accuracy becomes as follows: $\delta v \simeq (m/M)(v/100)$. The lowest electron energies that are still sufficiently larger than the molecular thermal kinetic energies to justify a definite interest in the accurate determination of the electron energy variations in collisions are in the range 0.1–1 eV ($v = 2 - 6 \times 10^5$ m sec^{-1}); for most molecules δv is thus in the 1–10^{-2} m sec^{-1} range.

According to eq. (1.1.1), the corresponding range of uncertainty in the determination of the spatial coordinate is 10^{-4} to 10^{-2} m. This means that the electron position is known only within a sphere having a diameter of this size. In order to be sure that the incident electron will pass through the disk-shaped slab of gas, it is necessary that the transverse dimensions of the gas sample be sufficiently larger than this electron uncertainty diameter. Then, when the conditions and requirements of the experiment are as described above, the diameter of the gas sample disk must be at least in the centimeters range.

The slab thickness d, on the other hand, must be chosen small enough so that the average velocity $\bar{v}'$ of the outgoing electrons, evaluated from the experimental data, will be very close to the incoming velocity v. The reason for this requirement can be easily understood on the basis that we want to define electron collision parameters which describe relatively elementary processes, mostly in order to use them thereafter in determining time-average properties of motion of electrons in gases. Mathematically, our goal will be to write specific differential equations describing this motion; that will be possible only if we can specify how vanishingly small changes of the average electron motion parameters take place as functions of the various motion and gas conditions. These statements will become clearer further on in this book.

Let us now divide the $\mathbf{v}'$ space into elementary cells, whose dimensions are significantly larger than the quantum uncertainties, but sufficiently small to provide a detailed description of the electron motion variations. "A detailed description" means here that conditions have been achieved such that, by changing the size of the cell, the mean number of electrons scattered into it is directly proportional to the cell volume.

In general, when the size of the gas sample has been properly chosen according to the preceding discussion, the $\mathbf{v}'$ elementary cells here required will exist. In fact, a resolution in velocity better than one hundredth of the mean velocity variation for each elastic electron-molecule collision at electron energies as low as 0.1–1 eV is usually quite adequate for the fine, practically continuous description of any significant electron collision process, either elastic or inelastic.

Having specified the main conditions to be satisfied in the experiment, we must now discuss how to handle the results. From what we have already stated, it appears clear that for our purpose the basic experiment is repeated a large number of times with constant initial conditions (same velocity $\mathbf{v}$ and same gas sample), in order to determine the fraction of the total number of outgoing electrons that is scattered into each $\mathbf{v}'$ elementary cell, and to interpret this as the scattering probability from $\mathbf{v}$ to $\mathbf{v}'$ (with the exclusion of the cell for which $\mathbf{v}' = \mathbf{v}$). If this fraction is divided by the cell volume, a scattering probability density is obtained, which may practically be regarded as a continuous function of $\mathbf{v}$ and $\mathbf{v}'$.

The geometrical symmetry conditions of the experiment indicate that this density may conveniently, and without loss of generality, be considered dependent on $\mathbf{v}$ and $\mathbf{v}'$, simply as a function of v, v', and ϑ, where ϑ is the angle between the scattered and the incident electron directions. Moreover, the scattering probability density may depend in general on the gas density, composition, and temperature and on the size of the disk-shaped gas sample (thickness d, cross-section diameter D). The scattering probability density will be indicated by $S(v, v', \vartheta, n_i, T_g, d, D)$, where n_i is the density of the ith type of particles and T_g the temperature of the gas sample.

The lack of definition of the scattering probability density function over the elementary cell volume at $\mathbf{v}' = \mathbf{v}$, a point which we have deliberately excluded, is removed by assuming that the function S is given there by a reasonable, continuous extrapolation from the nearby points, where it is well determined. Whereas no physical significance may be attributed to this value in any case, the assumption is convenient in order to write simpler mathematical relationships for the collision integral properties. Actually, we shall find that, in the evaluation of all the collision parameters of importance for the determination of electron swarm transport properties, S has to be multiplied by functions whose local values at $\mathbf{v'} = \mathbf{v}$ are zero, so that this point does not count at all in the determination of the parameters.

The quantities d and D are the only parameters that appear as independent variables in S and take into account the macroscopic geometry of the experiment. If a description of the collision phenomena of general physical validity is desired, the scattering probability density S must be replaced by other quantities in which d and D do not appear. For this purpose we must now establish the dependence of S on d and D.

Let us consider first a scattering experiment performed in a gas where the particles are neutral molecules. Examination of the results would show that, over large ranges of the most significant variables (in particular v, n_i, and d), satisfying all previously stated requirements, the following facts are observed. For the time being, and for reasons to be explained later, the region $\vartheta \simeq \pi/2$ is excluded from these considerations.

a. For most electrons the outgoing velocity $\mathbf{v'}$ is equal to the incoming velocity $\mathbf{v}$, within the errors predicted by the uncertainty principle. In other words, this means that the required small perturbation on the average electron motion results from the average between the unscattered path of most electrons and the scattered motion of a few of them, large perturbations being allowed as a common event of this kind of scattering.

b. The scattering probability density S is proportional to the total number of gas particles, that is, to $\Sigma_i n_i$, provided the relative composition is not changed.

c. The scattering probability density S is proportional to the gas sample thickness d and independent of its cross-section diameter D.

Moreover, it is observed that conditions b and c are most accurately satisfied at the lowest densities and at the lowest sample thicknesses d.

These same results can be very easily predicted on the basis of the classical theory of collisions. In fact, let us attach to each gas particle a finite volume, to be considered as the region over which the particle fields are strong enough to induce an observable change in the motion of an incoming electron (interaction region). When an electron crosses the interaction region of one gas particle, we shall say broadly that one collision has taken place. At the densities of significance in the most common and interesting experimental conditions, the average interparticle distances are much larger than the size of the interaction region of each gas particle. This implies that, when the gas slab is thin enough, the following situation may exist:

1. Most electrons do not make any collision during their flight within the gas sample.
2. A few electrons suffer one collision only and are scattered.

3. The number of electrons that undergo more than one collision is negligibly small.

It is immediately evident that such a physical model provides all the $\mathbf{v}'$ and S properties previously described and may thus be assumed to represent in classical terms, whenever experimental facts a–c are observed, the microscopic behavior of electron scattering in a neutral-molecule gas.

We have excluded the region $\vartheta \simeq \pi/2$ from the foregoing considerations, and the reason is now obvious. The electron scattered at right angles to the incident direction moves inside the slab along a path parallel to the largest size of the gas sample. Since this size may be large and cannot be reduced, because of the accuracy and the uncertainty principle requirements, it may happen with a significantly large probability (particularly at the highest densities) that the scattered electron will undergo a second collision, so that the final results do not satisfy the properties previously described. This difficulty may be easily overcome, however, by performing an experiment in which the angle between the slab plane and the incident electron direction is slightly different from $\pi/2$. We shall always assume (the reader may easily imagine how to integrate this experiment with the previous one for this purpose) that the measured scattering probability density S is given without any contribution of second collision events, also in the region $\vartheta \simeq \pi/2$.

When the gas density and composition and the sample dimensions are such that S satisfies, with reasonable approximation, the previously discussed properties a, b, and c, we say that the scattering experiment has been performed under "single-collision conditions." The reason for this designation follows from the simple interpretation given above. Since in a single-collision experiment S is proportional to d and independent of D, it is convenient to introduce a new function $S/d = s(v, v', \vartheta, n_i, T_g)$, which may be called the scattering probability density per unit path length.

If charged particles (ionized molecules and free electrons) are present in the gas at significant densities, single-collision conditions are not, in general, achievable. In this case, when the basic requirement of a sufficiently thin slab, such that the perturbation on the average electron motion will be small, is satisfied, the following features will appear under usual experimental conditions:

a. Except in a negligibly small fraction of cases, the velocities $\mathbf{v}'$ of outgoing electrons are close, but not equal, to the incoming velocity $\mathbf{v}$.

b. The scattering probability density S is independent of D but is no longer proportional to d; however, the mean changes of the most significant electron motion parameters, such as momentum and energy, which are obtained by integrating over velocity space the individual changes multiplied by the function S, are found to be proportional to d.

The result described under point *a* is simply predicted by the classical theory of collisions, if one assumes that the dimensions of the interaction regions of the charged particles are now larger than the interparticle distances, so that any incoming electron must necessarily intersect the perturbation regions of many particles. However, since the force field of each gas particle of this kind falls off significantly with distance, most perturbations are extremely weak, and the total effect on the electron motion remains small, as required.

Of particular significance is property b. In fact, it implies that the small perturbation effects on the momentum components and energy of the electron add up linearly. This may happen since, when d is increased, the average electron enters the added path length with a velocity very close to the original one, according to the small perturbation assumption, so that the same average changes of motion per unit length are expected, over both the original and the added paths. It is obvious that this is generally true only if d has been chosen many times larger than the interaction range of the charged particles, a condition easily satisfied in most experiments. In the next sections we shall make extensive use of this result.

Summarizing, we may say that in all cases of interest the scattering probability density S can be always regarded as independent of D, and such that the mean variations of electron motion parameters (e.g., momentum and energy), evaluated on its basis, are proportional to d. When the more restrictive conditions of single collisions are satisfied, not only the average motion variations but also the function S itself is proportional to d. In general, our future discussion will be related to S and its broadest properties, but, whenever of interest, the single-collision case and its characteristic function s will also be included.

It is very common to find that the scattering probability density plotted as a function of v' shows many isolated narrow peaks, located at practically the same v' values, independently of ϑ. For instance, this usually happens when the gas is made of molecules (Fig. 1.1.2). Each peak represents the statistical effect on the electron motion of a certain inelastic type of interaction between the electron and the gas particles; this interaction has established a transition (hereafter we shall label it j) between two definite states of a gas particle, whose energy difference approximately equals the electron loss of energy: $\frac{1}{2}m(v^2 - v'^2)$. This situation can be easily recognized in general by using well-known data on the energy levels of the gas particles, as determined and measured by quantum-mechanical computations and spectroscopic observations.

Using appropriate physical judgment, we can therefore write the function S as:

$$S = S_e + \sum_j S_j \qquad (1.1.2)$$

where each S_j characterizes the scattering probability density that can be

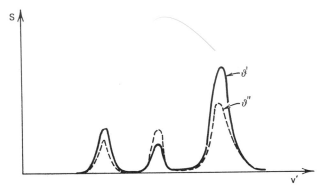

Fig. 1.1.2. Typical scattering probability density S of a molecular gas as a function of the speed v' of the outgoing electrons, for two different values of the scattering angle.

attributed to the excitation of a transition j by the incoming electron; and S_e is the remaining part of the distribution S, which can be interpreted as due to interactions that are only elastic. According to its definition, each distribution S_j is different from zero over a very small range of v' values only.

Since inelastic interactions require a close approach between the interacting particles, the S_j terms generally satisfy the single-collision conditions, so that (1.1.2) can be written as follows:

$$S = S_e + d \sum_j s_j \qquad (1.1.3)$$

When single-collision conditions dominate all processes, the above relationship becomes:

$$s = s_e + \sum_j s_j \qquad (1.1.4)$$

The scattering probability densities S and s, multiplied by the volume element $2\pi v'^2 \sin \vartheta \, dv' \, d\vartheta$, may be regarded as differential scattering probabilities, since they represent the average fraction of incident electrons that collisions scatter, elastically or inelastically, at an angle between ϑ and $\vartheta + d\vartheta$ with a speed between v' and $v' + dv'$. It is physically evident that, if one is interested in the macroscopically observable behavior of electron swarms, as we are, consideration must be given, not to the individual values of these differential scattering probabilities, but to their average properties from the electron motion point of view. In other words, we must introduce convenient integral parameters, where the scattering probabilities appear multiplied by the most significant motion quantities, and integrated over all angles ϑ or speeds v' or both. These parameters will be introduced and discussed in the next sections.

1.2   COLLISION PARAMETERS OF THE ELECTRON MOTION IN GASES

The integral collision parameters that we have to use in studying and describing the behavior of electron swarms will now be conveniently defined in terms of the results of the basic experiment discussed in Section 1.1. In fact, all these parameters can be derived from volume integrals over $\mathbf{v}'$ velocity space of the scattering probability density S or s, multiplied by physically significant functions (scalar, vector, or tensor) of the incoming and scattered electron velocities $\mathbf{v}$ and $\mathbf{v}'$. Four cases will be considered: the velocity functions are, respectively, 1, $\mathbf{v}' - \mathbf{v}$, $v'^2 - v^2$, and $(\mathbf{v}' - \mathbf{v})(\mathbf{v}' - \mathbf{v})$. We shall now examine separately each of these integrals, and define by means of them a convenient set of electron collision parameters.

A. The simplest case results from the integral of s over the entire velocity space $\mathbf{v}'$ or, in other words, from the integral of the differential scattering probability per unit path length $s \cdot 2\pi v'^2 \sin \vartheta \, dv' \, d\vartheta$ over all possible values of ϑ and v':

$$P_t(v, n_i, T_g) = 2\pi \int_0^\infty v'^2 \, dv' \int_0^\pi s(v, v', \vartheta, n_i, T_g) \sin \vartheta \, d\vartheta \qquad (1.2.1)$$

This quantity, being derived from s, is then defined on the basis of measurements performed under single-collision conditions only. If we count as one collision each observable perturbation effect on the electron motion, which the classical theory of collisions attributes to the individual action of one gas particle alone, then P_t measures the average number of collisions that an electron of constant speed v suffers per unit path length in a gas with component densities n_i and temperature T_g.

It follows that P_t may be properly called the total electron collision probability, implicit reference being made to a path length equal to unity. According to definition, the probability P_t exists only for gas conditions such that a single-collision type of experiment may be performed, as discussed in Section 1.1. Thus we can define P_t for the motion of an electron in a neutral molecular gas, but this is impossible in a gas where charged particles are present in significantly large densities.

When the distribution s is expressed as the sum of elastic and inelastic terms in the form (1.1.4), the total collision probability can be written similarly as the sum of corresponding terms, one for each s component, that is, one for each type of process:

$$P_t = P_e + \sum_j P_j \qquad (1.2.2)$$

Clearly P_e provides the collision probability for elastic collisions only, whereas P_j is the probability that an electron of constant speed v will excite, over a unit path length, a transition j in a gas particle. The probability P_j will be a function of v, T_g, and the density of gas particles in the initial energy level of the transition.

The quantity:

$$\Pi(v, v', n_i, T_g) = 2\pi v'^2 \int_0^\pi s(v, v', \vartheta, n_i, T_g) \sin \vartheta \, d\vartheta \qquad (1.2.3)$$

which appears in expression (1.2.1), is also useful sometimes when electron swarm properties are discussed using a microscopic picture. It represents the probability density per unit path length that the incident electron will be scattered into a range dv' at speed v'. In terms of Π, eq. (1.2.1) can be written as follows:

$$P_t = \int_0^\infty \Pi \, dv' \qquad (1.2.4)$$

As done for P_t, here too we can split Π, on the basis of relationship (1.1.4), into an elastic term (Π_e) and several inelastic terms (Π_j).

From the quantity P_t a set of related collision parameters can be derived according to simple definitions; we shall introduce the following ones:

a. the electron mean free path:

$$l_t = \frac{1}{P_t} \qquad (1.2.5)$$

b. the electron collision frequency:

$$\nu_t = v P_t = \frac{v}{l_t} \qquad (1.2.6)$$

c. the electron mean free time:

$$\tau_t = \frac{1}{\nu_t} = \frac{1}{v P_t} = \frac{l_t}{v} \qquad (1.2.7)$$

B. The integral over the entire velocity space $\mathbf{v}'$ of the scattering probability density S, multiplied by the change $(\mathbf{v}' - \mathbf{v})$ of the electron velocity, is now introduced as the fundamental quantity of a second important family of related collision parameters. If we indicate by $\overline{\delta \mathbf{v}}$ the quantity defined above, since it provides the average change of the electron velocity in passing through the gas, we have:

$$\overline{\delta \mathbf{v}} = \int_{\mathbf{v}} (\mathbf{v}' - \mathbf{v}) S(v, v', \vartheta, d, n_i, T_g) \, d\mathbf{v}' \qquad (1.2.8)$$

or:

$$\overline{\delta \mathbf{v}} = 2\pi \mathbf{v} \int_0^\infty v'^2 \, dv' \int_0^\pi \left(\frac{v'}{v} \cos \vartheta - 1 \right) S(v, v', \vartheta, d, n_i, T_g) \sin \vartheta \, d\vartheta \qquad (1.2.9)$$

Expression (1.2.9) is easily obtained from the preceding one, simply taking into account the fact that the distribution S is symmetrical around the $\mathbf{v}$ vector direction.

The quantity $-m\overline{\delta\mathbf{v}}$ is the corresponding loss of momentum and is proportional to d, according to the general properties of the distribution S, stated in Section 1.1. It is now convenient to introduce the parameter:

$$P_m(v, n_i, T_g) = \frac{m|\overline{\delta\mathbf{v}}|/d}{mv} = \frac{|\overline{\delta\mathbf{v}}|}{vd}$$

$$= \frac{2\pi}{d}\int_0^\infty v'^2\,dv'\int_0^\pi\left(1 - \frac{v'}{v}\cos\vartheta\right)S(v, v', \vartheta, d, n_i, T_g)\sin\vartheta\,d\vartheta$$

(1.2.10)

which gives the mean fraction of momentum lost over a unit path length by an electron of constant speed v, as the result of collisions. It follows that P_m can properly be called electron collision probability for momentum transfer and, by definition, has a broader range of existence than P_t, since it may be used in all cases, and also when the gas is such that single-collision conditions cannot be achieved. Analogously to P_t, P_m can be split into an elastic collision term P_{me} and several inelastic collision terms P_{mj}.

An electron mean free path l_m, collision frequency ν_m, and mean free time τ_m for momentum transfer can be derived from P_m using eqs. (1.2.5), (1.2.6), and (1.2.7), where the index t is now replaced by m.

C. The integral over the entire $\mathbf{v}'$ velocity space of S multiplied by the change $(v'^2 - v^2)$ of the square electron speed, that is:

$$\overline{\delta v^2} = 2\pi\int_0^\infty (v'^2 - v^2)v'^2\,dv'\int_0^\pi S(v, v', \vartheta, d, n_i, T_g)\sin\vartheta\,d\vartheta \quad (1.2.11)$$

is another important parameter, since $-m\,\overline{\delta v^2}/2d$ provides the mean energy lost over a unit path length by an electron of constant speed v. This energy loss is independent of d, according to the general S property stated in Section 1.1, and is usually measured in units of the excess of the incident electron kinetic energy $u = mv^2/2$ over the gas particle mean energy $3kT_g/2$ (k being the Boltzmann constant). This fractional loss is:

$$P_u(v, n_i, T_g) = \frac{-m\overline{\delta v^2}/2d}{mv^2/2 - 3kT_g/2} = -\left(v^2 - \frac{3kT_g}{m}\right)^{-1}\frac{\overline{\delta v^2}}{d}$$

$$= \left(v^2 - \frac{3kT_g}{m}\right)^{-1}\frac{2\pi}{d}\int_0^\infty (v^2 - v'^2)v'^2\,dv'$$

$$\times \int_0^\pi S(v, v', \vartheta, d, n_i, T_g)\sin\vartheta\,d\vartheta \quad (1.2.12)$$

The quantity P_u, here defined, measures the mean fraction of excess kinetic energy that the electron will lose, moving at constant speed v, over a unit path length, and as such it may be called the electron collision probability for energy exchange. Like all other collision probabilities, P_u can also be written as the sum of an elastic collision term P_{ue} and of several inelastic collision terms P_{uj}.

Used in the literature much more than P_u, however, is the so-called G-factor, defined by the ratio:

$$G = \frac{P_u}{P_m} \tag{1.2.13}$$

This factor provides the average fractional loss of excess electron kinetic energy per collision. Its popularity is due mainly to the fact that in an atomic gas, where only perfectly elastic collisions between electrons and gas particles are possible, this G-factor is simply equal to $2m/M$, independently of the electron velocity and of the gas temperature and density. We shall prove this point and discuss it further in the following sections.

Frequent use is also made of the related parameter, the collision frequency for energy exchange ν_u, defined by eq. (1.2.6), where the index t is replaced by u.

D. Finally we consider the integral of S/d multiplied by the tensor $(\mathbf{v'} - \mathbf{v})(\mathbf{v'} - \mathbf{v})$. The result is a tensor, independent of d, that we shall indicate as $\boldsymbol{\delta}$ (v, n_i, T_g). Since the distribution S is symmetrical around the $\mathbf{v}$ vector direction, it appears convenient to express it in an orthogonal frame, where $\mathbf{v}$ provides the direction of a coordinate axis (we shall label as $\parallel$ the component along this axis). In this frame the tensor $\boldsymbol{\delta}$ structure becomes:

$$\boldsymbol{\delta} = \begin{vmatrix} \delta_\parallel & 0 & 0 \\ 0 & \tfrac{1}{2}\delta_\perp & 0 \\ 0 & 0 & \tfrac{1}{2}\delta_\perp \end{vmatrix} \tag{1.2.14}$$

where:

$$\delta_\parallel = \frac{2\pi}{d} \int_0^\infty v'^2 \, dv' \int_0^\pi (v - v' \cos \vartheta)^2 S(v, v', \vartheta, d, n_i, T_g) \sin \vartheta \, d\vartheta \tag{1.2.15}$$

$$\delta_\perp = \frac{2\pi}{d} \int_0^\infty v'^4 \, dv' \int_0^\pi S(v, v', \vartheta, d, n_i, T_g) \sin^3 \vartheta \, d\vartheta \tag{1.2.16}$$

The quantity $v\boldsymbol{\delta}$ shows up in the kinetic description of electron swarm behavior as a diffusion coefficient in velocity space; therefore, it is often called the diffusion-in-velocity tensor.

The two components $\delta_\parallel$ and $\delta_\perp$ are not independent when all previously defined collision parameters are given. In fact, from the simple velocity relation:

$$(v - v' \cos \vartheta)^2 + (v' \sin \vartheta)^2 = v^2 + v'^2 - 2vv' \cos \vartheta$$

$$= 2v(v - v' \cos \vartheta) - (v^2 - v'^2) \quad (1.2.17)$$

we obtain (see eqs. 1.2.10 and 1.2.12) that the trace of the $\boldsymbol{\delta}$ tensor is completely determined by P_m and P_u probabilities:

$$\delta = \delta_{\parallel} + \delta_{\perp} = 2v^2 P_m - \left(v^2 - \frac{3kT_g}{m} \right) P_u$$

$$= \left[2v^2 - \left(v^2 - \frac{3kT_g}{m} \right) G \right] P_m \quad (1.2.18)$$

In most cases, as when elastic collisions with gas particles are dominant, $G \ll 1$, so that (1.2.18) further simplifies to the approximate relationship:

$$\delta = \delta_{\parallel} + \delta_{\perp} = 2v^2 P_m \quad (1.2.19)$$

Then we shall consider $\delta_{\perp}$ only and introduce a new collision probability parameter as follows:

$$P_d (v, n_i, T_g) = \frac{\delta_{\perp}}{v^2} = \frac{2\pi}{v^2 d} \int_0^{\infty} v'^4 \, dv' \int_0^{\pi} S(v, v', \vartheta, d, n_i, T_g) \sin^3 \vartheta \, d\vartheta \quad (1.2.20)$$

This quantity might well be called the electron collision probability for deflection, since it provides the mean energy going into the right-angle component of the scattered velocity, measured as a fraction of the incident energy.

A rather obvious remark can be made regarding the above definitions of the probabilities P_m, P_u, and P_d. They have all been related to the general scattering function S; but, whenever the gas conditions are such that the experiment can be performed according to the single-collision requirements, the ratio S/d may be systematically replaced by the function s in all the integrals.

In this regard, it is particularly pertinent to remark here that, when these conditions hold, eq. (1.2.12) can be rewritten in a much simpler form, making use of the Π definition given by (1.2.3):

$$P_u = \left(v^2 - \frac{3kT_g}{m} \right)^{-1} \int_0^{\infty} (v^2 - v'^2) \Pi \, dv' \quad (1.2.21)$$

This formula is generally applicable to the inelastic collision terms P_{uj}; in this case, since s_j, and therefore Π_j, are different from zero over only a very small range of v' values, whereas $(v^2 - v'^2)$ is practically constant over this region, relationship (1.2.21) simplifies further. Taking into account that $(u - u')$ approximately equals the energy difference u_j between the initial and final particle states ($u_j > 0$ for excitation, $u_j < 0$ for de-excitation of the particle), we obtain:

$$P_{uj} \simeq (u - \tfrac{3}{2}kT_g)^{-1} u_j \int_0^\infty \Pi_j \, dv' = (u - \tfrac{3}{2}kT_g)^{-1} u_j P_j \qquad (1.2.22)$$

The electron collision parameters defined in this section depend on the gas composition and temperature. To extract from the collisional data significant information on the scattering properties of the individual gas particles, these dependences must be examined and understood. This will be done in Sections 1.3 and 1.4.

1.3   COLLISIONS OF ELECTRONS WITH SINGLE GAS PARTICLES

According to the classical theory of collisions, each interaction between an electron and a single gas particle in empty space provides after-collision kinetic conditions, whose average characteristics, measured in the collidants' center-of-mass system, depend only on the species and internal state of the gas particle and on the relative speed of the collidants. If such information concerning the characteristics of an elementary electron collision process is desired from our data, it will be most useful to choose for this purpose a set of convenient conditions under which to perform, at least ideally, the experiment.

On simple intuitive grounds, we may expect the following conditions to be the best ones for a simple and straightforward determination of the elementary collision properties: (*a*) single-collision conditions, as defined in Section 1.1, are satisfied; (*b*) all gas particles are identical and are in the same internal energy state; (*c*) all gas particles are at rest, that is, $T_g = 0$. In this case the scattering probability density *s* and the derived collision probabilities P_t, P_m, P_u, and P_d are directly proportional to the gas particle density n_g (see property *b* on p. 9).

It is then convenient to introduce a new set of more elementary collision quantities, by taking the ratio of each collision probability over the particle density. Hence we have:

$$q_t(v) = \frac{P_t(v, n_g, 0)}{n_g} \qquad (1.3.1)$$

and similar definitions for $q_m(v)$, $q_u(v)$, and $q_d(v)$. The energy loss factor *G* becomes:

$$G(v, n_g, 0) = \frac{P_u(v, n_g, 0)}{P_m(v, n_g, 0)} = \frac{q_u(v)}{q_m(v)} = \lambda(v) \qquad (1.3.2)$$

All these quantities are functions only of the incoming electron speed v, which coincides with the collidants' relative speed, since the gas particles are assumed at rest. Knowledge of the values of these collision parameters for each state of the various gas particles represents the most elementary information to be used for the evaluation of the average electron motion properties, dominated by electron-gas particle collisions.

All q quantities have the dimensions of an area, and may be interpreted as diametric cross sections of the gas particles from an electron collision point of view. In fact, let us imagine each particle as a solid field-free sphere of diametric cross section q, which scatters an incoming electron, regarded as a point particle, each time that the electron strikes the sphere surface. If we now assume that our usual experiment is performed, with electrons incident on a slab element of cross-section area A and thickness d, filled with the above gas of sphere particles, the total collision probability (per unit path length) results:

$$P_t = \frac{qN}{Ad} = \frac{qn_gAd}{Ad} = qn_g \qquad (1.3.3)$$

where N is the number of gas particles in the slab portion under consideration. Comparing eqs. (1.3.1) and (1.3.3), we see that the particle geometrical cross section q may be set equal to q_t when the total collision probability is desired.

If the spheres are smooth, perfectly elastic, and infinitely heavy, we can easily evaluate also the collision probability for momentum transfer and then the corresponding particle cross section for momentum transfer q_m (some authors call this also the diffusion cross section). In fact, the electrons are scattered isotropically and without energy losses ($v' = v$), whereas the gas particles, in spite of collisions, remain at rest because of their infinite masses. Then we can write:

$$s(v, v', \vartheta, n_g, 0) = \frac{\delta(v - v')}{4\pi v'^2} \frac{qN}{Ad} = \frac{\delta(v - v')}{4\pi v'^2} qn_g \qquad (1.3.4)$$

where δ is Dirac's unity impulse function. Substituting this scattering distribution into eq. (1.2.10), we obtain:

$$P_m(v, n_g, 0) = \frac{qn_g}{2} \int_0^\infty \delta(v - v') \, dv' \int_0^\pi (1 - \cos \vartheta) \sin \vartheta \, d\vartheta = qn_g \qquad (1.3.5)$$

Hence q_m can be set equal, from a momentum loss point of view, to the diametric cross section of a smooth spherical gas particle, perfectly elastic and infinitely heavy.

If eq. (1.3.4) is substituted in (1.2.20), we obtain:

$$P_d(v, n_g, 0) = \frac{qn_g}{2v^2} \int_0^\infty v'^2 \delta(v - v') \, dv' \int_0^\pi \sin^3 \vartheta \, d\vartheta = \tfrac{2}{3}qn_g \qquad (1.3.6)$$

In this case, then, the deflection collision cross section q_d (some authors call it also the viscosity cross section) is equal to two thirds of the geometrical cross section of the gas particles.

In the same ideal gas that we have just considered, the probability P_u and the fractional energy loss per collision λ would instead be zero, since no electron energy loss can take place; in fact, no kinetic energy can be transferred from the electron to particles of infinite mass, and all the energy stored during a collision time must be given back to the electron, since the process has been described as perfectly elastic. Let us take the opportunity to consider at this point a similar ideal gas, but one less far from reality; this is the case of a gas of smooth, perfectly elastic, spherical particles of finite mass M. The value of λ, which is now different from zero, is well known and is derived in all textbooks on collision theory. However, it is such a basic quantity for our future discussion that we will take time here to show briefly how it can be derived.

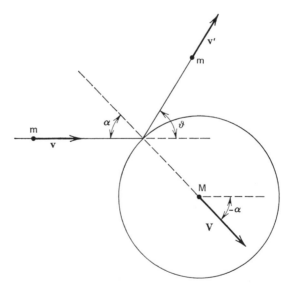

Fig. 1.3.1. The collision of a point-like electron of mass m and velocity **v** with a stationary, smooth, perfectly elastic, spherical gas particle of mass M.

As shown in Fig. 1.3.1, the electron is considered as a point particle of mass m and velocity **v**, which impinges on the stationary spherical gas particle of mass M with incident angle α ($0 \leqslant \alpha \leqslant \pi/2$). Whereas the electron is scattered along an angle ϑ ($0 \leqslant \vartheta \leqslant \pi$), the gas particle will recoil at angle $-\alpha$, since the acting force on the sphere can be directed only along the line connecting the impact point with the particle center of gravity, the surface being perfectly smooth in our assumption. The scattered electron and gas particle velocities are $\mathbf{v}'$ and $\mathbf{V}$, respectively.

The linear momentum and kinetic energy conservation equations are as follows:

$$mv = mv' \cos \vartheta + MV \cos \alpha \qquad (1.3.7)$$

$$0 = mv' \sin \vartheta - MV \sin \alpha \qquad (1.3.8)$$

$$mv^2 = mv'^2 + MV^2 \qquad (1.3.9)$$

This system enables us to obtain v', V, and ϑ in terms of v and α. Thus straightforward mathematics yields:

$$v' = v \left[1 - \frac{4mM}{(m + M)^2} \cos^2 \alpha \right]^{\frac{1}{2}} \qquad (1.3.10)$$

$$V = \frac{2m}{m + M} v \cos \alpha \qquad (1.3.11)$$

$$\tan \vartheta = \frac{\sin 2\alpha}{(m/M) - \cos 2\alpha} \qquad (1.3.12)$$

The probability that an electron, moving over a straight path of unit length, will strike a spherical particle of cross section q with an incident angle between α and $\alpha + d\alpha$ is $n_g q \sin 2\alpha \, d\alpha$. In fact, for each sphere, $2q \sin \alpha \, d\alpha$ is the area of the surface element between the cones α and $\alpha + d\alpha$, and this has to be multiplied by $\cos \alpha$, so to obtain the projection area over a plane perpendicular to the incident electron direction.

From the foregoing considerations it follows easily that the differential scattering probability is in this case:

$$s \cdot 2\pi v'^2 \sin \vartheta \, dv' \, d\vartheta$$

$$= \delta \left\{ v' - v \left[1 - \frac{4mM}{(m + M)^2} \cos^2 \alpha \right]^{\frac{1}{2}} \right\} dv' \cdot n_g q \sin 2\alpha \, d\alpha \quad (1.3.13)$$

where the second member is written as a function of α instead of ϑ, the relationship between these two quantities being given by eq. (1.3.12). Substituting (1.3.13) into eq. (1.2.12), where now $T_g = 0$, we have:

$$P_u = n_g q \int_0^\infty \left(1 - \frac{v'^2}{v^2} \right) dv' \int_0^{\pi/2} \delta \left\{ v' - v \left[1 - \frac{4mM}{(m + M)^2} \cos^2 \alpha \right]^{\frac{1}{2}} \right\} \sin 2\alpha \, d\alpha$$

$$(1.3.14)$$

Interchanging the order of integration and performing the simple integration with respect to v', we obtain:

$$P_u = n_g q \frac{4mM}{(m+M)^2} \int_0^{\pi/2} \cos^2 \alpha \sin 2\alpha \, d\alpha = \frac{2mM}{(m+M)^2} n_g q \quad (1.3.15)$$

When the gas particles are atoms or molecules, $m \ll M$, and eq. (1.3.15) becomes:

$$P_u = \frac{2m}{M} n_g q \quad (1.3.16)$$

It can easily be recognized that under these conditions P_m is given by the infinite mass value (1.3.5). The loss factor λ, defined by (1.3.2), is then:

$$\lambda = \frac{2m}{M} \quad (1.3.17)$$

independently of v.

Equation (1.3.17) has a broader range of validity than that resulting from the above demonstration. In fact, when the gas particles are atoms, and collisions are perfectly elastic, the fractional energy loss has, with good approximation, value (1.3.17), independently of the shape and internal structure of the atom. This can easily be shown as follows.

Equations (1.3.7)–(1.3.12) are always valid, provided collisions are perfectly elastic. If $m \ll M$, (1.3.12) gives:

$$\vartheta \simeq \pi - 2\alpha \quad (1.3.18)$$

which, substituted into eq. (1.3.10), provides:

$$v' \simeq v\left(1 - \frac{4m}{M}\sin^2\frac{\vartheta}{2}\right)^{1/2}$$

$$\simeq v\left[1 - \frac{2m}{M}(1 - \cos\vartheta)\right]^{1/2} \quad (1.3.19)$$

The differential scattering probability for this case can thus be written in the form:

$$s \cdot 2\pi v'^2 \sin\vartheta \, dv' \, d\vartheta$$

$$= \delta\left\{v' - v\left[1 - \frac{2m}{M}(1 - \cos\vartheta)\right]^{1/2}\right\} dv' \cdot s_\vartheta(\vartheta) \sin\vartheta \, d\vartheta \quad (1.3.20)$$

where s_ϑ, the factor of s which provides the angular scattering distribution for an incident electron of velocity v, can be any function of ϑ.

When relationship (1.3.20) is substituted into eq. (1.2.12), the integration order interchanged, and the simple integration with respect to v' performed, we obtain:

$$P_u = \frac{2m}{M} \int_0^\pi (1 - \cos \vartheta)\, s_\vartheta(\vartheta) \sin \vartheta \, d\vartheta \qquad (1.3.21)$$

With the same scattering probability (1.3.20) and the approximation $m \leqslant M$, eqs. (1.2.1) and (1.2.10) provide similarly:

$$P_t = \int_0^\pi s_\vartheta(\vartheta) \sin \vartheta \, d\vartheta \qquad (1.3.22')$$

$$P_m = \int_0^\pi (1 - \cos \vartheta) s_\vartheta(\vartheta) \sin \vartheta \, d\vartheta \qquad (1.3.22'')$$

Substituting (1.3.22") into (1.3.21) yields:

$$P_u = \lambda P_m = \lambda n_g q_m \qquad (1.3.23)$$

where λ is the value (1.3.17). This demonstrates our statement that λ equals $2m/M$ any time that collisions are perfectly elastic.

We saw in Section 1.2 that each collision probability can be written as the sum of an elastic collision term and of several inelastic collision terms. Similarly we can here define separate cross sections for elastic collisions (as q_e and q_{me}) and for inelastic collisions (as q_j and q_{mj}), dividing the corresponding elastic and inelastic probabilities by the density n_g. Of particular importance is the excitation cross section q_j for the j transition; conservation of energy requires that, when the gas particle gains energy in the transition, $q_j(v)$ be zero for all velocities below $\sqrt{2u_j/m}$, where u_j, as previously defined, is the energy difference between the initial and final states of the particle.

Let us consider now the collision probability for energy exchange P_u in an atomic or molecular gas of identical equal-state particles at zero temperature, when the presence of inelastic collisions between electrons and gas particles must be allowed for. In this case the elastic contribution is given by (1.3.23) and the inelastic ones by (1.2.22), where $T_g = 0$, so that we obtain:

$$P_u = \frac{2m}{M} P_{me} + \sum_j \frac{u_j}{u} P_j \qquad (1.3.24)$$

Actually, the expression for the inelastic collisions contribution in (1.3.24) is approximate, in that the kinetic energy transferred to the molecules during these collisions is not considered at all ($u - u' = u_j$ was assumed on p. 17). If $u \gg u_j$ for a given transition, it is easy to see that we can take into account the neglected transfer of kinetic energy to a molecule by simply adding a term $(2m/M)P_{mj}$ to the above expressions for P_u. If this can be assumed for all inelastic collisions, P_u is given with a better approximation by the formula:

$$P_u = \frac{2m}{M}\left(P_{me} + \sum_j P_{mj}\right) + \sum_j \frac{u_j}{u} P_j$$

$$= \frac{2m}{M} P_m + \sum_j \frac{u_j}{u} P_j$$

$$= \left(\frac{2m}{M} q_m + \frac{1}{u}\sum_j u_j q_j\right) n_g \qquad (1.3.25)$$

Since in general $(2m/M)\sum_j P_{mj}$ is much smaller than at least one of the other two terms in formula (1.3.25) and hence contributes very little to P_u, we shall always use (1.3.25) instead of (1.3.24) also when the assumption $u \gg u_j$ made above is not truly satisfied.

Correspondingly the fractional energy loss per collision λ becomes:

$$\lambda = \frac{2m}{M} + \frac{1}{u q_m}\sum_j u_j q_j \qquad (1.3.26)$$

According to what was stated in the Introduction, we are concerned only with the motion of slow electrons, whose energies are less than those required for dissociation, excitation of electronic states, and ionization of atoms and molecules in their ground states. In this energy range, if the gas particles lie in their normal thermal states, only transitions to rotationally and vibrationally molecular excited states are possible. Therefore, the inelastic collision terms in eq. (1.3.26) are present in a molecular gas but disappear when the gas is atomic, as in the case of rare gases, where λ reduces to the elastic value $2m/M$.

Collision cross sections, like q_t, q_m, q_d, and q_j, and the fractional energy loss λ are basic quantities, which characterize the collision action of a gas particle on an electron; as such they find use in all cases when, in an ionized gas or plasma environment, only one particle at a time interacts with each free electron (this corresponds to the "single-collision conditions" discussed previously for our basic experiment). The cross sections are measured in units of square centimeters, in units of square angstroms, or in atomic units of $\pi a_0{}^2 = 0.88 \times 10^{-16}$ cm^2 (where $a_0 = 0.53 \times 10^{-8}$ cm is the radius of the first Bohr orbit of the hydrogen atom). Energy loss factors are pure numbers, but very often they are given in units of the classical value (1.3.17) for elastic collisions.

Under the conditions assumed in this section, collision probabilities in a gas mixture, or in a gas where the particles are distributed among different energy levels, can be easily evaluated, since each collision is an independent process, and the overall probability is thus the sum of all component probabilities. Therefore, if we indicate with a subscript i the collision values for the i-component species, the following most important relationships will hold:

$$P_t = \sum_i n_i q_{ti} \qquad (1.3.27)$$

$$P_m = \sum_i n_i q_{mi} \qquad (1.3.28)$$

$$P_u = \sum_i \lambda_i n_i q_{mi} \qquad (1.3.29)$$

1.4 ELECTRON COLLISIONS IN A FINITE-TEMPERATURE MOLECULAR GAS

The quantities q and λ that we introduced in Section 1.3 are, by definition, derived from the results of experiments performed in ideal gases consisting entirely of identical, equal-state particles at rest. We recall that our basic experiment, since its description in Section 1.1, has also been defined assuming ideal conditions, its only purpose being the proper definition of useful electron collision parameters; therefore, the unrealistic nature of an identical-particle gas does not present in principle any conceptual difficulty. However, it is evident that for our future analysis of experimental data we shall also need to derive relations between all these elementary quantities and the collision probabilities and related parameters, which are actually measured in a real gas at nonzero temperatures.

These relationships will be derived in this section for the most common case of an electron making single collisions in a monatomic or in a mono-molecular gas, un-ionized and in thermodynamic equilibrium at a temperature below a few hundred degrees Kelvin. Here the finite temperature plays a double role: the molecules have a random kinetic motion, and they are distributed among different internal energy states.

We assume first that in the gas slab of our experiment all the particles are still identical, but that they move with a common constant velocity **V**. Now we consider the quantity vs, which provides the density in velocity space of the scattering rate suffered by an electron of constant speed v, since it is the product of the scattering probability density per unit path length, multiplied by the total length of the electron path during a unit time interval (at constant v). This scattering rate density must be the same, whether measured in a conventional laboratory frame of reference (O origin in Fig. 1.4.1) or in a system moving at constant velocity **V** (O' origin) where gas particles appear at rest before collisions. Since in the two reference systems the velocity space volume elements $d\mathbf{v}'$ and $d(\mathbf{v}' - \mathbf{V})$ are equal, we can write the equal rate density condition as follows:

$$vs(v, v', \vartheta, n_g, \mathbf{V}) = gs(g, g', \vartheta_r, n_g, 0) \qquad (1.4.1)$$

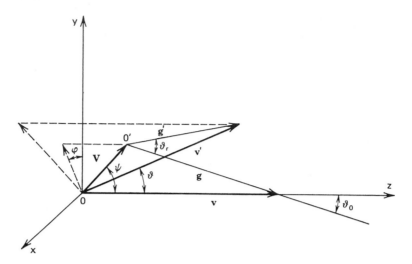

Fig. 1.4.1. Relationship between the electron velocity vectors in a laboratory frame (origin O, vectors **v** and **v′**) and in a system moving at constant velocity **V** (origin O', vectors **g** and **g′**).

where, for writing simplicity, the relative velocity vectors have been called **g**:

$$\mathbf{g} = \mathbf{v} - \mathbf{V} \qquad \therefore \qquad \mathbf{g}' = \mathbf{v}' - \mathbf{V} \tag{1.4.2}$$

and ϑ_r is the angle between the **g** and **g′** vectors.

We are now in a position to consider a gas of particles having kinetic energies according to thermodynamic equilibrium at temperature T_g. In this case the fraction of particles whose velocities lie between **V** and **V** + d**V** is given by a spherically symmetrical distribution $F(V)$, which more specifically is the Maxwellian distribution:

$$F_M(V)\,d\mathbf{V} = \left(\frac{M}{2\pi k T_g}\right) \exp\left(-\frac{MV^2}{2kT_g}\right) d\mathbf{V} \tag{1.4.3}$$

Then the density distribution of the electron scattering rate in a gas of identical particles at temperature T_g, on the basis of (1.4.1), will be:

$$vs(v, v', \vartheta, n_g, T_g) = \int_V gs(g, g', \vartheta_r, n_g, 0)F(V)\,d\mathbf{V} \tag{1.4.4}$$

where the integration is performed over all **V** values.

When single-collision conditions are satisfied, eqs. (1.2.1), (1.2.10), (1.2.12), and (1.2.20), which define the various collision probabilities P, may be written in the common form:

$$P(v, n_g, T_g) = \int_{\mathbf{v}'} \chi(\mathbf{v}, \mathbf{v}')s(v, v', \vartheta, n_g, T_g) \, d\mathbf{v}' \tag{1.4.5}$$

where $\chi(\mathbf{v}, \mathbf{v}')$ is different for each collision probability parameter, and integration is performed over the entire $\mathbf{v}'$ space. Substituting (1.4.4) into (1.4.5) yields:

$$P(v, n_g, T_g) = \frac{1}{v} \int_{\mathbf{v}'} d\mathbf{v}' \int_{\mathbf{V}} g\chi(\mathbf{v}, \mathbf{v}')s(g, g', \vartheta_r, n_g, 0)F(V) \, d\mathbf{V} \tag{1.4.6}$$

Interchanging the integration order, we obtain, since $d\mathbf{v}' = d\mathbf{g}'$ at constant $\mathbf{V}$:

$$P(v, n_g, T_g) = \frac{1}{v} \int_{\mathbf{V}} gF(V) \, d\mathbf{V} \int_{\mathbf{g}} \chi(\mathbf{v}, \mathbf{g}' + \mathbf{V})s(g, g', \vartheta_r, n_g, 0) \, d\mathbf{g}' \tag{1.4.7}$$

where the integration has to be performed over the entire $\mathbf{g}'$ space. It is important to recall that $\mathbf{g}$ is a function of $\mathbf{V}$ according to eq. (1.4.2).

Let us now consider separately each collision probability.

For the total collision probability P_t we have:

$$\chi(\mathbf{v}, \mathbf{v}') = 1 \tag{1.4.8}$$

The use of this formula in (1.4.7) gives, with the help of (1.3.1) and (1.4.5):

$$P_t(v, n_g, T_g) = \frac{1}{v} \int_{\mathbf{V}} gP_t(g, n_g, 0)F(V) \, d\mathbf{V} \tag{1.4.9}$$

or:

$$P_t(v, n_g, T_g) = \frac{n_g}{v} \int_{\mathbf{V}} gq_t(g)F(V) \, d\mathbf{V} \tag{1.4.10}$$

The volume element $d\mathbf{V}$ is given by:

$$d\mathbf{V} = V^2 \sin \psi \, dV \, d\psi \, d\varphi \tag{1.4.11}$$

where ψ and φ are shown in Fig. 1.4.1. From the same figure we derive the following relationship:

$$g^2 = v^2 + V^2 - 2vV \cos \psi \tag{1.4.12}$$

The variable g may replace the angular variable ψ; a differential relationship between them is easily obtained from the same eq. (1.4.12):

$$g \, dg = vV \sin \psi \, d\psi \tag{1.4.13}$$

The volume element $d\mathbf{V}$ then becomes:

$$d\mathbf{V} = \frac{Vg}{v} \, dV \, dg \, d\varphi \tag{1.4.14}$$

The integration range 0 to π for the variable ψ becomes the integration range $|v - V|$ to $v + V$ for the variable g.

Equation (1.4.10) thus becomes:

$$P_t(v, n_g, T_g) = \frac{n_g}{v^2} \int_0^\infty F(V)V \, dV \int_{|v-V|}^{v+V} g^2 q_t(g) \, dg \int_0^{2\pi} d\varphi$$

$$= \frac{2\pi n_g}{v^2} \int_0^\infty F(V)V \, dV \int_{|v-V|}^{v+V} g^2 q_t(g) \, dg \qquad (1.4.15)$$

When the gas particles are atoms or molecules with thermal energies (temperatures up to a few hundred degrees Kelvin only), $F = F_M$ is appreciably different from zero over a range of V values much lower than the electron speed values of usual interest. In fact molecules have velocities centered around $\sqrt{kT_g/M}$, and we are usually interested in electrons with energies of the same order as those of molecules, or larger, that is, with velocities of the order of $\sqrt{kT_g/m}$, or higher. In this case a simple but significant result is obtained by substituting for $q_t(g)$ the first terms of its Taylor series expansion around the v value as follows:

$$q(g) = q(v) + (g - v)\frac{dq}{dv} \qquad (1.4.16)$$

Within the same approximation the lower $|v - V|$ limit in the g integration may be conveniently replaced by $v - V$. Thus eq. (1.4.15) reduces to simple known integrals and provides the final result:

$$P_t(v, n_g, T_g) = n_g\left[1 + \frac{kT_g}{Mv^2}\left(1 + 2\frac{d \ln q_t}{d \ln v}\right)\right]q_t(v) \qquad (1.4.17)$$

under the assumptions previously made regarding the molecule and electron velocities $kT_g/Mv^2 \ll 1$. The dependence of cross sections on velocity is usually such that the following relationship also is satisfied:

$$\frac{kT_g}{Mv^2}\frac{d \ln q}{d \ln v} \ll 1 \qquad (1.4.18)$$

Thus eq. (1.4.17) says that, in a gas of equal-state particles with thermal energies, the dependence of the total collision probability on the gas temperature is usually very weak, so that in most cases it can be neglected:

$$P_t(v, n_g, T_g) \simeq n_g q_t(v) = P_t(v, n_g, 0) \qquad (1.4.19)$$

A molecular gas in thermodynamic equilibrium, however, is not an assembly of all equal-state molecules. In fact, at temperatures of a few hundred degrees Kelvin, most molecules are in the ground state, but an appreciable fraction of them may lie in rotationally excited states. Then, rigorously, the collision probability for a real gas has to be computed substituting (1.4.17) in (1.3.27); this yields the following formula for a gas at temperature T_g and with total molecular density n_g:

$$P_t(v, n_g, T_g) = n_g \sum_J \alpha_J(T_g)\left[1 + \frac{kT_g}{Mv^2}\left(1 + 2\frac{d \ln q_t{}^J}{d \ln v}\right)\right]q_t{}^J(v) \quad (1.4.20)$$

where $\alpha_J(T_g)$ is the fraction of molecules that, in thermodynamic equilibrium at temperature T_g, are in the level having rotational quantum number J, and $q_t{}^J$ is the total electron collision cross section of a molecule in this Jth rotational level.

Expressions of the α_J fractional populations appropriate for the various molecular structures can be found in most textbooks treating molecules and their spectra [for a clear and comprehensive exposition of this subject see Herzberg (1945, 1950)]. Here we shall report only the expressions for diatomic and linear triatomic molecules, which represent the most frequent cases considered in this book. If the diatomic molecule is homonuclear (like H_2, N_2, O_2), or if the linear triatomic molecule has a center of symmetry on the internuclear axis (like $CO_2 = {}^{16}O - {}^{12}C - {}^{16}O$), the fractional population is given by:

$$\alpha_J = \frac{(I + a)(2J + 1) \exp(-E_J/kT_g)}{\sum_J (I + a)(2J + 1) \exp(-E_J/kT_g)} \quad (1.4.21)$$

Here I is the nuclear spin of the atom outside the symmetry center (see the values in Table 1.4 1), a is 0 for all the para terms (which are the even J terms in some gases, like ${}^1H_2$ and ${}^{16}O_2$, and the odd J terms in other gases, like ${}^2D_2$, ${}^{14}N_2$, and ${}^{12}C{}^{16}O_2$) and 1 for all the ortho terms, and E_J is the energy of the Jth level. This energy is given to sufficient accuracy by:

$$E_J = J(J + 1)B_0 \quad (1.4.22)$$

where B_0 is the so-called rotational constant (E_J and B_0 are usually given in electron volts).

Not always in experiments will the population distribution have the values given by (1.4.21). In fact, transitions from para to ortho states take place very infrequently, so that, when a gas like hydrogen is cooled from room to low temperatures, equilibrium is established only in the two separate assemblies of para and ortho states, whereas for a time large compared to the duration of usual experiments the ratio of the two assemblies remains unchanged at the high-temperature value. When desirable, however, appropriate catalysts, such as O_2 and NO, accelerate to practical times the conversion to total equilibrium [Farkas (1935)].

In the case of heteronuclear diatomic molecules and of linear molecules without a symmetry center, distribution (1.4.21) still holds, except for the factor $(I + a)$, which has to be deleted.

In the discussion above we have assumed that all the molecules are in the lowest vibrational state $(v = 0)$. At very high temperatures, however, excitation to higher levels may take place and must be properly included by considering these additional (J, v) terms, whose energy level is still given by eq. (1.4.22), provided the value B_0 of the rotational constant is replaced by the value B_v appropriate to the vibrational state under consideration.

Table 1.4.1   Spin Values of Some Common Atoms in $\hbar$ Units

Atom:	^{1}H	^{2}D	^{12}C	^{14}N	^{16}O
Spin:	$\tfrac{1}{2}$	1	0	1	0

Equation (1.4.20) can be written in the form:

$$P_t(v, n_g, T_g) = n_g Q_t(v, T_g) \tag{1.4.23}$$

where:

$$Q_t(v, T_g) = \sum_J \alpha_J(T_g) \left[1 + \frac{kT_g}{Mv^2}\left(1 + 2\,\frac{d \ln q_t^J}{d \ln v}\right)\right] q_t^{J}(v) \tag{1.4.24}$$

This reduces to:

$$Q_t(v, T_g) \simeq Q_t^0 = \sum_J \alpha_J(T_g) q_t^{J}(v) \tag{1.4.25}$$

when the less accurate eq. (1.4.19) is used for P_t. The quantity Q_t may thus be regarded as an average cross section of the molecules for electron scattering. Analogously to P_t and q_t, Q_t can be written also as the sum of an elastic collision term Q_e and of several inelastic collision terms Q_j.

When the gas is atomic, all atoms are in the ground state at the temperatures mentioned; in this case, within the approximation of eq. (1.4.25), Q_t coincides with this state q_t, and no appreciable temperature dependence appears. Electron collisions in rare gases are a typical example of this situation.

Let us consider now the collision probability for momentum transfer P_m. We have:

$$\chi(\mathbf{v}, \mathbf{v'}) = \frac{\mathbf{v} \cdot (\mathbf{v} - \mathbf{v'})}{v^2}$$

$$= \frac{\mathbf{v} \cdot (\mathbf{g} - \mathbf{g'})}{v^2} \tag{1.4.26}$$

Thus we find, from (1.4.7) and the use of eqs. (1.2.8)–(1.2.10):

$$P_m(v, n_g, T_g) = \frac{1}{v} \int_{\mathbf{V}} gF(V) \, d\mathbf{V} \frac{\mathbf{v} \cdot}{v^2} \int_{\mathbf{g'}} (\mathbf{g} - \mathbf{g'}) s(g, g', \vartheta_r, n_g, 0) \, dg'$$

$$= \frac{1}{v^2} \int_{\mathbf{V}} g \frac{\mathbf{v} \cdot \mathbf{g}}{v} P_m(g, n_g, 0) F(V) \, d\mathbf{V}$$

$$= \frac{n_g}{v^2} \int_{\mathbf{V}} g(v - V \cos \psi) q_m(g) F(V) \, d\mathbf{V} \qquad (1.4.27)$$

If expressions (1.4.12) and (1.4.14) are substituted in eq. (1.4.27) for $\cos \psi$ and $d\mathbf{V}$, and integration over φ is performed, the collision probability for momentum transfer becomes:

$$P_m(v, n_g, T_g) = \frac{\pi n_g}{v^4} \int_0^\infty F(V) V \, dV \int_{|v-V|}^{v+V} g^2(v^2 + g^2 - V^2) q_m(g) \, dg \quad (1.4.28)$$

When the gas particles are equal-state atoms or molecules with thermal kinetic energies, expansion (1.4.16) for q_m can be used. By neglecting powers of kT_g/Mv^2 higher than unity, the following result is finally obtained:

$$P_m(v, n_g, T_g) = n_g \left[1 + \frac{kT_g}{Mv^2} \left(2 + 3 \frac{d \ln q_m}{d \ln v} \right) \right] q_m(v) \qquad (1.4.29)$$

Since eq. (1.4.29) has the same structure as (1.4.17), everything that has been said regarding the total collision probabilities and cross sections may be applied also to the collision probabilities and cross sections for momentum transfer. In particular, we can introduce an average molecular cross section for momentum transfer Q_m, according to the same basic formula for these gas conditions [see eq. (1.4.23)]:

$$Q(v, T_g) = \frac{1}{n_g} P(v, n_g, T_g) \qquad (1.4.30)$$

A formula corresponding to eq. (1.4.24) is easily derived for Q_m:

$$Q_m(v, T_g) = \sum_J \alpha_J(T_g) \left[1 + \frac{kT_g}{Mv^2} \left(2 + 3 \frac{d \ln q_m{}^J}{d \ln v} \right) \right] q_m{}^J(v) \quad (1.4.31)$$

When the kT_g/Mv^2 term can be neglected, this reduces to:

$$Q_m(v, T_g) \simeq Q_m{}^0 = \sum_J \alpha_J(T_g) q_m{}^J(v) \qquad (1.4.32)$$

which is formally identical to eq. (1.4.25). Here too Q_m can be written as the sum of Q_{me} and Q_{mj} cross sections.

Several authors, instead of using the average cross sections Q, prefer pressure-normalized, equivalent collision probabilities $\mathscr{P}$, derived from the corresponding P quantities similarly to eqs. (1.4.23) and (1.4.30):

$$P(v, n_g, T_g) = p_0 \mathscr{P}(v, T_g) \tag{1.4.33}$$

Here the reduced pressure $p_0 (= 273.16 p/T_g$, p being the actual gas pressure in torr) is the particle density in units of 3.54×10^{16} cm^{-3}, the standard concentration of molecules in a cubic centimeter at 0°C and 1 torr pressure. Thus the collision probabilities $\mathscr{P}$ in (cm · torr)$^{-1}$ units are equal to 3.12 times the average cross sections Q in atomic units.

Since frequent use will be made in this book of the collision frequency concept, we write here explicitly also the relationship between v and Q or $\mathscr{P}$, which is readily found by substituting (1.4.30) and (1.4.33) in the generalized eq. (1.2.6):

$$\nu = vP = n_g vQ = p_0 v \mathscr{P} \tag{1.4.34}$$

Next we consider the collision probability for energy exchange P_u, for which we have:

$$\chi(\mathbf{v}, \mathbf{v}') = \frac{v^2 - v'^2}{v^2 - 3kT_g/m} \tag{1.4.35}$$

Since:

$$v^2 - v'^2 = |\mathbf{g} + \mathbf{V}|^2 - |\mathbf{g}' + \mathbf{V}|^2$$
$$= (g^2 - g'^2) + 2\mathbf{V} \cdot (\mathbf{g} - \mathbf{g}') \tag{1.4.36}$$

eq. (1.4.7) for P_u reads:

$$
\begin{aligned}
P_u(v, n_g, T_g) &= \left(v^2 - \frac{3kT_g}{m} \right)^{-1} \frac{1}{v} \int_{\mathbf{V}} g F(\mathbf{V}) \, d\mathbf{V} \\
&\quad \times \left[\int_{\mathbf{g}} (g^2 - g'^2) s(g, g', \vartheta_r, n_g, 0) \, d\mathbf{g}' \right. \\
&\quad \left. + 2\mathbf{V} \cdot \int_{\mathbf{g}} (\mathbf{g} - \mathbf{g}') s(g, g', \vartheta_r, n_g, 0) \, d\mathbf{g}' \right] \\
&= \left(v^2 - \frac{3kT_g}{m} \right)^{-1} \frac{1}{v} \int_{\mathbf{V}} g[g^2 P_u(g, n_g, 0) \\
&\quad + 2\mathbf{V} \cdot \mathbf{g} P_m(g, n_g, 0)] F(\mathbf{V}) \, d\mathbf{V} \\
&= \left(v^2 - \frac{3kT_g}{m} \right)^{-1} \frac{1}{v} \int_{\mathbf{V}} g[g^2 P_u(g, n_g, 0) \\
&\quad + (v^2 - V^2 - g^2) P_m(g, n_g, 0)] F(\mathbf{V}) \, d\mathbf{V}
\end{aligned}
\tag{1.4.37}
$$

Introducing the fractional energy loss λ and the collision cross section q_m, defined according to eqs. (1.3.2) and (1.3.1), respectively, substituting for $d\mathbf{V}$ its expression (1.4.14) and integrating over φ we obtain eq. (1.4.37) in the form:

$$P_u(v, n_g, T_g) = \left(v^2 - \frac{3kT_g}{m}\right)^{-1} \frac{2\pi n_g}{v^2} \int_0^\infty F(V)V \, dV$$

$$\times \int_{|v-V|}^{v+V} g^2[g^2\lambda(g) + v^2 - V^2 - g^2]q_m(g) \, dg \quad (1.4.38)$$

When gas particles are equal-state atoms or molecules with thermal kinetic energies, eq. (1.4.38) becomes, in the first kT_g/Mv^2 approximation, analogously to eqs. (1.4.17) and (1.4.29):

$$P_u(v, n_g, T_g) = n_g\left[\lambda(v) - \frac{2kT_g}{Mv^2}\left(4 + \frac{d\ln q_m}{d\ln v}\right)\right]\left(1 - \frac{3kT_g}{mv^2}\right)^{-1} q_m(v) \quad (1.4.39)$$

In this derivation we have assumed that $\lambda(v)$ is a slow function of v much smaller than unity, which may be considered as having the same order of magnitude of kT_g/Mv^2. This last condition is best satisfied when the incoming electron kinetic energy is not much larger than the kinetic energies of the particles, that is, in the case for which we are chiefly interested in applying the gas particle motion corrections.

If eq. (1.3.26) is substituted for λ, and a velocity-independent v_m collision frequency is assumed $[d\ln q_m/d\ln v = -1]$ to simplify our formulas, we obtain:

$$P_u(v, n_g, T_g) = \frac{2m}{M} n_g q_m(v) + (u - \tfrac{3}{2}kT_g)^{-1}n_g \sum_j u_j q_j(v) \quad (1.4.40)$$

When the gas is composed of molecules in thermodynamic equilibrium at temperature T_g, the collision probability for energy exchange will be given by a sum of terms like (1.4.40), one for each rotational J level. Thus we have, introducing an average collision cross section for energy exchange Q_u according to eq. (1.4.30):

$$Q_u(v, n_g, T_g) = \frac{2m}{M} \sum_J \alpha_J q_m{}^J + (u - \tfrac{3}{2}kT_g)^{-1} \sum_J \alpha_J \sum_j^{(J)} u_j q_j \quad (1.4.41)$$

where $\sum_j^{(J)}$ means that the sum is made only over the transitions whose initial level has the rotational quantum number J. Let us introduce the previously defined cross section Q_j for the excitation, by an electron of speed v in a gas at temperature T_g, of a transition j from the rotational J level. According to eq. (1.4.25), Q_j is related to q_j by the formula:

$$Q_j(v, T_g) \simeq Q_j{}^0 = \alpha_J(T_g)q_j(v) \quad (1.4.42)$$

Using these Q_j^0 cross sections and the average cross section Q_m^0 for momentum transfer, we find that eq. (1.4.41) becomes:

$$Q_u(v, n_g, T_g) = \frac{2m}{M} Q_m^0 + (u - \tfrac{3}{2}kT_g)^{-1} \sum_j u_j Q_j^0 \qquad (1.4.43)$$

When (1.4.43) is divided by Q_m^0 the G-factor defined by eq. (1.2.13) is obtained. Therefore, this factor can be written as:

$$G(v, T_g) = \frac{2m}{M} + \frac{\sum\limits_j u_j Q_j^0}{(u - \tfrac{3}{2}kT_g)Q_m^0} \qquad (1.4.44)$$

For the reason explained in Section 1.3, the inelastic collision terms in eq. (1.4.44) disappear in the case of a slow electron moving through an atomic gas. When this happens, G and λ coincide, and both are simply equal to the constant value $2m/M$.

The last collision probability we shall consider is P_d. In this case we have:

$$\chi(\mathbf{v}, \mathbf{v}') = \frac{v'^2}{v^2} \sin^2 \vartheta \qquad (1.4.45)$$

The quantity $v'^2 \sin^2 \vartheta$ can be written as the sum of the two diagonal components along the axes x and y (Fig. 1.4.1) of the tensor $(\mathbf{v}' - \mathbf{v})(\mathbf{v}' - \mathbf{v}) = (\mathbf{g}' - \mathbf{g})(\mathbf{g}' - \mathbf{g})$. Then eq. (1.4.7) provides:

$$P_d(v, n_g, T_g) = \frac{1}{v^3} \int_{\mathbf{V}} gF(V) \, d\mathbf{V} \int_{\mathbf{g}'} \sum_{i=x,y} (\mathbf{g}' - \mathbf{g})_i (\mathbf{g}' - \mathbf{g})_i s(g, g', \vartheta_r, n_g, 0) \, d\mathbf{g}'$$

$$= \frac{1}{v^3} \int_{\mathbf{V}} g \sum_{i=x,y} \delta_{ii}(g, n_g, 0) F(V) \, d\mathbf{V} \qquad (1.4.46)$$

where the $\boldsymbol{\delta}$ tensor defined in Section 1.2 has been introduced. This tensor in the $\mathbf{V}$-relative velocity frame has the structure given by eq. (1.2.14); transformation from this frame to the absolute laboratory frame provides:

$$\sum_{i=x,y} \delta_{ii} = \frac{1 + \cos^2 \vartheta_0}{2} \delta_\perp + \sin^2 \vartheta_0 \, \delta_\| \qquad (1.4.47)$$

In Fig. 1.4.1 we recognize that ϑ_0 can be substituted for by ψ, according to the formula:

$$\sin \vartheta_0 = \frac{V}{g} \sin \psi \qquad (1.4.48)$$

Cross sections q_m and q_d and the loss factor λ can replace $\delta_\|$ and $\delta_\perp$ according to eqs. (1.2.18) and (1.2.20), so that (1.4.47) becomes:

$$\sum_{i=x,y} \delta_{ii}(g, n_g, 0) = n_g\{(g^2 - \tfrac{3}{2}V^2 \sin^2 \psi)q_d(g)$$

$$+ [2 - \lambda(g)]V^2 \sin^2 \psi \, q_m(g)\} \qquad (1.4.49)$$

Equations (1.4.12), (1.4.14), and (1.4.49), when substituted into eq. (1.4.46), yield, after integration over φ:

$$P_d(v, n_g, T_g) = \frac{\pi n_g}{2v^6} \int_0^\infty F(V)V\, dV$$

$$\times \int_{|v-V|}^{v+V} g^2 \Big\{ [\tfrac{3}{2}(v^2 - V^2)^2 + (v^2 - 3V^2)g^2 + \tfrac{3}{2}g^4]q_d(g)$$

$$- \Big[2 - \frac{\lambda}{g}\Big][(v^2 - V^2)^2 - 2(v^2 + V^2)g^2 + g^4]q_m(g) \Big\}\, dg$$

(1.4.50)

Since usually $\lambda \ll 1$, in a gas of equal-state molecules with thermal kinetic energies the same considerations that yielded eqs. (1.4.17) and (1.4.29) now provide:

$$P_d(v, n_g, T_g) = n_g\Big[1 + \frac{kT_g}{Mv^2}\Big(3 + 4\frac{d\ln q_d}{d\ln v}\Big)\Big]q_d(v)$$

$$+ n_g \frac{4kT_g}{Mv^2} q_m(v)$$

(1.4.51)

Equation (1.4.51) has the same structure as eqs. (1.4.17) and (1.4.29), so that everything which has been said for P_t and P_m applies also to P_d. Here too an average collision cross section Q_d could be easily derived, according to definition (1.4.30) and after having considered the contributions of all molecular rotational levels.

In summarizing the results of this section, it is important to remark that we have been able to replace all electron collision probabilities P_t, P_m, P_u, and P_d, in an un-ionized monatomic or monomolecular gas at thermal energies, by properly defined average particle cross sections. For a given gas they are functions only of the electron speed v and of the gas temperature T_g. The determination of the cross sections Q_m^0 and Q_j^0, or of the corresponding G-factor, is the major goal of the methods and techniques that will be discussed in this book. For this reason, in one of the following sections we shall express the most important electron swarm properties that we can measure in terms of these quantities.

However, when the swarm properties are measured in a typical ionized gas environment, as is frequently the case, the gas cannot be rigorously considered in thermodynamic equilibrium, since there will also be atoms or molecules in nonthermal excited states, as well as ions. In a weakly ionized gas these excited atoms and molecules usually constitute such a very small fraction of the total existing particles that their contribution can be entirely neglected. Although ions constitute a very small fraction too, their effect can be quite appreciable, since the electrostatic forces due to the unneutralized charges are of very long range, so that the number of distant elastic interactions between electrons and gas ions may be comparatively large.

Therefore, in these cases extra terms have to be added to the collision probabilities discussed, to take into account the contributions from ion collisions (collisions with neutral atoms or molecules and with ions are normally independent processes, and thus probabilities can simply be added). Since in general the microscopic description of these interactions is not of the single-collision type at the ion densities at which these terms usually become important, the ion collision contributions exist and can be added to all probabilities except P_t. For the same reason these ion terms cannot be given as functions of physically significant collision cross sections, as we have done above for the case of neutral atoms and molecules.

From the foregoing discussion it is apparent that, for the purpose of measuring electron collision parameters, the ionized gas environment must be chosen so to have a negligible contribution from the ions when our goal is to determine neutral molecule collision cross sections, and a comparable or dominant contribution from the ions when their collision parameters constitute the quantity we would like to measure.

1.5   THEORETICAL VALUES OF COLLISION PARAMETERS

We said in the Introduction that we do not intend to discuss here theories regarding electron collisions in gases. Therefore this section is devoted, not to theories, but only to the presentation of a few theoretical formulas whose validity is sufficiently general to justify their use either for a preliminary evaluation of the expected values of collision parameters, or for a guide in the analytical elaboration of the experimental data.

A. Elastic collisions

It has been known for a long time [Massey (1956)] that q_e, the total cross section for elastic collisions between electrons and atoms or molecules, can be written, according to quantum theory, as follows:

$$q_e(v) = \frac{\lambda_e^2}{\pi} \sum_{l=0}^{\infty} (2l + 1) \sin^2 \eta_l \tag{1.5.1}$$

where:

$$\lambda_e = \frac{h}{mv} = \left(\frac{153.5}{u}\right)^{\frac{1}{2}} \text{Å}^2 \tag{1.5.2}$$

is the incident electron de Broglie wavelength (u in electron volts), and η_l is the scattering phase shift that the lth partial wave, introduced in the solution of the wave equation for the collision process, suffers for the action of the gas particle scattering potential. The integer l is a measure of the electron angular momentum about the scattering particle center.

Similarly q_{me}, the elastic collision cross section for momentum transfer, can be written as:

$$q_{me} = \frac{\lambda_e^2}{\pi} \sum_{l=0}^{\infty} (l + 1) \sin^2 (\eta_l - \eta_{l+1}) \tag{1.5.3}$$

Both eqs. (1.5.1) and (1.5.3) are obtained by substituting into (1.3.22′) and (1.3.22″) the angular scattering distribution:

$$s_3(\vartheta) = \frac{n_g \lambda_e^2}{8\pi} \left| \sum_{l=0}^{\infty} (2l + 1)[\exp (2j\eta_l) - 1]P_l(\cos \vartheta) \right|^2 \tag{1.5.4}$$

the P_l being Legendre polynomials. Distribution (1.5.4) neglects the effect of spin-orbit splitting on the description of the scattering; however, according to calculations by Hoeper, Franzen, and Gupta (1968), this procedure is legitimate, except in the vicinity of scattering resonances.

Calculations of the phase shift values η_l as a function of the wavelength λ_e for various atoms and molecules in the ground state have been carried out by many authors. Although considerable success has been obtained in explaining some typical features of the collision cross section versus electron velocity behavior, like the existence in certain gases (as in argon, krypton, and xenon and in some molecules) of a well-pronounced minimum, known as the Ramsauer-Townsend effect, significant discrepancies still exist among most of the cross sections computed by the different authors, particularly at the lowest velocities ($u < 1$ eV). For this reason, and since excellent, detailed reviews of these calculations can be found in the literature [Massey (1956), Burke and Smith (1962), Moiseiwitsch (1962), Drukarev (1965), Burke (1969], we shall not discuss these theoretical results here with the exception of the work of O'Malley, Rosenberg, and Spruch (1962) for atoms.

The work of these authors seems to the writer particularly significant, since it provides for the phase shifts and for the collision cross sections formulas that at the same time are relatively simple and have sufficiently general validity, relying on only a small number of parameters for each gas. The values of the parameters are not given by this theory, but a few experimental points for each gas, anywhere in the energy region from zero to several electron volts, are sufficient to determine them. Therefore this method makes it possible to compare data obtained over different electron energy ranges by different methods, and in particular to provide estimates of the collision cross-section amplitudes at very low energies from data measured at higher energies.

The theory of O'Malley, Rosenberg, and Spruch applies to atoms that do not possess a permanent electric quadrupole moment. For this case, the long-range $\propto \alpha/r^4$ polarization potential, due to the dipole moment induced in an atom of polarizability α by the electron field, is shown to provide the major scattering interaction at very low energies. By modifying the so-called effective range theory, largely used in nuclear physics, so as to take into account this long-range potential, O'Malley et al. were able to derive convenient low-energy expansions for the scattering phase shifts and for the collision cross sections, according to eqs. (1.5.1) and (1.5.3).

The "atomic effective range theory" provides the following formulas for the phase shifts at electron energies below a few electron volts:

$$\left.\begin{aligned} \frac{\lambda_e \tan \eta_0}{2\pi a_0} &= -A - 0.2840\alpha\sqrt{u} - 0.04902 A\alpha u \ln u + Bu \\[2mm] \frac{\lambda_e \tan \eta_1}{2\pi a_0} &= 0.05679\alpha\sqrt{u} - 0.07353 A_1 u \\[2mm] \frac{\lambda_e \tan \eta_l}{2\pi a_0} &= \frac{0.8518\alpha\sqrt{u}}{(2l+3)(2l+1)(2l-1)}, \quad l > 1 \end{aligned}\right\} \quad (1.5.5)$$

where u is in electron volts, the polarizability α is in units of a_0^3, and the parameters A ("the scattering length" in units of a_0), B, and A_1 are left to be determined by experiment. At very low electron energies, the last terms in the expansions for η_0 and η_1 become negligibly small, and the collision cross sections are thus functions of A alone. We shall refer to these formulas later, when discussing the experimental results. The Q_m collision cross sections determined at low energies from swarm experiments will thus be compared with the results of other experiments performed at higher energies using conventional electron-beam methods (Q_t and s_3).

When elastic collisions with charged particles are considered, the scattering effect of the large number of small perturbations, which take place at significant densities of these particles, can be evaluated assuming an approximate description of the interaction process. Binary interaction according to a Coulomb potential (Rutherford scattering) within a properly defined sphere of each charged particle, and no interaction outside, were assumed by Spitzer and Härm (1953) for this purpose. Simple formulas for the various collision probabilities are thus obtained when the gas is a plasma, namely, when it is made of electrons and of positive ions (only singly ionized particles are considered) in approximately equal concentrations (n_e), and when the electron (T_e) and ion (T_i) temperatures are comparable. In this case physical intuition suggests assuming as the most appropriate interaction sphere of the collision process the largest sphere inside which the numbers of electrons and positive ions remain essentially uncorrelated, even when the potential energy due to the mutual Coulomb forces of the charged particles is taken into account. The radius of this sphere is $(\varepsilon_0 kT_e/e^2 n_e)^{1/2}$, e being the electron charge and ε_0 the free space permittivity [a derivation of this result can be found, e.g., in Longmire (1963)]; this same quantity is also equal to the Debye length for a positive ion shielded by an electron cloud, namely, it is equal to the characteristic length of the radial exponential decay of the average potential of the shielded ion.

Each electron collision probability is the sum of a term relative to elastic collisions with positive ions and of a term relative to elastic collisions with the other electrons. On the assumption that each scattering particle is at rest, so that temperature may enter only in determining the interaction sphere dimensions, each kind of charged particles provides a contribution to the collision probability for momentum transfer given by the expression:

$$P_{me}{}^0 = \frac{e^4(M+m)n_e}{8\pi\varepsilon_0{}^2 Mm^2 v^4}\ln\left[1 + \frac{16\pi^2\varepsilon_0{}^3 kT_e}{e^6 n_e}\left(\frac{mM}{M+m}\right)^2 v^4\right] \qquad (1.5.6)$$

where M is here the mass of the scattering particle under consideration. Analogously the other collision probabilities are given by the expressions:

$$P_{ue}{}^0 = \frac{2m}{M+m}\,P_{me}{}^0 \qquad (1.5.7)$$

$$P_{de}{}^0 = \frac{e^4 n_e}{4\pi\varepsilon_0{}^2 m^2 v^4}\left\{\ln\left[1 + \frac{16\pi^2\varepsilon_0{}^3 kT_e}{e^6 n_e}\left(\frac{mM}{M+m}\right)^2 v^4\right] - 1\right\} \qquad (1.5.8)$$

Most plasmas are such that in eq. (1.5.8) we can neglect unity compared to the logarithmic term and write:

$$P_{de}{}^0 = \frac{2M}{M+m}\,P_{me}{}^0 \qquad (1.5.9)$$

The kinetic motions of charged particles can be taken into account as discussed in Section 1.4, since the Spitzer-Härm theory assumes binary interactions between colliding particles. Thus P_{me} can be computed by means of eq. (1.4.28), where the $n_g q_m(g)$ factor is replaced by $P_{me}{}^0$, here a function of g in place of v. Treating the logarithmic term of $P_{me}{}^0$ as a constant, since it varies only slowly, we easily derive the following result:

$$\begin{aligned}
P_{me} &= P_{me}{}^0(v)\pi\int_0^\infty F(V)V\,dV\int_{|v-V|}^{v+V}\left(1 + \frac{v^2 - V^2}{g^2}\right)dg \\
&= P_{me}{}^0(v)4\pi\int_0^v F(V)V^2\,dV
\end{aligned} \qquad (1.5.10)$$

Analogously from eq. (1.4.38) we obtain:

$$P_{ue} = \left(1 - \frac{3kT_g}{mv^2}\right)^{-1} P_{ue}{}^0 \left[4\pi \int_0^v F(V)V^2 \, dV - \frac{M}{m} v 4\pi \int_v^\infty F(V)V \, dV\right]$$

(1.5.11)

$$P_{ue} = \left(1 - \frac{3kT_g}{mv^2}\right)^{-1} P_{ue}{}^0 \left[4\pi \int_0^v F_M(V)V^2 \, dV - \frac{kT_{i,e}}{m} 4\pi v F_M(v)\right]$$ (1.5.12)

and from eq. (1.4.50) within the limits of approximation (1.5.9):

$$P_{de} = P_{de}{}^0 \left[4\pi \int_0^v F(V)V^2 \, dV - \frac{1}{v^2} \frac{4\pi}{3} \int_0^v F(V)V^4 \, dV + v \frac{8\pi}{3} \int_v^\infty F(V)V \, dV\right]$$

(1.5.13)

$$P_{de} = P_{de}{}^0 \left[\left(1 - \frac{kT_{i,e}}{Mv^2}\right) 4\pi \int_0^v F_M(V)V^2 \, dV + \frac{kT_{i,e}}{M} 4\pi v F_M(v)\right]$$ (1.5.14)

Equations (1.5.10), (1.5.11), and (1.5.13) are obtained without any assumption regarding the actual shape of the distribution $F(V)$, whereas (1.5.12) and (1.5.14) are the specialized formulas that apply when $F(V)$ is Maxwellian with temperature T_i or T_e. It should be noted that in the general case of a non-Maxwellian distribution the approximate use of eqs. (1.5.6) and (1.5.8) is allowed, but then we must regard temperature T_e only as a kinetic parameter that specifies the mean energy of the electrons ($3kT_e/2$).

For positive ions, $M \gg m$ and the V values over which $F_M(V)$ is substantially different from zero are such that $V \ll v$ for electron velocities of interest in gas discharge studies. In these cases the above equations simplify considerably; for instance, eq. (1.5.10) becomes:

$$P_{me} \simeq P_{me}{}^0 \simeq \frac{e^4 n_e}{8\pi\varepsilon_0{}^2 m^2 v^4} \ln\left(1 + \frac{16\pi^2\varepsilon_0{}^3 kT_e m^2 v^4}{e^6 n_e}\right)$$

(1.5.15)

When the scattering particles are electrons ($M = m$), the Maxwellian expressions lose most of their value for studies of electron swarm properties. In fact, in most cases the velocity distribution function of the electrons in a swarm is not Maxwellian, and the two problems of finding the distribution and evaluating the effects on the electron motion of the electron interactions must be solved together. More will be said about this in the next chapter.

B. Inelastic collisions with molecules

Still rather limited is our theoretical knowledge regarding the q_i and q_{mi} cross sections for inelastic scattering, which are of interest in studies of low-energy electron swarms. In fact, for electrons with energies below a few electron volts, rotational and vibrational excitations of molecules must be considered as possible inelastic processes. This problem is amenable to a theoretical formulation, whose analytical treatment is feasible only when the molecules possess a dipole or a quadrupole moment, so that the corresponding long-range interactions overcome all óther interactions. All the theoretical work done on the determination of the electron collision cross sections for rotational excitation of molecules has been recently reviewed and summarized by Takayanagi and Itikawa (1970).

We shall discuss first the basic theory, due to Gerjuoy and Stein (1955), which provides the cross sections for rotational excitation by electron collisions of a homonuclear diatomic neutral molecule (zero electric dipole moment, but nonvanishing quadrupole moment).

According to the usual selection rules for electric quadrupole transitions, an electron can excite only rotational transitions in which the quantum number J changes by 2 units. Therefore, if we have a molecule in the Jth level, an electron inelastic collision of the first kind can excite the molecule to the $(J + 2)$th level only; the electron will then lose, according to eq. (1.4.22), the energy:

$$u_{J,J+2} = e(E_{J+2} - E_J) = e(4J + 6)B_0 \qquad (1.5.16)$$

Similarly, an electron inelastic collision of the second kind can de-excite the molecule to the $(J - 2)$th level only, and the electron gains the corresponding energy:

$$-u_{J,J-2} = e(E_J - E_{J-2}) = e(4J - 2)B_0 \qquad (1.5.17)$$

Let us call $q_{J,J+2}$ and $q_{J,J-2}$ the two q_j cross sections for the transition $J \rightarrow J + 2$ and $J \rightarrow J - 2$. Obviously $q_{J,J+2} = 0$ for $u < u_{J,J+2}$, and $q_{J,J-2} = 0$ for $J = 0$ and 1. Gerjuoy and Stein have shown that, for rotational transitions between Σ states, calculations based on the Born approximation and on the use of the large-r form of the quadrupole potential yield:

$$q_{J,J+2} = \frac{8\pi \mathcal{Q}^2 a_0^2}{15} \frac{(J + 2)(J + 1)}{(2J + 3)(2J + 1)} \left(1 - \frac{u_{J,J+2}}{u}\right)^{1/2} \quad \text{for } u \geqslant u_{J,J+2}$$

$$(1.5.18)$$

$$q_{J,J-2} = \frac{8\pi \mathcal{Q}^2 a_0^2}{15} \frac{J(J - 1)}{(2J + 1)(2J - 1)} \left(1 - \frac{u_{J,J-2}}{u}\right)^{1/2} \quad \text{for } J \geqslant 2 \qquad (1.5.19)$$

where $\mathcal{Q}$ is the electric quadrupole moment of the molecule, in units of ea_0^2. Since the scattering distribution s_j is isotropic, the inelastic cross sections for momentum transfer $q_{mJ,J+2}$, $q_{mJ,J-2}$ have the same values (1.5.18) and (1.5.19). According to definition (1.4.42), $Q_{J,J+2}$ and $Q_{J,J-2}$ are equal to the corresponding cross sections (1.5.18) and (1.5.19) multiplied by the Jth population fraction α_J, given by eq. (1.4.21).

These cross sections, as well as those for momentum transfer and the corresponding ones for excitation of vibrational levels, are generally much smaller than the elastic ones, so that they may usually be neglected in the expressions for Q_t and Q_m. However, they play a dominant role in the expressions for the collision probability for energy exchange P_u and for the energy loss factor G.

Thus, considering only excitation and de-excitation of rotational levels, the formulas for P_u and G, given at the end of Section 1.4, become:

$$P_u = \frac{2m}{M} n_g Q_m{}^0 + (u - \tfrac{3}{2}kT_g)^{-1} n_g \sum_J (u_{J,J+2} Q^0_{J,J+2} + u_{J,J-2} Q^0_{J,J-2})$$

(1.5.20)

$$G = \frac{2m}{M} + \frac{\sum_J (u_{J,J+2} Q^0_{J,J+2} + u_{J,J-2} Q^0_{J,J-2})}{(u - \tfrac{3}{2}kT_g) Q_m{}^0}$$

(1.5.21)

Since the ground states of the most common molecules, such as H_2, N_2, and O_2, are Σ states, formulas (1.5.20) and (1.5.21) can be extensively used in all cases where the electron energy is lower than the vibrational excitation threshold, or where the cross section for vibrational excitation is small enough compared to the cross section for rotational excitation.

These formulas simplify considerably when the electron energy, while remaining such that vibrational excitation can be neglected, is, however, large enough that the assumption $u \gg u_{J,J+2}$, $|u_{J,J-2}|$ can be made for all J levels that contribute significantly to the overall cross section values. In this case the cross sections (1.5.18) and (1.5.19) become independent of energy, and we have:

$$\sum_J (u_{J,J+2} Q^0_{J,J+2} + u_{J,J-2} Q^0_{J,J-2})$$

$$= \frac{8\pi \mathcal{Q}^2 a_0{}^2}{15} 2eB_0 \sum_J \left[\frac{(J+2)(J+1)}{2J+1} - \frac{J(J-1)}{2J+1} \right] \alpha_J$$

$$= \frac{32\pi}{15} ea_0{}^2 \mathcal{Q}^2 B_0$$

(1.5.22)

Since in these cases we have typically $u \gg kT_g$, it appears appropriate to write the expression for G in the form:

$$G = \frac{2m}{M} \left(1 + \frac{\eta \mathcal{Q}^2}{v^2 Q_m{}^0} \right)$$

(1.5.23)

Here η is a constant given by:

$$\eta = \frac{32\pi}{15} \frac{M}{m} \frac{eB_0}{m} a_0{}^2 = 6.02 \times 10^{-6} B_0 M_a \quad (m^4 \ sec^{-2})$$

(1.5.24)

where B_0 is in electron volts, and M_a indicates the molecular mass in atomic units.

Corrections to the cross sections of Gerjuoy and Stein were obtained by Dalgarno and Moffett (1963), who introduced a multiplying factor due to asymmetric polarization forces; by Sampson and Mjolsness (1965), who considered the effect of including both symmetric and asymmetric polarization forces; by Geltman and Takayanagi (1966), who included short-range forces; and by Ardill and Davison (1968), who added exchange effects. More recently Lane and Geltman (1967) computed rotational cross sections using close-coupling calculations, Henry and Lane (1969) modified them to include both polarization and exchange; and Hara (1969) performed similar calculations, including polarization and exchange, but using the adiabatic theory [see Fano (1970) for the relationship between the close-coupling and the adiabatic approaches]. Detailed formulas and numerical results for the case of hydrogen can be found in the original papers. The necessity to resort to these more sophisticated calculations, since Born approximation fails beyond the threshold energy region, has been justified by Chang (1970).

Stabler (1963) extended the theory of Gerjuoy and Stein to the case of rotational excitation of homonuclear diatomic molecular ions. The corresponding cross sections for an initial Σ state of rotational quantum number J are as follows:

$$q_{J,J+2} = \left(\ln 2 - \frac{2}{3}\right) \frac{4\pi \mathscr{D}^2 h^2}{5mu} \frac{(J+2)(J+1)}{(2J+3)(2J+1)} \tag{1.5.25}$$

$$q_{J,J-2} = \left(\ln 2 - \frac{2}{3}\right) \frac{4\pi \mathscr{D}^2 h^2}{5mu} \frac{J(J-1)}{(2J+1)(2J-1)} \tag{1.5.26}$$

The excitation cross section rises discontinuously from zero to the finite value (1.5.25) at the threshold energy (1.5.16). Later Sampson (1965) established that Stabler's derivation of (1.5.25) and (1.5.26) is valid only if the corresponding P_u values are $\lesssim 10\%$ of the value (1.5.7) for elastic collisions.

Takayanagi (1966) computed, using the Born approximation, the cross sections for the rotational transitions $J \to J+1$ and $J \to J-1$ in diatomic molecules with a permanent electric dipole moment $\mathscr{M}$ [in atomic units (au): $ea_0 = 2.54 \times 10^{-18}$ esu cm $= 2.54$ debye]:

$$q_{J,J+1} = \frac{8\pi e \mathscr{R}_y \mathscr{M}^2 a_0^2}{3u} \frac{J+1}{2J+1} \ln \frac{\sqrt{u} + \sqrt{u - u_{J,J+1}}}{\sqrt{u} - \sqrt{u - u_{J,J+1}}}$$

$$\text{for } u \geqslant u_{J,J+1} \tag{1.5.27}$$

$$q_{J,J-1} = \frac{8\pi e \mathscr{R}_y \mathscr{M}^2 a_0^2}{3u} \frac{J}{2J+1} \ln \frac{\sqrt{u - u_{J,J-1}} + \sqrt{u}}{\sqrt{u - u_{J,J-1}} - \sqrt{u}}$$

$$\text{for } J \geqslant 1 \tag{1.5.28}$$

where $\mathscr{R}_y$ is the Rydberg constant (13.6 eV), and $u_{J,J+1}, u_{J,J-1}$ are the energy jumps at each collision, equal to:

$$u_{J,J+1} = e(E_{J+1} - E_J) = 2e(J + 1)B_0$$

$$-u_{J,J-1} = e(E_J - E_{J-1}) = 2eJB_0$$

(1.5.29)

The cross sections for rotational transitions in the case of two specific polar molecules, CN and HCl, were also computed by Itikawa and Takayanagi (1969a), using the close-coupling method; thus they have shown that the simple Born approximation [eqs. (1.5.27) and (1.5.28)] gives fairly good results (maximum error 20%), even when the molecule has a dipole moment as large as $\mathscr{M} \simeq 1$ au. More recently Itikawa (1971) has derived the corresponding formulas in the Born approximation also for the case of nonlinear polyatomic molecules with a permanent dipole moment.

Formulas (1.5.27) and (1.5.28) can be used for computing P_u and G. The equivalent of the continuous approximation (1.5.23) is here [Hake and Phelps (1967)]:

$$G = \frac{2m}{M} \left(1 + \frac{\eta \, \mathscr{M}^2}{v^{\frac{1}{2}} Q_m^{\,0}} \right)$$

(1.5.30)

where:

$$\eta' = \frac{28\sqrt{2}\,\pi}{3} \Gamma\!\left(\frac{7}{8}\right) \frac{M}{m} \frac{e\mathscr{R}_y}{m} \left(\frac{eB_0}{m}\right)^{\!\frac{3}{4}} \left(\frac{eB_0}{kT_g}\right)^{\!\frac{1}{8}} a_0^{\,2}$$

$$= 2.37 \times 10^5 B_0^{7/8} M_a \quad (\mathrm{m}^{11/2}\ \mathrm{sec}^{-7/2})$$

(1.5.31)

In the case of dipolar molecules which also show significant quadrupole interaction effects in electron collision processes, the additional term $\eta \mathscr{Q}^2/v^2 Q_m^{\,0}$ of (1.5.23) has to be added within the parentheses of (1.5.30).

Altshuler (1957), who was also concerned with molecules having a permanent electric dipole moment, indicated that in this case the excitation of rotational motion may be so efficient that the sum of cross sections for the inelastic collisions usually overcomes the elastic cross section. In particular, he calculated the average cross section for momentum transfer, treating the molecules as point-like, linear, fixed rotators, and obtained with the Born approximation the following result:

$$Q_m^{\,0} = \sum_j Q_{mj} = \frac{1}{3\pi} \frac{h^2 \mathscr{M}^2}{mu} = 31.8 \frac{\mathscr{M}^2}{u\ (\mathrm{eV})}$$

(1.5.32)

It is important to note that this cross section is independent of the gas temperature T_g.

The reader must be cautioned, however, that Altshuler's result is based on the assumption that $v' \simeq v$ in the P_m definition (1.2.10). Therefore (1.5.32) can be used only when the incident electron energy is sufficiently larger than most energy jumps u_j.

More recently it was shown that Altshuler's result (1.5.32) retains its validity for all polar rigid rotators, regardless of shape or symmetry [Crawford (1967)], and Wijnberg (1966) derived an analogous formula for quadrupole contribution in linear molecules:

$$Q_m^0 = \frac{16}{45} \pi a_0^2 \mathcal{Q}^2 \tag{1.5.33}$$

Other significant developments of the electron scattering theory for dipolar molecules are as follows:

1. Shimizu's (1963) solution for the case of a dipole of finite size.
2. Mittleman's and von Holdt's (1965) exact solutions for the case of a point dipole.
3. Takayanagi's and Itikawa's (1968) calculations for a two-center dipole potential. In this case, for fixed charge separation and electron energy, the Q_m cross section exhibits as a function of $\mathcal{M}$ a resonance-type behavior.
4. Crawford's (1968) and Itikawa's (1969) close-coupling calculations in H_2O and in CN and HCl, respectively. Here, too, the resulting Q_m cross sections exhibit resonance as a function of $\mathcal{M}$ (peak at $\mathcal{M} \simeq 1$ au) and larger values than the Born estimates by Altshuler.

Furthermore, theoretical analysis of the scattering process in the $\mathcal{M}$ region above 0.639 au (1.625 debye) should also take into account the existence of electron bound states; in fact 0.639 au is the minimum dipole moment required to bind temporarily an electron to a stationary (infinitely massive) dipole in the ground state [Mittleman and Myerscough (1966), Turner and Fox (1966)], whereas for a nonstationary dipole and rotationally excited states higher critical $\mathcal{M}$ values are predicted [Garrett (1971), Bottcher (1971)].

The theory of vibrational excitation of molecules by electrons is still at a more rudimentary and less satisfactory stage of development than that of rotational excitation.

Vibrational excitation of molecules possessing a permanent electric dipole moment arises, at least partially, through the interaction represented by the dipole matrix element. Using the Born approximation, Takayanagi (1966) derived a formula for this collision cross section in terms of the infrared transition probability between the same vibrational states; Breig and Lin (1965) included polarization interactions in their Born approximation calculations and found that the polarization term is comparable to the dipole one. More recently Itikawa and Takayanagi (1969b) performed close-coupling calculations in CO, including both interactions; the Breig-Lin result that the two interaction terms are comparable is confirmed. The possibility of using the same Born-approximated formula of Takayanagi also in the case of polyatomic molecules with no permanent dipole moment, like CO_2, has been contended by Claydon, Segal, and Taylor (1970); these authors, who have discussed the case of CO_2 molecules, have come to the conclusion that in this case the Born approximation is valid only to estimate the order of magnitude of the excitation cross section of some vibrational mode (e.g., antisymmetric stretching and bending modes of CO_2).

Born approximation analyses for homonuclear molecules were also performed by Takayanagi (1965), Breig and Lin (1965), and Truhlar and Rice (1970). Close-coupling calculations in H_2 were carried out by Henry (1970), who considered both pure vibrational excitations and simultaneous rotational-vibrational excitations; an important result, use of which will be made later, is that the total vibrational cross section, namely, the sum of the cross sections for pure vibrational excitations and for simultaneous rotational-vibrational excitations, is almost independent of the initial rotational state, so that it depends neither on the gas temperature nor on the ortho- and para-form concentrations of the hydrogen gas.

Furthermore, vibrational excitations may display a resonance behavior, which occurs because of a resonance interaction between the electron and the molecular potential, the electron being first captured in a temporary negative-ion state of the molecule. Theoretical calculations of vibrational cross sections in H_2 based explicitly on this assumption were performed by Bardsley, Herzenberg, and Mandl (1966) and by Abram and Herzenberg (1969).

Chapter 2

The Electron Distribution Function and the Transport Parameters

In this chapter we propose to derive the basic mathematical expressions that relate the main observable macroscopic properties of electron swarms in gases to the elementary collision parameters defined in Chapter 1. These formulas will be used later for the discussion of the experimental data obtained by various investigators.

Since an electron swarm consists of a large number of electrons having different positions and velocities, a convenient description must be chosen for handling this type of problem. Adequate for our purpose is the statistical approach, in the form known as the kinetic theory of gases. This theory is based on knowledge of the phase-space distribution function for each species of existing gas particle.

The principal concern of this book is with electron swarm experiments performed for the purpose of determining slow electron collision parameters. In these cases it can generally be assumed that distribution functions are known for all existing particles, except electrons. In fact, the kinetic status of the gas in these experiments is such that we can usually regard atoms and molecules as particles with simple known distributions (mostly uniform in space and Maxwellian in velocity).

The first part of this chapter will be devoted to a review of the most important equations that can be used for the determination of the electron distribution function. There is a vast literature on the subject. Therefore we shall present here only the basic developments which provide the electron distribution function relations currently assumed by authors in the interpretation of the experimental data in which we are interested. The expressions for the transport properties (diffusion coefficients, electrical and thermal conductivities, etc.) are derived in the second part of the chapter. The role of the various collision parameters in these formulas will also be discussed.

The first sections of this chapter follow in general the concise but lucid exposition of this subject made by Allis (1956) in *Handbuch der Physik*. Readers who desire more details and a larger bibliography are referred to the extensive literature, well organized and summarized by Shkarofsky, Johnston, and Bachynski (1966) in their book *The Particle Kinetics of Plasmas*.

2.1 THE BOLTZMANN EQUATION

Let us consider an electron swarm in a gas, under the action of a continuous Lorentz force field $\mathbf{E} + \mathbf{v} \times \mathbf{B}$, where the electric and magnetic fields $\mathbf{E}, \mathbf{B}$ are in general functions of time and position. The number of electrons that at time t lie within the volume $d\mathbf{r}$ centered at $\mathbf{r}$ in configuration space and whose velocities lie within the volume $d\mathbf{v}$ centered at $\mathbf{v}$ in velocity space is $f(\mathbf{r}, \mathbf{v}, t)\, d\mathbf{r}\, d\mathbf{v}$. The density distribution function f so defined can be computed by solving an integrodifferential equation, which expresses mathematically the property of continuity for the flow of electrons in phase space.

If the continuity equation includes a term that accounts specifically for the interchange of electrons among the various volume elements of velocity space due to collisions, it is called the Boltzmann equation. This equation is usually written in the form:

$$\frac{\partial f}{\partial t} + \mathbf{v} \cdot \nabla_r f - \frac{e}{m} (\mathbf{E} + \mathbf{v} \times \mathbf{B}) \cdot \nabla_v f = \bar{B}(\mathbf{r}, \mathbf{v}, t) \qquad (2.1.1)$$

where $\bar{B}$ identifies the Boltzmann term, which provides the electron distribution function rate of change with time due to collisions.

We shall consider in this book only experiments in which electron collisions involving a change in the number of interacting particles, such as attachment, recombination, and ionizing collisions [called by Shkarofsky et al. (1966) transformation collisions], provide a contribution to the Boltzmann term, which, for the purpose of determining the distribution function, can be neglected in comparison to all other elastic and inelastic scattering collisions. We shall make only one exception to this general assumption: transformation collisions will be considered again in establishing the electron density continuity equation, which is obtained in Section 2.5 by appropriate integration of the Boltzmann equation, since in this case the integrated contribution of all other collisions is identically zero. Neglecting transformation collisions removes from the distribution function expressions the coefficients of these processes, and the experiments in which this simplification is applicable are then the most appropriate ones for solving the basic problem considered in this volume: determination of the slow electron collision parameters introduced in Chapter 1.

When transformation collisions are ignored, the Boltzmann $\bar{B}$ term can easily be related to the local scattering probability density $S(\mathbf{v}, \mathbf{v}', d)$ defined in Section 1.1 and measured under the same conditions.* This relation will be now obtained.

The probability that an electron lying in an element $d\mathbf{v}$ at $\mathbf{v}$ is scattered in the element $d\mathbf{v}'$ at $\mathbf{v}'$ in time Δt is simply $S(\mathbf{v}, \mathbf{v}', v\,\Delta t)\,d\mathbf{v}'$. The Δt interval is chosen small enough so that $d = v\,\Delta t$ satisfies the assumption of a thin slab according to Section 1.1, but also large enough so that the corresponding mean changes of the average electron motion parameters, such as momentum and energy, are proportional to d (see pp. 10 and 11). The number of electrons leaving the element $d\mathbf{v}$ at $\mathbf{v}$ in time Δt will then be:

$$f(\mathbf{v}) \int_{\mathbf{v}'} S(\mathbf{v}, \mathbf{v}', v\,\Delta t)\,d\mathbf{v}'\,d\mathbf{v}$$

Analogously the number of electrons entering the element $d\mathbf{v}$ at $\mathbf{v}$ in time Δt due to collision scattering from all other elements $d\mathbf{v}''$ in the precollision velocity space $\mathbf{v}''$ will be:

$$\int_{\mathbf{v}''} f(\mathbf{v}'')S(\mathbf{v}'', \mathbf{v}, v''\Delta t)\,d\mathbf{v}\,d\mathbf{v}''$$

The difference between these two numbers of electrons has to be equal to $\bar{B}\,d\mathbf{v}\,\Delta t$, since $\bar{B}$ is the distribution function rate of change. Thus we have:

$$\bar{B}(\mathbf{v}) = \frac{1}{\Delta t}\left[\int_{\mathbf{v}''} f(\mathbf{v}'')S(\mathbf{v}'', \mathbf{v}, v''\,\Delta t)\,d\mathbf{v}'' - f(\mathbf{v}) \int_{\mathbf{v}'} S(\mathbf{v}, \mathbf{v}', v\,\Delta t)\,d\mathbf{v}'\right] \qquad (2.1.2)$$

As seen in Chapter 1, the total scattering probability density S can be written as the sum of various terms, one for each kind of collision. When gas conditions are such that for a certain type of collision $S(\mathbf{v}, \mathbf{v}', d)$ can be replaced by the single-collision expression $d\,s(\mathbf{v}, \mathbf{v}')$, then for that term eq. (2.1.2) becomes:

$$\bar{B}(\mathbf{v}) = \int_{\mathbf{v}''} v''f(\mathbf{v}'')s(\mathbf{v}'', \mathbf{v})\,d\mathbf{v}'' - vf(\mathbf{v}) \int_{\mathbf{v}'} s(\mathbf{v}, \mathbf{v}')\,d\mathbf{v}' \qquad (2.1.3)$$

---

* Note that, only for the purpose of simplifying notations, the functional dependence of S on n_i and T_g does not appear explicitly here. For the same reason, in the discussion that follows on the Boltzmann term we have always omitted indication of the functional dependence on $\mathbf{r}$ and t.

Since the last integral is the total collision probability [see (1.2.1)], eq. (2.1.3) can also be written in the form:

$$\bar{B}(\mathbf{v}) = \int_{\mathbf{v}''} \mathbf{v}'' f(\mathbf{v}'') s(\mathbf{v}'', \mathbf{v}) \, d\mathbf{v}'' - v f(\mathbf{v}) P_t(v) \tag{2.1.4}$$

In the next section we shall evaluate the Boltzmann term for all the kinds of collisions of interest in this book, using eq. (2.1.2) or (2.1.4), depending on the case.

2.2 EVALUATION OF THE BOLTZMANN TERM FOR TYPICAL CASES

A. Single-Process Scattering by Molecules at Rest

We assume here a molecular gas at zero temperature. All particles are then identical, equal-state molecules at rest. Moreover we assume a collision process such that the electron scattering probability density $s(v, v', \vartheta)$ along a certain direction ϑ is different from zero only within a very small range around a certain value v_s of the velocity v'. This usually happens when the molecule can undergo one type of collision process only, since classical momentum and energy conservation laws compel the outgoing electron to have a velocity v' uniquely related to the incoming velocity v and to the scattering angle ϑ. In Section 1.1 we stated that this situation takes place for inelastic collisions which excite a definite transition between two levels of a molecule (in this case v' is independent of ϑ). The same is demonstrated to be true by eq. (1.3.19) for the case of perfectly elastic collisions.

Then we write:

$$s(v, v', \vartheta) \cdot 2\pi v'^2 \sin \vartheta \, dv' \, d\vartheta$$
$$= \delta[v' - v_s(v, \vartheta)] \, dv' \cdot s_{\vartheta}(v, \vartheta) \sin \vartheta \, d\vartheta \tag{2.2.1}$$

In particular, for inelastic collisions with infinitely heavy molecules, such that their kinetic energy is not changed by the electron impact:

$$v_s = \left(v^2 - \frac{2u_j}{m} \right)^{\frac{1}{2}} = (v^2 - v_j^2)^{\frac{1}{2}} \tag{2.2.2}$$

where v_j is the speed corresponding to the kinetic energy u_j. Here we shall consider v_j^2 positive or negative with the same sign as u_j. For elastic collisions [eq. (1.3.19)] we have:

$$v_s = v \left[1 - \frac{2m}{M} (1 - \cos \vartheta) \right]^{\frac{1}{2}} \tag{2.2.3}$$

In most cases it is convenient to solve the Boltzmann equation by assuming that the distribution function can be written as the product of separate functions of polar coordinates in velocity space, or that it can be expanded in a sum of terms of this type. Thus we assume:

$$f(\mathbf{v''}) = f(v'')\Theta(\vartheta)\Phi(\varphi) \qquad (2.2.4)$$

The first term of eq. (2.1.4) can be written as:

$$I(\mathbf{v}) = \int_0^{2\pi} \Phi(\varphi)\, d\varphi \int_0^\pi \Theta(\vartheta) \sin \vartheta\, d\vartheta \int_0^\infty v''f(v'')s(v'', v, \vartheta)v''^2\, dv'' \qquad (2.2.5)$$

Here ϑ and φ denote, respectively, the polar and azimuthal angles in a reference system where the polar axis has been taken as directed along the vector $\mathbf{v}$.

Integral (2.2.5), where $s(v'', v, \vartheta)$ is given according to eq. (2.2.1), becomes:

$$I(\mathbf{v}) = \frac{1}{2\pi v^2}\int_0^{2\pi} \Phi(\varphi)\, d\varphi \int_0^\pi \Theta(\vartheta) \sin \vartheta\, d\vartheta \int_0^\infty v''^3 f(v'')$$

$$\times\, \delta[v - v_s(v'', \vartheta)]s_\vartheta(v'', \vartheta)\, dv'' \qquad (2.2.6)$$

When y and h are any two continuous well-behaved functions of x:

$$\int \delta[y(x)]h(x)\, dx = \int \delta(y)h[x(y)]\left(\left|\frac{dy}{dx}\right|\right)^{-1} dy = h(x)\left(\left|\frac{dy}{dx}\right|\right)^{-1}\bigg|_{x=x(0)} \qquad (2.2.7)$$

provided the domain of integration includes the $x(0)$ point. We have assumed that the integrals are performed along the directions of increasing x and y values, respectively; for this reason the absolute value of dy/dx appears in the y-domain integral. When this relationship is used, eq. (2.2.6) yields:

$$I(\mathbf{v}) = \frac{1}{2\pi v^2}\int_0^{2\pi} \Phi(\varphi)\, d\varphi \int_0^\pi v''^3 f(v'')s_\vartheta(v'', \vartheta)\left[\frac{\partial v_s(v'', \vartheta)}{\partial v''}\right]^{-1}\Theta(\vartheta)\sin \vartheta\, d\vartheta \qquad (2.2.8)$$

where now v'' is no longer an independent variable but is a function of v and ϑ as given by the equation:

$$v_s(v'', \vartheta) = v \qquad (2.2.9)$$

The assumption is now made that for charged particles in electric and magnetic fields the distribution function $f(\mathbf{v})$ can be written as:

$$f(\mathbf{v}) = f^0(v) + \frac{1}{v}\mathbf{v}\cdot\mathbf{f}^1(v) \qquad (2.2.10)$$

$$= f^0(v) + f^1(v)\cos \chi$$

where χ denotes the polar angle of the vector $\mathbf{v}$ relative to the direction ($\mathbf{z}$ axis) of the $\mathbf{f}^1(v)$ vector. Expression (2.2.10) is the result of taking the first two terms only of an expansion in spherical harmonics in velocity space; this approximation is sufficient for solving most of the problems in which we are interested (its limits will be established in Section 2.3).

Correspondingly we have:

$$f(\mathbf{v}'') = f^0(v'') + f^1(v'') \cos \chi'' \qquad (2.2.11)$$

where the polar angle χ'' of $\mathbf{v}''$ can be easily written as a function of ϑ, φ, and φ_z, the polar and azimuthal angles of $\mathbf{v}''$ and the azimuthal angle of $\mathbf{z}$ in the reference system with $\mathbf{v}$ polar axis:

$$\cos \chi'' = \cos \vartheta \cos \chi + \sin \vartheta \sin \chi \cos (\varphi - \varphi_z) \qquad (2.2.12)$$

When this is substituted into eq. (2.2.11), $f(\mathbf{v}'')$ becomes the sum of three terms like (2.2.4). Each term can be evaluated according to eq. (2.2.5), so that the Boltzmann term $\bar{B}(\mathbf{v})$ given by eq. (2.1.4) becomes, after some straightforward mathematics:

$$\bar{B}(\mathbf{v}) = \bar{B}^0(v) + \bar{B}^1(v) \cos \chi = \bar{B}^0(v) + \frac{1}{v} \mathbf{v} \cdot \bar{\mathbf{B}}^1(v) \qquad (2.2.13)$$

$$\bar{B}^0(v) = \frac{1}{v^2} \int_0^\pi v''^3 f^0(v'') s_\vartheta(v'', \vartheta) \left[\frac{\partial v_s(v'', \vartheta)}{\partial v''} \right]^{-1} \sin \vartheta \, d\vartheta - v f^0(v) P_t(v) \quad (2.2.14)$$

$$\bar{\mathbf{B}}^1(v) = \frac{1}{v^2} \int_0^\pi v''^3 f^1(v'') s_\vartheta(v'', \vartheta) \left[\frac{\partial v_s(v'', \vartheta)}{\partial v''} \right]^{-1} \cos \vartheta \sin \vartheta \, d\vartheta - v \mathbf{f}^1(v) P_t(v)$$

$$(2.2.15)$$

For inelastic collisions eqs. (2.2.2) and (2.2.9) yield:

$$\frac{\partial v_s(v'', \vartheta)}{\partial v''} = \frac{v''}{v_s(v'', \vartheta)} = \frac{v''}{v} \qquad (2.2.16)$$

whereas for elastic collisions eqs. (2.2.3) and (2.2.9) provide analogously:

$$\frac{\partial v_s(v'', \vartheta)}{\partial v''} = \frac{v_s(v'', \vartheta)}{v''} = \frac{v}{v''} \qquad (2.2.17)$$

When these are substituted into eqs. (2.2.14) and (2.2.15) and the first-order approximation $v'' = v$ is made, the following results are easily obtained for both types of collisions:

$$\bar{B}^0(v) = v f^0(v) \int_0^\pi s_\vartheta(v, \vartheta) \sin \vartheta \, d\vartheta - v f^0(v) P_t(v) = 0 \qquad (2.2.18)$$

$$\bar{\mathbf{B}}^1(v) = v \mathbf{f}^1(v) \int_0^\pi s_\vartheta(v, \vartheta) \cos \vartheta \sin \vartheta \, d\vartheta - v \mathbf{f}^1(v) P_t(v)$$

$$= v \mathbf{f}^1(v) \int_0^\pi s_\vartheta(v, \vartheta)(\cos \vartheta - 1) \sin \vartheta \, d\vartheta$$

$$= -v \mathbf{f}^1(v) P_m(v) = -n_g q_m(v) v \mathbf{f}^1(v) \qquad (2.2.19)$$

Since usually the spherically symmetrical part of the distribution $f^0(v)$ is the dominant term, whereas $\mathbf{f}^1(v)$ represents only a perturbation term due to the existing fields and boundary conditions, a further approximation is required for $\bar{B}^0(v)$ than for $\bar{\mathbf{B}}^1(v)$ in order to obtain a congruent setup of formulas.

Let us consider first the inelastic collision case. According to eqs. (2.2.2) and (2.2.9), v'' is given by:

$$v''^2 = v^2 + v_j{}^2 \tag{2.2.20}$$

and then is not a function of ϑ. In this case eq. (2.2.14) becomes [Holstein (1946)]:

$$\bar{B}^0(v) = \frac{v''^2}{v} f^0(v'') \int_0^\pi s_\vartheta(v'', \vartheta) \sin \vartheta \, d\vartheta - v f^0(v) P_j(v)$$

$$= \frac{n_g}{v} [q_j(\sqrt{v^2 + v_j{}^2})(v^2 + v_j{}^2) f^0(\sqrt{v^2 + v_j{}^2}) - q_j(v) v^2 f^0(v)] \tag{2.2.21}$$

where $P_j = n_g q_j$ replaces P_t since we are considering the inelastic collision process for the j transition alone.

Let us consider next the elastic collision case, for which we have:

$$\bar{B}^0(v) = \frac{1}{v^3} \int_0^\pi v''^4 f^0(v'') s_\vartheta(v'', \vartheta) \sin \vartheta \, d\vartheta - v f^0(v) P_t(v) \tag{2.2.22}$$

$$v''^2 = v^2 \left[1 - \frac{2m}{M} (1 - \cos \vartheta) \right]^{-1}$$

$$\simeq v^2 \left[1 + \frac{2m}{M} (1 - \cos \vartheta) \right] \tag{2.2.23}$$

Since v'' is very close to v, we may conveniently use the following Taylor series expansion for the integrand of eq. (2.2.22):

$$v''^4 f^0(v'') s_\vartheta(v'', \vartheta) = v^4 f^0(v) s_\vartheta(v, \vartheta) + (v''^2 - v^2) \frac{\partial}{\partial(v^2)} [v^4 f^0(v) s_\vartheta(v, \vartheta)]$$

$$= v^4 f^0(v) s_\vartheta(v, \vartheta) + \frac{2m}{M} (1 - \cos \vartheta) v^2 \frac{\partial}{\partial(v^2)} [v^4 f^0(v) s_\vartheta(v, \vartheta)] \tag{2.2.24}$$

The Boltzmann term $\bar{B}^0$ then becomes [Morse, Allis, and Lamar (1935)]:

$$\bar{B}^0(v) = \frac{2m}{M} \frac{1}{v} \frac{\partial}{\partial(v^2)} \left[v^4 f^0(v) \int_0^\pi (1 - \cos \vartheta) s_\vartheta(v, \vartheta) \sin \vartheta \, d\vartheta \right]$$

$$= \frac{2m}{M} \frac{1}{v} \frac{\partial}{\partial(v^2)} [n_g q_m(v) v^4 f^0(v)] \tag{2.2.25}$$

Thus eqs. (2.2.13), (2.2.21), (2.2.25), and (2.2.19) are the basic relationships for the Boltzmann collision term when the gas is composed of all identical molecules at rest and the different collision processes are individually considered.

B. Elastic Scattering by Charged Particles at Rest

In this case more than one particle at a time interacts appreciably with each electron; hence we must use eq. (2.1.2). Since most of these interactions are very weak, we are justified in assuming that only when $\mathbf{v}''$ is close to $\mathbf{v}$ is S significantly different from zero. In this case the following Taylor expansion can be conveniently used:

$$
\begin{aligned}
f(\mathbf{v}'')S(\mathbf{v}'', \mathbf{v}, v''\,\Delta t) &= f(\mathbf{v}'')S(\mathbf{v}'', \mathbf{v}'' - \Delta\mathbf{v}, v''\,\Delta t) \\
&= f(\mathbf{v})S(\mathbf{v}, \mathbf{v} - \Delta\mathbf{v}, v\,\Delta t) \\
&\quad + \Delta\mathbf{v} \cdot \mathbf{\nabla}_v[f(\mathbf{v})S(\mathbf{v}, \mathbf{v} - \Delta\mathbf{v}, v\,\Delta t)] \\
&\quad + \tfrac{1}{2}\Delta\mathbf{v}\,\Delta\mathbf{v} \cdot \mathbf{\nabla}_v\mathbf{\nabla}_v[f(\mathbf{v})S(\mathbf{v}, \mathbf{v} - \Delta\mathbf{v}, v\,\Delta t)] \quad (2.2.26)
\end{aligned}
$$

When this expansion and the new variable $\Delta\mathbf{v} = \mathbf{v}'' - \mathbf{v}$ are placed into the first integral of eq. (2.1.2), it appears immediately that the first expansion term cancels out the second integral of the equation. The Boltzmann term then becomes:

$$
\begin{aligned}
\bar{B}(\mathbf{v}) = \frac{1}{\Delta t}\bigg\{ &\int_{\Delta v}\Delta\mathbf{v} \cdot \mathbf{\nabla}_v[f(\mathbf{v})S(\mathbf{v}, \mathbf{v} - \Delta\mathbf{v}, v\,\Delta t)]\, d(\Delta\mathbf{v}) \\
&+ \frac{1}{2}\int_{\Delta v}\Delta\mathbf{v}\,\Delta\mathbf{v} \cdot \mathbf{\nabla}_v\mathbf{\nabla}_v[f(\mathbf{v})S(\mathbf{v}, \mathbf{v} - \Delta\mathbf{v}, v\,\Delta t)]\, d(\Delta\mathbf{v})\bigg\} \quad (2.2.27)
\end{aligned}
$$

Obvious changes of variables and simple mathematics yield:

$$
\begin{aligned}
\bar{B}(\mathbf{v}) = \mathbf{\nabla}_v \cdot vf(\mathbf{v})\frac{1}{d}&\int_{v'}(\mathbf{v} - \mathbf{v}')S(\mathbf{v}, \mathbf{v}', d)\, d\mathbf{v}' \\
&+ \tfrac{1}{2}\mathbf{\nabla}_v\mathbf{\nabla}_v \cdot vf(\mathbf{v})\frac{1}{d}\int_{v'}(\mathbf{v} - \mathbf{v}')(\mathbf{v} - \mathbf{v}')S(\mathbf{v}, \mathbf{v}', d)\, d\mathbf{v}' \quad (2.2.28)
\end{aligned}
$$

The equation that results when this expression for $\bar{B}(\mathbf{v})$ is substituted into the Boltzmann equation (2.1.1) is usually known as the Fokker-Planck equation.

Following Allis, we can write eq. (2.2.28) in polar coordinates. Giving proper consideration to the fact that the scattering function S is symmetrical around the $\mathbf{v}$ direction and that $\mathbf{v} - \mathbf{v}'$ is a small quantity, as already assumed previously, we obtain:

$$\bar{B}(\mathbf{v}) = \frac{1}{v^2}\frac{\partial}{\partial v}\left\{v^3 f(\mathbf{v})\frac{1}{d}\int_{\mathbf{v}'}\left[(v - v'\cos\vartheta) - \frac{1}{2v}v'^2\sin^2\vartheta\right]S(\mathbf{v}, \mathbf{v}', d)\,d\mathbf{v}'\right.$$

$$\left. + \frac{1}{2}\frac{\partial}{\partial v}v^3 f(\mathbf{v})\frac{1}{d}\int_{\mathbf{v}'}(v - v'\cos\vartheta)^2 S(\mathbf{v}, \mathbf{v}', d)\,d\mathbf{v}'\right\}$$

$$+ \frac{1}{4v\sin\vartheta}\frac{\partial}{\partial\vartheta}\left[\sin\vartheta\frac{\partial}{\partial\vartheta}f(\mathbf{v})\frac{1}{d}\int_{\mathbf{v}'}v'^2\sin^2\vartheta S(\mathbf{v}, \mathbf{v}', d)\,d\mathbf{v}'\right] \qquad (2.2.29)$$

The various integrals that appear in this expression are easily recognized as those used in Chapter 1 for the definition of the various collision probabilities; in particular, they are given by eqs. (1.2.10), (1.2.15), and (1.2.16). It is thus a simple matter to show that the Boltzmann term becomes:

$$\bar{B}(\mathbf{v}) = \frac{1}{v^2}\frac{\partial}{\partial v}\left\{(P_m - \tfrac{1}{2}P_d)v^4 f(\mathbf{v}) + \frac{\partial}{\partial v}\left[P_m - \frac{1}{2}\left(1 - \frac{3kT_g}{mv^2}\right)P_u - \tfrac{1}{2}P_d\right]v^5 f(\mathbf{v})\right\}$$

$$+ \frac{vP_d}{4\sin\vartheta}\frac{\partial}{\partial\vartheta}\left[\sin\vartheta\frac{\partial}{\partial\vartheta}f(\mathbf{v})\right] \qquad (2.2.30)$$

For electrons, the assumption of particles at rest would clearly contradict the existence of a continuous distribution function $f(\mathbf{v})$. A discussion of the electron-electron collision Boltzmann term will be postponed, therefore, until we examine the case of a gas with particles having agitation motion.

For positive ions, we make the assumption of particles at rest ($T_g = 0$) and use relationships (1.5.7) and (1.5.9); we obtain:*

$$\bar{B}(\mathbf{v}) = \frac{1}{2}\frac{1}{v^2}\frac{\partial}{\partial v}[P_u{}^i v^4 f(\mathbf{v})] + \frac{vP_d{}^i}{4\sin\vartheta}\frac{\partial}{\partial\vartheta}\left[\sin\vartheta\frac{\partial}{\partial\vartheta}f(\mathbf{v})\right] \qquad (2.2.31)$$

When expansion (2.2.10) is substituted, this yields:

$$\bar{B}^0(v) = \frac{1}{v}\frac{\partial}{\partial(v^2)}[P_u{}^i v^4 f^0(v)] \qquad (2.2.32)$$

$$\bar{\mathbf{B}}^1(v) = -\tfrac{1}{2}P_d{}^i v \mathbf{f}^1(v) + \frac{1}{v}\frac{\partial}{\partial(v^2)}[P_u{}^i v^4 \mathbf{f}^1(v)] \qquad (2.2.33)$$

---

* In order to simplify notations, hereafter $P_{me}{}^0$, $P_{ue}{}^0$, and $P_{de}{}^0$, when the scatterers are positive ions, will be indicated by the symbols $P_m{}^i$, $P_u{}^i$, and $P_d{}^i$, respectively, and the same collisions probabilities, when the scatterers are electrons, will be indicated by $P_m{}^e$, $P_u{}^e$, and $P_d{}^e$.

Neglecting m/M terms compared to unity, we obtain for the above expressions:

$$\bar{B}^0(v) = \frac{2m}{M} \frac{1}{v} \frac{\partial}{\partial(v^2)} \, [P_m{}^i v^4 f^0(v)] \tag{2.2.34}$$

$$\bar{\mathbf{B}}^1(v) = -\tfrac{1}{2} P_d{}^i v \mathbf{f}^1(v) = -P_m{}^i v \mathbf{f}^1(v) \tag{2.2.35}$$

It is important to observe that these terms are the same as those given by eqs. (2.2.25) and (2.2.19), respectively.

C. Scattering by a Gas at Temperature T_g

We evaluate first the Boltzmann terms due to scattering by neutral and ionized molecules only, that is, we neglect at present the electron-electron interaction contributions.

Since the gas is supposed to be at temperature T_g, the molecules will be distributed among different internal energy states as discussed in Section 1.4. The Boltzmann terms will be the sums of the contributions of the various different types of collisions. Introducing the average cross sections $Q_m{}^0$ and $Q_j{}^0$ defined in Section 1.4 and neglecting for the time being the agitational motions of the molecules, we can write the previously derived formulas as:

$$\bar{B}^0(v) = \frac{n_g}{v} \sum_j [Q_j{}^0(\sqrt{v^2 + v_j{}^2})(v^2 + v_j{}^2) f^0(\sqrt{v^2 + v_j{}^2}) - Q_j{}^0(v) v^2 f^0(v)]$$

$$+ \frac{2m}{M} \frac{1}{v} \frac{\partial}{\partial(v^2)} \, \{[n_g Q_m{}^0(v) + P_m{}^i(v)] v^4 f^0(v)\} \tag{2.2.36}$$

$$\bar{\mathbf{B}}^1(v) = -[n_g Q_m{}^0(v) + P_m{}^i(v)] v \mathbf{f}^1(v) \tag{2.2.37}$$

where n_g is the number of neutral molecules per unit volume. In deriving these formulas a few familiar assumptions, already discussed in Chapter 1, have been made: the gas is weakly ionized, so that the number of molecules out of thermodynamic equilibrium is negligibly small; each inelastic collision is given simply as the result of a pure u_j transition of an infinitely heavy molecule and of an independent simultaneous elastic collision with the actual molecule [see for analogy eq. (1.3.25)].

Recalling eqs. (1.4.30), (1.4.32), (1.4.42), and (1.5.15), which may be used over a very large range of practical experimental conditions, we can rewrite the above Boltzmann terms more simply as follows:

$$\bar{B}^0(v) = \frac{1}{v} \sum_j [P_j(\sqrt{v^2 + v_j{}^2})(v^2 + v_j{}^2) f^0(\sqrt{v^2 + v_j{}^2}) - P_j(v) v^2 f^0(v)]$$

$$+ \frac{2m}{M} \frac{1}{v} \frac{\partial}{\partial(v^2)} \, [P_m(v) v^4 f^0(v)] \tag{2.2.38}$$

$$\bar{\mathbf{B}}^1(v) = -P_m(v) v \mathbf{f}^1(v) \tag{2.2.39}$$

Let us now take into consideration the agitational motion of the molecules. In doing this we neglect effects which arise only from the fact that for each electron absolute velocity v there exists a corresponding range of relative collisional velocities and that the various collision probabilities are in general functions of this relative velocity. In fact, we saw in Chapter 1 that these effects are of the order of kT_g/Mv^2, a quantity negligibly small compared to unity in all cases that have potential interest for us. Only two steps remain then in the previous derivations of the formulas of this section, where the molecular motion has to be included: (a) the $(v''^2 - v^2)$ term, which was set equal to $(2m/M)(1 - \cos \vartheta)v^2$ in eq. (2.2.24); and (b) the $T_g = 0$ assumption made in deriving eq. (2.2.31) from (2.2.30).

We consider first the elastic electron-molecule collision case, that is, the correct evaluation of the $(v''^2 - v^2)$ term in eq. (2.2.24). If $\mathbf{V}$ is the velocity of a molecule, eq. (1.4.36) yields:

$$v''^2 - v^2 = (g''^2 - g^2) + 2\mathbf{V} \cdot (\mathbf{v}'' - \mathbf{v}) \qquad (2.2.40)$$

Whereas the first term will provide the contribution that already appears in eq. (2.2.24), the second term is expected to give an additional contribution. The latter will be of importance at the lowest electron velocities (thermal kinetic energies), that is, when V is of the order of $\sqrt{m/M}\, v$; in this case the second term of eq. (2.2.40), being of the order of $\sqrt{m/M}\, v^2$, is one magnitude larger than the first one [whose order is $(m/M)v^2$]. Hence to evaluate consistently the contribution of the $\mathbf{V} \cdot (\mathbf{v}'' - \mathbf{v})$ term we must retain not only the $(v''^2 - v^2)$ Taylor expansion term in eq. (2.2.24), but also the following one, where appears the factor $(v''^2 - v^2)^2 \simeq [2\mathbf{V} \cdot (\mathbf{v}'' - \mathbf{v})]^2$, whose magnitude is $(m/M)v^4$.

Let us evaluate this last contribution, which can be done simply. In fact it is obtained by integrating the expression:

$$\tfrac{1}{2}(v''^2 - v^2)^2 \frac{\partial^2}{\partial(v^2)^2} [v^4 f^0(v)s_\vartheta(v, \vartheta)]$$

$$\simeq 2[\mathbf{V} \cdot (\mathbf{v}'' - \mathbf{v})]^2 \frac{\partial^2}{\partial(v^2)^2} [v^4 f^0(v)s_\vartheta(v,\vartheta)] \qquad (2.2.41)$$

over a Maxwellian velocity distribution of molecules. This yields:

$$2\frac{kT_g}{M} |\mathbf{v}'' - \mathbf{v}|^2 \frac{\partial^2}{\partial(v^2)^2} [v^4 f^0(v)s_\vartheta(v, \vartheta)]$$

$$\simeq 4\frac{kT_g}{M}(1 - \cos \vartheta)v^2 \frac{\partial^2}{\partial(v^2)^2} [v^4 f^0(v)s_\vartheta(v, \vartheta)] \qquad (2.2.42)$$

Hence its contribution to the Boltzmann $\bar{B}^0$ term will be:

$$4\frac{kT_g}{M}\frac{1}{v}\frac{\partial^2}{\partial(v^2)^2}\left[v^4 f^0(v)\int_0^\pi (1-\cos\vartheta)s_g(v,\vartheta)\sin\vartheta\,d\vartheta\right]$$

$$=4\frac{kT_g}{M}\frac{1}{v}\frac{\partial^2}{\partial(v^2)^2}[n_g q_m(v)v^4 f^0(v)]\quad (2.2.43)$$

Instead of re-evaluating the $(v''^2 - v^2)$ term in the expansion, one may easily derive it indirectly. In fact, the Boltzmann $\bar{B}^0$ term will now have the form:

$$\bar{B}^0(v)=\frac{1}{v}\frac{\partial}{\partial(v^2)}[A(v)f^0(v)]+4\frac{kT_g}{M}\frac{1}{v}\frac{\partial^2}{\partial(v^2)^2}[n_g q_m(v)v^4 f^0(v)]$$

$$=\frac{1}{v}\frac{\partial}{\partial(v^2)}\left[A^1(v)f^0(v)+4\frac{kT_g}{M}n_g q_m(v)v^4\frac{\partial}{\partial(v^2)}f^0(v)\right]\quad (2.2.44)$$

where $A(v)$ and $A^1(v)$ indicate unknown functions of v. But $\bar{B}^0$ must vanish when electrons and molecules are in thermodynamic equilibrium, that is, when $f^0(v)$ is the Maxwell distribution at the gas temperature T_g:

$$f^0(v)=n_e\left(\frac{m}{2\pi kT_g}\right)^{3/2}\exp\left(-\frac{mv^2}{2kT_g}\right)\quad (2.2.45)$$

This determines $A^1(v)$:

$$A^1(v)=-4\frac{kT_g}{M}n_g q_m(v)v^4\frac{\partial\ln f^0(v)}{\partial(v^2)}$$

$$=\frac{2m}{M}n_g q_m(v)v^4\quad (2.2.46)$$

and the Boltzmann term becomes [Davydov (1935)]:

$$\bar{B}^0(v)=\frac{2m}{M}\frac{1}{v}\frac{\partial}{\partial(v^2)}\left\{n_g q_m(v)v^4\left[1+\frac{2kT_g}{m}\frac{\partial}{\partial(v^2)}\right]f^0(v)\right\}\quad (2.2.47)$$

The elastic electron-ion collision case can be handled much more straightforwardly. In fact, when expressions (1.5.10), (1.5.11), and (1.5.13) are substituted in eq. (2.2.30), letting $T_g \neq 0$, the quantity in the wavy brackets becomes:

$$\tfrac{1}{2}P_u{}^i v^4\left\{4\pi\int_0^v F(V)V^2\,dV f(\mathbf{v})\right.$$

$$\left.+\frac{8\pi}{3}\frac{M}{m}\left[\int_0^v F(V)V^4\,dV+v^3\int_v^\infty F(V)V\,dV\right]\frac{\partial f(\mathbf{v})}{\partial(v^2)}\right\}\quad (2.2.48)$$

When the usual assumptions for molecular ions under typical gas discharge conditions are made ($M \gg m$, $F = F_M$, $T_i = T_g \simeq$ a few hundreds degrees Kelvin), expression (2.2.48) reduces to:

$$\frac{1}{2} P_u^i v^4 \left[1 + \frac{2kT_g}{m} \frac{\partial}{\partial(v^2)} \right] f(\mathbf{v}) \tag{2.2.49}$$

In this case the $\bar{B}^0(v)$ Boltzmann term turns out to be the same as that given by eq. (2.2.32), except for the fact that the operator

$$\left[1 + \frac{2kT_g}{m} \frac{\partial}{\partial(v^2)} \right]$$

has to be placed in front of $f^0(v)$. According to eq. (2.2.47), the same thing happens in the case of electron-neutral collisions too.

Thus, when the agitational motion of the molecules is taken into account, (2.2.38) must be modified as follows:

$$\bar{B}^0(v) = \frac{1}{v} \sum_j [P_j(\sqrt{v^2 + v_j^2})(v^2 + v_j^2) f^0(\sqrt{v^2 + v_j^2}) - P_j(v) v^2 f^0(v)]$$

$$+ \frac{2m}{M} \frac{1}{v} \frac{\partial}{\partial(v^2)} \left\{ P_m(v) v^4 \left[1 + \frac{2kT_g}{m} \frac{\partial}{\partial(v^2)} \right] f^0(v) \right\} \tag{2.2.50}$$

This reduces again to eq. (2.2.38) when $u \gg kT_g$ over the important energy range of $f^0(v)$. It is a simple matter to verify that correspondingly eq. (2.2.39) is not changed by the agitational motion.

When the electron distribution function $f^0(v)$ is Maxwellian at gas temperature T_g [eq. (2.2.45)], the Boltzmann term $\bar{B}^0(v)$ must vanish, since collisions cannot change this distribution function. This happens for the elastic collision term of eq. (2.2.50), and the same must be true for the inelastic collision term. From this a useful relation between corresponding inelastic collision probabilities is easily derived.

Let us call P_j the probability for collisions of the first kind only (electron energy loss u_j), and P_{-j} the corresponding probability for collisions of the second kind (electron energy gain u_j). The summation of inelastic collision terms in eq. (2.2.50) then reads:

$$\sum_j [P_j(\sqrt{v^2 + v_j^2})(v^2 + v_j^2) f^0(\sqrt{v^2 + v_j^2}) - P_j(v) v^2 f^0(v)$$

$$+ P_{-j}(\sqrt{v^2 - v_j^2})(v^2 - v_j^2) f^0(\sqrt{v^2 - v_j^2}) - P_{-j}(v) v^2 f^0(v)] \tag{2.2.51}$$

Each term of this summation must vanish when $f^0(v)$ is given by eq. (2.2.45).

Thus we obtain:

$$P_j(\sqrt{v^2 + v_j^2})(v^2 + v_j^2) \exp\left(-\frac{mv_j^2}{2kT_g}\right) - P_j(v)v^2$$

$$+ P_{-j}(\sqrt{v^2 - v_j^2})(v^2 - v_j^2) \exp\left(\frac{mv_j^2}{2kT_g}\right) - P_{-j}(v)v^2 = 0 \quad (2.2.52)$$

This simply reduces to:

$$P_{-j}(v) = \left(1 + \frac{v_j^2}{v^2}\right) \exp\left(-\frac{mv_j^2}{2kT_g}\right) P_j(\sqrt{v^2 + v_j^2}) \qquad (2.2.53)$$

which is well known as the Klein-Rosseland relation [Mitchell and Zemansky (1934)].

Let us now consider again eq. (2.2.50), and make the assumption that for all inelastic collisions $v_j^2 \ll v^2$ over the important range of electron velocity. Hence a Taylor series expansion can be used, and we obtain:

$$P_j(\sqrt{v^2 + v_j^2})(v^2 + v_j^2)f^0(\sqrt{v^2 + v_j^2}) - P_j(v)v^2f^0(v)$$

$$= v_j^2 \frac{\partial}{\partial(v^2)} [P_j(v)v^2 f^0(v)] \quad (2.2.54)$$

neglecting terms in v_j^4 and higher. When the electron distribution function $f^0(v)$ is such that over the important energy range we have $u \gg kT_g$, eq. (2.2.38), which replaces (2.2.50) over this range, can be written with the help of (2.2.54) as follows:

$$\bar{B}^0(v) = \frac{1}{v}\frac{\partial}{\partial(v^2)}\left\{\left[\frac{2m}{M}P_m(v) + \sum_j \frac{u_j}{u}P_j(v)\right]v^4 f^0(v)\right\}$$

$$= \frac{1}{v}\frac{\partial}{\partial(v^2)}[G(v)P_m(v)v^4 f^0(v)] \qquad (2.2.55)$$

Here $G(v)$ has been introduced in accordance with the definition and formulas of Chapter 1, simplified by the assumption $u \gg kT_g$ [see, in particular, eq. (1.4.44)]:

$$G(v) = \frac{2m}{M} + \frac{1}{uP_m(v)}\sum_j u_j P_j(v) \qquad (2.2.56)$$

When the important electron energies are not very large in comparison to kT_g, eq. (2.2.50) must be used. Instead of eq. (2.2.55) we now obtain:

$$\bar{B}^0(v) = \frac{1}{v}\frac{\partial}{\partial(v^2)}\left\{v^4\left[\frac{2m}{M}P_m(v)\left\{1 + \frac{2kT_g}{m}\frac{\partial}{\partial(v^2)}\right\} + \sum_j \frac{u_j}{u}P_j(v)\right]f^0(v)\right\} \quad (2.2.57)$$

Frequently it is assumed that this equation also can be written in a form similar to eq. (2.2.55), as follows:

$$\bar{B}^0(v) = \frac{1}{v} \frac{\partial}{\partial(v^2)} \left\{ G(v)P_m(v)v^4 \left[1 + \frac{2kT_g}{m} \frac{\partial}{\partial(v^2)} \right] f^0(v) \right\} \qquad (2.2.58)$$

Here, however, $G(v)$ would be:

$$G(v) = \frac{2m}{M} + \frac{\sum\limits_j u_j P_j(v)}{\left\{ u + \left[\dfrac{\partial \ln f^0(v)}{\partial \ln (v^2)} \right] kT_g \right\} P_m(v)} \qquad (2.2.59)$$

which is not a true collision parameter, since the electron distribution function comes into its definition. Only as an approximation can this $G(v)$ be identified with the true fractional energy loss per collision defined in Chapter 1; the amount of difference is quite apparent if we compare, for instance, eq. (2.2.59) with eq. (1.4.44). In spite of this limitation, (2.2.58) is widely used in most discussions of experimental data under the assumption that $G(v)$ is the fractional energy loss per collision given by (1.4.44); as such it will also be employed hereafter in this book.

D. Electron-Electron Scattering

Finally we consider the role of electron-electron collisions. The contribution to the Boltzmann $\bar{B}(v)$ term should be obtained from eq. (2.2.30) with the appropriate expressions for the electron-electron collision probabilities. Here, however, these probabilities cannot be taken in general as given by (1.5.10), (1.5.11), and (1.5.13), since all these formulas were obtained assuming an isotropic distribution of scatterers $F(V)$, whereas for electron-electron collisions we must set $F(V) = f(\mathbf{v})$ and $f(\mathbf{v})$ is not an isotropic distribution if expansion (2.2.10) is considered.

Thus, by substituting (1.5.10), (1.5.11), and (1.5.13) into eq. (2.2.30), only the isotropic Boltzmann term $\bar{B}^0(v)$ is correctly obtained. In this case the quantity between wavy brackets in (2.2.30) is given again by expression (2.2.48), where now m, $f^0(v)$, and $P_m^e = P_u^e$ replace the corresponding ion quantities M, $F(V)$, and P_u^i. The resulting Boltzmann term is then conveniently written as:

$$\bar{B}^0 = \frac{1}{v} \frac{\partial}{\partial(v^2)} \left\{ v^4 P_m^e \left[I_0^0 + \frac{v}{3} (I_2^0 + J_{-1}^0) \frac{\partial}{\partial v} \right] f^0(v) \right\} \qquad (2.2.60)$$

where the I and J integrals are defined [Allis (1956)] as:

$$I_j^l(v) = \frac{4\pi}{n_e v^j} \int_0^v f^l(v) v^{2+j} \, dv \qquad (2.2.61)$$

$$J_j^l(v) = \frac{4\pi}{n_e v^j} \int_v^\infty f^l(v) v^{2+j} \, dv \qquad (2.2.62)$$

Various authors [Spitzer and Härm (1953), Allis (1956), Hwa (1958), Dreicer (1960), Shkarofsky (1963)], following essentially identical procedures, have computed the Boltzmann $\bar{\mathbf{B}}^1(v)$ term analogously, but using the appropriate $P_m{}^e$, $P_u{}^e$, and $P_d{}^e$ expressions for the nonisotropic distribution of scatterers (2.2.10). The result can be written in the form:

$$
\bar{\mathbf{B}}^1 = \frac{vP_m{}^e}{6}\left[v^2(I_2{}^0 + J_{-1}{}^0)\frac{\partial^2}{\partial v^2} + (3I_0{}^0 - I_2{}^0 + 2J_{-1}{}^0)\left(v\frac{\partial}{\partial v} - 1\right)\right]\mathbf{f}^1(v)
$$

$$
+ 4\pi v^4 P_m{}^e f^0(v)\,\mathbf{f}^1(v)
$$

$$
+ \frac{v^2 P_m{}^e}{10}\left[v(\mathbf{I}_3{}^1 + \mathbf{J}_{-2}{}^1)\frac{\partial^2}{\partial v^2} + \tfrac{1}{3}(5\mathbf{I}_1{}^1 - 3\mathbf{I}_3{}^1 + 2\mathbf{J}_{-2}{}^1)\frac{\partial}{\partial v}\right]f^0(v) \quad (2.2.63)
$$

Whenever electron-electron collisions are of importance, expressions (2.2.60) and (2.2.63) must be added to the $\bar{B}^0$ and $\bar{\mathbf{B}}^1$ equations, derived in the foregoing paragraphs for electron-neutral and electron-ion collisions only.

2.3 SOLUTIONS OF BOLTZMANN EQUATION

In Section 2.2 we assumed [eq. (2.2.10)] that the distribution function $f(\mathbf{v})$ is well represented by the first two terms alone of its expansion in spherical harmonics in velocity and that the spherically symmetric term f^0 is predominant since collisions tend to disorder any directional motion of the electrons. Correspondingly the Boltzmann collision term is given by eq. (2.2.13). Substitution of these expressions in the Boltzmann equation (2.1.1) yields after some mathematics:

$$
\frac{\partial f^0}{\partial t} + \frac{1}{v}\mathbf{v}\cdot\frac{\partial \mathbf{f}^1}{\partial t} + \frac{v}{3}\nabla_r\cdot\mathbf{f}^1 + \mathbf{v}\cdot\nabla_r f^0 - \frac{e}{m}\frac{1}{3v^2}\frac{\partial}{\partial v}(v^2\mathbf{E}\cdot\mathbf{f}^1)
$$

$$
- \frac{e}{m}\frac{1}{v}\mathbf{v}\cdot\left(\mathbf{E}\frac{\partial f^0}{\partial v} + \mathbf{B}\times\mathbf{f}^1\right) = \bar{B}^0 + \frac{1}{v}\mathbf{v}\cdot\bar{\mathbf{B}}^1 \quad (2.3.1)
$$

Separation of the scalar parts (spherical terms that are independent of the direction of $\mathbf{v}$) and of the vector parts (nonspherical terms that are given by functions of v multiplied by $\mathbf{v}$) provides two independent equations:

$$
\frac{\partial f^0}{\partial t} + \frac{v}{3}\nabla_r\cdot\mathbf{f}^1 - \frac{e}{m}\frac{1}{3v^2}\frac{\partial}{\partial v}(v^2\mathbf{E}\cdot\mathbf{f}^1) = \bar{B}^0 \quad\quad (2.3.2)
$$

$$
\frac{\partial \mathbf{f}^1}{\partial t} + v\nabla_r f^0 - \frac{e}{m}\left(\mathbf{E}\frac{\partial f^0}{\partial v} + \mathbf{B}\times\mathbf{f}^1\right) = \bar{\mathbf{B}}^1 \quad\quad (2.3.3)
$$

The distribution function components f^0 and $\mathbf{f}^1$ are thus obtained by solving this system of coupled equations.

Let us consider the case of an applied alternating electric field; a static magnetic field $\mathbf{B}$ may also be present. The electric field $\mathbf{E}$ is written in the exponential notation $\mathbf{E}_p \exp(j\omega t)$, and the frequency ω and all other significant conditions of the experiment are assumed to be such that the following approximations can be made:

a. A steady-state regime has been attained at the time of interest.

b. The spherically symmetrical component f^0 has negligible alternating variations at the frequency ω and its harmonics.

c. The distribution component $\mathbf{f}^1$ is sufficiently well represented by the first two terms alone of its complex exponential Fourier series in ωt:

$$\mathbf{f}^1 = \mathbf{f}_0^1 + \mathbf{f}_1^1 \exp(j\omega t) \tag{2.3.4}$$

d. The space variations of the various quantities are slow enough not to affect the velocity-dependent part of the spherically symmetric component f^0 of the distribution.

Margenau and Hartmann (1948), Allis (1956), Shkarofsky, Johnston, and Bachynski (1966), Johnston (1966), and others have provided more general analyses for cases in which one or more of the above assumptions are not adequate. The interested reader is referred to their treatments for a more detailed study of the problems.

When assumptions *a*, *b*, and *c* are made, eq. (2.3.2) yields:

$$\frac{v}{3}\nabla_r \cdot \mathbf{f}_0^1 - \frac{e}{m}\frac{1}{6v^2}\frac{\partial}{\partial v}(v^2\mathbf{E}_p \cdot \operatorname{Re}\mathbf{f}_1^1) = \bar{B}^0 \tag{2.3.5}$$

The real part of $\mathbf{f}_1^1$ comes in, since the product $\mathbf{E}\cdot\mathbf{f}^1$ cannot be handled as a conventional product using the exponential technique $\exp(j\omega t)$; this would erroneously provide no time-independent term due to the applied field. Actually, we must take the product of the two real functions, that is, the product of their real parts, and this yields the time-independent contribution $\frac{1}{2}\mathbf{E}_p \cdot \operatorname{Re}\mathbf{f}_1^1$.

When the same assumptions *a*, *b*, and *c* are made and expression (2.2.39) for $\bar{B}^1(v)$ is used, eq. (2.3.3) splits into the following dc and ac vector equations:

$$(\nu_m - \boldsymbol{\omega}_b \times)\mathbf{f}_0^1 = -v\nabla_r f^0 \tag{2.3.6}$$

$$(j\omega + \nu_m - \boldsymbol{\omega}_b \times)\mathbf{f}_1^1 = \frac{e\mathbf{E}_p}{m}\frac{\partial f^0}{\partial v} \tag{2.3.7}$$

Here the vector cyclotron frequency $\boldsymbol{\omega}_b$ has been used, according to the relation:

$$\boldsymbol{\omega}_b = \frac{e\mathbf{B}}{m} \tag{2.3.8}$$

and, for simplicity of notation, the collision frequency parameter ν_m has been preferred to the collision probability P_m. In accordance with the derivation of (2.2.39), ν_m indicates, now and in what follows, except when otherwise noted, the sum of the electron collision frequencies of neutral atoms or molecules and of positive ions. Since expression (2.2.39) for $\bar{\mathbf{B}}^1(v)$ does not include electron-electron collisions, however, (2.3.6) and (2.3.7) and their solutions are applicable only to the case of negligible effects of these collisions. The role of electron-electron collisions in the Boltzmann equation solutions will be discussed separately at the end of this section.

Equations (2.3.6) and (2.3.7) are solved by adopting the same technique. We shall discuss first the solution of (2.3.7), which is obtained by multiplying both sides by $(\nu_m + j\omega + \boldsymbol{\omega}_b \times)$. The result is:

$$[(\nu_m + j\omega)^2 - \boldsymbol{\omega}_b \times (\boldsymbol{\omega}_b \times)] \mathbf{f}_1^{\;1} = \frac{e}{m} (\nu_m + j\omega + \boldsymbol{\omega}_b \times) \mathbf{E}_p \frac{\partial f^0}{\partial v} \tag{2.3.9}$$

whence:

$$(\nu_m + j\omega) \mathbf{f}_{1\|}^{\;1} = \frac{e}{m} \mathbf{E}_{p\|} \frac{\partial f^0}{\partial v} \tag{2.3.10}$$

$$[(\nu_m + j\omega)^2 + \omega_b^{\;2}] \mathbf{f}_{1T}^{\;1} = \frac{e}{m} (\nu_m + j\omega + \boldsymbol{\omega}_b \times) \mathbf{E}_{pT} \frac{\partial f^0}{\partial v} \tag{2.3.11}$$

respectively, for the components parallel (index $\|$) and normal (index T) to the magnetic field.

If field $\mathbf{E}_{pT}$ is circularly polarized, a simple expression for $\mathbf{f}_{1T}^{\;1}$ is obtained from eq. (2.3.11). In fact, we have in this case:

$$\boldsymbol{\omega}_b \times \mathbf{E}_{pT} = \pm j\omega_b \mathbf{E}_{pT} \tag{2.3.12}$$

where the upper and lower signs hold, respectively, for right-hand and left-hand polarization around the magnetic field. Then eq. (2.3.11) simplifies to:

$$(\nu_m + j\omega \mp j\omega_b) \mathbf{f}_{1T}^{\;1} = \frac{e}{m} \mathbf{E}_{pT} \frac{\partial f^0}{\partial v} \tag{2.3.13}$$

If $\mathbf{E}_{pT}$ is linearly polarized, it can be split into two equal and opposite circularly polarized fields. Then $\mathbf{f}_{1T}^{\;1}$ is correspondingly given by the vector sum of the two resulting circular components according to (2.3.13); however, since these rotating vectors have different phases and amplitudes, their sum will not be aligned with the direction of $\mathbf{E}_{pT}$, and we will have a component along $\mathbf{E}_{pT}$ given by the algebraic sum of the circular components, and another one perpendicular to $\mathbf{E}_{pT}$ given by j times their algebraic difference.

These results are most conveniently written in tensor form as follows:

$$\mathbf{f}_1^{\ 1} = \frac{e}{m}\,\mathfrak{T}\cdot\mathbf{E}_p\,\frac{\partial f^0}{\partial v} \tag{2.3.14}$$

where the time tensor $\mathfrak{T}$ written with the z axis along $\mathbf{B}$ is:

$$\mathfrak{T} = \begin{vmatrix} \frac{1}{2}(\mathfrak{T}_C + \mathfrak{T}_\ni) & \frac{1}{2}j(\mathfrak{T}_C - \mathfrak{T}_\ni) & 0 \\[2mm] -\frac{1}{2}j(\mathfrak{T}_C - \mathfrak{T}_\ni) & \frac{1}{2}(\mathfrak{T}_C + \mathfrak{T}_\ni) & 0 \\[2mm] 0 & 0 & \mathfrak{T}_\| \end{vmatrix} \tag{2.3.15}$$

Here $\mathfrak{T}_C$, $\mathfrak{T}_\ni$, and $\mathfrak{T}_\|$ are the expressions previously found for the circularly polarized and parallel field cases, given in (2.3.13) and (2.3.10):

$$\left.\begin{aligned} \mathfrak{T}_C &= [\nu_m + j(\omega - \omega_b)]^{-1} \\ \mathfrak{T}_\ni &= [\nu_m + j(\omega + \omega_b)]^{-1} \\ \mathfrak{T}_\| &= [\nu_m + j\omega]^{-1} \end{aligned}\right\} \tag{2.3.16}$$

Equation (2.3.6) is solved similarly, and the results are combined according to (2.3.4), so that the solution for $\mathbf{f}^1$ can be written as follows:

$$\mathbf{f}^1 = -v\mathfrak{T}_0\cdot\nabla_r f^0 + \frac{e}{m}\,\mathfrak{T}\cdot\mathbf{E}\frac{\partial f^0}{\partial v} \tag{2.3.17}$$

Here $\mathfrak{T}_0$ is tensor $\mathfrak{T}$, when the position $\omega = 0$ has been made. It can be easily verified that, when the applied field is dc and steady-state conditions have been attained, this same eq. (2.3.17) holds provided we set $\omega = 0$.

We note that eq. (2.3.17) provides a valid indication of the order of magnitude of the ratio $|f^1/f^0|$; in fact, if we call L a length appropriate to characterize the expected spatial gradients $[L^{-1} \sim |\nabla_r \ln f|]$, (2.3.17) states that $|f^1/f^0|$ will have an order of magnitude not larger than the quantity $(v/\nu_m L) + (eE/m\nu_m v)$. Successive higher-order terms not considered in expansion (2.2.10) are of the order of powers of the above quantity, so that neglecting them was legitimate if $v/\nu_m L \ll 1$ and $eE/m\nu_m v \ll 1$ over the entire range of important electron velocities. These conditions can be restated by saying that the current first-order theory is applicable provided spatial gradients are such that the fractional change of f in a mean free path is much less than 1, and provided the energy gain per mean free path (eEv/ν_m) by an electron is much less than its kinetic energy $(mv^2/2)$ [Bernstein (1969)]. A substantially different line of approach, which would be appropriate if the last condition should fail and the two-term expansion (2.2.10) could not be adopted, has been proposed by Cavalleri and Sesta (1968), (1969) [see also Cavalleri (1969b)]; the same results have also been obtained recently by Paveri-Fontana (1970) and by Braglia (1970b), following more traditional lines of development based on Boltzmann and Fokker-Planck expressions.

The spherically symmetrical part f^0 of the electron distribution is obtained by substituting expression (2.3.17) for $\mathbf{f}_0{}^1$ and $\mathbf{f}_1{}^1$ into (2.3.5), and neglecting all space gradients on the basis of assumption d. This yields:

$$- \frac{e^2}{m^2} \frac{1}{6v^2} \frac{\partial}{\partial v}\left(\mathbf{E}_p \cdot \mathrm{Re}\ \mathfrak{T} \cdot \mathbf{E}_p v^2 \frac{\partial f^0}{\partial v}\right) = \bar{B}^0 \qquad (2.3.18)$$

In general, (2.3.18) is an integrodifferential equation in the unknown $f^0(v)$, where $\bar{B}^0$ is given as a function of $f^0(v)$ from eq. (2.2.50) or (2.2.58). The additional term (2.2.60) for electron-electron collisions is here neglected, since, as previously noted, the effects of these collisions will not be considered except at the end of this section. The left-hand side of (2.3.18) can be evaluated making use of tensor expression (2.3.15). This yields:

$$\frac{e^2}{m^2} \frac{1}{6v^2} \frac{\partial}{\partial v}\left\{[\mathrm{Re}\ \mathfrak{T}_{\|} \cdot E_{p\|}{}^2 + \tfrac{1}{2} \mathrm{Re}\ (\mathfrak{T}_C + \mathfrak{T}_{\supset}) \cdot E_{pT}{}^2]v^2 \frac{\partial f^0}{\partial v}\right\} + \bar{B}^0 = 0$$

$$(2.3.19)$$

It is convenient to introduce a velocity-dependent effective ac electric field, with a rms value given by:

$$E_e{}^2 = \frac{1}{2}\left[\frac{E_{p\|}{}^2}{1 + (\omega/\nu_m)^2} + \frac{1}{2}\frac{E_{pT}{}^2}{1 + (\omega - \omega_b)^2/\nu_m{}^2} + \frac{1}{2}\frac{E_{pT}{}^2}{1 + (\omega + \omega_b)^2/\nu_m{}^2}\right]$$

$$(2.3.20)$$

When this is introduced into (2.3.19), where the tensor components are given by expressions (2.3.16), the following basic equation for the determination of $f^0(v)$ is obtained:

$$\frac{e^2}{m^2} \frac{1}{3v^2} \frac{\partial}{\partial v}\left(\frac{v^2 E_e{}^2}{\nu_m} \frac{\partial f^0}{\partial v}\right) + \bar{B}^0 = 0 \qquad (2.3.21)$$

Before discussing possible solutions of eq. (2.3.21), it is helpful to consider more carefully the effective field concept. According to definition (2.3.20) and in cases where only electron-neutral collisions are of importance, this field E_e is a function of the applied electric field components, of the gas composition, which specifies the velocity-dependent collision frequency law

$v_m/p_0 = v\mathcal{P}_m$, and of the quantities ω/p_0 and ω_b/p_0 (or B/p_0). Moreover the following special cases are of particular interest for later use.

a. When no static magnetic field is applied ($\omega_b = 0$) or when the static magnetic and ac electric fields are parallel ($E_T = 0$), E_e becomes:

$$E_e^{\,2} = \frac{E_p^{\,2}}{2[1 + (\omega/v_m)^2]} \tag{2.3.22}$$

b. It can easily be verified that, when a dc electric field E is applied and space gradients are neglected, the steady-state distribution function f^0 is given again from eq. (2.3.21), the effective field E_e being now:

$$E_e^{\,2} = E_\parallel^{\,2} + \frac{E_T^{\,2}}{1 + (\omega_b/v_m)^2} \tag{2.3.23}$$

This expression is obtained from the same eq. (2.3.20) by setting $\omega = 0$ and replacing the rms value of the ac field by the dc field value. In fact, factor $\tfrac{1}{2}$, which appears in front of the product $\mathbf{E}_p \cdot \mathrm{Re}\,\mathbf{f}_1^1$, comes from the average of $\cos^2 \omega t$ and must be replaced by unity in the dc case.

c. The effective field E_e is equal to the dc or to the rms ac value E of the applied field under the following circumstances: (1) in general, when $v_m^{\,2} \gg (\omega + \omega_b)^2$; (2) in the special case of parallel electric and magnetic fields ($E_T = 0$), when $v_m^{\,2} \gg \omega^2$, independently of the ω_b value. Therefore, always $E_e = E$ when $\omega = \omega_b = 0$; in the very frequent case when v_m is proportional to the gas pressure, the electron-neutral collisions being dominant, adequately high pressures will be required in order to have $E_e = E$ for given ω and ω_b values.

d. The effective field E_e is proportional to the electron collision frequency v_m: (1) in the special cases of no magnetic field or of parallel electric and magnetic fields ($E_T = 0$), when $v_m^{\,2} \ll \omega^2$, independently of the ω_b value; (2) in the special case of orthogonal electric and magnetic fields ($E_\parallel = 0$), when $v_m^{\,2} \ll (\omega - \omega_b)^2$; (3) in the most general case, when both conditions, $v_m^{\,2} \ll \omega^2$ and $v_m^{\,2} \ll (\omega - \omega_b)^2$, are satisfied. Thus, at cyclotron resonance ($\omega = \omega_b$) and close to it, we never have $E_e \propto v_m$ except for parallel fields. In all cases, for given ω and ω_b values, fairly low pressures will be required to satisfy the above conditions.

e. When we have orthogonal dc electric and magnetic fields ($E_\parallel = 0$), at low pressures and high magnetic fields ($v_m^{\,2} \ll \omega_b^{\,2}$) the effective field, in addition to being proportional to v_m, is proportional to the ratio E/B.

We shall now briefly discuss the most useful solutions of eq. (2.3.21), starting from the particularly simple case of an electron swarm moving through an atomic gas, for which we can set $P_j = 0$ over the energy range of interest. It is a simple matter to show that in this case eqs. (2.2.50) and (2.3.21), after a straightforward first integration with physically significant boundary conditions, yield:

$$\frac{e^2}{3m^2} \frac{E_e^2}{v_m} \frac{\partial f^0}{\partial v} + \frac{m}{M} v v_m \left(1 + \frac{kT_g}{mv} \frac{\partial}{\partial v} \right) f^0 = 0 \qquad (2.3.24)$$

A second integration provides the solution for f^0:

$$f^0(v) = C \exp \left(-\int_0^v \frac{mv \, dv}{kT_g + e^2 M E_e^2 / 3m^2 v_m^2} \right) \qquad (2.3.25)$$

Here the normalizing constant C can be determined if the electron density is known; for this purpose we use the simple relationship:

$$n_e = 4\pi \int_0^\infty f^0 v^2 \, dv \qquad (2.3.26)$$

Exactly in the same way f^0 is determined when $\bar{B}^0$ can be written in form (2.2.58). The solution is here:

$$f^0(v) = C \exp \left(-\int_0^v \frac{mv \, dv}{kT_g + 2e^2 E_e^2 / 3mGv_m^2} \right) \qquad (2.3.27)$$

Setting $G = 2m/M$, we obtain eq. (2.3.25) again.

In the limit of very low electric fields, when condition:

$$E_e^2 \ll \frac{3mk}{2e^2} Gv_m^2 T_g = E_{et}^2 \qquad (2.3.28)$$

is satisfied, the distribution f^0 becomes Maxwellian and corresponds to an average electron energy:

$$\bar{u} = \frac{2\pi m}{n_e} \int_0^\infty f^0 v^4 \, dv = \tfrac{3}{2} kT_g = \bar{u}_g \qquad (2.3.29)$$

When G and v_m can be regarded as constant, namely, as independent of the electron velocity, or when G is constant and $E_e \propto v_m$ (this happens at fairly low pressures, as specified under points d and e of the foregoing discussion on E_e behavior), the distribution is also Maxwellian. In both cases the electron average energy is:

$$\bar{u} = \bar{u}_g + \frac{e^2}{mG} \left(\frac{E_e}{v_m} \right)^2 \qquad (2.3.30)$$

The Maxwellian distribution written as a function of the energy $\bar{u}$ is:

$$f^0(u) = n_e \left(\frac{3m}{4\pi \bar{u}} \right)^{3/2} \exp \left(-\frac{3u}{2\bar{u}} \right) \qquad (2.3.31)$$

Alternatively, we may replace $\bar{u}$ in expression (2.3.31), introducing the electron temperature T_e or the corresponding energy $kT_e = \hat{u}$, according to the classical formula:

$$\bar{u} = \tfrac{3}{2}kT_e = \tfrac{3}{2}\hat{u} \qquad (2.3.32)$$

Distribution (2.3.31) may thus be rewritten as:

$$f^0(u) = n_e \left(\frac{m}{2\pi\hat{u}}\right)^{3/2} \exp\left(-\frac{u}{\hat{u}}\right) \qquad (2.3.33)$$

and it can easily be verified that the value of v which gives the maximum of $v^2 f^0(v)$, that is, the most probable electron speed, is the one corresponding to $\hat{u}$ ($v = \sqrt{2\hat{u}/m}$).

When G is constant, E_e equal to the applied field strength E (high-pressure case; see point c of the discussion on E_e behavior), and the collision probability P_m constant ($\nu_m \propto v$), eq. (2.3.27) can be written in the form:

$$f^0(x) = C \exp\left[-\int_0^x \left(1 + \frac{\alpha}{x}\right)^{-1} dx\right] \qquad (2.3.34)$$

where x is a new, more convenient variable:

$$x = \frac{mv^2}{2kT_g} = \frac{u}{kT_g} \qquad (2.3.35)$$

and α is a velocity-independent parameter:

$$\alpha = \frac{e^2 E^2}{3k^2 T_g^2 G P_m^2} \qquad (2.3.36)$$

The integration is elementary and yields [Kelly, Margenau, and Brown (1957)]:

$$f^0(x) = C\left(1 + \frac{x}{\alpha}\right)^\alpha e^{-x} \qquad (2.3.37)$$

Other cases can be handled similarly, when G and P_m are constant. For instance, when the effective field is given by (2.3.22), eq. (2.3.27) takes the form:

$$f^0(x) = C \exp\left[-\int_0^x \left(1 + \frac{\alpha}{x + \beta}\right)^{-1} dx\right] \qquad (2.3.38)$$

where the velocity-independent parameters α and β are now:

$$\alpha = \frac{e^2 E_p^2}{6k^2 T_g^2 G P_m^2} \qquad (2.3.36')$$

$$\beta = \frac{m\omega^2}{2kT_g P_m^2} \qquad (2.3.39)$$

Integration yields [Kelly et al. (1957)]:

$$f^0(x) = C\left(1 + \frac{x}{\alpha + \beta}\right)^\alpha e^{-x} \tag{2.3.40}$$

It is a simple matter to verify that the same solution holds in a few common cases when the electric field is orthogonal to the magnetic field. For the dc case it is always so; α and β are here:

$$\alpha = \frac{e^2 E_T{}^2}{3k^2 T_g{}^2 G P_m{}^2} \tag{2.3.36''}$$

$$\beta = \frac{m\omega_b{}^2}{2kT_g P_m{}^2} \tag{2.3.39'}$$

In the ac case it is required that $v_m{}^2 \ll (\omega + \omega_b)^2$, a condition usually well satisfied when cyclotron resonance is observed; the distribution function is given again from eq. (2.3.40) if we let:

$$x = \frac{mv^2}{2kT'} = \frac{u}{kT'} \tag{2.3.35'}$$

$$T' = T_g + \frac{e^2 E_{pT}{}^2}{6mk(\omega + \omega_b)^2 G} \tag{2.3.41}$$

$$\alpha = \frac{e^2 E_{pT}{}^2}{12k^2 T'^2 G P_m{}^2} \tag{2.3.36'''}$$

$$\beta = \frac{m(\omega - \omega_b)^2}{2kT' P_m{}^2} \tag{2.3.39''}$$

Expanding the integral of (2.3.38) and neglecting terms of order x^3 and higher, we can write f^0 also in the approximate form:

$$f^0(x) = C \exp\left[-\frac{\beta}{\alpha + \beta} x - \frac{\alpha}{2(\alpha + \beta)^2} x^2\right] \tag{2.3.42}$$

which is generally more convenient for the computation of transport parameters [Gurevich (1956), Kelly et al. (1957)].

The assumptions of a constant energy loss factor and of a constant collision frequency or probability are seldom satisfied in actual gases. More useful solutions are obtained if we assume that these quantities are simple functions of velocity, so that the integration which appears in eq. (2.3.27) can be easily performed whenever the effective electric field is equal to the applied field, or proportional to the collision frequency, and the electron thermal motion is negligible compared to the field-induced random motion:

$$\bar{u}_g \ll \frac{e^2 E_e^{\;2}}{mG\nu_m^{\;2}} \tag{2.3.43}$$

Let us take, for instance, that G is a function of some power of velocity $G \propto v^g$, and $E_e \propto \nu_m$. Then we obtain [Bekefi and Brown (1958)]:

$$f^0(v) = C \exp(-bv^l) \tag{2.3.44}$$

independently of the $\nu_m(v)$ law. Here:

$$l = 2 + g \tag{2.3.45}$$

$$b = \frac{3}{2l} \frac{G}{v^g} \left(\frac{m\nu_m}{eE_e}\right)^2 \tag{2.3.46}$$

are two velocity-independent parameters.

Assume now that both G and ν_m are functions of some power of velocity $G \propto v^g$, $\nu_m \propto v^h$, and that the effective field is equal to the applied one. The distribution function is given again by the same form (2.3.44), where now [Bekefi and Brown (1958)]:

$$l = 2 + g + 2h \tag{2.3.45'}$$

$$b = \frac{3}{2l} \frac{G}{v^g} \left(\frac{\nu_m}{v^h}\right)^2 \left(\frac{m}{eE}\right)^2 \tag{2.3.46'}$$

The general expression (2.3.44) includes two of the most common and simple distributions: $l = 2$ is the Maxwellian distribution; $l = 4$ is the Druyvesteyn distribution, which was originally introduced [Druyvesteyn and Penning (1940)] when determining the distribution function for the case of an applied dc field in a gas characterized by $g = 0$, $h = 1$. The average electron energy of distribution (2.3.44), computed according to definition (2.3.29), comes out to be:

$$\bar{u} = \frac{\Gamma(5/l)}{2\Gamma(3/l)} mb^{-2/l} \tag{2.3.47}$$

where Γ signifies a gamma function.

Among the many other situations in which eq. (2.3.27) can be easily integrated under the conditions stated above, we mention here only the case in which G is given by (1.5.23). We note that this expression for G has been derived under assumptions that are consistent with (2.3.43). Thus eq. (2.3.27) becomes [Gerjuoy and Stein (1955)]:

$$f^0(v) = C \exp\left[-\frac{3m^3}{e^2 M} \int_0^v \left(1 + \frac{\eta n_g \mathcal{Q}^2}{v\nu_m}\right) \frac{\nu_m^{\;2}}{E_e^{\;2}} v\, dv\right] \tag{2.3.48}$$

When $\nu_m \propto v^h$ and E_e is equal to the strength E of the applied field, we obtain:

$$f^0(v) = C \exp\left\{-\frac{3}{2(1+h)}\frac{m}{M}\left(\frac{m}{eE}\right)^2\left(\frac{\nu_m}{v^h}\right)^2\left[1 + 2\frac{\eta\mathcal{Q}^2}{v^2 Q_m^0(v)}\right]v^{2(1+h)}\right\} \quad (2.3.49)$$

whereas, when $E_e \propto \nu_m$, the result is:

$$f^0(v) = C \exp\left\{-\frac{3}{2}\frac{m}{M}\left(\frac{m\nu_m}{eE_e}\right)^2\left[1 + \frac{2}{1-h}\frac{\eta\mathcal{Q}^2}{v^2 Q_m^0(v)}\right]v^2\right\} \quad (2.3.50)$$

Let us consider now the case of a more accurate evaluation, when we wish to use for $\bar{B}^0$ the rigorous expression (2.2.50) instead of approximation (2.2.58). If the Klein-Rosseland relation (2.2.53) is used, eq. (2.3.21) can be conveniently integrated in the form [Frost and Phelps (1962)]:

$$[\alpha(x) + x]x\frac{\partial f^0}{\partial x} + x^2 f^0 + \frac{M}{2mP_m(x)}\sum_j\int_x^{x+x_j}P_j(x')$$

$$\times\,[f^0(x') - e^{-x_j}f^0(x' - x_j)]x'\,dx' = 0 \quad (2.3.51)$$

Here we have introduced the normalized energy variable x, already defined in this section [eq. (2.3.35)], and:

$$\alpha = \frac{e^2 M E_e^2}{6mk^2 T_g^2 P_m^2} \quad (2.3.52)$$

which corresponds to definition (2.3.36) with $G = 2m/M$; α must be considered here a function of x through the velocity dependence of P_m and E_e.

Equation (2.3.51), as well as previous f^0 solutions, shows that the electron distribution is a function of the applied fields and of the gas pressure,* only in so far as they enter into the velocity-dependent function E_e/P_m. Recalling the discussion on the meaning of the effective field according to definition (2.3.20), we can state that, if electron-neutral collisions only are important, the distribution function will be completely defined when the gas is specified and when the values of the proper variables $E_{p\parallel}/p$, E_{pT}/p, ω/p, ω_b/p (or B/p), and T_g, or any equivalent combination of them, are given.

Following previous developments by Sherman (1960) and Boyer, Frost, and Phelps (1962) have shown how $f^0(x)$ can be determined numerically by convenient computer techniques. For this purpose they first make the transformation:

$$f^0(x) = c\varphi(x)\gamma(x) \quad (2.3.53)$$

---

* In order to simplify notations from now on in this chapter and in the following one, p will always mean reduced pressure p_0 (see Chapter 1) except when otherwise noted.

where [see eq. (2.3.34)]:

$$\gamma(x) = \exp\left[-\int_0^x \left(1 + \frac{\alpha}{x}\right)^{-1} dx\right] \tag{2.3.54}$$

It is assumed that the new unknown function $\varphi(x)$ can be set equal to unity for sufficiently large x values, that is, for $x \geqslant \delta$, where δ is conveniently chosen to satisfy this statement. The x range from zero to δ is, on the other hand, divided into N equidistant intervals Δx, and φ is determined at the $(N - 1)$ points $x_h = h\Delta x$ (h being an integer) by solving with a computer the linear equation system:

$$\varphi(x_h) = 1 + \frac{M}{2m} \Delta x \left[a_h + \sum_{k=1}^{N-1} b_{hk}\gamma(x_k)\varphi(x_k)\right] \tag{2.3.55}$$

Here:

$$a_h = \sum_j \int_\delta^{\delta+x_j} xP_j(x)\gamma(x)\{\hat{\Gamma}(\delta) - \hat{\Gamma}[\max(x_h, x - x_j)]\}\, dx \tag{2.3.56}$$

where:

$$\hat{\Gamma}(x) = \int_0^x [x(\alpha + x)P_m(x)\gamma(x)]^{-1}\, dx \tag{2.3.57}$$

and $\hat{\Gamma}[\min(a, b)]$ means that the argument of $\hat{\Gamma}$ is taken as the lesser of the values a and b, and similarly for $\hat{\Gamma}[\max(a, b)]$. Also:

$$\begin{aligned}b_{hk} = \varepsilon_{hk}x_k \sum_j P_j(x_k)\{\hat{\Gamma}(x_k) - \hat{\Gamma}[\max(x_h, x_k - x_j)]\} \\ - \sum_j \varepsilon_{hk}{}^j(x_k + x_j)P_j(x_k + x_j)e^{-x_j} \\ \times \{\hat{\Gamma}[\min(\delta, x_k + x_j)] - \hat{\Gamma}[\max(x_h, x_k)]\}\end{aligned} \tag{2.3.58}$$

where:

$$\varepsilon_{hk} = \begin{cases} 0 & \text{for } k \leqslant h \\ 1 & \text{for } k > h \end{cases} \tag{2.3.59}$$

$$\varepsilon_{hk}{}^j = \begin{cases} 0 & \text{for } k \leqslant h - x_j/\Delta x \\ 1 & \text{for } k > h - x_j/\Delta x \end{cases} \tag{2.3.60}$$

Clearly, this procedure can be easily handled by a computer. Later we shall mention cases in which a computer has actually been used. The less satisfactory feature of this procedure lies with the assumption $\varphi(x \geqslant \delta) = 1$.

When the applied field is large enough to satisfy the condition $\bar{u} \gg \bar{u}_g$, we shall have $x_j \gg 1$ over the entire significant energy range of the distribution function. In this case we may neglect in eq. (2.3.51) the term $e^{-x_j}f^0(x - x_j)$, which represents collisions of the second kind; for this case simpler and faster methods of solving (2.3.51) have been developed. One of them [Frost and

Phelps (1962)] relies on the fact that eq. (2.3.51) without the above term is such that only $f^0(x')$ for $x' > x$ is required to calculate $f^0(x)$. The equation can then be solved by backward prolongation, after having assumed that for sufficiently high energy f^0 is given by the distribution function (2.3.25) (where the kT_g term can be neglected), since at energies much greater than the highest inelastic threshold the elastic collision losses become dominating over inelastic ones. The Gauss-Seidel method of solving integrodifferential equations has also been used [Crompton, Gibson, and McIntosh (1969)].

All of the f^0 solutions that we have discussed so far are based on eq. (2.3.21), which was derived neglecting space gradients in the original equations. When space gradients cannot be neglected, much more complicated equations for the dc and ac cases are found.

Without going into any general analysis of this problem, we shall only provide some information on cases of later interest. Thus we consider the electron swarm in an applied uniform dc electric field in the absence of any static magnetic field; substituting (2.3.17) with $\omega = \omega_b = 0$ into (2.3.2), we can easily show the steady-state f^0 distribution to be:

$$\frac{1}{2v^2}\frac{\partial}{\partial v}\left\{G\nu_m v^3\left[f^0 + \left\{\frac{kT_g}{m} + \frac{2}{3G}\left(\frac{eE}{m\nu_m}\right)^2\right\}\frac{1}{v}\frac{\partial f^0}{\partial v}\right]\right\}$$

$$+ \frac{1}{3}\frac{\partial}{\partial z}\left[\frac{eE}{m\nu_m}v\frac{\partial f^0}{\partial v} + \frac{1}{v^2}\frac{\partial}{\partial v}\left(\frac{eE}{m\nu_m}v^3 f^0\right)\right] + \frac{v^2}{3\nu_m}\nabla_r^2 f^0 = 0 \quad (2.3.61)$$

where field $\mathbf{E}$ has been assumed to be directed along the negative z-axis direction.

When this equation holds, it can be immediately seen that, also in the absence of the electric field ($E = 0$), the distribution f^0 cannot be Maxwellian at gas temperature T_g. In fact, the first term:

$$\frac{1}{2v^2}\frac{\partial}{\partial v}\left[G\nu_m v^3\left(f^0 + \frac{kT_g}{mv}\frac{\partial f^0}{\partial v}\right)\right]$$

which we recognize to be $\bar{B}^0$ [eq. (2.2.58)], the isotropic part of the distribution function rate of change due to collisions alone, must be different from zero because of the presence of the last term $(v^2/3\nu_m)\nabla_r^2 f^0$. In general $\nabla_r^2 f^0$ will be negative, since the space variation of f^0 is dominated by the configuration of the electron density, which typically decreases with increasing slope from a maximum at the center of the swarm to zero at the boundary, where diffusing electrons are collected by the container walls. This requires that $\bar{B}^0$ be positive, that is, that collisions alone bring more electrons into each velocity volume element than out of it; furthermore, because of factor v^2/ν_m in front of $\nabla_r^2 f^0$, the relative weight of this effect is a growing function of the electron velocity for most gases. Consequently the distribution f^0 will appear to be contracted toward lower energies with respect to the Maxwellian at temperature T_g.

Physically, this effect follows from the fact that the faster electrons diffuse more rapidly to the container walls, and this may cause a significant reduction of the electron average energy compared to that of the gas molecules. For this reason the effect is commonly called diffusion cooling [Biondi (1954)]. The perturbation on f^0 due to diffusion cooling is most important at the lowest collision frequencies, because of the role that ν_m plays in the various terms of (2.3.61); therefore, diffusion cooling is usually significant only at pressures below 1 torr and in gases that display particularly low collision frequency values in the thermal range, such as neon and argon [see Chapter 4].

Let us now consider again the complete eq. (2.3.61); a simple criterion can be obtained for the conditions under which the gradient terms can be neglected in comparison to the collision and field terms representing energy loss or gain [Parker (1963)]. For this purpose we consider the case of velocity-independent ν_m and G parameters and substitute for f^0 the approximate Maxwellian form (2.3.31), where n_e has been assumed to be the only position-dependent quantity. When this is done, the ratio $\mathscr{R}$ of the three gradient terms to either the energy loss collision term, or to the sum of the collision term and the field term representing energy gain, can be expressed as:

$$\mathscr{R} = \frac{\dfrac{eE}{m\nu_m}\left(1 - \dfrac{mv^2}{\bar{u}}\right)\dfrac{\partial n_e}{\partial z} + \dfrac{v^2}{3\nu_m}\nabla_r^2 n_e}{\frac{2}{3}G\nu_m\left(1 - \dfrac{mv^2}{2\bar{u}}\right)n_e} \tag{2.3.62}$$

Integration over velocity space of the numerator of the above ratio multiplied by f^0 must be identically zero, since zero is the result of integration of the collision and field terms of eq. (2.3.61), as can be easily verified. Hence:

$$\nabla_r^2 n_e = \frac{3eE}{2\bar{u}}\frac{\partial n_e}{\partial z} \tag{2.3.63}$$

which, when substituted into (2.3.62), yields:

$$\mathscr{R} = \frac{2eE}{3mG\nu_m^2}\frac{1}{n_e}\frac{\partial n_e}{\partial z} \tag{2.3.64}$$

or, by using (2.3.30):

$$\mathscr{R} = \frac{2}{3}(\bar{u} - \bar{u}_g)\frac{1}{eE}\frac{\partial \ln n_e}{\partial z} \tag{2.3.65}$$

These formulas show that gradient terms are most important at high fields, where average electron energies are large in comparison to thermal ones.

Specific solutions of eq. (2.3.61) have been given by Parker (1963) for the cases of a velocity-independent collision frequency and of a velocity-independent collision cross section. Assuming that electrons are provided by a point source at the coordinate origin, which emits monoenergetic electrons in an unbounded region, Parker has shown that the distribution function can be written as the sum of energy modes, each of which decays with distance from the source. The lowest of these modes provides the far-distant distribution, since the higher ones decrease more rapidly with distance. Thus, far from the source and for average electron energies large in comparison to thermal ones [so that gradient effects are most important, according to (2.3.65)] the following first-order solution is provided by Parker for the constant collision frequency case and for a unity electron emission rate:

$$f^0(\rho, z, u) = \frac{\nu_m}{2z}\left(\frac{3m}{4\pi\mathscr{U}}\right)^{5/2}\left(1 + \frac{u}{eEz}\right)$$

$$\times \exp\left\{-\frac{3u}{2\mathscr{U}}\left[1 + \left(\frac{\rho}{2z}\right)^2\right]\right\}\exp\left(-\frac{3eE}{8\mathscr{U}}\frac{\rho^2}{z}\right) \quad (2.3.66)$$

Here ρ is the distance from the z axis,

$$\mathscr{U} = \frac{e^2}{mG}\left(\frac{E}{\nu_m}\right)^2 \quad (2.3.67)$$

is the average electron energy in the corresponding gradient-free case [see eq. (2.3.30), where the thermal energy term $\bar{u}_g$ is assumed negligibly small], G is constant, and the initial energy of the electrons emitted from the source has been set equal to zero.

Approximate expressions for the density and the average energy can be computed by substituting distribution (2.3.66) into (2.3.26) and (2.3.29), respectively. The results are:

$$n_e \simeq \frac{3m\nu_m}{8\pi\mathscr{U}}\frac{1 + \mathscr{U}/eEz}{z[1 + (\rho/2z)^2]^{3/2}}\exp\left(-\frac{3eE}{8\mathscr{U}}\frac{\rho^2}{z}\right) \quad (2.3.68)$$

$$\bar{u} \simeq \frac{1 + 2\mathscr{U}/3eEz}{1 + (\rho/2z)^2}\mathscr{U} \quad (2.3.69)$$

For electrons on the axis ($\rho = 0$) and far distant from the source ($z \to \infty$), the average energy goes to the expected gradient-free value $\mathscr{U}$; on-axis electrons closer to the source have energies larger by the factor $1 + 2\mathscr{U}/3eEz$; off-axis electrons far from the source have energies lower by the factor $1 + (\rho/2z)^2$. These results will be used in Chapter 3 for the discussion of the Townsend method of determining electron dc transport coefficients.

For problems in which the time dependence of f^0 has to be considered, we must use, not the steady-state equation (2.3.61), but its equivalent, which is obtained without dropping time derivatives; it can be easily verified that in this case eq. (2.3.61) is modified only in that the right-hand term is no longer zero, but $\partial f^0/\partial t$. The solution of the resulting equation has been thoroughly discussed by Parker and Lowke (1969) [see also Skullerud (1969)]. If a one-dimensional δ-function pulse of zero-energy electrons is assumed, the far-distance part of the time-evolving f^0 distribution is given in the first approximation by:

$$f^0(v, z, t) = \frac{f_e^0(v)}{2\sqrt{\pi \omega_2 t}} \exp\left[-\frac{(z - \omega_1 t)^2}{4\omega_2 t}\right] \qquad (2.3.70)$$

Here $f_e^0(v)$ is the time- and space-independent distribution (2.3.27), and parameters ω_1 and ω_2 are:

$$\omega_1 = -\frac{4\pi}{3n_e}\frac{eE}{m}\int_0^\infty \frac{v^3}{\nu_m}\frac{\partial f_e^0}{\partial v}\,dv \qquad (2.3.71)$$

$$\omega_2 = \frac{4\pi}{3n_e}\int_0^\infty \frac{v^4}{\nu_m}\left[f_e^0 + \left(3\omega_1\frac{\nu_m}{v^2} + \frac{eE}{mv}\frac{\partial}{\partial v}\right)\varphi(v)\right]dv \qquad (2.3.72)$$

where:

$$\varphi(v) = -\frac{mf_e^0(v)}{eE}\int_0^v \frac{v_\alpha}{1 + [\bar{u}_g/\mathscr{U}(v_\alpha)]}$$

$$\times \left[1 + \frac{\nu_m(v_\alpha)}{v_\alpha^3 f_e^0(v_\alpha)}\int_0^{v_\alpha} v_\beta^{\,2}\left(\frac{3m}{eE}\omega_1 + \frac{v_\beta}{\nu_m(v_\beta)}\frac{\partial}{\partial v_\beta}\right)f_e^0(v_\beta)\,dv_\beta\right]dv_\alpha \qquad (2.3.73)$$

Here $\mathscr{U}$ is defined according to eq. (2.3.67). Parker and Lowke also provide solutions of higher-order approximation than (2.3.70), and solve [Lowke and Parker (1969)] with the same approach the problem in which inelastic collisions are taken into account in the original equations, using for $\bar{B}^0$ its rigorous expression (2.2.50) instead of (2.2.58); furthermore they extend the basic analysis to the three-dimensional case. Detailed expressions for all these cases and useful suggestions on how to proceed in actual computations can be found in the papers by these authors.

Now we shall briefly indicate how electron-electron collisions influence the isotropic distribution component f^0. These collisions require introducing the additional $\bar{B}^0$ term (2.2.60) into the right-hand side of eq. (2.3.2) and the additional $\bar{\mathbf{B}}^1$ term (2.2.63) into the right-hand side of (2.3.3). The presence of these contributions makes the distribution functions f^0 look more Maxwellian than those derived in their absence; according to Allis (1956), this statement can be justified as follows.

Equation (2.3.21) is considered, wherein $\bar{B}^0$ is taken as the sum of (2.2.50) and (2.2.60). In this way the simplification is made that the electrons, only insofar as they act as a scattering medium, have an isotropic distribution that allows setting $\bar{B}^1 = 0$. If, furthermore, integrals (2.2.61) and (2.2.62) are computed assuming as a first approximation that this distribution is Maxwellian at a temperature T_e, (2.2.60) simplifies further and, in terms of the electron-electron collision frequency $\nu_m{}^e = vP_m{}^e$, reduces to:

$$\bar{B}^0 = \frac{1}{v}\frac{\partial}{\partial v^2}\left[v^3 I_0{}^0 \nu_m{}^e\left(1 + \frac{kT_e}{mv}\frac{\partial}{\partial v}\right)f^0\right] \qquad (2.3.74)$$

When this additional term is introduced, eq. (2.3.24) becomes:

$$\frac{e^2}{3m^2}\frac{E_e{}^2}{\nu_m}\frac{\partial f^0}{\partial v} + \frac{m}{M}v\nu_m\left(1 + \frac{kT_g}{mv}\frac{\partial}{\partial v}\right)f^0 + \tfrac{1}{2}vI_0{}^0\nu_m{}^e\left(1 + \frac{kT_e}{mv}\frac{\partial}{\partial v}\right)f^0 = 0 \quad (2.3.75)$$

The operator $[1 + (kT/mv)(\partial/\partial v)]$ yields zero when applied to a Maxwellian distribution at temperature T; therefore the last term acts in the direction of forcing the distribution toward a Maxwellian distribution at temperature T_e. Since $\nu_m{}^e$ is proportional to the electron density, this effect actually

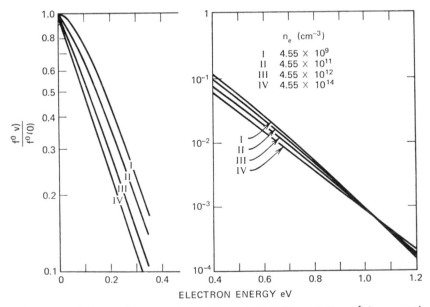

Fig. 2.3.1. Evolution of f^0 for a fixed applied electric field ($E_e = 10$ V cm^{-1}) in a potassium-seeded helium plasma (potassium particle density = 1%, $T_g = 1250°$K, $p = 1$ atm) at different ionization degrees. As the electron density increases, the distribution approaches the Maxwellian form [after Viegas and Kruger (1969)].

becomes of importance only in a highly ionized gas and may therefore be neglected in most of the experiments described in this book, which refer to weakly ionized gases.

By numerical iterative solution of the above-mentioned equations [(2.3.21), wherein $\bar{B}^0$ is the sum of (2.2.50) and (2.2.60)], Viegas and Kruger (1969) computed the distribution f^0 for various cases of actual interest, demonstrating its evolution toward the Maxwellian form when the electron density increases so as to make $\nu_m{}^e \gg \nu_m$ (Fig. 2.3.1).

2.4 TRANSPORT COEFFICIENTS

The experimental determination of the distribution function of the electrons in an ionized medium is a rather difficult job, the accuracy of the determination is seldom adequate for the derivation of the electron collision parameters, and the interpretation of the results is often questionable. It is much better for our purpose to measure certain macroscopic parameters, which are state or transport quantities defined as appropriate integral values taken over the distribution function.

Any state quantity that we shall consider can be defined as the integral in velocity space of a certain scalar function $X(v)$ weighted over the distribution function. Since we have assumed that the distribution function is well represented by the two terms alone of eq. (2.2.10), we find that this integral can be computed from the expression:

$$\int_{\mathbf{v}} X(v) f(\mathbf{v}) \, d\mathbf{v} = 4\pi \int_0^\infty X(v) f^0(v) v^2 \, dv = n_e \bar{X} \qquad (2.4.1)$$

The overimposed bar on a velocity-dependent quantity indicates, here and hereafter, the mean value of the quantity taken over $f^0(v)$, according to the standard definition:

$$\bar{X} = \frac{4\pi}{n_e} \int_0^\infty X(v) f^0(v) v^2 \, dv \qquad (2.4.2)$$

The transport quantities are, on the other hand, the integrals in velocity space of the vector expressions $\mathbf{v}X(v)$, weighted over the distribution function; in this case we have:

$$\int_{\mathbf{v}} \mathbf{v} X(v) f(\mathbf{v}) \, d\mathbf{v} = \frac{4\pi}{3} \int_0^\infty X(v) \mathbf{f}^1(v) v^3 \, dv \qquad (2.4.3)$$

When $X = 1$, eq. (2.4.1) yields the electron density:

$$n_e = 4\pi \int_0^\infty f^0(v) v^2 \, dv \qquad (2.4.4)$$

and eq. (2.4.3) the electron flow:

$$\mathbf{\Gamma}_e = \frac{4\pi}{3} \int_0^\infty \mathbf{f}^1(v)v^3 \, dv \tag{2.4.5}$$

When $X = u = mv^2/2$, eq. (2.4.1) yields the electron kinetic energy density:

$$\rho_e = 2\pi m \int_0^\infty f^0(v)v^4 \, dv = n_e \dot{u} \tag{2.4.6}$$

and eq. (2.4.3) the electron kinetic energy flux:

$$\mathbf{q}_e = \frac{2\pi}{3} m \int_0^\infty \mathbf{f}^1(v)v^5 \, dv \tag{2.4.7}$$

Except when otherwise noted, we shall consistently neglect electron-electron collision effects; then expression (2.3.17) can be substituted for $\mathbf{f}^1$ into eqs. (2.4.5) and (2.4.7). In the first case we obtain:

$$\mathbf{\Gamma}_e = -\frac{4\pi}{3} \int_0^\infty (\mathfrak{T}_0 \cdot \nabla_r f^0)v^4 \, dv + \frac{4\pi}{3} \frac{e}{m} \int_0^\infty (\mathfrak{T} \cdot \mathbf{E}) \frac{\partial f^0}{\partial v} v^3 \, dv$$

$$= -\nabla_r \cdot \left[\frac{4\pi}{3} \int_0^\infty \mathfrak{T}_0 f^0 v^4 \, dv \right] + \left[\frac{4\pi}{3} \frac{e}{m} \int_0^\infty \mathfrak{T} \frac{\partial f^0}{\partial v} v^3 \, dv \right] \cdot \mathbf{E} \tag{2.4.8}$$

with the understanding that in the last equation the space gradient has to be considered as a partial derivative operating on the distribution function factor alone of the expression to which it is applied.

Equation (2.4.8) is conveniently rewritten in the form:

$$\mathbf{\Gamma}_e = -\nabla_r \cdot (Dn_e) - \frac{1}{e} \mathbf{\sigma} \cdot \mathbf{E} \tag{2.4.9}$$

where D is the diffusion tensor:

$$D = \frac{4\pi}{3n_e} \int_0^\infty \mathfrak{T}_0 f^0 v^4 \, dv = \overline{\tfrac{1}{3}v^2 \mathfrak{T}_0} \tag{2.4.10}$$

and $\mathbf{\sigma}$ is the ac complex conductivity tensor:

$$\mathbf{\sigma} = \mathbf{\sigma}_r + J\mathbf{\sigma}_i = -\frac{4\pi}{3} \frac{e^2}{m} \int_0^\infty \mathfrak{T} \frac{\partial f^0}{\partial v} v^3 \, dv = \frac{e^2 n_e}{m} \langle \mathfrak{T} \rangle \tag{2.4.11}$$

In expression (2.4.11) the brackets indicate a particular kind of mean value, like (2.4.2), but obtained with a different weighting of the f_0 distribution, namely:

$$\langle X \rangle = -\frac{4\pi}{3n_e} \int_0^\infty X(v) \frac{\partial f^0}{\partial v} v^3 \, dv \tag{2.4.12}$$

When the applied electric field is the dc E field, eq. (2.4.9) is preferably written in one of the forms:

$$\boldsymbol{\Gamma}_e = -\boldsymbol{\nabla}_r \cdot (D n_e) - n_e \boldsymbol{\mu} \cdot \mathbf{E} \tag{2.4.13}$$

$$\boldsymbol{\Gamma}_e = -\boldsymbol{\nabla}_r \cdot (D n_e) + n_e \mathbf{w} \tag{2.4.14}$$

where $\boldsymbol{\mu}$ is the mobility tensor, and $\mathbf{w}$ the drift velocity vector. The mobility tensor is given by:

$$\boldsymbol{\mu} = \frac{\boldsymbol{\sigma}_{dc}}{e n_e} = \frac{e}{m} \langle \mathfrak{T}_0 \rangle \tag{2.4.15}$$

Here $\boldsymbol{\sigma}_{dc}$ is the conductivity (2.4.11) at the zero frequency limit.

In the absence of a static magnetic field all of the above tensors become scalar quantities. Equations (2.4.9) and (2.4.13) can be written in the form:

$$\boldsymbol{\Gamma}_e = -\boldsymbol{\nabla}_r (D n_e) - \frac{1}{e} \sigma \mathbf{E} \tag{2.4.16}$$

$$\boldsymbol{\Gamma}_e = -\boldsymbol{\nabla}_r (D n_e) - n_e \mu \mathbf{E} \tag{2.4.17}$$

and the transport coefficients become [Huxley (1937), Margenau (1946)]:

$$D = \frac{4\pi}{3 n_e} \int_0^\infty \frac{v^4}{v_m} f^0 \, dv = \frac{1}{3} \left(\frac{\overline{v^2}}{v_m} \right) \tag{2.4.18}$$

$$\sigma = \sigma_r + j\sigma_i = -\frac{4\pi}{3} \frac{e^2}{m} \int_0^\infty \frac{v^3}{v_m + j\omega} \frac{\partial f^0}{\partial v} \, dv = \frac{e^2 n_e}{m} \left\langle \frac{v_m - j\omega}{v_m^2 + \omega^2} \right\rangle \tag{2.4.19}$$

$$\mu = \frac{w}{E} = \frac{\sigma_{dc}}{e n_e} = -\frac{4\pi}{3 n_e} \frac{e}{m} \int_0^\infty \frac{v^3}{v_m} \frac{\partial f^0}{\partial v} \, dv = \frac{e}{m} \left\langle \frac{1}{v_m} \right\rangle \tag{2.4.20}$$

More general expressions for μ, to be used when $eE/mv_m v$ is not much less than unity, so that approximation (2.2.10) fails, follow from the developments of Cavalleri and Sesta (1968, 1969) [see Section 2.3], and can be found in their papers and in Braglia (1970a, 1970b).

In the presence of a static magnetic field, according to expression (2.3.15) and to definitions (2.4.10), (2.4.11), and (2.4.15), all the transport parameters will have tensor components parallel to the magnetic field, $D_\|$, $\sigma_\|$, and $\mu_\| = w_\|/E_\|$, equal to the above expressions for D, σ, and μ in the absence of a magnetic field. However, their components in the plane at right angles to the field will be different.

In this plane the coefficient D_T for the diffusive flow along the gradient of the pressure-like quantity $D_T n_e$ is called the transverse diffusion coefficient. Substitution into eq. (2.4.10) of the diagonal term of $\mathfrak{T}_0$, as given from

(2.3.15) and (2.3.16), yields [Huxley (1937)]:

$$D_T = \frac{4\pi}{3n_e} \int_0^\infty \frac{\nu_m v^4}{\nu_m{}^2 + \omega_b{}^2} f^0 \, dv = \frac{1}{3}\left(\overline{\frac{\nu_m v^2}{\nu_m{}^2 + \omega_b{}^2}}\right) \qquad (2.4.21)$$

The coefficient $D_\perp$ for the diffusive flow in the same plane, but at right angles to the gradient of the pressure-like quantity $D_\perp n_e$, is called the perpendicular diffusion coefficient. Substitution into eq. (2.4.10) of the off-diagonal term of $\mathfrak{T}_0$, as given from (2.3.15) and (2.3.16), yields:

$$D_\perp = \frac{4\pi}{3n_e}\omega_b \int_0^\infty \frac{v^4}{\nu_m{}^2 + \omega_b{}^2} f^0 \, dv = \frac{\omega_b}{3}\left(\overline{\frac{v^2}{\nu_m{}^2 + \omega_b{}^2}}\right) \qquad (2.4.22)$$

In a similar way we can introduce the transverse coefficients, ac conductivity σ_T and mobility μ_T, for the flow along the transverse electric field direction, and the perpendicular coefficients, ac conductivity $\sigma_\perp$ and mobility $\mu_\perp$, for the flow perpendicular to both the electric and the magnetic fields. The appropriate expressions, derived as for the diffusion coefficients above, are:

$$\sigma_T = \sigma_{Tr} + j\sigma_{Ti} = \frac{e^2 n_e}{2m}\left\langle [\nu_m + j(\omega + \omega_b)]^{-1} + [\nu_m + j(\omega - \omega_b)]^{-1}\right\rangle$$

$$= \frac{e^2 n_e}{m}\left\langle \frac{\nu_m(\nu_m{}^2 + \omega^2 + \omega_b{}^2) - j\omega(\nu_m{}^2 + \omega^2 - \omega_b{}^2)}{[\nu_m{}^2 + (\omega + \omega_b)^2][\nu_m{}^2 + (\omega - \omega_b)^2]}\right\rangle \qquad (2.4.23)$$

$$\sigma_\perp = \sigma_{\perp r} + j\sigma_{\perp i} = j\frac{e^2 n_e}{2m}\left\langle [\nu_m + j(\omega + \omega_b)]^{-1} - [\nu_m + j(\omega - \omega_b)]^{-1}\right\rangle$$

$$= \frac{e^2 n_e}{m}\omega_b\left\langle \frac{\nu_m{}^2 - \omega^2 + \omega_b{}^2 - 2j\omega\nu_m}{[\nu_m{}^2 + (\omega + \omega_b)^2][\nu_m{}^2 + (\omega - \omega_b)^2]}\right\rangle \qquad (2.4.24)$$

$$\mu_T = \frac{w_T}{E_T} = \frac{\sigma_{dcT}}{en_e} = \frac{e}{m}\left\langle \frac{\nu_m}{\nu_m{}^2 + \omega_b{}^2}\right\rangle \qquad (2.4.25)$$

$$\mu_\perp = \frac{w_\perp}{E_T} = \frac{\sigma_{dc\perp}}{en_e} = \frac{e}{m}\omega_b\left\langle \frac{1}{\nu_m{}^2 + \omega_b{}^2}\right\rangle \qquad (2.4.26)$$

When relation (2.3.17) is substituted for $\mathbf{f}^1$ into eq. (2.4.7), an expression for the kinetic energy flux similar to (2.4.8) is obtained. In future applications, we shall be concerned only with the nonalternating part of the electron heat transfer, also in the presence of applied ac fields; therefore we are going to consider the energy flux due to $f_0{}^1$ alone. In doing this, however, we include the probable presence of a dc electric field $\mathbf{E}$, since this field is usually found, being generated by the formation of electrostatic space charges inside the ionized medium. Thus we obtain:

$$\mathbf{q}_e = -\nabla_r \cdot (\Theta n_e) - n_e \mathfrak{M} \cdot \mathbf{E} \tag{2.4.27}$$

where:

$$\Theta = \frac{2\pi m}{3n_e} \int_0^\infty \mathfrak{T}_0 f^0 v^6 \, dv = \frac{m}{6} \overline{v^4 \mathfrak{T}_0} \tag{2.4.28}$$

$$\mathfrak{M} = -\frac{2\pi e}{3n_e} \int_0^\infty \mathfrak{T}_0 \frac{\partial f^0}{\partial v} v^5 \, dv = \frac{e}{2} \langle v^2 \mathfrak{T}_0 \rangle \tag{2.4.29}$$

and the space gradient is understood to operate on the distribution function factor alone.

Since we like to have explicit evidence of the dependence of the kinetic energy flow on the gradient of the average electron energy $\bar{u}$, in accordance with the common physical description of the heat transfer processes, eq. (2.4.27) is written more conveniently as [Shkarofsky (1961)]:

$$\mathbf{q}_e = -\Theta \cdot \nabla_r n_e - n_e \mathfrak{D} \cdot \nabla_r \bar{u} - n_e \mathfrak{M} \cdot \mathbf{E} \tag{2.4.30}$$

where:

$$\mathfrak{D} = \frac{\partial \Theta}{\partial \bar{u}} = \frac{2\pi m}{3n_e} \int_0^\infty \mathfrak{T}_0 \frac{\partial f^0}{\partial \bar{u}} v^6 \, dv \tag{2.4.31}$$

is called the thermal diffusivity tensor. Alternatively, some authors prefer using the thermal conductivity tensor:

$$\mathscr{K} = \tfrac{3}{2} k n_e \mathfrak{D} \tag{2.4.32}$$

which is more convenient in order to write (2.4.30) in terms of the electron temperature, as defined by (2.3.32); in fact, we have:

$$\mathbf{q}_e = -\Theta \cdot \nabla_r n_e - \mathscr{K} \cdot \nabla_r T_e - n_e \mathfrak{M} \cdot \mathbf{E} \tag{2.4.30'}$$

In the absence of a static magnetic field all of the above tensors become scalar quantities; the energy flux equation and the transport coefficients are then:

$$\mathbf{q}_e = -\Theta \nabla_r n_e - n_e \mathfrak{D} \nabla_r \bar{u} - n_e \mathfrak{M} \mathbf{E} \tag{2.4.33}$$

$$\mathbf{q}_e = -\Theta \nabla_r n_e - \mathscr{K} \nabla_r T_e - n_e \mathfrak{M} \mathbf{E} \tag{2.4.33'}$$

$$\Theta = \frac{m}{6} \left(\frac{\overline{v^4}}{v_m} \right) \tag{2.4.34}$$

$$\mathfrak{D} = \frac{2}{3k n_e} \mathscr{K} = \frac{\partial \Theta}{\partial \bar{u}} \tag{2.4.35}$$

$$\mathfrak{M} = \frac{e}{2} \left\langle \frac{v^2}{v_m} \right\rangle \tag{2.4.36}$$

In the presence of a static magnetic field, we must consider parallel, transverse, and perpendicular coefficients, as we have done for the other transport coefficients. For later use we report here the detailed expressions for $\mathfrak{M}_T$ and $\mathfrak{M}_\perp$ only:

$$\mathfrak{M}_T = \frac{e}{2} \left\langle \frac{v^2 \nu_m}{\nu_m^2 + \omega_b^2} \right\rangle \tag{2.4.37}$$

$$\mathfrak{M}_\perp = \frac{e}{2} \omega_b \left\langle \frac{v^2}{\nu_m^2 + \omega_b^2} \right\rangle \tag{2.4.38}$$

Another measurable quantity, which has to be considered because of its close relation to electron collision parameters, is the power of incoherent spontaneous radiation of frequency ω, emitted along a certain direction and in one of two available polarizations from a unit volume of an ionized medium. This power per unit solid angle and per unit radian frequency interval is denoted by j_ω and can be obtained from eq. (2.4.1) if X is taken as the corresponding power radiated by an electron of velocity v.

According to Kirchhoff's law, the emitted power j_ω for a tenuous ionized medium of refractive index close to unity and of low absorptivity (attenuation coefficient α) can be written as:

$$j_\omega = 2\alpha B_\omega = \sqrt{\mu_0/\varepsilon_0}\, \sigma_{er} B_\omega \tag{2.4.39}$$

where σ_{er} is the real part of an equivalent conductivity of the ionized medium at the frequency ω, introduced here as a suitable quantity defined, for any propagation direction and polarization, in terms of the corresponding α by the same well-known relation for isotropic media: $2\alpha = \sqrt{\mu_0/\varepsilon_0}\, \sigma_r$, and B_ω is the so-called Planck function for the polarized black-body radiation from matter at temperature T_r. In the Rayleigh-Jeans limit ($\hbar\omega \ll kT_r$), which is appropriate for all radio frequencies up to the microwave region:

$$B_\omega = \frac{\varepsilon_0 \mu_0}{8\pi^3} \omega^2 kT_r = \frac{kT_r}{2\pi\lambda_0^2} \tag{2.4.40}$$

where the wavelength in free space λ_0 ($= 2\pi/\omega\sqrt{\varepsilon_0\mu_0}$) has been introduced in expression (2.4.40).

In the low-absorptivity case, if we retain first-order loss terms only, the above-introduced σ_{er} for any propagation direction and polarization will be given by a linear combination of the real parts of the conductivity tensor components, readily derivable from eqs. (2.4.19), (2.4.23), and (2.4.24). It follows that σ_{er} can be written in the form [Bunkin (1957), Bekefi (1966)]:

$$\sigma_{er} = \frac{e^2 n_e}{m} \langle R_\omega(\nu_m) \rangle \tag{2.4.41}$$

where $R_\omega(\nu_m)$ is an appropriate function of the collision frequency ν_m.

When the medium is in thermodynamic equilibrium, the electron distribution function f^0 is Maxwellian, according to eq. (2.3.33), and $T_r = T_e$. Since for a Maxwell distribution:

$$\frac{\partial f^0}{\partial v} = -\frac{3mv}{2\bar{u}} f^0 \qquad (2.4.42)$$

from definitions (2.4.2) and (2.4.12) it follows:

$$\langle X \rangle = \frac{\overline{uX}}{\bar{u}} \qquad (2.4.43)$$

Therefore:

$$\langle R_\omega \rangle = \frac{m}{3kT_e} \overline{(v^2 R_\omega)} \qquad (2.4.44)$$

Substituting all the above relationships into (2.4.39), we find:

$$j_\omega = \frac{e^2 \mu_0 \sqrt{\varepsilon_0 \mu_0}}{24\pi^3} \omega^2 n_e \overline{(v^2 R_\omega)} = n_e \overline{Y} \qquad (2.4.45)$$

Expression (2.4.45) shows, then, that the power Y radiated by an electron of velocity v can be identified as:

$$Y = \frac{e^2 \mu_0 \sqrt{\varepsilon_0 \mu_0}}{24\pi^3} \omega^2 v^2 R_\omega \qquad (2.4.46)$$

Since this power cannot depend on the particular distribution function that

Table 2.4.1 Expressions of the R_ω Function for Typical Propagation Cases

Propagation Vector **K** vs. Magnetic Field **B**	Wave Polarization	R_ω
B = 0	Any	$\dfrac{\nu_m}{\nu_m^2 + \omega^2}$
K ∥ **B**	Right circular	$\dfrac{\nu_m}{\nu_m^2 + (\omega - \omega_b)^2}$
K ∥ **B**	Left circular	$\dfrac{\nu_m}{\nu_m^2 + (\omega + \omega_b)^2}$
K ⊥ **B**	Linear $\mathbf{E}_r \parallel \mathbf{B}$	$\dfrac{\nu_m}{\nu_m^2 + \omega^2}$
$\widehat{\mathbf{KB}} = \vartheta$	Elliptical $\mathbf{E}_r \perp \mathbf{B}$	$\simeq \dfrac{1 + \cos^2 \vartheta}{2} \dfrac{\nu_m}{\nu_m^2 + (\omega - \omega_b)^2}$ (resonant term only)

was used for its derivation, expression (2.4.45) establishes a general formula for the computation of the emitted power j_ω.

In view of the relative simplicity of the experiments to which we shall apply the foregoing results, we indicate in Table 2.4.1 the R_ω expressions for various typical cases of electromagnetic wave propagation [Bekefi, Hirshfield, and Brown (1961)]; in the Table $\mathbf{E}_r$ designates the electric field of the radiated wave.

2.5   TRANSPORT EQUATIONS

By appropriate integrations over velocity space of the spherical component (2.3.2) of the Boltzmann equation, it is possible to derive useful equations for the transport parameters introduced in Section 2.4.

Multiplying (2.3.2) by $4\pi v^2 \, dv$, integrating from zero to infinity, and noting that the field-heating term (the last one on the left-hand side) and the collision term ($\bar{B}^0$) do not contribute to a net gain or loss of electrons leads to the familiar continuity equation:

$$\frac{\partial n_e}{\partial t} + \nabla_r \cdot \mathbf{\Gamma}_e = 0 \tag{2.5.1}$$

where the electron flow $\mathbf{\Gamma}_e$ is defined by (2.4.5).

Until now we have disregarded transformation collisions (see Section 2.1) that change the number of particles, such as ionization and attachment processes. However, these processes can be easily introduced into the continuity equation by the addition of an electron source term equal to the time variation of the electron density $\partial n_e/\partial t$ in the absence of the electron flow ($\mathbf{\Gamma}_e = 0$). Thus the continuity equation can be written in more general form as:

$$\frac{\partial n_e}{\partial t} + \nabla_r \cdot \mathbf{\Gamma}_e = (\bar{\nu}_i - \bar{\nu}_a)n_e \tag{2.5.2}$$

where $\bar{\nu}_i$ and $\bar{\nu}_a$ are, for the case under consideration, the ionization and attachment mean frequencies, respectively [according to definition (2.4.2)].

When experiments are performed in an active discharge or in an afterglow, a large number of positive ions may be present and recombination then becomes an additional important electron loss process. In this case, if electrons and positive ions are present in equal numbers, the continuity equation (2.5.2) has to be modified as follows:

$$\frac{\partial n_e}{\partial t} + \nabla_r \cdot \mathbf{\Gamma}_e = (\bar{\nu}_i - \bar{\nu}_a)n_e - \alpha n_e^2 \tag{2.5.3}$$

where α is the recombination coefficient.

Important energy transport equations are obtained when eq. (2.3.2) is multiplied by $(\frac{1}{2}mv^2)(4\pi v^2\,dv)$ and integrated from zero to infinity.

We consider first the case of an applied ac electric field, for which eq. (2.3.2) can be replaced adequately by the equation:

$$\frac{\partial f^0}{\partial t} + \frac{v}{3}\mathbf{V}_r\cdot\mathbf{f}_0^1 - \frac{e}{m}\frac{1}{6v^2}\frac{\partial}{\partial v}(v^2\mathbf{E}_p\cdot\text{Re }\mathbf{f}_1^1) = \bar{B}^0 \qquad (2.5.4)$$

which is the same as (2.3.5) except for the term $\partial f^0/\partial t$, which takes into account slow variations of f^0 with time. If we multiply and integrate (2.5.4) as stated above, we obtain:

$$\frac{\partial(n_e\bar{u})}{\partial t} + \mathbf{V}_r\cdot\mathbf{q}_e - \frac{\pi}{3}\frac{e^2}{m}\mathbf{E}_p\cdot\int_0^\infty v^2\frac{\partial}{\partial v}\left(\text{Re }\mathfrak{T}\cdot\mathbf{E}_p v^2\frac{\partial f^0}{\partial v}\right)dv$$

$$= \pi m\int_0^\infty v^2\frac{\partial}{\partial v}\left[Gv_m v^3\left(1 + \frac{kT_g}{mv}\frac{\partial}{\partial v}\right)f^0\right]dv \qquad (2.5.5)$$

having used definition (2.4.7) for $\mathbf{q}_e$, expression (2.3.14) for $\mathbf{f}_1^1$, and form (2.2.58) for $\bar{B}^0$. If the field and the collision terms are integrated by parts, with the physically appropriate boundary conditions, and use is made of (2.4.11) for the definition of $\boldsymbol{\sigma}$, the following result is obtained:

$$\frac{\partial(n_e\bar{u})}{\partial t} = \frac{1}{2}\mathbf{E}_p\cdot\text{Re }\boldsymbol{\sigma}\cdot\mathbf{E}_p - \mathbf{V}_r\cdot\mathbf{q}_e - 2\pi m\int_0^\infty Gv_m v^4\left(1 + \frac{kT_g}{mv}\frac{\partial}{\partial v}\right)f^0\,dv$$

$$(2.5.6)$$

This equation can be easily reduced to the form:

$$\frac{\partial(n_e\bar{u})}{\partial t} = \sigma_{Tr}E_T^2 + \sigma_r E_{\parallel}^2 - \mathbf{V}_r\cdot\mathbf{q}_e - 2\pi m\int_0^\infty Gv_m v^4\left(1 + \frac{kT_g}{mv}\frac{\partial}{\partial v}\right)f^0\,dv$$

$$(2.5.7)$$

where E_T and $E_{\parallel}$ are the rms values of the applied field components.

Equation (2.5.7) will be used in the following chapters under conditions for which the distribution f^0 is very nearly Maxwellian. In this case substitution of (2.4.42) into the collision term of (2.5.7) makes this term equal to:

$$-n_e\overline{Gv_m}u\left(1 - \frac{\bar{u}_g}{\bar{u}}\right)$$

Then eq. (2.5.7) can be rewritten as:

$$\frac{\partial(n_e\bar{u})}{\partial t} = \sigma_{Tr}E_T^2 + \sigma_r E_{\parallel}^2 - \mathbf{V}_r\cdot\mathbf{q}_e - \frac{n_e}{\tau_r}(\bar{u} - \bar{u}_g) \qquad (2.5.8)$$

where:

$$\tau_r = \frac{\bar{u}}{G\nu_m u} = \frac{\bar{u}}{\nu_u u} \tag{2.5.9}$$

may properly be regarded as the electron relaxation time through energy exchange in collisions. Since for Maxwellian distributions relationship (2.4.43) holds, τ_r is more conveniently written in the form:

$$\tau_r = \frac{1}{\langle G\nu_m \rangle} = \frac{1}{\langle \nu_u \rangle} \tag{2.5.10}$$

When the applied field is dc and the induced electron flow is small enough that we can neglect the transport of energy due to the direct drift action of the applied field, with a procedure equivalent to the one used for the ac case we arrive at an energy transport equation quite similar to (2.5.7):

$$\frac{\partial(n_e \bar{u})}{\partial t} = en_e(\mu_T E_T^2 + \mu E_{\parallel}^2) - \nabla_r \cdot \mathbf{q}_e$$

$$- 2\pi m \int_0^\infty G\nu_m v^4 \left(1 + \frac{kT_g}{mv}\frac{\partial}{\partial v}\right) f^0 \, dv \tag{2.5.11}$$

In the case when the collision frequency for energy exchange $\nu_u \, (= G\nu_m)$ is velocity independent, this equation becomes:

$$\frac{\partial(n_e \bar{u})}{\partial t} = en_e(\mu_T E_T^2 + \mu E_{\parallel}^2) - \nabla_r \cdot \mathbf{q}_e - \nu_u n_e(\bar{u} - \bar{u}_g) \tag{2.5.12}$$

When space uniformity is assumed and steady-state energy conditions are attained, eq. (2.5.12) is conveniently written as:

$$\nu_u = \frac{e(\mu_T E_T^2 + \mu E_{\parallel}^2)}{\bar{u} - \bar{u}_g} \tag{2.5.13}$$

This expression is quite interesting since it relates the collision frequency ν_u, assumed to be velocity independent, to other quantities that are measurable parameters of the moving electron swarm.

2.6   DERIVATION OF COLLISION PARAMETERS FROM ELECTRON SWARM PROPERTIES

All of the transport coefficients defined in Section 2.4 are explicit functions of usually known parameters and of the collision frequency $\nu_m(v)$. If the distribution function f^0 were known in each case, from an appropriately large set of values of one or more transport coefficients for a given gas the collision frequency $\nu_m(v)$ could be derived if unknown.

However, as we saw in Section 2.3, f^0 is in turn a function of $v_m(v)$, so that this distribution must be substituted in the transport coefficient expressions in one of its explicit forms, including $v_m(v)$, or left as a general function of $v_m(v)$, and the collision frequency determined by solving self-consistently the resulting set of appropriate integral equations. Since in the Boltzmann equation we find the presence of additional electron collision parameters, like the energy loss factor $G(v)$ or the inelastic collision probabilities, we are required to know them too, or to consider them as unknown, to be determined from the integral equations set at the same time as $v_m(v)$.

The techniques for determining electron collision parameters are the major items of interest in this book. In this section we provide formulas for the cases in which substitution of explicit and simple solutions for f^0 and the use of appropriate laws for the velocity dependence of the collision parameters yield transport coefficient expressions that are particularly useful for the purpose of deriving from them the collision parameter values. Moreover, we define and introduce here secondary transport parameters which are combinations of those defined in Section 2.4, and which happen to be those

Table 2.6.1   Combinations of Field Characteristics and of Collision Parameter Laws that Yield a Distribution Function of the Form $f^0 \propto \exp(-bv^l)$

E Field	G	v_m	l	b	$\bar{u}$
1. Dc, ac, $E_e^2 \ll E_{et}^2$	Any	Any	2	$\dfrac{3m}{4\bar{u}}$	eq. (2.3.29)
2. Ac, $\omega^2 \gg v_m^2$, $(\omega - \omega_b)^2 \gg v_m^2$	Const.	Any	2	$\dfrac{3m}{4\bar{u}}$	eq. (2.3.30)
3. Ac, $\mathbf{E} \parallel \mathbf{B}$ $\omega^2 \gg v_m^2$	Const.	Any	2	$\dfrac{3m}{4\bar{u}}$	eq. (2.3.30)
4. Dc, ac, $\mathbf{E} \perp \mathbf{B}$ $(\omega - \omega_b)^2 \gg v_m^2$	Const.	Any	2	$\dfrac{3m}{4\bar{u}}$	eq. (2.3.30)
5. Ac, $E_e^2 \gg E_{et}^2$ $\omega^2 \gg v_m^2$, $(\omega - \omega_b)^2 \gg v_m^2$	$\propto v^g$	Any	$2 + g$	eq. (2.3.46)	eq. (2.3.47)
6. Ac, $\mathbf{E} \parallel \mathbf{B}$, $E_e^2 \gg E_{et}^2$ $\omega^2 \gg v_m^2$	$\propto v^g$	Any	$2 + g$	eq. (2.3.46)	eq. (2.3.47)
7. Dc, ac, $\mathbf{E} \perp \mathbf{B}$, $E_e^2 \gg E_{et}^2$ $(\omega - \omega_b)^2 \gg v_m^2$	$\propto v^g$	Any	$2 + g$	eq. (2.3.46)	eq. (2.3.47)
8. Dc, ac, $E_e^2 \gg E_{et}^2$ $(\omega + \omega_b)^2 \ll v_m^2$	$\propto v^g$	$\propto v^h$	$2 + g + 2h$	eq. (2.3.46')	eq. (2.3.47)
9. Dc, ac, $\mathbf{E} \parallel \mathbf{B}$, $E_e^2 \gg E_{et}^2$ $\omega^2 \ll v_m^2$	$\propto v^g$	$\propto v^h$	$2 + g + 2h$	eq. (2.3.46')	eq. (2.3.47)

actually measured in the experiments to be discussed in the following chapters.

Most of this analysis is based on the simple f^0 solution (2.3.44), which includes the Maxwell distribution case ($l = 2$); we have summarized in Table 2.6.1 the combinations of field characteristics and of collision parameter laws that yield this f^0 form, according to the results of Section 2.3. The field E_{et}, which appears in the table, is defined by (2.3.28) and represents the effective field that induces an electron random motion equal to the thermal one; in general, E_e and E_{et} are functions of the electron velocity, and the stated E_e versus E_{et} conditions must hold over the entire range of significant electron velocities.

A. Use of Direct Current Transport Parameters when no Magnetic Field is Present

We discuss first the transport coefficients that are usually considered in an electron swarm moving under the action of a dc electric field when no static magnetic field is applied. These coefficients, according to eq. (2.4.17), are the following two: the electron mobility μ or the equivalent electron drift velocity w [eq. (2.4.20)], and the electron diffusion D [eq. (2.4.18)]. It is also essential to consider the related ratio D/μ, which is a very important experimentally measurable quantity, usually called the characteristic energy, since:

$$\frac{D}{\mu} = \frac{m}{3e} \frac{\overline{(v^2/v_m)}}{\langle 1/v_m \rangle} = \frac{2}{3e} \frac{\overline{(u/v_m)}}{\langle 1/v_m \rangle} \tag{2.6.1}$$

In fact, when v_m is velocity independent:

$$\frac{D}{\mu} = \frac{2\bar{u}}{3e} \tag{2.6.2}$$

since, when X is constant:

$$\langle X \rangle = X \tag{2.6.3}$$

The same result (2.6.2), known as the Einstein relation, is obtained for any $v_m(v)$ law when distribution f^0 is Maxwellian, since in this case, as we have seen before:

$$\langle X \rangle = \frac{\overline{uX}}{\bar{u}} \tag{2.4.43}$$

According to Table 2.6.1, this situation will always occur at very low electric fields (case 1). When eq. (2.6.2) holds, D/μ is a measure of the electron average energy in electron volts ($= \bar{u}/e$): hence the name characteristic energy for this ratio. The ratio of the characteristics energy to its thermal equilibrium value kT_g/e [from (2.6.2) when $\bar{u} = 3kT_g/2$] is the dimensionless

Townsend's energy factor:

$$k_1 = \frac{e}{kT_g} \frac{D}{\mu} \qquad (2.6.4)$$

hence:

$$k_1 = 39.8 \frac{D}{\mu} \quad \text{(eV)} \quad \text{at } 20°C \qquad (2.6.5)$$

Frequent use of this factor is found in the literature, particularly by authors of the Australian school.

In the present ($\omega = \omega_b = 0$) case the effective electric field is equal to the applied one, and, according to the results of Section 2.3, in a given gas and at a constant temperature the distribution f^0 will be a function of E/p only. From (2.4.18), (2.4.20), and (2.6.1) it follows that μp (or w), Dp, and D/μ or k_1 will likewise be functions of E/p alone. Clearly this result applies when the contribution of electron-ion collisions is negligible and only electron-neutral collisions are relevant; this condition is adopted throughout this section, except when otherwise noted.

Frequent use can also be found in the literature of two other parameters, related to the previous set of dc, zero magnetic field coefficients. They are as follows [Frost and Phelps (1962)]:

a. An effective momentum transfer collision frequency ν_m^*, defined in accordance with eq. (2.4.20) by the relation:

$$\nu_m^* = \frac{e}{m\mu} = \frac{1}{\langle 1/\nu_m \rangle} \qquad (2.6.6)$$

b. An effective energy exchange collision frequency ν_u^*, defined as the power input per electron due to the electric field divided by the characteristic energy excess over its thermal equilibrium value:

$$\nu_u^* = \frac{e\mu E^2}{e(D/\mu) - kT_g} \qquad (2.6.7)$$

When the characteristic energy is given by eq. (2.6.2), we have:

$$\nu_u^* = \frac{3}{2} \frac{e\mu E^2}{\bar{u} - \bar{u}_g} = \tfrac{3}{2}\nu_u \qquad (2.6.8)$$

where the last equality holds when ν_u is velocity independent [see eq. (2.5.13)].

Both quantities, ν_m^*/p and ν_u^*/p, are functions of E/p. Because of their intimate relationships to the corresponding electron collision frequencies, as they appear from the above equations, in non-polar gases ν_m^*/p will usually be more sensitive to changes in the elastic collision probability, whereas

v_u^*/p will be so to changes in the inelastic collision probabilities. For this reason these quantities are particularly significant when collision data have to be derived from experimentally measured parameters.

Let us write the collision frequency v_m in the form:

$$v_m(u) = v_m(\bar{u})\psi(u/\bar{u}) \tag{2.6.9}$$

where $\bar{u}$, according to our standard notations, is the average energy of the electron swarm whose properties are being measured in the experiment, and ψ represents the normalized law of dependence of the collision frequency on the electron energy [$\psi(1) = 1$]. Since the velocity-dependent part of the distribution function f^0 has dimensions of v^{-3}, it can always be written in the quite general form:

$$f^0(u) = Kn_e \frac{\varphi(u/\bar{u})}{\bar{u}^{3/2}} \tag{2.6.10}$$

where φ provides the distribution shape, and K is a dimensionless constant. When (2.6.9) and (2.6.10) are substituted into (2.4.20), (2.4.18), and (2.6.1), we can write:

$$\mu p = \frac{w}{E/p} = \mathscr{A}\frac{e}{m[v_m(\bar{u})/p]} \tag{2.6.11}$$

$$Dp = \mathscr{A}\mathscr{C}\frac{2}{3m}\frac{\bar{u}}{v_m(\bar{u})/p} \tag{2.6.12}$$

$$\frac{D}{\mu} = \mathscr{C}\frac{2\bar{u}}{3e} \tag{2.6.13}$$

where:

$$\mathscr{A} = -\frac{2}{3}\frac{\int_0^\infty (\partial\varphi/\partial y)(y^{3/2}/\psi)\,dy}{\int_0^\infty \varphi y^{1/2}\,dy} = \frac{2}{3}\frac{\int_0^\infty \varphi(\partial/\partial y)(y^{3/2}/\psi)\,dy}{\int_0^\infty \varphi y^{1/2}\,dy} \tag{2.6.14}$$

(y being the normalized energy $u/\bar{u}$), and:

$$\mathscr{A}\mathscr{C} = \frac{\int_0^\infty (\varphi y^{3/2}/\psi)\,dy}{\int_0^\infty \varphi y^{3/2}\,dy} \tag{2.6.15}$$

The quantities $\mathscr{A}$ and $\mathscr{C}$ depend only on the distribution shape φ and on the collision frequency law ψ. It can easily be verified that $\mathscr{A}$ and $\mathscr{C}$ are unity when the collision frequency is energy independent ($\psi = 1$). Moreover, from (2.6.6) we have the relation:

$$\frac{v_m^*}{p} = \frac{1}{\mathscr{A}}\frac{v_m(\bar{u})}{p} \tag{2.6.16}$$

Expressions (2.6.11)–(2.6.15) are useful when we can employ the same ψ and φ normalized behaviors over a wide range of experimental conditions, including a significant range of values of the average electron energy $\bar{u}$. In fact, in these cases, $\mathscr{A}$ and $\mathscr{C}$ are constant.

According to Table 2.6.1, in the present dc no-magnetic field situation this happens when $\nu_m \propto v^h$, $G \propto v^g$, and $E^2 \gg E_t^2$. In fact, we have:

$$\psi = y^{h/2} \tag{2.6.17}$$

$$\varphi \propto \exp\left\{-\left[\frac{\Gamma(5/l)}{\Gamma(3/l)} y\right]^{l/2}\right\} \tag{2.6.18}$$

where $l = 2 + g + 2h$.

For these conditions $\mathscr{A}$ and $\mathscr{C}$ are easily computed; the result is [Skharofski, Bachynski, and Johnston (1961), Gilardini (1963)]:

$$
\begin{aligned}
\mathscr{A}(l, h) &= \frac{3 - h}{3} \frac{\Gamma[(3 - h)/l]}{\Gamma(3/l)} \left[\frac{\Gamma(5/l)}{\Gamma(3/l)}\right]^{h/2} \\
&= \frac{\Gamma[(3 + l - h)/l]}{\Gamma[(3 + l)/l]} \cdot \left[\frac{\Gamma(5/l)}{\Gamma(3/l)}\right]^{h/2} \tag{2.6.19}
\end{aligned}
$$

$$
\begin{aligned}
\mathscr{C}(l, h) &= \frac{3}{3 - h} \frac{\Gamma(3/l)}{\Gamma(5/l)} \frac{\Gamma[(5 - h)/l]}{\Gamma[(3 - h)/l]} \\
&= \frac{\Gamma[(3 + l)/l]}{\Gamma[(3 + l - h)/l]} \frac{\Gamma[(5 - h)/l]}{\Gamma(5/l)} \tag{2.6.20}
\end{aligned}
$$

In Fig. 2.6.1 $\mathscr{A}$ and $\mathscr{C}$ are plotted as a function of l for various values of h. From (2.6.13), (2.3.47), and (2.3.46') we obtain:

$$\frac{D}{\mu} \propto \bar{u} \propto b^{-2/l} \propto \left(\frac{E}{p}\right)^{4/l}$$

Since the power law (2.6.17) means $\nu_m(\bar{u})/p \propto \bar{u}^{h/2}$, we have for the mobility from (2.6.11):

$$\mu p \propto \left(\frac{\nu_m}{p}\right)^{-1} \propto \bar{u}^{-h/2} \propto \left(\frac{E}{p}\right)^{-2h/l} \propto \left(\frac{D}{\mu}\right)^{-h/2}$$

Therefore, by plotting for a given gas the experimental data of μp versus E/p or D/μ, and of D/μ versus E/p, both on logarithmic scales, we can easily verify whether the above-assumed dependences are satisfied (the data must fall on straight lines) and can derive the values of h and g [the slope of the μp versus E/p, of the μp versus D/μ, and of the D/μ versus E/p straight lines are, respectively,

$$-2h/l = -2h/(2 + g + 2h), \quad -h/2, \quad \text{and} \quad 4/l = 4/(2 + g + 2h)].$$

Thereafter, $\mathscr{A}$ and $\mathscr{C}$ can be computed so that (2.6.13) gives $\bar{u}$; (2.6.11), the collision frequency ν_m at this energy; and (2.3.47), together with (2.3.45′) and (2.3.46′), the loss factor G at the same energy. When G is known, it is sufficient to measure one transport parameter only as a function of E/p; customarily this is μp versus E/p. In this case, after having determined h as explained above, for an arbitrarily chosen E/p point we compute $\nu_m(\bar{u})$ by means of (2.6.11), and hence the energy $\bar{u}$ by means of (2.3.47) and (2.3.46′).

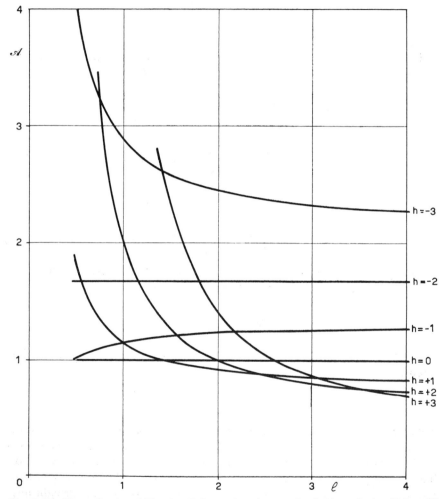

Fig. 2.6.1. Normalized mobility (coefficient $\mathscr{A}$) and normalized D/μ ratio (coefficient $\mathscr{C}$), plotted as a function of the parameters of the distribution function $f^0 \propto \exp(-bv^l)$ and of the collision frequency $\nu_m \propto v^h$. Vertical axes can be read also as normalized conductivity ratio and normalized radiation temperature, respectively if the sign of h is changed.

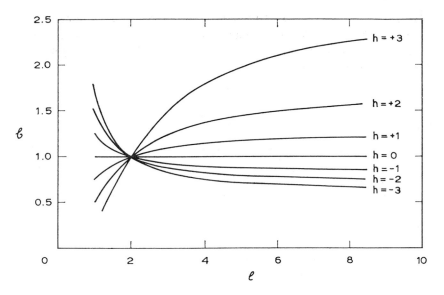

Figure 2.6.1. (contd.)

Another interesting situation takes place at very low fields, when $E^2 \ll E_t{}^2$. In this case (see Table 2.6.1) the distribution is Maxwellian, since electrons are in thermal equilibrium with the gas. Assuming an inverse power series representation of p/ν_m in terms of the electron velocity:

$$\frac{p}{\nu_m} = \frac{1}{v\mathscr{P}_m} = \sum_j b_j v^{-j} \tag{2.6.21}$$

and introducing the notation:

$$\gamma_p{}^q = \frac{\Gamma[(p+q)/2]}{\Gamma(p/2)} \tag{2.6.22}$$

eq. (2.4.20) yields [Pack and Phelps (1961)]:

$$\mu p = \frac{e}{m} \sum_j \gamma_5{}^{-j} b_j \left(\frac{2\hat{u}}{m}\right)^{-j/2} = \sum_j B_j \left(\frac{2\hat{u}}{m}\right)^{-j/2} = \sum_j B_j \left(\frac{2kT_g}{m}\right)^{-j/2} \tag{2.6.23}$$

We recall that $\sqrt{2\hat{u}/m}$ is the most probable speed of the distribution. Assuming that $\nu_m \propto v^h$, instead of using a power series, is formally a special case of the above, and (2.6.23) takes the particular form:

$$\mu p = \frac{e}{m} \gamma_5{}^{-h} b_h \left(\frac{2\hat{u}}{m}\right)^{-h/2} = \frac{e}{m} \gamma_5{}^{-h} \frac{p}{\nu_m(\hat{u})} \tag{2.6.23'}$$

When from experiment we have μp at several gas temperatures T_g, it is usually possible to choose a reasonable set of values of j and to determine the coefficients B_j from the experimental data. The electron collision probability is thus given as a function of velocity by:

$$\mathscr{P}_m^{-1} = \frac{m}{e} \sum_j \frac{B_j}{\gamma_5^{-j}} v^{1-j} \qquad (2.6.24)$$

If the electron diffusion coefficient is measured in the absence of any applied field, the above discussion and formulas can be used again, since Dp is equal to $(kT_g/e)\mu p$, that is, Dp and μp differ by a known constant only.

When electron-positive ion collisions are dominating (only singly ionized particles are considered, since this is the actual case in the experiments we shall discuss later on), the mobility can be computed from (2.4.20), where ν_m is equal to $vP_m{}^i$, $P_m{}^i$ being given by (1.5.15). In expression (1.5.15) the electron velocity v appears twice, but the dependence of the logarithmic term on velocity is weak, so that no significant error is made if we replace in it the quantity mv^2 by its average value $3kT_e$. Then we may write:

$$\nu_m{}^i = vP_m{}^i = A \frac{n_e}{v^3} \ln\left(1 + B \frac{T_e{}^3}{n_e}\right) = A \frac{n_e \mathscr{L}}{v^3} \qquad (2.6.25)$$

with:

$$A = \frac{e^4}{8\pi\varepsilon_0{}^2 m^2} = 4.03 \times 10^5 \quad (\text{m}^6 \text{ sec}^{-4})$$

$$B = \frac{144\pi^2\varepsilon_0{}^3 k^3}{e^6} = 1.53 \times 10^{14} \quad (\text{m}^\circ\text{K})^{-3} \qquad (2.6.26)$$

In this expression the symbol:

$$\mathscr{L} = \ln\left(1 + B \frac{T_e{}^3}{n_e}\right) \qquad (2.6.27)$$

has been introduced for future writing convenience*; within the parentheses unity may often be neglected, since the second term is usually much larger.

Substituting (2.6.25) into (2.4.20), which, we recall, implies neglecting any effect due to electron-electron collisions, yields:

$$\mu = \frac{e}{mAn_e\mathscr{L}} \langle v^3 \rangle \qquad (2.6.28)$$

* The $\mathscr{L}$ expression (2.6.27) derives from the assumption that the interaction sphere radius for electron collisions is equal to the Debye length appropriate to the potential screening of a positive ion by electrons only. Consideration of collision times justifies this assumption [Delcroix (1965)], but use of the Debye length of an ion shielded by both electrons and positive ions is sometimes used in the literature; in this case $\mathscr{L} = \ln [1 + BT_e{}^3 T_i/n_e(T_e + T_i)]$.

and, for the case of a Maxwell distribution:

$$\mu = \frac{8e}{\sqrt{\pi}\ mAn_e\mathscr{L}} \left(\frac{2kT_e}{m}\right)^{3/2} \tag{2.6.29}$$

The corresponding dc conductivity is [see eq. (2.4.20)]:

$$\sigma_{\mathrm{dc}} = en_e\mu = \frac{8e^2}{\sqrt{\pi}\ mA\mathscr{L}} \left(\frac{2kT_e}{m}\right)^{3/2} \tag{2.6.30}$$

Spitzer and Härm (1953) investigated how eq. (2.6.29) must be modified when electron-electron collisions are of importance. For this purpose the expression of the distribution component $\mathbf{f}^1$ to be substituted into (2.4.5) is obtained from (2.3.3), where the additional electron-electron term (2.2.63), properly simplified for the case of Maxwellian distribution f^0, is included in $\bar{\mathbf{B}}^1$. Along these lines Spitzer and Härm found that μ and σ_{dc} are still given by eqs. (2.6.29) and (2.6.30) except for a multiplying numerical factor, equal to 0.5816. Since eq. (2.3.3) with term (2.2.63) cannot be solved in a closed form for $\mathbf{f}^1$, an approximate way of solving it and of deriving the Spitzer–Härm result is to expand the $\mathbf{f}^1$ components in terms of generalized Laguerre or Sonine polynomials, and to calculate the expansion coefficients from a set of linear nonhomogeneous equations, derived by substituting expansions in eq. (2.3.3) and properly applying the orthogonality properties of these polynomials [Shkarofsky (1961)]; different approximations are thus possible, depending on the number of expansion terms which are considered. Identical results are obtained if, following the general theory of Chapman-Enskog [see for details Chapman and Cowling (1952) and Landshoff (1949, 1951)], which applies to states not far from that of thermodynamic equilibrium, one solves the Boltzmann equation by a method of successive approximations; the distribution function, normalized to the Maxwellian one, is directly expanded in terms of Laguerre or Sonine polynomials, and from the Boltzmann equation the system of linear nonhomogeneous equations to be solved in succession is derived.

More recently, the Spitzer-Härm factor has been recomputed, replacing the crude approximation of a truncated Coulomb potential used by these authors (see Section 1.5) with more physically significant models of the interaction that takes place in distant encounters between charged particles. For this purpose both the so-called unified theory of Kihara and Aono (1963) and Liboff's (1959) model of a shielded Coulomb potential have been used; the common result [Itikawa (1963), Kruger, Mitchner, and Daybelge (1968), Viegas (1971)] is that the multiplying factor depends on $\mathscr{L}$. In Fig. 2.6.2 the ratio of Viegas' second approximation to the Spitzer-Härm value is plotted as a function of $\mathscr{L}$ for the various interaction models.

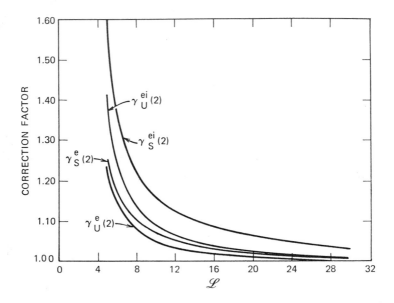

Fig. 2.6.2. Ratio $[\gamma(2)]$ of Viegas second approximation to the Spitzer-Härm value of the dc electrical conductivity in the case of dominating electron-ion and electron-electron collisions, computed with different models and plotted as a function of $\mathscr{L}$. Subscripts S and U indicate that γ has been computed using the shielded Coulomb potential or the unified theory, whereas superscripts e and ei indicate, respectively, that the shielding characteristic length is taken equal to $(\varepsilon_0 k T_e/e^2 n_e)$ (Debye length for an ion shielded by electrons only) or $1/\sqrt{2}$ this value (total Debye length for an ion shielded by a two-component electron-ion isothermal plasma) [after Viegas (1971)].

Finally we consider the case of a mixture of gases. Collision frequencies are additive, so that, if p_i is the partial pressure of the ith component, we have:

$$\nu_m = v \sum_i p_i \mathscr{P}_{mi} = \sum_i p_i \left(\frac{\nu_m}{p}\right)_i \tag{2.6.31}$$

If collision frequencies of the components $(\nu_m/p)_i$ are assumed to be velocity independent, we obtain the formula known as Blanc's law:

$$\frac{1}{\mu} = \sum_i \frac{p_i}{(\mu p)_i} \tag{2.6.32}$$

which follows at once since $\nu_m/p = e/m(\mu p)$ for each component.

The same formula (2.6.32) holds also when the collision frequencies $(\nu_m/p)_i$ are not velocity independent, but all are governed by the same dependence law on velocity, and the electrons are in thermal equilibrium with the gas ($E^2 \ll E_t{}^2$). Another case of practical concern is that of a binary mixture with isothermal electrons, whose collision probabilities are constant in gas 1 and proportional to v^{-2} in gas 2. In this case eq. (2.4.20) yields:

$$\frac{1}{\mu} = \frac{p_1}{(\mu p)_1} \left[1 - z + z^2 e^z \int_z^\infty u^{-1} \exp(-u)\, du \right]^{-1} \simeq \frac{p_1}{(\mu p)_1} (1 + \beta z) \qquad (2.6.32')$$

where z indicates the ratio $p_2 \mathscr{P}_{m2}(\hat{u}_g)/p_1 \mathscr{P}_{m1}$, and β changes from 1 when $z \ll 1$, to $\frac{4}{7}$ over the $0.2 \leqslant z \leqslant 7$ region, and up to $\frac{1}{2}$ when $z > 10$ [Christophorou, Hurst, and Hendrick (1966)].

Of particular interest also is the conductivity of a partially ionized gas, in which electron-neutral, electron-ion, and electron-electron collisions have all to be considered for the determination of the overall conductivity. In this case use of the mixture rule (2.6.32) for a two-component gas (neutral atoms or molecules and ions) yields results that may be in error by as much as a factor of 2 [see, e.g., Kruger, Mitchner, and Daybelge's (1968) calculations for atmospheric-pressure argon]. Better results are obtained with an empirical mixture rule proposed by Frost (1961), in which the conductivity expression (2.4.20) is used, but the distribution function is assumed to be Maxwellian and the collision frequency is taken as the sum of the electron-neutral collision frequency and of a modified electron-ion collision frequency:

$$\nu_m{}^i = 0.952A \, \frac{n_e \mathscr{L}}{v^2} \left(\frac{m}{2kT_e} \right)^{1/2} \qquad (2.6.25')$$

such that the resultant conductivity reduces to the exact Spitzer-Härm value [0.5816 times the value given by (2.6.30)] in the limit of a fully ionized plasma. As shown by Schweitzer and Mitchner (1966), Kruger et al. (1968), and Viegas and Kruger (1969) for several experimental electron-neutral collision frequencies, Frost's formula yields, in general, remarkably accurate results.

When the electrical conductivity of a Maxwellian plasma in the presence of both electron-neutral collisions and charged particle interactions has to be computed more accurately than Frost's mixture rule permits, converging approximations can be adopted, following the expansion procedures mentioned previously for the case of dominant collisions between charged particles. Among the different schemes that have actually been used, the most significant one appears to be that developed by Devoto (1966, 1967b) [see also Li and Devoto (1968)], which is based on the Chapman-Enskog method and uses a shielded Coulomb potential for the evaluation of interactions between charged particles. In all these calculations electron-neutral collisions appear in the pertinent equations as a set of properly defined average cross sections,

which add to corresponding average cross sections for charged particle interactions. Approximations, however, have usually to be carried to higher
orders than the second order, which is sufficient for the case of interactions
between charged particles; in fact, for dependence $v_m \propto v^h$, the necessary
order of approximation for a reasonable level of accuracy increases with
increasing h [this demonstrates why this technique is most appropriate for
the case of dominant electron-ion collisions, when $h = -3$ according to
eq. (2.6.25)].

B. Use of Alternating Current Transport Parameters when no Magnetic Field is Present

Measurements of alternating current transport coefficients, most of which
have been performed at radio and microwave frequencies, represent a large
source of data for the determination of collision parameters. In most cases
two main quantities are considered: the ac conductivity ratio ρ, defined as
[Phelps, Fundingsland, and Brown (1951)]:

$$\rho = -\frac{\sigma_r}{p\sigma_i} \tag{2.6.33}$$

and the radiation temperature T_r, defined as [Bekefi, Hirshfield, and Brown
(1961)]:

$$T_r = \frac{8\pi^3 B_\omega}{\varepsilon_0 \mu_0 k \omega^2} = \frac{8\pi^3}{\mu_0 \sqrt{\varepsilon_0 \mu_0}\, k\omega^2}\frac{j_\omega}{\sigma_{er}} \tag{2.6.34}$$

The last definition was derived from eq. (2.4.40), where Planck's function
B_ω has been replaced by the measurable emitted power j_ω according to
Kirchhoff's relationship (2.4.39).

When use is made of eqs. (2.4.19) for σ, (2.4.41) for σ_{er}, and (2.4.45) for
j_ω, the above quantities become:

$$\omega\rho = \frac{1}{p}\frac{\left\langle \dfrac{v_m}{v_m^2 + \omega^2} \right\rangle}{\left\langle \dfrac{1}{v_m^2 + \omega^2} \right\rangle} = \frac{\left\langle \dfrac{v_m/p}{(v_m/p)^2 + (\omega/p)^2} \right\rangle}{\left\langle \dfrac{1}{(v_m/p)^2 + (\omega/p)^2} \right\rangle} \tag{2.6.35}$$

$$T_r = \frac{2}{3k}\frac{\overline{uR_\omega}}{\langle R_\omega \rangle} \tag{2.6.36}$$

Hence:

$$T_r = \frac{2}{3k}\frac{\left(\dfrac{v_m u}{v_m^2 + \omega^2} \right)}{\left\langle \dfrac{v_m}{v_m^2 + \omega^2} \right\rangle} \tag{2.6.37}$$

when the no-field expression of R_ω is chosen according to Table 2.4.1. It appears clear that the last form of (2.6.35) has been finalized for application in the case in which electron-neutral collisions only are important; this formula also justifies the introduction of pressure in the definition of ρ [eq. (2.6.33)]. The validity of eq. (2.6.37) is not restricted to the case of a tenuous plasma [Plantinga (1961)].

When ν_m is velocity independent:

$$\omega\rho = \frac{\nu_m}{p} \tag{2.6.38}$$

$$T_r = \frac{2\bar{u}}{3k} = T_e \tag{2.6.39}$$

the electron temperature T_e being defined by relationship (2.3.32). It then appears that, like the electron mobility and the characteristic energy in the dc case, the conductivity ratio and the radiation temperature provide here an appropriate measure of the collision frequency and of the electron energy, respectively. Both quantities, when we may neglect electron-ion collisions, are independent of the swarm electron densities, and this is one of the reasons for their choice.

When the distribution is Maxwellian, as at very low applied field intensities, eq. (2.6.39) holds for any collision frequency law; this is easily proved by substituting relationship (2.4.43) into (2.6.36). We have, on the other hand, no simple general expression for the conductivity ratio in a Maxwellian electron swarm, except for special cases, two of which will now be discussed.

a. In the microwave-frequency and low-pressure (or density) limit $\omega^2 \gg \nu_m{}^2$, independently of the electron energy distribution, the conductivity ratio (2.6.35) becomes:

$$\omega\rho = \frac{\langle \nu_m \rangle}{p} \tag{2.6.40}$$

If we assume a collision frequency law $\nu_m \propto v^h$ and consider a Maxwellian distribution, from (2.6.40) we obtain:

$$\omega\rho = \gamma_5{}^h \frac{\nu_m(\hat{u})}{p} = \mathscr{B}(h) \frac{\nu_m(\bar{u})}{p} = \mathscr{B}(h) \left(\frac{2\bar{u}}{m}\right)^{\!\!\frac{1}{2}} \mathscr{P}_m(\bar{u}) \tag{2.6.41}^{\cdot}$$

where $\gamma_5{}^h$ is defined by (2.6.22), and:

$$\mathscr{B}(h) = (\tfrac{2}{3})^{h/2}\gamma_5{}^h \tag{2.6.42}$$

In the case of electron-positive ion collisions, $\nu_m \propto v^{-3}$ according to (2.6.25); substitution of this into (2.6.41) and (2.6.42) yields:

$$\omega\rho = \frac{\sqrt{6}\,A\mathscr{L}}{\sqrt{\pi}} \frac{n_e}{p} \left(\frac{2\bar{u}}{m}\right)^{\!\!-\frac{3}{2}} = \frac{4A\mathscr{L}}{3\sqrt{\pi}} \frac{n_e}{p} \left(\frac{2kT_e}{m}\right)^{\!\!-\frac{3}{2}} = A'\mathscr{L} \frac{n_e}{p} T_e^{-\frac{3}{2}} \tag{2.6.43}$$

where $A' = 1.78 \times 10^{-6}$ m³ sec⁻¹ °K$^{3/2}$. We note that, at constant temperature, this $\omega\rho$ is essentially proportional to n_e/p, since the influence of limited
variations of n_e in the $\mathscr{L}$ term [eq. (2.6.27)] is usually negligible. Furthermore,
Landshoff's (1949), Spitzer's (1952), and Robinson's and Bernstein's (1962)
results on the conductivities of highly ionized gases show that eq. (2.6.43)
is valid also when electron-electron collision effects are included in the analysis.

According to (2.6.40), in a mixture of gases $\omega\rho$ will be given by the weighted
sum of the same quantity $\omega\rho$ for each component gas, the weight being the
fractional pressure or density of the gas in the mixture.

When electron collisions with both neutral atoms or molecules and ions
are important, $\omega\rho$ is given by the sum of (2.6.41) for the former and (2.6.43)
for the latter.

When the $\omega^2 \gg \nu_m{}^2$ limit is not satisfied, ρ may conveniently be expressed
in terms of the limit value (2.6.41); we shall indicate this here as $\bar{\rho}$. The
result [Phelps et al. (1951), Phelps (1960), Bakshi, Haskell, and Papa
(1967)] can be written in the form:

$$\rho = \bar{\rho}\,\frac{D^h_{(3+h)/2}[\omega/\nu_m(\hat{u})]}{D^h_{3/2}[\omega/\nu_m(\hat{u})]} = \bar{\rho}\,\frac{D^h_{(3+h)/2}[\gamma_5{}^h/p\bar{\rho}]}{D^h_{3/2}[\gamma_5{}^h/p\bar{\rho}]} \qquad (2.6.44)$$

where:

$$D_i{}^h(x) = \frac{1}{\Gamma(i+1)}\int_0^\infty \frac{y^i\exp(-y)}{y^h + x^2}\,dy \qquad (2.6.45)$$

For integer values of h between -3 and $+3$ these integrals can be related to
the tabulated Dingle integrals [Dingle, Arndt, and Roy (1957); see Appendix
B]. Figure 2.6.3 is a plot of $\rho/\bar{\rho}$ versus $1/p\bar{\rho}$ for different values of h.

b. We consider again the microwave-frequency and low-pressure limit
$\omega^2 \gg \nu_m{}^2$, but assume now a power series representation of ν_m/p as a function
of the electron velocity:

$$\frac{\nu_m}{p} = v\mathscr{P}_m = \sum_j a_j v^j \qquad (2.6.46)$$

The conductivity ratio becomes [Phelps et al. (1951)]:

$$\omega\rho = \frac{\langle\nu_m\rangle}{p} = \sum_j \gamma_5{}^j a_j \left(\frac{2\hat{u}}{m}\right)^{j/2} \qquad (2.6.47)$$

In the case of an electron swarm in thermal equilibrium with the gas,
measurements of ρ with a sufficiently low field not to perturb this equilibrium
afford the possibility of determining the thermal energy values of the electron
collision frequency by means of the above formulas. The common procedures
[Phelps et al. (1951), Gould and Brown (1954), Anderson and Goldstein

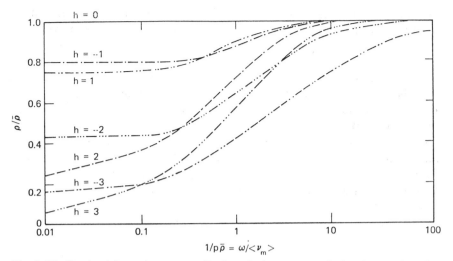

Fig. 2.6.3. Conductivity ratio ρ, normalized to the low-pressure limit value $\bar{\rho}$, plotted as a function of $1/p\bar{\rho}$ for different values of the power-law exponent h [after Bakshi, Haskell, and Papa (1967)].

(1955)] are as follows:

1. First the conductivity ratio is measured at sufficiently low pressures to yield $\bar{\rho}$; then the pressure is increased, whereas the gas temperature and the frequency ω are kept constant, so that measurements provide the curve of $\rho/\bar{\rho}$ versus $1/p\bar{\rho}$. The value of h is found from the best fit between the experimental points and the theoretical curves of Fig. 2.6.3; the value of ν_m at $\hat{u} = kT_g$ is then computed from $\bar{\rho}$, using eq. (2.6.41).

2. The gas temperature is changed, and ρ is measured as a function of the electron temperature ($T_e = T_g$), the measuring signal frequency being constant and such that $\omega^2 \gg \nu_m^2$. With reference to eq. (2.6.47), we choose an appropriate set of values of j so that the experimental $\omega\rho$ versus T_e curve can be represented by the expansion:

$$\omega\rho = \sum_j A_j \left(\frac{2kT_e}{m}\right)^{j/2} = \sum_j A_j \left(\frac{2\hat{u}}{m}\right)^{j/2} \tag{2.6.48}$$

where the values of the coefficients A_j are determined for the best fit. The electron collision probability is then given as a function of velocity in the thermal energy range by:

$$\mathscr{P}_m = \sum_j \frac{A_j}{\gamma_5{}^j} v^{j-1} \tag{2.6.49}$$

3. At a constant frequency such that $\omega^2 \gg \nu_m^2$ and at a constant temperature, the conductivity ratio ρ is measured as a function of n_e/p. Should

we find a straight-line relationship, the following interpretation can be adopted: the intercept at zero n_e/p is the value due to electron-neutral collisions; the linearly dependent part of ρ represents the electron-ion collision contribution, according to (2.6.43). Then the slope makes it possible to determine the quantity $A\mathscr{L} \simeq A \ln (BT_g{}^3/n_e)$, which has to be consistent with the theoretical values [see (2.6.26)]; otherwise, different A and B values are derived from the experimental data, and a new empirical expression, still in the form (2.6.25), is obtained for $\nu_m{}^i$.

4. Let us consider again the simultaneous presence of relevant contributions to $\omega\rho$ from both electron-neutral and electron-ion collisions. If a strong field is suddenly applied to the ionized medium, the electron temperature begins growing, and ρ will change correspondingly, its initial time derivative being:

$$\omega \frac{d\rho}{dt} = \frac{h}{2} \mathscr{B}(h) \frac{\nu_m(\bar{u}_g)}{p} \frac{1}{T_g} \frac{dT_e}{dt} \tag{2.6.50}$$

for the contribution of electron-neutral collisions, and:

$$\omega \frac{d\rho}{dt} = -\frac{2A(\mathscr{L} - 2)}{\sqrt{\pi}} \frac{n_e}{p} \left(\frac{2kT_g}{m}\right)^{-3/2} \frac{1}{T_g} \frac{dT_e}{dt} \tag{2.6.51}$$

for the contribution of electron-ion collisions*. These expressions have been derived assuming that the electron energy distribution remains Maxwellian during the initial transient; (2.6.50) is obtained from (2.6.41), and (2.6.51) from (2.6.43), use being made of (2.6.27) for $\mathscr{L}$ in the approximate form, which applies when unity can be neglected. The rate of change (2.6.50) is positive if h is positive, whereas the rate of change (2.6.51) is negative; then the experimental conditions may be adjusted until the total rate of change is zero. These conditions make it possible to determine the quantity $A(\mathscr{L} - 2)$ if the electron-neutral collision frequency is known; in fact, by equating to zero the sum of (2.6.50) and (2.6.51), we obtain:

$$A(\mathscr{L} - 2) = \frac{\sqrt{\pi}}{4} h \mathscr{B}(h) \left(\frac{2kT_g}{m}\right)^{3/2} \frac{\nu_m(\bar{u}_g)}{n_e} \tag{2.6.52}$$

Here too the experimental values of the quantity $A(\mathscr{L} - 2)$ can be compared with the theoretical ones, or used to derive new values for the coefficients A and B.

In the microwave and low-pressure limit $\omega^2 \gg \nu_m{}^2$, interesting solutions exist also when the applied field is not bounded to be very low. In this case,

---

*The $(\mathscr{L} - 2)$ factor becomes $(\mathscr{L} - \frac{5}{3})$ when the $\mathscr{L}$ expression given in the footnote of page 96 is used [Anderson and Goldstein (1955)].

according to (2.3.20), the effective field E_e is simply related to the rms value of the applied field E:

$$E_e = \frac{\nu_m}{\omega} E \tag{2.6.53}$$

and the distribution f^0, which we know to be a function, in a given gas and at a constant temperature, of E_e/ν_m alone, becomes here a function of E/ω only. The conductivity ratio is given by (2.6.40) and the radiation temperature by:

$$T_r = \frac{2}{3k} \frac{\overline{\nu_m u}}{\langle \nu_m \rangle} \tag{2.6.54}$$

which is obtained from (2.6.37) with the position $\omega^2 \gg \nu_m^2$. In a given gas and at a constant temperature $\omega\rho$ and T_r are thus functions of E/ω only.

If the collision frequency and the distribution function can be written in forms (2.6.9) and (2.6.10), the conductivity ratio is again given by (2.6.41), where now:

$$\mathscr{B} = -\frac{2}{3} \frac{\int_0^\infty (\partial\varphi/\partial y)\psi y^{3/2}\, dy}{\int_0^\infty \varphi y^{1/2}\, dy} = \frac{2}{3} \frac{\int_0^\infty \varphi(\partial/\partial y)(\psi y^{3/2})\, dy}{\int_0^\infty \varphi y^{1/2}\, dy} \tag{2.6.55}$$

Likewise (2.6.54) yields:

$$T_r = \mathscr{F} \frac{2\bar{u}}{3k} \tag{2.6.56}$$

where:

$$\mathscr{BF} = \frac{\int_0^\infty \varphi\psi y^{3/2}\, dy}{\int_0^\infty \varphi y^{3/2}\, dy} \tag{2.6.57}$$

If $\nu_m \propto v^h$, $G \propto v^g$, and $E_e^2 \gg E_{et}^2$, the functions ψ and φ are given, respectively, by eqs. (2.6.17) and (2.6.18), where now $l = 2 + g$. Recalling eqs. (2.6.14) and (2.6.15) for $\mathscr{A}$ and $\mathscr{C}$, we can readily demonstrate that:

$$\mathscr{B}(l, h) = \mathscr{A}(l, -h) \tag{2.6.58}$$

$$\mathscr{F}(l, h) = \mathscr{C}(l, -h) \tag{2.6.59}$$

so that the diagrams of Fig. 2.6.1 can be used also for the present microwave case.

Since $\omega\rho \propto \nu_m(\bar{u})/p \propto \bar{u}^{h/2} \propto T_r^{h/2}$, a plot of experimental data for $\omega\rho$ versus T_r on a logarithmic scale for a given gas and at a constant temperature must be a straight line; if this condition is found to be sufficiently well satisfied, the slope yields $h/2$. Moreover, recalling (2.3.47), (2.3.45), (2.3.46), and (2.6.53), we have $T_r \propto \bar{u} \propto b^{-2/(2+g)} \propto (E/\omega)^{4/(2+g)}$, so that g can be obtained from the slope of the logarithmic scale plot of T_r versus E. The collision frequency is then derived from the experimental ρ values by means of

(2.6.41), and the loss factor G from the experimental T_r values by means of (2.6.56), (2.3.47), (2.3.46), and (2.6.53).

More simple and convenient solutions may be used when, in addition to satisfying the microwave and low-pressure limit $\omega^2 \gg \nu_m^2$ as in the above discussion, we can further assume G to be velocity independent. We recall that this assumption is always exactly satisfied in weakly ionized atomic gases, where $G = 2m/M$. The simplifications that occur in the constant-G case are based largely on the fact that, when over the most significant part of the electron distribution function f_0 we can take $\omega^2 \gg \nu_m^2$ and $G \simeq$ constant, this distribution becomes Maxwellian, the electron temperature being [eqs. (2.3.30) and (2.6.53)]:

$$T_e = T_g + \frac{2e^2E^2}{3km\omega^2 G} = T_r \qquad (2.6.60)$$

The formulas that we previously derived for the case of a Maxwellian distribution can thus be applied; more specifically, we shall now review the basic methods that have actually been used for the derivation of the values of the electron collision parameters.

a. When G is known, as in atomic gases, the conductivity ratio is measured as a function of the applied electric field and the results are plotted as a $\omega\rho$ versus T_e curve, the temperature T_e being computed from (2.6.60). With a best-fit procedure this curve is approximated by means of a power expansion like (2.6.48), so that $\mathscr{P}_m$ is then determined from (2.6.49).

b. When G is unknown, the same procedure is used for determining $\mathscr{P}_m$, except that T_e is not computed from (2.6.60); in its place the radiation temperature T_r, which is here equal to T_e, is measured. Furthermore the unknown, constant value of G can be determined according to equation (2.6.60):

$$G = \frac{2e^2E^2}{3m\omega^2 k \, \Delta T} \qquad (2.6.61)$$

where:

$$\Delta T = T_r - T_g \qquad (2.6.62)$$

Formato and Gilardini (1962) showed that eq. (2.6.61) can also be used if we should observe, in the absence of the electric field, non-equilibrium thermodynamic conditions, such as an electron distribution function different from the Maxwellian distribution at the equilibrium temperature T_g. In this case, however, if T_r' and ρ' are the values of these parameters in the absence of the electric field and T_r'' and ρ'' the values in the presence of the field, ΔT has to be taken as:

$$\Delta T = (T_r'' - T_g) - \frac{\rho'}{\rho''}(T_r' - T_g) \qquad (2.6.63)$$

c. When G is unknown, its value in the thermal energy range can be determined as well without measuring radiation temperatures. For this purpose [Bekefi and Brown (1958)] the conductivity ratio is measured both as a function of T_g using very low fields and as a function of E^2 at a constant gas temperature T_{g0}. The factor G is determined from any pair of T_g and E^2 values which yield the same ρ; the formula to be used is again (2.6.61), where now:

$$\Delta T = T_g - T_{g0} \qquad (2.6.64)$$

d. Another procedure for the determination of an unknown G, based on measurements of conductivity ratio values alone, was devised and used by Fundingsland, Faire, and Penico (1954). The conductivity ratio is measured both with a very low field, which does not disturb thermal equilibrium at T_g, and with a larger field of intensity E, and we call these two values ρ_0 and ρ_∞, respectively; furthermore we measure the initial time derivative of the conductivity ratio when the field E is suddenly applied ($\dot\rho_0$). Assuming the collision frequency law $\nu_m \propto v^h$, we have $\omega\rho \propto \bar{u}^{h/2} \propto T_e^{h/2}$, and eq. (2.6.60) can be written as:

$$\frac{T_e - T_g}{T_g} = \left(\frac{\rho_\infty}{\rho_0}\right)^{2/h} - 1 = \frac{e^2 E^2}{m\omega^2 G \bar{u}_g} \qquad (2.6.65)$$

If the time variation of the electron density, as well as any space non-uniformity, is neglected, the initial time derivative of the average electron energy is given by the simplified form of (2.5.8):

$$\left.\frac{d\bar{u}}{dt}\right|_{t=0} = \frac{\sigma_r}{n_e} E^2 = \frac{e^2 E^2}{m\omega} \, p\rho_0$$

which provides:

$$\dot\rho_0 = \frac{h}{2}\frac{\rho_0}{\bar{u}_g}\left.\frac{d\bar{u}}{dt}\right|_{t=0} = \frac{h}{2}\frac{e^2 E^2}{m\omega\bar{u}_g} \, p\rho_0{}^2 \qquad (2.6.66)$$

Elimination of the field E^2 between (2.6.66) and (2.6.65) yields:

$$\frac{2\dot\rho_0}{\omega p\rho_0{}^2} = hG\left[\left(\frac{\rho_\infty}{\rho_0}\right)^{2/h} - 1\right] \qquad (2.6.67)$$

Since the product hG occurs only as a factor on the right-hand side of the equation, whereas h occurs also in the exponent within the bracket, it is possible from any two sets of ρ_0, ρ_∞, and $\dot\rho_0$ values to eliminate the product hG and thus determine the value of the exponent h alone. The loss factor and the electron collision frequency in the thermal range are then computed, substituting the value of h so determined, together with the experimental data, into (2.6.67) and (2.6.41), respectively.

After having thoroughly discussed the two most important groups of cases, namely, the thermal-equilibrium, low-field group and the microwave, low-pressure $\nu_m^2 \ll \omega^2$ group, we shall now briefly reconsider the general case, in which the field strength and the frequency have arbitrary values; we assume again that $\nu_m \propto v^h$ and $G = $ constant. Numerical evaluations of ρ can be made in this case using (2.6.35) and the distribution expression (2.3.27); the results obtained in this way by Haskell, Papa, and Bakshi (1967) are plotted in Figs. 2.6.4–2.6.9 as $\rho/\bar{\rho}$ versus $1/p\bar{\rho}$ [$\bar{\rho}$ being given by the limit (2.6.41) at $\hat{u} = kT_g$] for different values of the field strength parameter:

$$a = \frac{e^2 E^2}{m\omega^2 G \bar{u}_g} \left(\frac{\gamma_5^h}{p\bar{\rho}}\right)^2 \tag{2.6.68}$$

and for integer values of h between $+3$ and -3. As pointed out before, $\rho/\bar{\rho} = 1$ when $h = 0$ and when $1/p\bar{\rho} \gg 1$; the $a = 0$ curves coincide with the corresponding ones of Fig. 2.6.3. The $h = 1$ curves can also be computed using the f^0 expression (2.3.42).

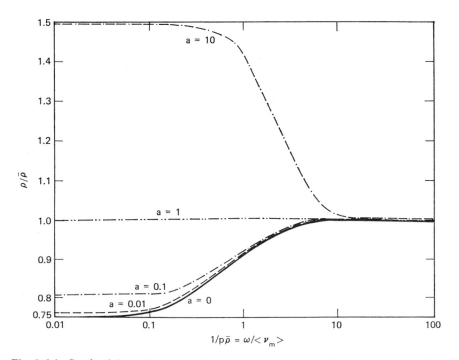

Fig. 2.6.4. Conductivity ratio ρ, normalized to the low-pressure limit value $\bar{\rho}$, plotted as a function of $1/p\bar{\rho}$ for $h = 1$ and for different values of the field strength parameter a [after Haskell, Papa, and Bakshi (1967)].

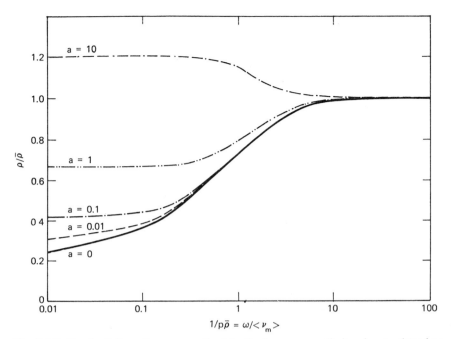

Fig. 2.6.5. Conductivity ratio ρ, normalized to the low-pressure limit value $\bar{\rho}$, plotted as a function of $1/p\bar{\rho}$ for $h = 2$ and for different values of the field strength parameter a [after Haskell, Papa, and Bakshi (1967)].

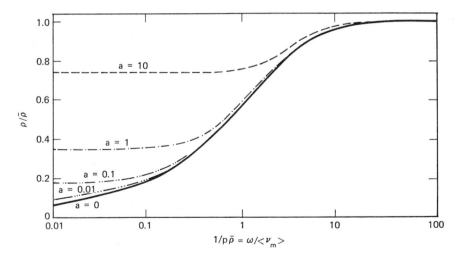

Fig. 2.6.6. Conductivity ratio ρ, normalized to the low-pressure limit value $\bar{\rho}$, plotted as a function of $1/p\bar{\rho}$ for $h = 3$ and for different values of the field strength parameter a [after Haskell, Papa, and Bakshi (1967)].

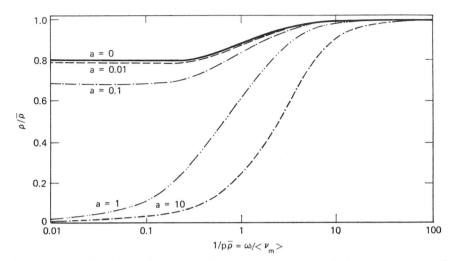

Fig. 2.6.7. Conductivity ratio ρ, normalized to the low-pressure limit value $\bar{\rho}$, plotted as a function of $1/p\bar{\rho}$ for $h = -1$ and for different values of the field strength parameter a [after Haskell, Papa, and Bakshi (1967)].

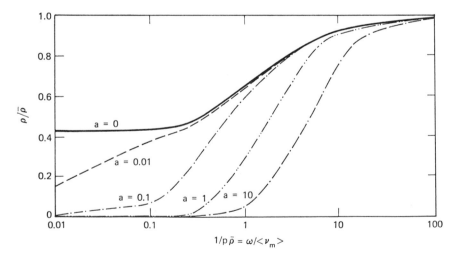

Fig. 2.6.8. Conductivity ratio ρ, normalized to the low-pressure limit value $\bar{\rho}$, plotted as a function of $1/p\bar{\rho}$ for $h = -2$ and for different values of the field strength parameter a [after Haskell, Papa, and Bakshi (1967)].

110

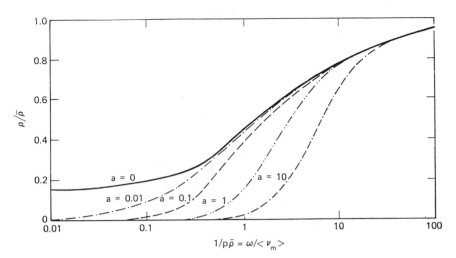

Fig. 2.6.9. Conductivity ratio ρ, normalized to the low-pressure limit value $\bar{\rho}$, plotted as a function of $1/p\bar{\rho}$ for $h = -3$ and for different values of the field strength parameter a [after Haskell, Papa, and Bakshi (1967)].

The ac conductivity that we have considered in the previous paragraphs relates the applied field to the electron current at the same frequency. More refined solutions of the Boltzmann equation than those discussed in Section 2.3 provide information on the ac currents at harmonic frequencies: odd harmonics result from the assumption of a velocity-dependent collision frequency, and as such they exist also in the case where all quantities of importance are spatially uniform; even harmonics, on the other hand, can be due only to the presence of gradients in the electric field, electron density, or temperature or to the presence of a superposed small dc field. For the purpose of evaluating the harmonic currents and for the sake of taking collisions correctly into account, a complete expansion of the distribution function in spherical harmonics in velocity space and in Fourier harmonics in ωt must be introduced into the Boltzmann equation; thereafter the current for each harmonic frequency is computed by substituting in (2.4.5) the appropriate $\mathbf{f}^1$ harmonic component. A similar analysis may also be performed for the currents at the sum and difference frequencies produced by the nonlinear mixing of two applied fields of different frequencies.

Numerous papers have been published on these subjects, and the reader is referred to the comprehensive review by Sodha and Kaw (1969). We mention here only two contributions, which provide valid and interesting results for two of the most typical cases:

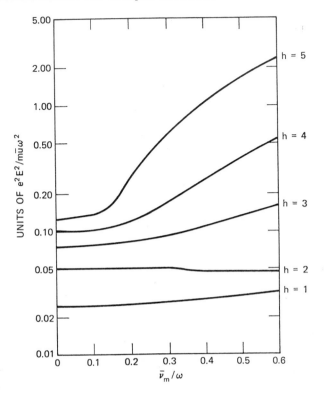

Fig. 2.6.10. Third harmonic/fundamental current ratio (in units of $e^2E^2/m\bar{u}\omega^2$) as a function of the mean collision frequency for different values of the power-law exponent h [after Krenz (1965)].

1. Krenz (1965) assumes a power-law dependence for the electron collision frequency and computes the ratio of the third harmonic current to the in-phase fundamental one; his results for a Maxwell distribution are plotted in Fig. 2.6.10.

2. Shkarofsky (1968) gives adequate formulas for the evaluation of the currents generated at difference, sum, and second harmonic frequencies by the gradients of two applied electric fields, consideration being given also to the presence of the appropriate self-consistent electric fields at these frequencies.

In all these cases the generated currents and powers are functions of the electron collision frequencies and of their dependences on velocity; it is conceivable, then, that data on the collision parameters can be obtained from measurements of these nonlinear effects. However, this possibility has not yet been pursued adequately either theoretically or experimentally; therefore, no further discussion of it will be given in this book.

It seems appropriate at this point to discuss also a group of experiments in which energy exchange and flow parameters are determined. For this purpose an electron swarm, typically an isothermal plasma, is heated by a pulsed microwave field and either the time behavior of the electron energy when the field is suddenly removed, or the space flow of the electron energy outside the heated region, is observed. In these cases the electron energy behavior is described by the transport equation (2.5.8); the coefficients of importance are the energy relaxation time τ_r and the energy transport coefficients Θ, $\mathfrak{D}$, and $\mathfrak{M}$.

We discuss the relaxation time τ_r first, and shall consider two cases of relevant interest.

a. We assume that G is velocity independent, and that experiments are conducted so as to satisfy the low-pressure limit condition $\nu_m^2 \ll \omega^2$. Hence the distribution function f^0 is Maxwellian at the end of a sufficiently long heating pulse, and after removal of the pulse it will relax to a new Maxwellian, the one corresponding to thermal equilibrium between electrons and molecules. Hence we may well take f^0 to be very nearly Maxwellian also during transients, and eq. (2.5.10) for τ_r to be applicable. This equation, together with (2.6.40), yields:

$$p\tau_r = \frac{1}{G}\frac{p}{\langle \nu_m \rangle} = \frac{1}{G}\frac{1}{\omega\rho} \tag{2.6.69}$$

so that, if we know G, this is equivalent to measuring and elaborating the conductivity ratio ρ; conversely, if we measure both $p\tau_r$ and $\omega\rho$, we can determine G [Anderson and Goldstein (1955)].

b. The heating pulse power is sufficiently weak to justify again assuming a Maxwellian distribution, the deviations from thermal equilibrium being small. If we consider homonuclear diatomic molecules, we can assume for G expression (1.5.23), and the following result is obtained:

$$p\tau_r = \frac{M}{2m}\left(\frac{\langle \nu_m \rangle}{p} + 2.13 \times 10^{17}B_0 M_a \mathscr{Q}^2 \langle v^{-1} \rangle\right)^{-1}$$

$$= \frac{M}{2m}(\omega\rho + 2.91 \times 10^{13}B_0 M_a \mathscr{Q}^2 T_e^{-\frac{1}{2}})^{-1} \tag{2.6.70}$$

If we consider polar diatomic molecules, for which (1.5.30) applies, the corresponding result is:

$$
p\tau_r = \frac{M}{2m} \left(\frac{\langle \nu_m \rangle}{p} + 8.38 \times 10^{27} B_0^{7/8} M_a \mathscr{M}^2 \langle v^{-5/2} \rangle \right)^{-1}
$$

$$
= \frac{M}{2m} (\omega\rho + 2.54 \times 10^{18} B_0^{7/8} M_a \mathscr{M}^2 T_e^{-5/4})^{-1} \qquad (2.6.71)
$$

Hence comparison of these theoretical expressions with experimental data may allow the determination of quadrupole and dipole moments of diatomic molecules [Mentzoni and Row (1963), Mentzoni and Donohoe (1968)].

Considering now the energy transport coefficients, we restrict the discussion to cases in which the electron distribution function can be taken as Maxwellian. In these cases, using eq. (2.4.43), we can establish from (2.4.28) and (2.4.29) a general relationship which provides Θ as a function of $\mathfrak{M}$:

$$
\Theta = \frac{\bar{u}}{3} \langle v^2 \mathfrak{T}_0 \rangle = \frac{2\bar{u}}{3e} \mathfrak{M} = \frac{\hat{u}}{e} \mathfrak{M} \qquad (2.6.72)
$$

When a power-law dependence $\nu_m \propto v^h$ is assumed, (2.4.36) yields:

$$
\mathfrak{M}p = \tfrac{5}{2} \gamma_7^{-h} \frac{e\hat{u}}{m} \frac{p}{\nu_m(\hat{u})} \qquad (2.6.73)
$$

and the thermal diffusivity $\mathfrak{D}$ given by (2.4.35) becomes:

$$
\mathfrak{D}p = \frac{2}{3e} \frac{d}{d\hat{u}} (\hat{u}\mathfrak{M}p) = \tfrac{5}{6}(4 - h)\gamma_7^{-h} \frac{\hat{u}}{m} \frac{p}{\nu_m(\hat{u})} \qquad (2.6.74)
$$

In the case of dominating electron-ion collisions, the collision frequency is given by (2.6.25), and we obtain:

$$
\mathfrak{M} = \frac{16e}{\sqrt{\pi} \, An_e \mathscr{L}} \left(\frac{2kT_e}{m} \right)^{5/2} = 4\mu k T_e \qquad (2.6.75)
$$

Neglecting the dependence of $\mathscr{L}$ on T_e, we further obtain:

$$
\mathfrak{D} = \frac{112}{3\sqrt{\pi} \, An_e \mathscr{L}} \left(\frac{2kT_e}{m} \right)^{5/2} \qquad (2.6.76)
$$

Also for these parameters Spitzer and Härm (1953) calculated corrections to include electron-electron collision effects; these corrections are simply multiplying factors for the above equations, namely, 0.4652 for $\mathfrak{M}$ and 0.2938 for $\mathfrak{D}$.* As for the dc conductivity, Itikawa (1963) and Kruger,

---

* Attention must be paid to the fact that Spitzer and Härm discuss the thermal conductivity coefficient at constant p, whereas we have defined this parameter at constant n_e; therefore, in the notation of Spitzer and Härm our factor $\mathfrak{D}$ is given by $(5\delta_T + 2\delta_E)/7$.

Mitchner, and Daybelge (1968) recomputed the multiplying factors, using either the Kihara-Aono theory or Liboff's shielded potential model; here, too, factors become functions of $\mathscr{L}$.

C. Use of Direct Current Transport Parameters in a Static Magnetic Field

As we shall see in Chapter 3, the mobility parameter, which is the most convenient one to use when comparing calculated and experimental results of the electron swarm motion in a static magnetic field, is not the transverse mobility μ_T or the perpendicular mobility $\mu_\perp$, but their ratio [Jory (1965) and references therein]. More specifically, we define as magnetic mobility μ_M and magnetic drift velocity w_M the quantities:

$$\mu_M = \frac{w_M}{E} = \frac{1}{B}\frac{\mu_\perp}{\mu_T} = \frac{1}{B}\frac{w_\perp}{w_T} \tag{2.6.77}$$

Making use of (2.4.25) and (2.4.26) we obtain:

$$\mu_M = \frac{e}{m}\frac{\langle 1/(v_m^2 + \omega_b^2)\rangle}{\langle v_m/(v_m^2 + \omega_b^2)\rangle} \tag{2.6.78}$$

or:

$$\mu_M = \mu\frac{\langle 1/(v_m^2 + \omega_b^2)\rangle}{\langle 1/v_m\rangle\langle v_m/(v_m^2 + \omega_b^2)\rangle} \tag{2.6.79}$$

where μ is the mobility (2.4.20) computed for the same distribution function. When v_m is velocity independent:

$$\mu_M = \mu \tag{2.6.80}$$

and this justifies the above definition. The ratio $\Psi = \mu_M/\mu$ has been termed the magnetic deflection coefficient [Frost and Phelps (1962)].

Other experimentally observable parameters that are worth consideration in the presence of a magnetic field are the transverse diffusion coefficient D_T [see eq. (2.4.21)] and the D/μ ratios:

$$\frac{D_T}{\mu_\parallel} = \frac{2}{3e}\frac{\overline{[v_m u/(v_m^2 + \omega_b^2)]}}{\langle 1/v_m\rangle} \tag{2.6.81}$$

$$\frac{D_\parallel}{\mu_T} = \frac{2}{3e}\frac{\overline{(u/v_m)}}{\langle v_m/(v_m^2 + \omega_b^2)\rangle} \tag{2.6.82}$$

Of particular interest are the limit magnetic field cases. At low magnetic fields (LMF), defined as those for which $\omega_b^2 \ll v_m^2$ over the relevant velocity range of the electron distribution, the transverse mobility and diffusion coefficients and the D/μ ratios given above approach the no-magnetic-field values, whereas the magnetic mobility becomes:

$$\mu_M = \frac{e}{m} \frac{\langle 1/\nu_m^2 \rangle}{\langle 1/\nu_m \rangle} = \frac{\langle 1/\nu_m^2 \rangle}{\langle 1/\nu_m \rangle^2} \mu \qquad (2.6.83)$$

Under these conditions we shall always have $\Psi \geqslant 1$. In the strong magnetic field (SMF) limit, defined by the opposite condition $\omega_b^2 \gg \nu_m^2$, we have:

$$\mu_M = \frac{1}{B^2 \mu_T} = \frac{e}{m\langle \nu_m \rangle} = \frac{\mu}{\langle \nu_m \rangle \langle 1/\nu_m \rangle} \qquad (2.6.84)$$

$$D_T = \frac{\nu_m \nu^2}{3\omega_b^2} \ll D \qquad (2.6.85)$$

$$\frac{D_T}{\mu_\parallel} = \frac{2}{3e\omega_b^2} \frac{\overline{\nu_m u}}{\langle 1/\nu_m \rangle} \ll \frac{2\bar{u}}{3e} \qquad (2.6.86)$$

$$\frac{D_\parallel}{\mu_T} = \frac{2\omega_b^2}{3e} \frac{\overline{(u/\nu_m)}}{\langle \nu_m \rangle} \gg \frac{2\bar{u}}{3e} \qquad (2.6.87)$$

According to the results of Section 2.3 and from the definitions of the various coefficients here considered, it follows that when electron-neutral collisions only are important the quantities $\mu_M p$ (or w_M), $D_T p$, $D_T/\mu_\parallel$, and $D_\parallel/\mu_T$ in a given gas and at a constant temperature are all functions of E/p and B/p (or E/B) only. In the LMF limit the effective field is equal to the applied one, and both the electron energy distribution and the above quantities become functions of E/p only. In the SMF limit, on the other hand, when fields are orthogonal, E_e/ν_m is equal to E/ω_b, so that the electron energy distribution becomes a function of E/ω_b (or E/B) only, and the same must happen for $\mu_M p$, $\omega_b^2 D_T/p$, $\omega_b^2 D_T/\mu_\parallel p^2$, and $D_\parallel p^2/\mu_T \omega_b^2$. From the foregoing it is apparent that the case in which the coefficient Ψ is really valuable for the purpose of deriving electron collision parameters is the LMF one, since here the condition of equal distribution functions for μ_M and μ is automatically satisfied, f^0 being the same at equal E/p values independently of the presence or the absence of the magnetic field.

Useful formulas, similar to the ones derived for the no-magnetic-field case, could be easily obtained from the above expressions for the LMF and SMF limits. In particular, when $\nu_m \propto \nu^h$, $G \propto \nu^g$, and $E^2 \gg E_t^2$, the magnetic deflection coefficient in the LMF limit, given by (2.6.83), becomes:

$$\Psi = \frac{\mathscr{A}(l, 2h)}{\mathscr{A}^2(l, h)} \qquad (2.6.88)$$

with $l = 2 + g + 2h$. This quantity is plotted in Fig. 2.6.11 as a function of l for various values of h.

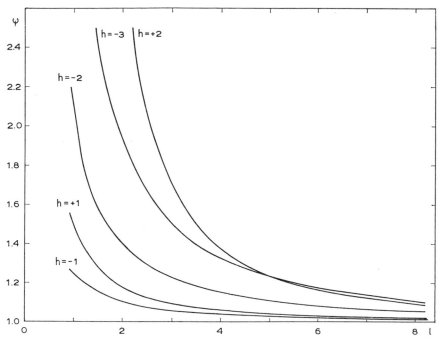

Fig. 2.6.11. Magnetic deflection coefficient Ψ, plotted according to (2.6.88) as a function of the parameters of the distribution function $f^0 \propto \exp\left(-bv^l\right)$ and of the collision frequency $v_m \propto v^h$.

The expressions for the SMF limit with orthogonal dc fields are strongly related to the previously discussed formulas for the microwave-frequency, low-pressure limit $\omega^2 \gg v_m^2$. In fact, since E_e/v_m is equal to E/ω in the microwave case [eq. (2.6.53)] and to E/ω_b in the SMF limit, the distribution function will be the same for the same values of the ratios E/ω and E/ω_b in the two cases. Furthermore, $\langle v_m \rangle / p$ is equal to $\omega\rho$ in the microwave case [see eq. (2.6.40)] and to $e/m\mu_{MP}$ in the SMF case [see eq. (2.6.84)]; the curve of $\omega\rho$ versus E/ω must then coincide with the curve of $e/m\mu_{MP}$ (or $\omega_b w_T/w_{\perp}p$) versus E/ω_b. Therefore the experimental data of these two quantities may be discussed and elaborated together.

For the other quantities useful formulas are obtained when the electron energy distribution is Maxwellian, as happens at very low **E** fields or at any field strength if G is velocity independent (Table 2.6.1, case 4). Recalling relationship (2.4.43) for Maxwellian distributions, we obtain:

$$D_T = \frac{\hat{u}\langle v_m \rangle}{m\omega_b^2} \tag{2.6.89}$$

$$\frac{D_T}{\mu_{\|}} = \frac{\hat{u}}{e} \frac{\langle v_m \rangle}{\omega_b^2 \langle 1/v_m \rangle} = \left(\frac{\hat{u}}{e}\right)^2 \left(\frac{D_{\|}}{\mu_T}\right)^{-1} \tag{2.6.90}$$

where the electron energy $\hat{u}$ is [eq. (2.3.30)]:

$$\hat{u} = \tfrac{2}{3}\bar{u} = \tfrac{2}{3}\left[\bar{u}_g + \frac{m}{G}\left(\frac{E}{B}\right)^2\right] \tag{2.6.91}$$

Equations (2.6.41), (2.6.43), and (2.6.47) may conveniently be used as appropriate expressions to be substituted for the average collision frequency $\langle\nu_m\rangle$ in eq. (2.6.89). For the diffusion/mobility ratios we consider only as a situation of future interest the case $\nu_m \propto v^h$, which yields:

$$\frac{D_T}{\mu_\parallel} = \frac{\hat{u}}{e}\frac{\gamma_5^h}{\gamma_5^{-h}}\left[\frac{\nu_m(\hat{u})}{\omega_b}\right]^2 = \left(\frac{\hat{u}}{e}\right)^2\left(\frac{D_\parallel}{\mu_T}\right)^{-1} \tag{2.6.92}$$

On more general grounds it is also worth remarking that, according to (2.4.43), any time that a distribution is Maxwellian μ and D are related by the generalized Einstein relation, which follows from (2.4.10) and (2.4.15):

$$D = \frac{\hat{u}}{e}\mu = \frac{2\bar{u}}{3e}\mu \tag{2.6.2'}$$

There are experimental cases in which the observable parameter is not the mobility, but the dc conductivity. If n_e is known or can be reliably estimated, we derive the corresponding mobility [eqs. (2.4.25) and (2.4.26)], and the problem can be treated as discussed above; if we prefer making no assumption regarding n_e, we must adopt methods based on mobility or conductivity ratios only. Worth mentioning here is the method of Mullaney and Dibelius (1961). These authors measure σ_{dc} without magnetic field and σ_{dcT} in the magnetic field under conditions that do not change n_e; furthermore, they assume ν_m to be velocity independent, so that its value can be computed from the expression:

$$\nu_m = \frac{\omega_b}{\sqrt{(\mu/\mu_T) - 1}} = \frac{\omega_b}{\sqrt{(\sigma_{dc}/\sigma_{dcT}) - 1}} \tag{2.6.93}$$

which is easily derived from eqs. (2.4.20) and (2.4.25).

D. Use of Alternating Current Transport Parameters in a Static Magnetic Field

Let us consider first the case of an ac electric field, circularly polarized in a plane perpendicular to a static magnetic field. Equation (2.4.11) gives the conductivity in terms of the time tensor $\mathfrak{T}$, whose analytical expressions are (2.3.15) and (2.3.16); from these it can be seen that the conductivity for a circularly polarized field, orthogonal to the static magnetic field, is given by the same expression as that for the no-magnetic-field case (2.4.19), provided ω is replaced by $(\omega - \omega_b)$ or $(\omega + \omega_b)$ for right or left circular polarization, respectively. At this point it can easily be shown also that the conductivity ratio and the radiation temperature for this case are given by expressions (2.6.35) and (2.6.37) after the above substitutions have been made. Since the results and procedures previously described can be used, no further discussion along this specific line will follow here.

On the other hand, the resonant behavior of the conductivity of the right circularly polarized field suggests new solutions for the problem of determining from experiments the values of the collision parameters. In fact, the real part of this conductivity:

$$\sigma_{C_r} = \frac{e^2 n_e}{m} \left\langle \frac{\nu_m}{\nu_m{}^2 + (\omega - \omega_b)^2} \right\rangle \tag{2.6.94}$$

shows a typical cyclotron-resonance behavior, so that from the resonance line shape (normalized to unity at the peak):

$$\sigma_{C_r}^* = \frac{\langle \nu_m / [\nu_m{}^2 + (\omega - \omega_b)^2] \rangle}{\langle 1/\nu_m \rangle} \tag{2.6.95}$$

collision frequency data can be derived.

In particular, when ν_m is velocity independent, the resonant line half-width $\Delta\omega_{0.5}$ at half-maximum points ($\sigma_{C_r}^* = 0.5$) is equal to ν_m, and the resonant line half-width $\Delta\omega_I$ at the inflection (maximum slope) points is equal to $0.577\nu_m$. If the distribution function f^0 is Maxwellian and $\nu_m \propto v^h$, eq. (2.6.95) becomes:

$$\sigma_{C_r}^* = \frac{\Gamma[(5 + h)/2]}{\Gamma[(5 - h)/2]} D_{(3+h)/2}^h \left[\frac{\omega - \omega_b}{\nu_m(\hat{u})} \right] \tag{2.6.96}$$

where the function $D_i{}^h$ is defined by (2.6.45). In this case the line half-width $\Delta\omega_{0.5}$ can be written as:

$$\Delta\omega_{0.5} = \frac{\nu_m(\hat{u})}{X(h)} \tag{2.6.97}$$

and an analogous expression holds for $\Delta\omega_I$; the coefficients $X(h)$ and $X_I(h)$ were computed numerically by Fehsenfeld (1963) and are given in Table 2.6.2.

If h is known or can be properly assumed, measurements of $\Delta\omega_{0.5}$ or $\Delta\omega_I$ provide in this way $\nu_m(\hat{u})$. If h is unknown, one may measure the line widths at different heights of the resonance curve; since each line width is related to $\nu_m(\hat{u})$ by an expression similar to (2.6.97), but with different $X(h)$, it is apparent that the ratios of the line widths of the same curve measured at different heights will be functions of h alone and this parameter can thus be obtained from the experimental data. Thereafter $\nu_m(\hat{u})$ is determined directly from (2.6.97) or from any similar expression for other line widths.

Table 2.6.2   Coefficients $X(h)$, $X_I(h)$, and $Y(h)$, Which Relate the Collision Frequency $\nu_m(\hat{u})$ to the Cyclotron Resonance Half-Widths $\Delta\omega_{0.5}$ and $\Delta\omega_I$ and to the Displacement $\delta\omega_s$ of the σ_{Tr} Maximum, Respectively

h	X	X_I	Y
-2	3.073	7.899	2.155
-1	1.5975	3.318	1.446
0	1.0000	1.732	1.000
$+1$	0.7973	1.726	0.6917
$+2$	0.8943	6.165	0.4640

A similar procedure was followed by Meyerand and Flavin (1964), who adopted, however, a slightly different collision frequency law: $\nu_m(v) - \nu_m(0) \propto v^h$, which has no null at zero velocity. Thus these authors derive from the line width ratios ($\Delta\omega_I$, which corresponds to the most accurate section of the resonance curve to locate experimentally, is taken as the basis for comparison) the two collision parameters h and $\nu_m(0)/\nu_m(\hat{u})$; then $\nu_m(\hat{u})$ is obtained from $\Delta\omega_I$. Tice and Kivelson (1967) extended this technique, using for ν_m a two-term velocity expansion ($Av^{h_1} + Bv^{h_2}$) and determining the four unknown constants (A, B, h_1, h_2) by the best fitting of the experimental line shapes with the theoretical ones.

In more general cases, when the distribution is not Maxwellian, the line half-width at half-maximum points is obtained by solving the equation:

$$\left\langle \frac{\nu_m}{\nu_m{}^2 + \Delta\omega^2} \right\rangle = \frac{1}{2}\left\langle \frac{1}{\nu_m} \right\rangle \tag{2.6.98}$$

Another significant and useful result is obtained when the ratio of the resonance peak to the area under the resonance curve is computed. In fact, (2.6.95) yields for this ratio:

$$\frac{\langle 1/\nu_m \rangle}{\langle \int_{-\infty}^{+\infty} \{\nu_m/[\nu_m{}^2 + (\omega - \omega_b)^2]\}\, d(\omega - \omega_b)\rangle} = \frac{1}{\pi}\left\langle \frac{1}{\nu_m} \right\rangle \tag{2.6.99}$$

Kelly, Margenau, and Brown (1957) worked out formulas and performed computations for the high-frequency conductivity near cyclotron resonance also for the case of velocity-independent G and P_m parameters, using the distribution function (2.3.42), which is the appropriate one when the strength of the applied electric field is such as to heat electrons over their thermal equilibrium value. Details can be found in their paper.

The incoherent spontaneous radiation near electron cyclotron resonance also displays interesting behavior. In terms of the radiation temperature appropriate to the emission of a resonant wave with its electric field perpendicular to the applied **B** field, we have [see eq. (2.6.36) and Table 2.4.1]:

$$T_r = \frac{2}{3k} \frac{\overline{\{\nu_m u/[\nu_m^2 + (\omega - \omega_b)^2]\}}}{\langle \nu_m/[\nu_m^2 + (\omega - \omega_b)^2]\rangle} \tag{2.6.100}$$

When the distribution function is Maxwellian, or when ν_m is velocity independent, (2.6.100) yields [see property (2.4.43)]:

$$T_r = \frac{2}{3k} \bar{u} \tag{2.6.39}$$

In this case no resonance is observed. But, if the distribution deviates significantly from a Maxwellian one, and the collision frequency is strongly dependent on velocity, calculations show that peaks or dips may occur at $\omega = \omega_b$. It is clear that, any time $\nu_m(v)$ is taken as known, by inverting numerically relation (2.6.100) between the observable $T_r(\omega_b)$ and the unknown $f^0(v)$, the latter can be determined [Fields, Bekefi, and Brown (1963); Wright (1966); Noon, Blaszuk, and Holt (1968)]. In particular Wright and Bekefi (1971) suggest using for this purpose a modified Maxwellian distribution:

$$f^0(v) = \begin{cases} \exp\left(-\frac{mv^2}{2kT_0}\right) \sum_{n=0}^{N} C_n v^{2n} & \text{for} \quad v \leqslant v_c \\ 0 & \text{for} \quad v \geqslant v_c \end{cases} \tag{2.6.101}$$

with no f^0 discontinuity at $v = v_c$. Since in this case the problem consists in determining the N coefficients C_n, which minimize the mean-square difference between the measured and the calculated T_r temperatures, the linear dependence of $f^0(v)$ on these coefficients greatly simplifies the fitting procedure. Wright and Bekefi have found that in practice from two to four C_n coefficients are sufficient for deriving a satisfactory solution.

Most microwave techniques adopted for the determination of the electron collision parameters, more specifically $\nu_m(v)$, rely on measurements of the transverse ac conductivity σ_T, and make use of formula (2.4.23) to establish the appropriate conditions of measurement. We note first that near cyclotron resonance σ_T is approximately equal to $\sigma_C/2$; hence the methods based on measurements of the σ_{C_r} resonance line widths can be applied as well if we observe the σ_{T_r} resonance curve. Three other methods, all of which make use of σ_T measurements and have been experimentally tested, will now be described.

A. The method of Hirshfield and Brown (1958) is based on the determination of the magnetic field at which the imaginary part of σ_T is zero, when this conductivity component crosses from negative to positive values near cyclotron resonance:

$$\left\langle \frac{\nu_m{}^2 + \omega^2 - \omega_b{}^2}{[\nu_m{}^2 + (\omega + \omega_b)^2][\nu_m{}^2 + (\omega - \omega_b)^2]} \right\rangle = 0 \qquad (2.6.102)$$

This equation is, in general, difficult to evaluate. However, if we limit the frequency and pressure ranges so that $\omega^2 \gg \nu_m{}^2$, it becomes:

$$1 - 2\omega \, \delta\omega \left\langle \frac{1}{\nu_m{}^2} \right\rangle = 0 \qquad (2.6.103)$$

where $\delta\omega = \omega_b - \omega$. For a Maxwellian distribution and $\nu_m \propto v^h$ this yields:

$$\nu_m{}^2(\hat{u}) = 2\gamma_5^{-2h} \omega \, \delta\omega \qquad (2.6.104)$$

which is an appropriate formula for computing $\nu_m(\hat{u})$.

B. The method of Narasinga Rao, Verdeyen, and Goldstein (1961) is based on the behavior of σ_{Tr}, the real part of the transverse conductivity. In fact, for a Maxwellian distribution and $\nu_m \propto v^h$, $\omega\sigma_{Tr}/n_e$ can be written, according to (2.4.23), as a function of $\nu_m(\hat{u})/\omega$, ω_b/ω, and h, and each $\omega\sigma_{Tr}/n_e$ versus $\nu_m(\hat{u})/\omega$ curve displays a maximum (see Fig. 2.6.12 for $h = 0$). Thus, at any fixed frequency, we have a relationship between the applied magnetic field and the $\nu_m(\hat{u})$ value for which σ_{Tr} is a maximum. According to the calculations of Shkarofsky, Johnston, and Bachynski (1966), who use the near-resonance approximation $\sigma_{Tr} \simeq \sigma_{Cr}/2$, this relationship is conveniently written, in terms of the displacement from cyclotron resonance, as:

$$|\delta\omega_s| = |\omega_b - \omega| = \frac{\nu_m(\hat{u})}{Y(h)} \qquad (2.6.105)$$

Values of the coefficients $Y(h)$ are given in Table 2.6.2. If we measure $\delta\omega_s$ by determining the magnetic field at which $\partial\sigma_{Tr}/\partial\nu_m$ vanishes, $\nu_m(\hat{u})$ can then be computed by means of (2.6.105).

C. A method proposed by Bruce, Crawford, and Harp (1968) considers the transient response of an ionized gas when a very short electric field pulse is applied perpendicularly to the static magnetic field. If E is the intensity of the applied dc electric field and τ its pulse duration, we may consider the applied waveform to be represented by $E\tau \, \delta(t)$ and all the electrons instantaneously brought up to the velocity:

$$v(0) = \frac{eE\tau}{m} \qquad (2.6.106)$$

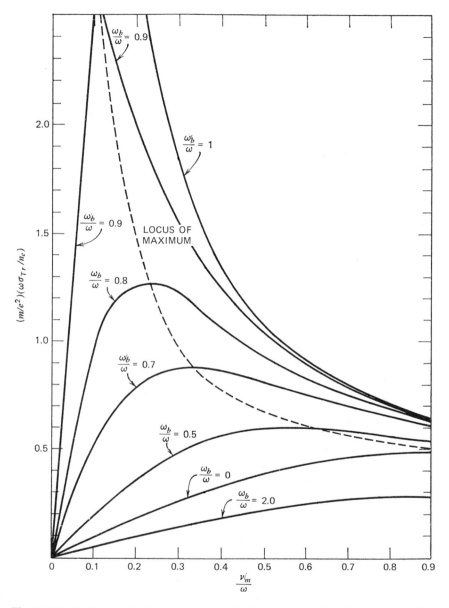

Fig. 2.6.12. Real part of the transverse conductivity σ_T as a function of the collision frequency ν_m (independent of the electron velocity), for different values of the cyclotron frequency ω_b [after Narasinga Rao, Verdeyen, and Goldstein (1961)].

Obviously this is correct only provided the velocity (2.6.106) is much larger than the electron thermal velocities before the pulse is applied, and provided τ is much shorter than the mean free time $1/\nu_m$. The transient response of the current density component J in the electric field direction can be obtained by applying the Laplace formalism ($\mathfrak{L}$ and $\mathfrak{L}^{-1}$ being the transform and anti-transform operators), as follows:

$$\mathfrak{L}[J] = \sigma_T \mathfrak{L}[E_T \, \delta(t)] = \sigma_T E_T \tag{2.6.107}$$

Substituting (2.4.23) for σ_T, when ν_m is velocity independent we obtain:

$$
\begin{aligned}
J(t) &= \frac{e^2 n_e}{2m} E_T \mathfrak{L}^{-1}\left(\frac{1}{p + j\omega_b + \nu_m} + \frac{1}{p - j\omega_b + \nu_m}\right) \\
&= \frac{e^2 n_e}{m} E_T \exp\left(-\nu_m t\right) \cos \omega_b t \\
&= e n_e v(0) \exp\left(-\nu_m t\right) \cos \omega_b t \tag{2.6.108}
\end{aligned}
$$

This current then oscillates at the cyclotron frequency; the initial time derivative of its envelope, normalized to unity at $t = 0$, is:

$$\dot{J}(0) = -\nu_m \tag{2.6.109}$$

so that ν_m is determined from the observable $\dot{J}(0)$ or from any radiation field proportional to $J(t)$. When ν_m is a function of velocity, $-\dot{J}(0)$ can be interpreted as the value of ν_m at the common electron velocity $v(0)$.

The foregoing description of the method of Bruce, Crawford and Harp was based on the application of a short dc electric field pulse. It can easily be verified, however, that application of a short pulse of ac electric field at the cyclotron frequency yields the same results; in fact, application of such a field accelerates electrons continuously along a spiral in a plane perpendicular to **B**, and the final value at the pulse end is again (2.6.106) provided $\omega_b \tau \gg 1$.

Experiments on the electron energy flow in the presence of a static magnetic field require discussion of the appropriate energy transport coefficients. As in the case of no magnetic field, we consider a Maxwellian distribution and derive similar formulas for $\mathfrak{M}_T$, $\mathfrak{D}_T$, $\mathfrak{M}_\perp$, and $\mathfrak{D}_\perp$. In the high-field and low-pressure limit $\omega_b{}^2 \gg \nu_m{}^2$, (2.4.37) and (2.4.38) yield:

$$\mathfrak{M}_T = \frac{e}{2\omega_b{}^2} \langle v^2 \nu_m \rangle \tag{2.6.110}$$

$$\mathfrak{M}_\perp = \frac{e}{2\omega_b} \langle v^2 \rangle = \frac{5\hat{u}}{2B} \tag{2.6.111}$$

When a power-law dependence $v_m \propto v^h$ is assumed, (2.6.110) leads to the result:

$$\mathfrak{M}_T = \tfrac{5}{2}\gamma_7^{\ h} \frac{\hat{u}}{B} \frac{v_m(\hat{u})}{\omega_b} \tag{2.6.112}$$

Hence:

$$\mathfrak{M}_T = \frac{\gamma_7^{\ h}}{\gamma_7^{\ -h}} \left[\frac{v_m(\hat{u})}{\omega_b} \right]^2 \mathfrak{M} \tag{2.6.112'}$$

when use is made of expression (2.6.73) for $\mathfrak{M}$. The corresponding formulas for the thermal diffusivity components are:

$$\mathfrak{D}_T = \frac{2}{3e} \frac{d}{d\hat{u}} (\hat{u}\mathfrak{M}_T) = \tfrac{5}{6}(4 + h)\gamma_7^{\ h} \frac{\hat{u}}{eB} \frac{v_m(\hat{u})}{\omega_b} \tag{2.6.113}$$

$$\mathfrak{D}_T = \frac{4 + h}{4 - h} \frac{\gamma_7^{\ h}}{\gamma_7^{\ -h}} \left[\frac{v_m(\hat{u})}{\omega_b} \right]^2 \mathfrak{D} \tag{2.6.113'}$$

$$\mathfrak{D}_\perp = \frac{2}{3e} \frac{d}{d\hat{u}} (\hat{u}\mathfrak{M}_\perp) = \frac{10\hat{u}}{3eB} \tag{2.6.114}$$

The ratios of the transverse to the perpendicular components of the $\mathfrak{M}$ and $\mathfrak{D}$ tensors are in these cases given by the simple formulas:

$$\frac{\mathfrak{M}_T}{\mathfrak{M}_\perp} = \gamma_7^{\ h} \frac{v_m(\hat{u})}{\omega_b} \tag{2.6.115}$$

$$\frac{\mathfrak{D}_T}{\mathfrak{D}_\perp} = \left(1 + \frac{h}{4}\right)\gamma_7^{\ h} \frac{v_m(\hat{u})}{\omega_b} \tag{2.6.116}$$

All these formulas suggest simple experimental methods for the determination of $v_m(\hat{u})$.

In the case of dominating electron-ion collisions, substitution of the v_m expression (2.6.25) yields:

$$\mathfrak{M}_T = \frac{2eA\mathscr{L}n_e}{3\sqrt{\pi}\,\omega_b^{\ 2}} \left(\frac{2kT_e}{m}\right)^{-\frac{1}{2}} \tag{2.6.117}$$

which, neglecting the dependence of $\mathscr{L}$ on T_e, provides:

$$\mathfrak{D}_T = \frac{2A\mathscr{L}n_e}{9\sqrt{\pi}\,\omega_b^{\ 2}} \left(\frac{2kT_e}{m}\right)^{-\frac{1}{2}} \tag{2.6.118}$$

Electron-electron collision effects may be taken into account by multiplying expressions (2.6.117) and (2.6.118), respectively, by 2.150 and 3.404, which are the inverse of Spitzer's and Härm's (1953) factors previously reported for $\mathfrak{M}$ and $\mathfrak{D}$ (no-magnetic-field case).

When the $\omega_b{}^2 \gg \nu_m{}^2$ limit is not satisfied, the following expressions are obtained from (2.4.37) and (2.4.38):

$$\mathfrak{M}_T = \overline{\mathfrak{M}}_{T'} \left[\frac{\omega_b}{\nu_m(\hat{u})} \right]^2 D^h_{(5+h)/2} \left[\frac{\omega_b}{\nu_m(\hat{u})} \right]$$

$$= \overline{\mathfrak{M}}_T \left(\gamma_7{}^h \frac{\overline{\mathfrak{M}}_\perp}{\overline{\mathfrak{M}}_T} \right)^2 D^h_{(5+h)/2} \left[\gamma_7{}^h \frac{\overline{\mathfrak{M}}_\perp}{\overline{\mathfrak{M}}_T} \right] \qquad (2.6.119)$$

$$\mathfrak{M}_\perp = \overline{\mathfrak{M}}_\perp \left[\frac{\omega_b}{\nu_m(\hat{u})} \right]^2 D^h_{5/2} \left[\frac{\omega_b}{\nu_m(\hat{u})} \right]$$

$$= \overline{\mathfrak{M}}_\perp \left(\gamma_7{}^h \frac{\overline{\mathfrak{M}}_\perp}{\overline{\mathfrak{M}}_{T'}} \right)^2 D^h_{5/2} \left[\gamma_7{}^h \frac{\overline{\mathfrak{M}}_\perp}{\overline{\mathfrak{M}}_{T'}} \right] \qquad (2.6.120)$$

$$\frac{\mathfrak{M}_T}{\mathfrak{M}_\perp} = \frac{(\overline{\mathfrak{M}}_T/\overline{\mathfrak{M}}_\perp) D^h_{(5+h)/2}[\gamma_7{}^h(\overline{\mathfrak{M}}_\perp/\overline{\mathfrak{M}}_T)]}{D^h_{5/2}[\gamma_7{}^h(\overline{\mathfrak{M}}_\perp/\overline{\mathfrak{M}}_T)]} \qquad (2.6.121)$$

where the bar over $\mathfrak{M}_T$ and $\mathfrak{M}_\perp$ indicates values (2.6.112) and (2.6.111), and $D_i{}^h(x)$ is defined by (2.6.45). In the limit $B \to 0$, $\mathfrak{M}_T$ approaches the value given by (2.6.73) and $\mathfrak{M}_\perp$ approaches zero.

When required, the values of $\mathfrak{D}_T$ and $\mathfrak{D}_\perp$ for the general case can be derived from the foregoing $\mathfrak{M}_T$ and $\mathfrak{M}_\perp$ expressions by performing the appropriate differentiations, as indicated from (2.6.113) and (2.6.114). A worthwhile general statement regarding these coefficients can be easily demonstrated from the foregoing formulas without any tedious development: the ratios $\mathfrak{D}_T/\mathfrak{D}$ and $\mathfrak{D}_T/\mathfrak{D}_\perp$ are functions only of $\omega_b/\nu_m(\hat{u})$ and of the parameter h.

Chapter 3

Experimental Methods of Measuring Transport Parameters

In this chapter we describe in detail the experimental methods by which the electron swarm transport parameters defined in Chapter 2 can be properly measured. In Section 3.1 the discussion is restricted to a few general remarks regarding the conditions appropriate for the experiments, in order to derive from them the desired electron collision parameters, and regarding the common techniques employed in all the methods. The transport parameters are individually treated in the successive sections; for each of them we outline first the various methods of measurements, and then we analyze their secondary effects, validity limits, and error sources. For each method at least one typical, modern apparatus and the relevant measuring techniques are described; different features of other experiments based on the same method are mentioned only when sufficiently meaningful.

Among the many experimental techniques that we discuss in this chapter for the purpose of determining electron swarm properties, we shall not consider the use of electric probes for electron density and distribution function measurements, because there are many more possibilities of incorrect measurements and misleading interpretations in the use of electric probes than with the other techniques treated here. The main advantage of the electric probe is its capability for local parameter measurements, but this implies also its most significant drawback, namely, the perturbation of local swarm characteristics.

Even if we are not basically concerned with probe techniques, their use will occasionally be mentioned. For adequate knowledge of this subject the reader is referred to the reviews by Chen (1965) and Schott (1968).

3.1 COMMON FEATURES OF THE EXPERIMENTAL METHODS

According to the general statements made in the Introduction of this book, we intend to deal here only with those experimental methods of measuring transport parameters that are appropriate for the purpose of deriving the elastic and inelastic, nontransformation electron collision parameters at energies below about 1 eV. Hence we shall not be concerned with cases in which transport parameters are measured in field-sustained discharges. In fact, in these cases most of the swarm electrons have energies well above 1 eV; furthermore, the distribution function f^0, a knowledge of which is necessary for deriving the collision parameters, is strongly dependent on the electron excitation and ionization cross sections, so that the determination of the collision parameters at low energies is significantly influenced by any error or assumption made concerning these higher-energy cross sections. Measurements in field-sustained discharges are therefore of no real value for the present concern.

For the same reasons we shall also disregard cases in which an electron swarm moves under the action of electric fields not so strong as to produce or to maintain gas breakdown, but large enough to generate significant excitation and ionization of the gas.

The experiments that will be discussed in the following sections of this chapter have a few common characteristics. In all of them there is a chamber, which, except in a very few high-pressure cases, is first evacuated by means of a very high-vacuum pumping system, and then filled with the gas under investigation at a pressure that must be accurately measured. High-vacuum techniques and pressure-measuring equipment are very well treated in excellent textbooks, to which the interested reader is referred for an appropriate knowledge of these subjects [see, in particular, Dushman (1962), Pirani and Yarwood (1961), Guthrie (1963), Steinhertz (1963), Roberts and Vanderslice (1963), Leck (1964), Redhead, Hobson, and Kornelsen (1968)]. Residual gas pressures as low as a few units of 10^{-9} torr are attained after high-temperature bakeout in some of the cleanest chambers, as typically used in microwave experiments. Such extremely low residual pressures, however, are rarely mandatory, since values of 10^{-5} to 10^{-7} torr are generally quite sufficient when a chamber is afterwards filled with a gas at much higher pressures (>0.1 torr); in any case it is most important to verify that the rate of rise of the background pressure is such that during the time of the experiment the results of concern have not changed beyond the measuring errors. Conventional mercury and oil manometers and McLeod gauges have been extensively used for pressure measurements, but in most of the experiments performed during the sixties preference was given to better devices, such as the bakable null-reading manometer of Alpert, Matland, and McCoubrey (1951), the precision capsule gauge of Crompton and Elford (1957), and other commercial mechanical gauges (metallic diaphragm or quartz-spiral manometers). Measurements from 5 up to 250 torr with an error of less than 0.1 % have been performed at Canberra with a Texas quartz-spiral gauge, calibrated against a C.E.C. type G-201 primary pressure standard.

In a large number of experiments the electron swarm moves under the action of a dc electric field; this field must be constant and accurately known. In general the high voltage to be applied between the chamber electrodes is obtained from highly regulated and well-filtered power supplies; in a few cases preference has been given to batteries. The use of digital voltmeters for accurate measurements of these applied voltages has gained much favor in recent years; some workers, however, still perform measurements by means of wire-wound resistors (preferably immersed in an insulating oil bath, whose temperature is recorded for appropriate corrections of the resistance value) in series with a microammeter.

In most experiments dealing with measurements of dc transport parameters very low dc currents have to be measured. They are typically in the 10^{-11} to 10^{-13} A range and must be measured by means of electrometers in conjunction with very high values of resistance; here, too, a digital voltmeter has been conveniently used by some workers for reading with high accuracy the electrometer output. Commercial vibrating-reed, vacuum-tube, and MOS field-effect transistor electrometers have been found adequate for the present use; of these, however, the vibrating-reed electrometer is generally preferred, since it satisfies the most stringent requirements in respect to input impedance and drift characteristics.

Alternating current transport parameters are most often determined at microwave frequencies. Although we assume that the reader is familiar with these techniques in general, we recommend the book by Heald and Wharton (1965) as the most valuable comprehensive reference on the specific subject of microwave techniques applied to measurements of plasma properties.

The other general features of the experiments we are going to discuss, such as the occasional presence of a static magnetic field and the use of a thermostatic bath to control the chamber temperature, are not particularly critical, so that it is rather easy to satisfy the requirements by simply following all good practices of laboratory work. Specific features of the experiments, on the other hand, will be discussed in detail for each method separately in the following sections of this chapter.

3.2 DRIFT VELOCITY FROM TRANSIT TIME MEASUREMENTS: GENERAL DESCRIPTION OF THE EXPERIMENTAL METHODS

The electron drift velocity w is usually determined by measuring in the gas under investigation the average transit time of the electrons of a swarm across a known distance (drift distance) under the action of an applied dc electric field E. In all experiments that are adequate to provide accurate significant data, two plane parallel boundaries, separated by d, can be easily recognized: they enclose the drift region over which the electron transit time is measured. The electric field is perpendicular to these drift region boundaries and is uniform. The gas is here homogeneous, unionized, and in a thermodynamic, equilibrium state.

Two basically different approaches can be adopted. The most popular one, which has been used since 1936 [Bradbury and Nielsen (1936)] with only one significant exception [Nolan and Phelps (1965)], has provided about all known reliable data. It requires releasing in the drift region a very short burst of electrons and observing the electron motion in a constant applied dc field. In most of these experiments the electron burst, generated in a time much shorter than the electron transit time, is released or injected in the drift region at the cathode boundary (the lower one in Fig. 3.2.1); the transit

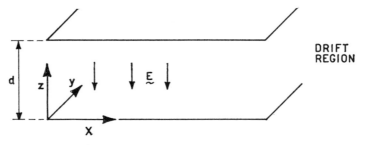

Fig. 3.2.1. Drift region and related coordinate system.

time is determined from the arrival time of the electron packet at the anode boundary. In a few other experiments electrons are simultaneously generated throughout the drift region, and the time they take to clear out of the region is measured: this is equal to the transit time of the most distant electrons, those which have to cross the entire drift distance d. In all these cases the electron injection process is repeated at a rate such that at each new cycle all previous swarm electrons have been completely cleared out of the drift region.

The other approach is much older; most of the measurements performed with this technique date back to the 1920's, immediately after its successful introduction by Loeb (1922). A continuous source of electrons at one boundary feeds the drift region, where the electrons move under the influence of an alternating electric field; electrons will reach the other boundary and give rise to a current with a nonzero dc component only when the half-period of the applied field waveform, during which electrons are accelerated toward that boundary, is sufficiently long. Thus it is possible to measure the electron mobility by observing the frequency of the applied field at which the dc current vanishes.

We shall now examine successively in more detail the appropriate techniques for (A) releasing repetitive, short electron bursts in the drift region;

(B) measuring transit times by observing pulses of electrons drifting in a constant electric field; and (C) deriving electron mobilities from measurements of the dc component of the electron flow across the drift region in an alternating electric field.

A. Let us consider first the problem of the pulsed electron source to be used for injecting through the cathode boundary a very short pulse of electrons into the drift region. In order to obtain accurate results with this method, at each new cycle a very thin packet of electrons, preferably uniform in density over the entire area of the tube, must be produced at the source boundary, the generation time being as short as possible. This goal is achieved by using one of two basically different approaches: (a) electrons are generated

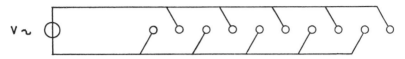

Fig. 3.2.2. Schematic representation of the electron shutter.

continuously by an appropriate source, but are admitted in the drift region through an electron shutter, which is normally closed and opens for a very short time interval only; or (b) short pulses of electrons are periodically released from a properly driven cathode, or are generated in the gas by bursts of well-collimated ionizing radiation, along a line perpendicular to the field. The electron source boundary of the drift region can be identified, in case a, with the shutter plane, and, in case b, with the cathode or with the ionizing particle plane. Now we shall explain these two techniques in more detail.

The electron shutter (Fig. 3.2.2), a derivation of the electron filter originally conceived in 1926 by Loeb (1955) for the purpose of capturing selectively all traversing electrons, consists of a plane grid of parallel wires, of which alternate wires are connected together. A steady flow of electrons, emitted by a heated filament or by an illuminated photocathode, or generated by volume ionization from a radioactive source, will pass freely across the shutter when no bias voltage is applied between the two halves of the grid, but it will be largely collected by the grid when an adequate bias is applied. Two modes of operation can be used: (1) an ac signal is applied between the two halves of the grid, so that the stream of electrons will be divided into a series of pulses, electrons being admitted only each time the ac signal goes through a zero [Bradbury and Nielsen (1936)]; (2) the grid is biased to absorption and made periodically transmitting by applying a series of short rectangular voltage pulses, such as to reduce the fields between wires to zero [Phelps,

Pack, and Frost (1960)]. It is appropriate, when using the electron shutter, to apply in the region between the electron source and the shutter itself the same dc electric field as in the drift region, so that electrons will enter this region already having the correct form of energy distribution.

For the purpose of releasing directly short pulses of electrons at the source boundary, a photocathode, upon which is periodically incident an ultra-violet radiation pulse, was introduced by Hornbeck (1951) and has since become the most frequently adopted solution. A grid acting as the positive electrode of a short-duration discharge has also been used for injecting electron bursts in the drift region [Breare and Von Engel (1963), Chanin and Steen (1964)]; this solution is of particular value when the gas under investigation may contaminate the photocathode. Other solutions are provided by the use of pulsed X-rays and of well-collimated beta-particles or alpha-particles; these sources were introduced, respectively, by Herreng (1942), by Stevenson (1952), and by Colli and Facchini (1952). Sources of alpha-particles have been used since then in a few transit time experiments: each alpha-particle creates a bunch of electrons, say about 10^5 in a typical case, along a line perpendicular to the electric field. Attention must be paid, however, to the following facts: the electron density over the source boundary plane is highly nonuniform, the pulse rate has to be low when a high degree of collimation is desired, and the time interval between successive bursts is not constant.

Let us consider finally the case in which we desire to generate a burst of electrons over the entire drift volume. Here too direct ionization of the gas by an alpha-particle crossing the entire drift region can be used [Klema and Allen (1950), Fischer-Treuenfeld (1965)]; however, the foregoing remarks on the use of alpha-particle ionization can be repeated here. Another approach makes use of a pulsed X-ray beam: the X-rays may ionize the gas directly, or generate Compton electrons by striking an appropriate target, like a gold-backed window [Hudson (1944), Kirschner and Toffolo (1952)]. Ionization from fission fragments generated by neutrons striking a uranium layer target has also been adopted [Nagy, Nagy, and Dési (1960)].

B. Under the influence of an applied constant and uniform dc electric field bursts of electrons flow across the drift region to a collector, which is also the positive electrode (anode) of the applied voltage. Six basic techniques can be used for measuring the transit time of the electrons across the drift region.

a. The collector is a large plate, which represents also the drift region anode boundary; an electron packet is released at the cathode boundary (source). The resulting current induced at the collector by the drifting electron swarm is observed; the magnitude of this current, while the electrons are traversing the drift region, is constant and equal to:

$$i = \frac{ewN}{d} \tag{3.2.1}$$

where N is the total number of electrons, and both diffusion and space-charge effects are neglected. The current rises to this value when the electron packet is released (time origin $t = 0$) and drops to zero at time t_d, when the packet is swept out by the collector. The current pulse length is then equal to the electron transit time:

$$t_d = \frac{d}{w} \tag{3.2.2}$$

and the drift velocity is calculated accordingly:

$$w = \frac{d}{t_d} \tag{3.2.3}$$

Because of the finite size of the electron packet, in the actual experiments the current pulse is not exactly rectangular; then it is customary to identify as transit time t_d the time interval between the two instants at which the current is half of its maximum value. The method as described was devised by Herreng (1942) and improved by Hornbeck (1951), and has been extensively used by Bowe (1960) [Bowe and Langs (1966)].

The use of a variable drift distance, adopted by some investigators, makes possible differential measurements, thereby allowing the removal of end effects like those due to the nonideal characteristics of electron injection. Thus Herreng (1942), who used an X-ray beam source, by moving the beam known distances from the collector and making differential transit time measurements between similar points on the current decline curves, was able to avoid the consequences of the usually ill-defined edges of the X-ray beam. Later Chanin and Steen (1964) used a movable collector and performed differential transit time measurements, so as to remove the contribution of the anomalous initial portion of the transit, where electrons have not yet reached the energy distribution appropriate to the swarm in the applied field. In these cases a convenient procedure consists of (i) plotting the measured transit time as a function of the drift distance, (ii) verifying that at least for the largest distances a straight line can be correctly drawn through the experimental points, and (iii) determining the drift velocity as the reciprocal of the straight-line slope.

Levine and Uman (1964) slightly modified the basic method by feeding the collector current to a large resistance at the input of a cathode follower preceding a low-noise amplifier, so as to improve the signal strength. The time constant, which is appropriate to this resistance and to the total parallel capacitance C (drift chamber plus input circuit of the cathode follower), is much larger than the pulse duration; hence the voltage waveform at the input to the cathode follower becomes approximately a ramp function (for the duration of the current pulse in the chamber):

$$v = \frac{1}{C} \int_0^t i \, dt = \frac{ewN}{dC} t = \frac{eN}{C} \frac{t}{t_d} \quad \text{for } 0 \leqslant t \leqslant t_d \qquad (3.2.4)$$

The transit time is then assumed by Levine and Uman to be equal to the duration of the sloping portion of the output voltage pulse, as displayed on an oscilloscope (Fig. 3.2.3). Similar conditions hold in the case of drift

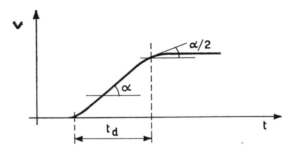

Fig. 3.2.3. Output voltage waveform in Levine's and Uman's (1964) experiment and definition of the transit time t_d.

velocity measurements at high pressures performed by Grünberg (1967). Rigorously, since the waveform is now the integral of the current pulse, the two previously defined half-maximum current instants, which determine t_d, should now be identified as the time instants at which the waveform slope attains half of its maximum value.

b. As in case *a*, the collector plate represents also the drift region anode boundary, but now a burst of electrons is generated simultaneously across the entire drift region. It is easy to see that, for a very short burst of electrons (compared to their transit time), here too the current at the collector drops to zero only when the electrons generated at the cathode boundary have reached the collector; the electron drift velocity is thus obtained. The current waveform, however, is no longer rectangular; this would be immaterial, except for the fact that the zero current point is now badly defined, the current steadily decreasing with time. Some improvement is obtained [Nagy, Nagy, and Dési (1960)] when ionization is produced by fission fragments due to the neutron bombardment of a uranium layer deposited on a negative electrode at the drift region cathode boundary; most electrons are generated near this electrode and thus the current has a more rectangular shape, as in case *a*.

A spark technique has been introduced in the most recent experiments, to detect with better definition the presence of electrons in the drift region. The experiment is performed as follows. The electron burst source (an alpha-particle in the scheme of Fig. 3.2.4) establishes the time origin $t = 0$ and triggers, after a variable delay, a high-voltage pulse, which is superimposed on the steady applied voltage so as to initiate an avalanche discharge. The time

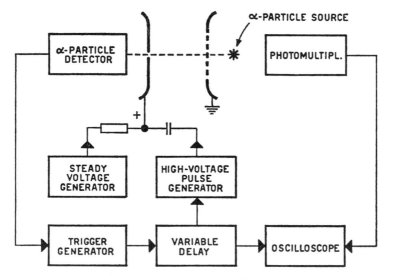

Fig. 3.2.4. Block diagram of Fischer-Treuenfeld's (1965) method.

at which the high-voltage pulse is applied is changed, and the strength of the avalanche discharge is determined by the maximum height of a photo-multiplier current output, which measures the light emitted by the discharge. The high-voltage pulse polarity is opposite to the steady voltage polarity, so that the largest avalanche amplification effect takes place for the electrons closest to the collector; in particular, then, this effect is very large when the last electrons—those which have crossed the entire gap—are about to leave the drift volume when the high-voltage pulse is applied. The photomultiplier current maximum plotted versus the triggering time of the high-voltage pulse will accordingly show a sharp drop to zero at time t_d, which corresponds to the transit time of the electrons across the entire gap length (eq. 3.2.2).

In the form described the method is due to Fischer-Treuenfeld (1965), who also worked out the formulas that apply to his experimental conditions. He considers a one-dimensional geometry, in which the electron density* n

---

* From now on, the subscript e for the electron quantities (density, flow, etc.) will be dropped, except when there is a possibility of ambiguity.

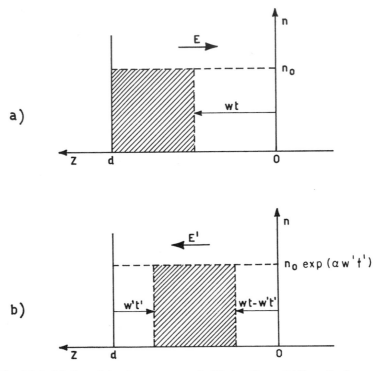

Fig. 3.2.5. Motion of the electron swarm in Fischer-Treuenfeld's method.

depends only on z, the distance from the right electrode of Fig. 3.2.4, and t. The alpha-particle creates at time $t = 0$ a uniform electron density n_0 across the gap; later on at time t, if diffusion is neglected, the density becomes (Fig. 3.2.5a):

$$n(z, t) = \begin{cases} n_0 & \text{for } 0 \leqslant t < d/w, \; wt < z < d \\ 0 & \text{for other } z \text{ and } t \end{cases} \tag{3.2.5}$$

We assume now that at time t the high-voltage pulse of opposite polarity is applied; the electrons will reverse their motion and, neglecting transients, will move away from the collector with drift velocity w', which is appropriate to the new electric field. At the same time gas ionization takes place and the electron density increases as $\exp(\alpha w't')$, t' being the time passed since the high-voltage pulse was applied and α the first Townsend coefficient. Then we have:

$$n(z, t') = n(z + w't', t) \exp(\alpha w't')$$

and the current density across the gap is (Fig. 3.2.5b):

$$i(t, t') = \frac{ew'}{d} \int_0^{d-w't'} n(z, t') \, dz$$

$$= \frac{ew'}{d} \exp(\alpha w't') \int_0^{d-w't'} n(z + w't', t) \, dz$$

$$= \frac{ew'}{d} \exp(\alpha w't') \int_{w't'}^{d} n(z, t) \, dz \qquad (3.2.6)$$

Substituting distribution (3.2.5), we obtain:

$$i(t, t') = \begin{cases} ew'\left(1 - \dfrac{wt}{d}\right) n_0 \exp(\alpha w't') & \text{for } 0 \leqslant w't' \leqslant wt \\[3mm] ew'\left(1 - \dfrac{w't'}{d}\right) n_0 \exp(\alpha w't') & \text{for } wt \leqslant w't' \leqslant d \end{cases} \qquad (3.2.7)$$

It is easy to verify that:

(1) when $0 \leqslant wt \leqslant d - (1/\alpha)$, the current as a function of t' has the t-independent maximum:

$$i_{max} = \frac{ew'n_0}{\alpha d} \exp(\alpha d - 1) \quad \text{at } w't' = d - \frac{1}{\alpha} \qquad (3.2.8)$$

(2) when $d - (1/\alpha) \leqslant wt \leqslant d$, the current has the maximum:

$$i_{max} = ew'n_0\left(1 - \frac{wt}{d}\right) \exp(\alpha wt) \quad \text{at } w't' = wt \qquad (3.2.9)$$

The photomultiplier current is proportional to the discharge current, and the behavior of its maximum versus time is then given from (3.2.8) and (3.2.9). If we normalize this maximum current i_p to unity at $t = 0$ and the time to the transit time t_d (eq. 3.2.2), introducing the new time variable $\tau = t/t_d$, we obtain:

$$i_p(\tau) = \begin{cases} 1 & \text{for } 0 \leqslant \tau \leqslant 1 - (1/\alpha d) \\ \alpha d(1 - \tau) \exp[1 - \alpha d(1 - \tau)] & \text{for } 1 - (1/\alpha d) \leqslant \tau \leqslant 1 \end{cases} \qquad (3.2.10)$$

This behavior is shown in Fig. 3.2.6. By fitting the observed current maxima plotted versus the actual time to the behavior predicted above, the transit time t_d is determined; the drift velocity is given again from (3.2.3).

A similar technique was also reported by Fischer (1966). This author, however, instead of determining the strength of the avalanche discharge following the high-voltage pulse, measures only its sparking efficiency,

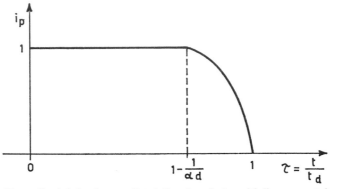

Fig. 3.2.6. Normalized behavior predicted for the photomultiplier current in Fischer-Treuenfeld's method.

defined as the number of actual sparks divided by the number of particle triggers. This efficiency, which is expected to decay as $1 - \exp(-N)$, here becomes the appropriate observable parameter, since the author operates the chamber according to the normal usage of nuclear spark chambers, namely, at very low electron density levels. It seems, however, that these conditions, insofar as they allow only a statistical measure of the disappearance of the last electron, are not particularly suitable for accurate drift velocity measurements; furthermore, one cannot see any real argument in support of the author's basic choice of determining the transit time from the delay corresponding to a sparking efficiency of 50 % (0.7 electron present).

Following a different concept, English and Hanna (1953) modified the original method by introducing a grid in the drift region. Each alpha-particle creates in the cathode-grid region a column of electrons parallel to the field; when this column passes through the grid in the collector region, an induced current pulse appears at the collector, so that the drift velocity is given by the ratio of the alpha-particle ionization range to this pulse rise time.

In addition to the discussed operations with very short bursts of electrons, long ionizing pulses, lasting more than the electron transit time, have also been used [Hudson (1944), Kirschner and Toffolo (1952)]. In this case the electron current across the gap increases steadily until a regime value is attained; this occurs when the electrons generated at the cathode at the beginning of the ionizing pulse have reached the anode (this time is t_d). The appropriate general solutions for the current behavior can be found in the paper by Kirschner and Toffolo (1952), to which the interested reader is referred for more information on this now obsolete technique.

c. At the cathode boundary a packet of electrons is injected into the drift region; at the anode boundary a properly driven electron shutter admits or does not admit the packet to the collector. There are three different modes of operation, which will now be described.

Bradbury and Nielsen (1936) adopted two equal shutters: one for the production of the electron pulses, the other for the control of the collector current (Fig. 3.2.7); on both shutters they apply the same sinusoidal voltage between the two halves of the grids. Since electrons flow across a shutter only when

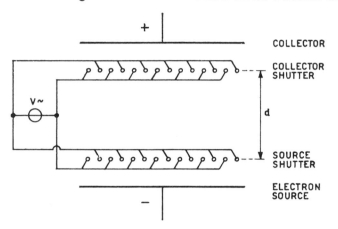

Fig. 3.2.7. Schematic diagram of the Bradbury-Nielsen (1936) apparatus.

the ac voltage applied to the grids is approximately zero, the current across the chamber will reach the collector only when the applied voltage frequency is such that the electron transit time between shutters t_d is equal to the time interval between two zeros of the voltage. Therefore, if the voltage frequency is varied and the collector current versus frequency is recorded, sharp maxima of this current will be observed any time that the voltage frequency f is an integral multiple of $1/2t_d$. From a knowledge of the frequencies producing at least two successive maxima (orders K and $K + 1$):

$$f_1 = \frac{K}{2t_d} \; ; \qquad f_2 = \frac{K + 1}{2t_d} \qquad (3.2.11)$$

the drift velocity (eq. 3.2.3) can easily be derived:

$$w = 2d(f_2 - f_1) \qquad (3.2.12)$$

A typical curve for this case is shown in Fig. 3.2.8.

Instead of using a sinusoidal voltage, Phelps, Pack, and Frost (1960) applied to both shutter grids a dc bias and a superimposed sequence of narrow rectangular pulses with a uniform rate; the pulse sequence applied to the collector shutter can be delayed by a variable amount with respect to

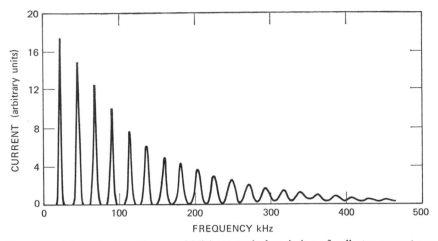

Fig. 3.2.8. Method of Bradbury and Nielsen: typical variation of collector current as a function of frequency. The curve was obtained in hydrogen at $E/p = 0.08$ V cm^{-1} torr^{-1}, $p = 500$ torr, $T_g = 293°$K [after Lowke (1963)].

the one applied to the source shutter. Since the bias voltage reduces the transmitted current to a very small value, whereas the pulse amplitude is such as to bring the grid field back to zero and thus restore the full electron flow across the shutter, the collector current will rise to a sharp maximum only when the delay between the two pulse sequences equals the electron transit time from the source to the collector shutters. For comparable performance the amplitudes of the applied pulses are much smaller than the large sine waves required by Bradbury and Nielsen, and this simplifies minimizing field distortions in the drift region.

In order to correct for end effects, the use of two drift distances d_1 and d_2 is generally preferred. This can be accomplished either by changing the actual spacing between the shutters or by using two collector shutters at different distances from the source. Figure 3.2.9 shows a simplified schematic of the system, as it was used for this purpose at Westinghouse [Pack and Phelps (1961)]; in this case the need of a source shutter is eliminated by using the light of a periodically pulsed ultraviolet source on a photocathode. In all these cases, if the collector current waveforms are not badly distorted by diffusion effects, the drift velocity is given by the ratio of the spacing $d_2 - d_1$ between the two collector shutters to the difference of the corresponding pulse delay times (t_1, t_2), at which the collector current peaks occur:

$$w = \frac{d_2 - d_1}{t_2 - t_1} \qquad (3.2.13)$$

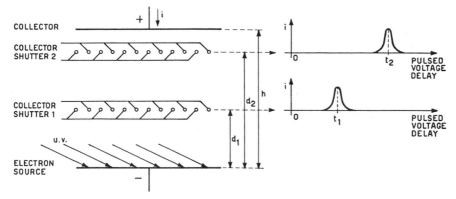

Fig. 3.2.9. Simplified schematic of the Westinghouse method and collector current as a function of the time delay between the light pulse and the applied grid voltage pulse for each shutter. The widths of the pulses shown are not to scale [after Pack and Phelps (1961)].

Also, errors caused by waveform distortions due to an unsymmetrical light pulse can be eliminated by means of these differential measurements.

A third mode of operation can also be considered when the Westinghouse structure is used. This is called the zero-bias or rejection mode of operation [Pack and Phelps (1961)], since no dc bias, but pulses only, are applied to the shutter grids. When a pulse is applied to a shutter grid, its electron transmission is reduced. Then, if a sequence of voltage pulses makes the grid transmission periodically reduced at the same rate as the pulsed light source, the collector current will be a minimum when the time location of the pulse sequence is such that the electron packets coming from the photocathode arrive at the shutter position in coincidence with the voltage pulses. If t_1 is here the delay of the pulse sequence applied to the first shutter (the reference time origin being the uv light pulses), for which the collector current attains a minimum, and t_2 is the delay of the pulse sequence applied to the second shutter, which also produces a collector current minimum, then the drift velocity is again computed from eq. (3.2.13).

This mode of operation has the advantage that there is no field between the grid wires, except during the very short pulse duration, so that the perturbing effects of the grid fields are reduced in comparison to the other modes of operation.

It is worth remarking that the statement that a field between the wire grids reduces the shutter transmission is not always true. Pack, Voshall, and Phelps (1962) found that argon, krypton, and xenon show an anomalous behavior, probably due to the existence of a Ramsauer minimum, above a certain E/p for each gas. In these gases, in fact, the grid transmission first increases with the applied bias voltage up to a maximum and then decreases as the bias is increased further. It is rather obvious how this effect can be taken into account, when the bias and pulse amplitude voltages are chosen for maximum and reduced transmission.

d. A gas discharge tube of the Geiger-Müller type is mounted to the collector anode electrode and communicates with the drift chamber through a small centrally located aperture. The tube is triggered by the arrival of one of the electrons of a short pulse injected into the drift region at the cathode boundary. The time location of the Geiger signal yields the transit time of the triggering electron [time-of-flight (TOF) method]. Because of the long dead times of the Geiger tube no other electron of the same swarm can be detected.

In its most complete form the method makes use of a periodically pulsed source of electrons and of repeated observation of the arrival times at successive bursts, so that the electron transit time distribution is obtained. For this purpose each electron of the swarm must have the same probability of triggering the tube, and it can be easily understood that this will be the case only when conditions are such that the average number of electrons detected for each burst is much less than 1. If diffusion effects are not too large, the time of the distribution peak can be regarded as the arrival time of the average electron, which moves through the chamber with drift velocity.

When measurements are performed at fairly high pressures, diffusion becomes small and the distribution very narrow [see eq. (3.3.4) of Section 3], so that it is difficult, if not even impossible, to measure the arrival swarm distribution with any significant precision. In these cases the drift velocity is simply determined from eq. (3.2.3), where t_d is the time between the emission of the electron bursts and the Geiger pulses, as they appear, practically all with the same delay, when directly displayed on an oscilloscope.

This method, conceived by Stevenson (1952), has been improved [Bortner, Hurst, and Stone (1957), Hurst, O'Kelly, Wagner, and Stockdale (1963), Hurst and Parks (1966)] and most extensively used by Hurst and his co-workers at the Oak Ridge National Laboratory. In the latest experiments electrons are generated on a photocathode by the light of a pulsed uv source; this light, detected by a photodiode, provides the time origin for the electron transit time measurements. A time-to-amplitude converter and a channel pulse height analyzer, or a high quality time-of-flight analyzer, can be used for measuring the transit time distribution. Only good counter gases can fill the Geiger and the drift chamber; in order to overcome this limitation, in some experiments the Geiger has been replaced with a differentially pumped electron multiplier [Wagner, Davis, and Hurst (1967)].

e. Comunetti and Huber (1960) devised a different solution, in which the function of the collector anode electrode is performed by a composite structure (Fig. 3.2.10): a fine mesh grid at the end of the drift space allows electrons to enter into a very thin region, where these electrons, under the

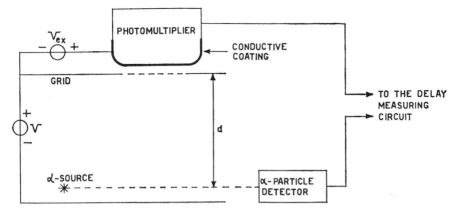

Fig. 3.2.10. Schematic diagram of Comunetti's and Huber's (1960) apparatus.

action of a much larger field, slightly below breakdown, excite gas molecules; the generated light is detected by a photomultiplier. The quartz entrance window of this multiplier is very close to the transmission grid, and its coating is conductive, so that all the electrons are collected, and optically transparent, so that the emitted radiation is not significantly attenuated. The large field for excitation is obtained by applying an adequately large voltage between the grid and the conductive coating of the photomultiplier.

Except for a very small fixed delay, which can be easily determined and taken into account, the detected light can be considered as a good replica of the electron current across the grid. The drift velocity of a short burst of electrons can then be determined by measuring the transit time between its generation and the maximum of the detected light. In the experiment by Comunetti and Huber electrons are generated by ionization from an alpha-particle collimated in a plane parallel to the anode grid.

f. A method for observing a short pulse of electrons, while drifting between a source and a collector, was proposed and tested in hydrogen by Breare and Von Engel (1963, 1964). The method relies on the fact that, at sufficiently high values of E/p, some electrons will excite gas molecules to electronic levels, resulting in the emission of radiation. As shown in Fig. 3.2.11, only the light emitted on a given plane is detected, since the swarm path viewed is restricted by means of an appropriate optical collimator in front of a photomultiplier detector. In this way at the highest pressures (> 20 torr in hydrogen) the photomultiplier sees the arrival of the radiating electrons of the swarm front; hence the transit time across a drift region is easily determined by using two photomultipliers located at the region boundaries.

On the other hand, since the light emission efficiency decreases with pressure, only very few electrons in each swarm generate photomultiplier

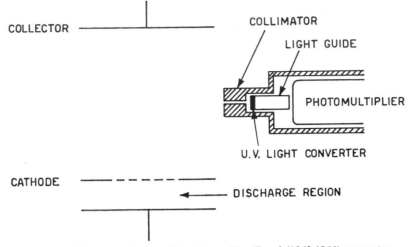

Fig. 3.2.11. Schematic diagram of the Breare-Von Engel (1963,1964) apparatus.

output pulses at the lowest pressures (e.g., $\simeq 2$ torr in hydrogen). In this case each detected photomultiplier pulse indicates the arrival time of a single electron, whose location can be anywhere in the swarm; hence observation of a large number of successive drifting swarms yields the distribution of electron transit times and makes possible the determination of the drift velocity from the time position of the distribution peak.

In the particular experiment of Breare and Von Engel, the electron burst is generated by a pulsed discharge, or by a continuous discharge and a gate electrode. The light emitted by the drifting electrons, which is mostly in the ultraviolet region, is converted into visible light by a fluorescent substance (sodium salicylate) deposited at the end of a glass rod in front of the photomultiplier.

C. We consider now the essential features of the approach based on applying an alternating electric field between two parallel plate electrodes, one of which is a steady source of electrons, and on measuring the dc component of the resulting electron current.

Following Nolan and Phelps (1965), one may conveniently use a square-wave voltage. In this case, if V is the peak value of this voltage of frequency f applied over a gap of separation d, $\mu V/d$ will be the drift velocity and $(\mu V/d)(1/2f)$ the maximum distance from the source, attained by the most favored electrons. Only when this distance is equal to or larger than d, will a dc component of electron current flow. It is a simple matter to detect this current by means of a dc meter in the circuit; leaving constant the voltage amplitude, its frequency f is increased till the dc current vanishes (let f_0 be this frequency), so that the electron mobility can be computed by means of the relationship:

$$\frac{\mu V}{d} \frac{1}{2f_0} = d$$

$$\mu = \frac{2f_0 d^2}{V} \tag{3.2.14}$$

With an applied square voltage the measured μ corresponds to a well-defined $E/p = V/dp$ value; this is not the case, however, when a different voltage waveform is used, as in the early experiments by Loeb (1922, 1924) and by Wahlin (1923, 1924, 1926, 1931). These authors applied sinusoidal voltages and computed mobility assuming a μ value independent of E/p; at a constant frequency f the voltage is decreased till the dc current vanishes (let V_0 be the peak value of this voltage), so that eq. (3.2.14) is replaced by:

$$\frac{\mu V_0}{d} \int_0^{\frac{1}{2}f} \sin(2\pi f t)\, dt = \frac{\mu V_0}{d} \frac{1}{\pi f} = d$$

$$\mu = \frac{\pi f d^2}{V_0} \tag{3.2.15}$$

In the experiment of Nolan and Phelps the drift distance d can be changed; if the field V/d is kept constant and d is plotted versus $1/2f_0$, the resulting curve must be a straight line with the slope equal to the drift velocity $\mu V/d$. End effects are thus significantly reduced. Furthermore, it can be easily verified that in the region $f < f_0$ the dc current is a linear function of $(f_0 - f)$, so that f_0 can be accurately determined by the intercept with the f axis of the best straight line drawn through the experimental points.

3.3 DRIFT VELOCITY DETERMINATIONS FROM TRANSIT TIME MEASUREMENTS: APPROXIMATION LIMITS AND RELATED ERRORS

In Section 3.2, for the purpose of describing the different experimental methods for determining electron drift velocities from transit time measurements, we used, explicitly or implicitly, several convenient simplifying assumptions. When a more exact treatment is desired and the validity limits of the theory are requested, reconsideration of these assumptions is mandatory.

The most relevant assumptions that underlie the preceding analysis are the following:

   a. The diffusive spreading of an electron swarm as it drifts under the action of the dc field has negligible effects on the parameters of the signal waveform, from which the drift velocity is deduced.

   b. Electron densities are so low that space-charge effects and Coulomb mutual interactions do not modify appreciably the swarm motion.

   c. Volume losses of electrons (as attachment) and volume generation (as ionization) in the drifting swarm are neglected.

   d. The duration of each electron burst generation is infinitely short.

   e. A plane parallel geometry and a perfectly uniform drift field are assumed.

   f. The energy distribution of the electrons approximates closely the equilibrium kinetic distribution throughout the drift region and within the electron swarm.

   g. The repetition rate is slow enough to allow each electron pulse to be cleared out of the drift chamber before the next one is initiated.

Assumptions *b* and *g* are usually well satisfied in all the experiments without particular effort; only in Chanin's and Steen's (1964) experiment in cesium might space-charge effects have been important, according to computer calculations by Ward (1966). More significant limitations on the accuracy of the measurements, however, may come from the other assumptions. Therefore we shall now discuss their validity limits and evaluate the errors that occur when these limits are not satisfied well enough.

   A. In the case of drift velocity measurements the errors due to diffusion are the most important ones and hence will be discussed first. This analysis will rest heavily on Lowke's (1962) work, which is an extension of a previous investigation by Duncan (1957).

   As we saw in Chapter 2, the differential equation governing the motion of an electron swarm under the combined action of an applied uniform dc field along the z direction and of diffusion, when all other assumptions previously listed are satisfied, is:

$$\frac{\partial n}{\partial t} = -\nabla \cdot \mathbf{\Gamma} = -\nabla \cdot [\mathbf{w}n - \nabla(Dn)]$$

$$= -w \frac{\partial n}{\partial z} + D \nabla^2 n \tag{3.3.1}$$

Since we have assumed a plane parallel geometry, the electron density n is a function of t and z only, and (3.3.1) can be rewritten as:

$$\frac{\partial n}{\partial t} = D \frac{\partial^2 n}{\partial z^2} - w \frac{\partial n}{\partial z} \tag{3.3.2}$$

The first case we shall consider is that of an electron source located at $z = 0$, which generates uniformly at time $t = 0$ an infinitely short burst of N electrons per unit area of the xy plane; the resulting electron packet is allowed to move in space according to eq. (3.3.2) without any wall limitation. The appropriate solution of the equation is then:

$$n = \frac{N}{2\sqrt{\pi Dt}} \exp\left[-\frac{(z - wt)^2}{4Dt}\right] \tag{3.3.3}$$

As a function of z the density has the shape of a normal Gauss curve. Although the form does not change with time, the variance $\sqrt{2Dt}$ does and

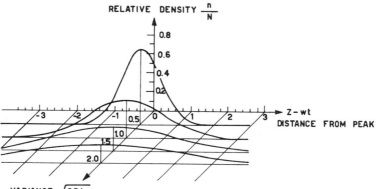

Fig. 3.3.1. Relative density of a drifting and diffusing electron swarm.

then the density distribution decreases in height and extends farther along the z axis as time goes on (Fig. 3.3.1). Moreover, the peak of the distribution moves along the z axis with drift velocity w.

Therefore, whereas the electron drift velocity can be obtained by making measurements of the velocity of the electron distribution peak, the pertinent accuracy will depend on the width of the distribution. This width Δz can be defined as the distance between the half-maximum density points:

$$\exp\left[-\frac{(\Delta z/2)^2}{4Dt}\right] = \frac{1}{2}$$

and then is given by:

$$\Delta z = 4\sqrt{\ln 2 \cdot Dt}$$

When $t = t_a$, the arrival time at the end of the drift region [eq. (3.2.2)], the fractional distribution width becomes:

$$\frac{\Delta z}{d} = 4\sqrt{\ln 2(D/wd)} \tag{3.3.4}$$

It appears that in drift time measurements the relative error due to diffusion will decrease when the dimensionless quantity D/wd decreases. This quantity can also be written as $D/\mu V$, where $V = Ed$ is the voltage applied across the drift region; for a given E/p, the characteristic energy D/μ is fixed, and then to improve accuracy we need larger voltages, that is, larger pd [$= V/(E/p)$] values. To do so at high values of E/p, however, may require values of pd approaching or exceeding that for electrical breakdown.

For a quantitative evaluation of the error occurring when there is some kind of electrode at $z = d$, a more refined analysis is required and methods a, c, d, and e described in Section 2 have to be separately considered. This will be done in the next paragraphs, but after reconsideration of the basic equation (3.3.3) to include also the possibility of significant deviations from the assumption of a uniform equilibrium energy distribution within the electron pulse. In fact, the correct energy distribution for the present problem is given by (2.3.70), and the electron density as a function of z and t must be rigorously derived by integrating (2.3.70) according to relationship (2.3.26). When this is done, we obtain the same expression (3.3.3) if we identify ω_1 as w and ω_2 as D; ω_1 defined by (2.3.71) is effectively equal to w defined by (2.4.20), but ω_2 defined by (2.3.72) is equal to D defined by (2.4.18) only when $\varphi(v)$ is negligible. Equation (2.3.73) indicates that, at a given E/p, $\varphi(v)$ is proportional to $1/p$; therefore expression (3.3.3) will be correct only in the high-pressure limit, whereas in general it will be better to use a similar expression:

$$ n = \frac{N}{2\sqrt{\pi D_L t}} \exp\left[-\frac{(z - wt)^2}{4 D_L t}\right] \tag{3.3.3'} $$

where D_L ($= \omega_2$) is a new coefficient for diffusion parallel to the drift field, called the longitudinal diffusion coefficient and computed by means of (2.3.72). Parker and Lowke (1969), who derived eq. (3.3.3'), also determined, by considering $f^0(v, z, t)$ solutions of higher-order approximation than (2.3.70), the corresponding higher-order shape characteristics of the pulse; such further refinements are useless here, however, since our present purpose is only to evaluate the transit time errors due to the diffusive spreading of the electron swarm. Since the above analysis shows that diffusion parallel to the electric field takes place according to the diffusion coefficient D_L, it is also appropriate to replace whenever possible in future developments the basic equations (3.3.1) and (3.3.2) with the new forms:

$$ \frac{\partial n}{\partial t} = -w \frac{\partial n}{\partial z} + D \nabla_t^2 n + D_L \frac{\partial^2 n}{\partial z^2} \tag{3.3.1'} $$

$$ \frac{\partial n}{\partial t} = D_L \frac{\partial^2 n}{\partial z^2} - w \frac{\partial n}{\partial z} \tag{3.3.2'} $$

where ∇_t^2 is the transverse part of the ∇^2 operator.

The presence of a plane collector or of an absorbing shutter imposes the boundary condition $n = 0$ at $z = d$, and this can be satisfied for all times by simply introducing into distribution (3.3.3') an image term of appropriate strength, which represents a negative density packet of electrons generated at $z = 2d$. The density distribution is then given by:

$$n = \frac{N}{2\sqrt{\pi D_L t}} \left\{ \exp\left[-\frac{(z - wt)^2}{4D_L t} \right] - \exp\left[\frac{wd}{D_L} - \frac{(z - 2d - wt)^2}{4D_L t} \right] \right\} \quad (3.3.5)$$

The resulting current density across the drift region is equal to:

$$i = \frac{ew}{d} \int_{-\infty}^{d} n \, dz \quad (3.3.6)$$

The lower limit of integration is set equal to $-\infty$ instead of zero to compensate for the fact that distribution (3.3.5) does not satisfy the condition $n = 0$ at $z = 0$ for all $t > 0$, as it should in the presence of a cathode or of an absorbing shutter at the electron source location. On the other hand, the use of this condition would introduce mathematical difficulties into the solution, without significantly improving the physical value of the final result we are seeking here.

Substituting distribution (3.3.5) into eq. (3.3.6) yields, after integration:

$$i = \frac{ewN}{2d} \left\{ \left[1 + \mathrm{erf}\left(\frac{d - wt}{2\sqrt{D_L t}} \right) \right] - \left[1 - \mathrm{erf}\left(\frac{d + wt}{2\sqrt{D_L t}} \right) \right] \exp\left(\frac{wd}{D_L} \right) \right\} \quad (3.3.7)$$

where:

$$\mathrm{erf}\, x = \frac{2}{\sqrt{\pi}} \int_{0}^{x} \exp(-u^2) \, du \quad (3.3.8)$$

The maximum current occurs abruptly at $t = 0$ and is equal to the value given by eq. (3.2.1).

For the arrival time t_a of the electron burst at the collector the method based on the observation of the collector-induced current uses the time at which the current (3.3.7) drops to half of its maximum value. Equation (3.2.2) suggests writing:

$$t_a = \frac{d}{w} (1 - r) \quad (3.3.9)$$

so that r represents the fractional change in pulse length due to the combined perturbation effects of the electron diffusion and of the collector boundary presence. If $|r| \ll 1$, as it should be in a well-designed experiment, from eq. (3.3.7) and the arrival time definition the following equation for computing r is obtained:

$$\mathrm{erf}[(\sqrt{wd/D_L})(r/2)] - [1 - \mathrm{erf}(\sqrt{wd/D_L})] \exp\left(\frac{wd}{D_L} \right) = 0 \quad (3.3.10)$$

The following approximations can be made:

$$\text{erf } x = \begin{cases} \dfrac{2}{\sqrt{\pi}}\, x & \text{when } |x| \ll 1 \\[2mm] 1 - \dfrac{1}{\sqrt{\pi}}\dfrac{e^{-x^2}}{x} & \text{when } x > 2 \end{cases} \tag{3.3.11}$$

Since, under the usual experimental conditions:

$$|r| \ll 2\sqrt{D_L/wd} < 1 \tag{3.3.12}$$

introduction of the above approximations into eq. (3.3.10) leads to the solution:

$$r = \frac{D_L}{wd} \tag{3.3.13}$$

Therefore the drift velocity computed from transit time measurements using formula (3.2.3) is not exactly the true drift velocity w, but is related to this velocity as follows:

$$\frac{d}{t_d} = \frac{w}{1 - r} \simeq w(1 + r) \tag{3.3.14}$$

According to eq. (3.3.13), the value of the drift velocity computed from (3.2.3) is then too high, and the relative error is D_L/wd.

Since the error D_L/wd can be written as $D_L/\mu V$, we can state that the error is given by the ratio of the characteristic energy D_L/μ of the electron swarm in electron volts to the drift region voltage in volts. This form is useful for computations. For instance, for measurements at an energy of 1 eV with an error less than 1%, it is required to use a voltage larger than 100 V; if for a particular gas the 1-eV energy is attained at $E/p = 10$ V cm^{-1} torr^{-1}, then the experiment has to be performed under conditions such that $pd > 10$ cm torr.

If the r value (3.3.13) is substituted into (3.3.9), we can immediately verify that the slope of the transit time versus drift distance curve remains equal to the reciprocal of the drift velocity, independently of diffusion effects. Here we have another good reason for the use of tubes with a variable drift distance.

Let us consider now the case of an absorbing shutter in front of the collector. The shutter is made periodically open for a very small interval of time τ, so that the electron density distribution (3.3.5) is not modified by this process. The number N_c of electrons per unit area, which at time t pass across the shutter, located at $z = d$, and arrive at the collector, is then:

$$N_c = \tau\Gamma_{z=d} = \tau\left(wn - D_L\frac{\partial n}{\partial z}\right)_{z=d} \tag{3.3.15}$$

Because of the boundary condition $n = 0$ at $z = d$:

$$N_c = -\tau D_L \left(\frac{\partial n}{\partial z}\right)_{z=d} \tag{3.3.16}$$

hence:

$$N_c = \frac{\tau N d}{2t\sqrt{\pi D_L t}} \exp\left[-\frac{(d - wt)^2}{4 D_L t}\right] \tag{3.3.17}$$

The collector current density efN_c, where f is the pulse repetition frequency, will be maximum when:

$$\frac{d(fN_c)}{dt} = 0 \tag{3.3.18}$$

The pulse sequence delay t_d at which this happens can be written as:

$$t_d = \frac{d}{w}(1 - r) \tag{3.3.19}$$

and the assumption $|r| \ll 1$ may be made again.

When f and τ are constant, as in Phelps, Pack, and Frost transmission experiments, solution of eq. (3.3.18) with (3.3.17) and (3.3.19) yields:

$$r = 3\frac{D_L}{wd} \tag{3.3.20}$$

A more sophisticated expression for r, to be used when the assumption $|r| \ll 1$ cannot be made, is given by Whealton and Woo (1969). The delay difference for two drift spacings is:

$$t_2 - t_1 = \frac{d_2}{w}\left(1 - 3\frac{D_L}{wd_2}\right) - \frac{d_1}{w}\left(1 - 3\frac{D_L}{wd_1}\right) = \frac{d_2 - d_1}{w} \tag{3.3.21}$$

This equation is the same as (3.2.13); hence we can state that in this case the determination of the drift velocity should be subject to a much smaller error than in the single-spacing case.

When the method of Bradbury and Nielsen is used, the applied signal frequency f replaces time t in (3.3.17) according to the relationship [see eq. (3.2.11)]:

$$t = \frac{K}{2f} \tag{3.3.22}$$

Moreover $\tau \propto 1/f$ and $N \propto 1/f$, since both shutters are open for a fixed fraction of the applied signal period. As a function of frequency the maximum current will then occur when:

$$\frac{d(fN_c)}{df} = \frac{d}{df}\left\{\sqrt{f}\exp\left[-\frac{f}{2KD_L}\left(d - \frac{Kw}{2f}\right)^2\right]\right\} = 0 \tag{3.3.23}$$

Let us write the corresponding frequency as:

$$f = K \frac{w}{2d} (1 + r) \qquad (3.3.24)$$

where r is a quantity much less than unity in accordance with eq. (3.2.11). Solution of (3.3.23), neglecting terms in r^2, yields:

$$r = \frac{D_L}{wd} \qquad (3.3.25)$$

The drift velocity can be computed from the frequencies f_1 and f_2 of two successive maxima, according to:

$$w = 2d \frac{f_2 - f_1}{1 + r} \simeq 2d(f_2 - f_1)(1 - r) \qquad (3.3.26)$$

Therefore r represents the fractional error made when eq. (3.2.12) is used for the computation of the drift velocity from the measured frequencies of current maxima.

When the rejection mode of operation of a double-shutter system is used, the grids in front of the collector are transparent most of the time and the zero density boundary moves to the collector position $z = h$. The number N_c of electrons collected by a shutter grid, pulsed at time t and located at $z = d$, is given again by eq. (3.3.15) if zero electron transmission during the pulse is assumed. The density n to be substituted into this equation is given by (3.3.5), where d has now been replaced by h. The resulting expression for N_c turns out to be much more complicated than (3.3.17).

Since in this case the minimum collector current condition is given by the same eq. (3.3.18) and position (3.3.19) can again be made, it is possible to derive from the known expression of N_c a solution for the fractional error r, equivalent to (3.3.20). The result is:

$$r = \left\{ 3 - \frac{1 - \left[1 - 2 \dfrac{h(h-d)^2}{d^3} \dfrac{wd}{D_L} \right] \exp \left[- \dfrac{h(h-d)}{d^2} \dfrac{wd}{D_L} \right]}{1 + \dfrac{h-d}{d} \exp \left[- \dfrac{h(h-d)}{d^2} \dfrac{wd}{D_L} \right]} \right\} \frac{D_L}{wd} \qquad (3.3.27)$$

In Fig. 3.3.2 we have plotted $r(wd/D_L)$ as a function of d/h for $D_L/wd = 0.1$ and $D_L/wd = 0.01$.

In general, then, the quantity rd is not independent of d, as it was in the transmission mode of operation; therefore the error does not disappear, as in (3.3.21), when the delay difference for two drift spacings is calculated. This may imply a larger error when the rejection mode of operation is adopted

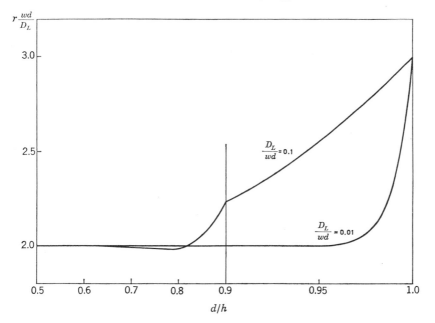

Fig. 3.3.2. Plot of the fractional error r to be expected when the Westinghouse method, the rejection mode of operation, is used.

and eq. (3.2.13) is used. The actual value for this error can be easily computed for each individual case by means of (3.3.27). However, Fig. 3.3.2 shows that there is a large range of d/h values over which rd is practically independent of d, so that, when the experiment is performed accordingly, the error for the rejection mode may be as small as for the transmission mode, double-spacing case.

Finally, when the methods of Stevenson and Hurst (TOF) and of Comunetti and Huber are used, the measured distribution of the arrival times of the electrons at the collector will be proportional to the electron flow $(\Gamma)_{z=d}$ and hence to $(\partial n/\partial z)_{z=d}$, as given in eq. (3.3.17). It follows that the treatment of this case will be the same as that previously presented for the Phelps, Pack, and Frost transmission method: the time of the peak of the distribution can be written as (3.3.19), where r is given by (3.3.20). Use of (3.3.27) appears, on the other hand, more appropriate for computing the fractional error r in the case of the experiment of Breare and Von Engel.

The discussion of the diffusion effects has been based entirely on eq. (3.3.5), which, as already remarked, does not satisfy the condition $n = 0$ at $z = 0$ for all $t > 0$. In other words we have thoroughly neglected the fact that some of the electrons will diffuse backward from the drifting swarm and will be swept out by the cathode or by the closed source shutter. Because of the difficulty of handling the problem rigorously, Lowke performed only an approximate evaluation of the effect and found that the fractional error due to back diffusion is given by the general expression:

$$r = 2\frac{D_L}{wd} \tag{3.3.28}$$

This contribution simply adds to the r values previously obtained in the analysis of the other diffusion effects for the various methods. Back diffusion might also be responsible for the initial small drop in the induced collector current observed by Chanin and Steen (1964) in cesium.

Lowke also investigated the contribution to the error due to the finite time that the collector shutter is open in actual experimental conditions, when the method of Bradbury and Nielsen is used. For this problem he performed numerical computations and showed how the predicted error decreases when the shutter is open for an increasing fraction of the period.

A similar effect is expected if the generation of the burst of electrons takes a finite length of time. This condition has not been analytically investigated, and no data are available from the literature. However, we may argue that for this reason previous values of r yield overestimates of the actual fractional errors.

It is also worth remarking that this situation is most important when the Bradbury-Nielsen method is used, since reducing the shutter-open time requires in this case a larger voltage on the grids, and many obvious good reasons (breakdown, field distortions, etc.) forbid increasing this voltage too much. On the other hand, light pulses and gates applied to grids, as required by the other methods, do not suffer, in general, from the same limitations and can usually be made sufficiently short to justify neglecting the finite duration of the shutter-open time.

All preceding considerations apply only to the methods in which an electron pulse is produced at the cathode boundary of the drift region. The effect of the electron diffusion in the alternative case, when at $t = 0$ electrons are produced throughout the drift region [see method b in Section 3.3], was discussed by Fischer-Treuenfeld (1965) for the conditions of his experiment. The approach that he used for this evaluation is based on some crude and questionable approximations; however, the experimental data fit well with the predicted shape of the curve of photomultiplier current maximum versus time, and this provides some confidence concerning the value of the theoretical analysis.

Fischer-Treuenfeld assumes that the current across the gap is still given by eq. (3.2.6), but that the preionization density instead of the distribution (3.2.5) has the distribution which results from the drift and diffusion motion of an electron swarm, which initially has a uniform density n_0 for $z < d$, and 0 for $z \geqslant d$. Solving eq. (3.3.2′) with this initial condition yields, over the range of interest $0 < z < d$:

$$n(z, t) = \frac{n_0}{2}\left[1 + \text{erf}\left(\frac{d + wt - z}{2\sqrt{D_L t}}\right)\right] \qquad (3.3.29)$$

An analytical expression like (3.2.10) cannot be obtained using this distribution, and the problem must be handled numerically.

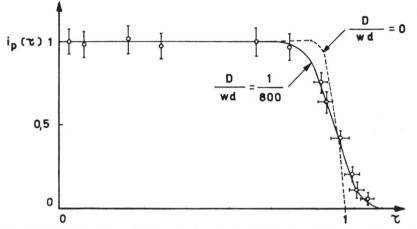

Fig. 3.3.3. Fischer-Treuenfeld's method: calculated behavior of the normalized photomultiplier current for $\alpha d = 9.5$ and $D_L/wd = 1/800$ and experimental points for nitrogen at $E/p = 1$ V cm^{-1} torr^{-1} and $T_g = 293°$K [after Fischer-Treuenfeld (1965)].

Here the maximum current $i_p(\tau)$ becomes a function of both dimensionless quantities αd and D_L/wd. In Fig. 3.3.3 is shown the computed behavior of $i_p(\tau)$ for $\alpha d = 9.5$ and $D_L/wd = 1/800$, compared to the no-diffusion case $(D_L/wd = 0)$; a set of experimental data for nitrogen at $E/p = 1$ V cm^{-1} torr^{-1} can be fitted well over the curve which includes the effects of diffusion. Fischer-Treuenfeld's data have all been handled in this way, so that his transit time determinations are already corrected for diffusion effects.

Diffusion effects may be of importance also in the experiments with alternating fields, since they make less definite the cutoff point at which current vanishes. Figure 3.3.4 shows a typical curve of the dc current component versus frequency, where one can observe around the breaking point the rounding off of the curve caused by diffusion.

B. Let us evaluate now the effect of attachment when the experiment is performed in gases, where this electron loss process takes place. In this case eq. (3.3.1) is modified by the addition to its right-hand side of the term $-\bar{\nu}_a n$,

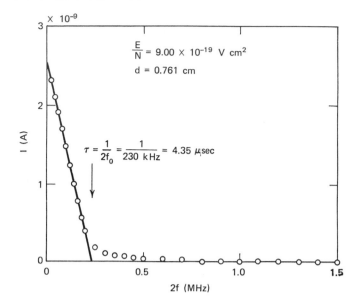

Fig. 3.3.4. Typical curve of average anode current versus twice the square-wave frequency, when the alternating field method is used. The rounding off of the curve in the region around the breaking point is caused by diffusion [after Nolan and Phelps (1965)].

where $\bar{\nu}_a$ is the attachment frequency; it is easy to see that the density distribution $n(z, t)$ for the various cases is now given simply by the distribution derived in the absence of attachment multiplied by the factor $\exp(-\bar{\nu}_a t)$. In the following paragraphs more definite conclusions will be drawn from this result for all methods that have actually been applied to measurements in attaching gases.

Since the collector current (3.3.7) must also be multiplied by the factor $\exp(-\bar{\nu}_a t)$, the pulse top, when the induced current is observed, will be falling down with the constant slope $\bar{\nu}_a$ on a current logarithmic scale. Hence the attachment frequency can be determined directly from measurements of this slope, whereas the electron transit time is given by the half-current points, the reference unity current being provided by the linearly extrapolated pulse top (Fig. 3.3.5a).

When the integrated current pulse is observed, as many workers do, the negative slope of the pulse top implies a continuously decreasing rate of rise of the observed integrated current (Fig. 3.3.5b). In this case the exact determination of the transit time of the electrons is more difficult, since the sudden drop in the rate of rise, which is usually observed when the electron burst reaches the collector, takes place here on a curve that already shows a

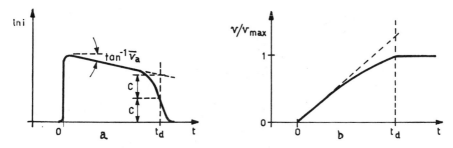

Fig. 3.3.5. Effects of electron attachment on the transit time measurements: (a) collector current waveform as induced by the drifting electron pulse; (b) voltage waveform when capacitive integration of the current is performed.

continuously diminishing slope due to the attachment process. When the effect is observed to be particularly large, as in water vapor at E/p between 15 and 30 V cm^{-1} torr^{-1}, according to Ryzko (1966) it is better to determine $\bar{v}_a$ from the pulse front shape and to compute w as the ratio $\bar{v}_a/\eta$, where η is the probability of attachment per unit length, known independently, for instance, from the integrated maximum value of the current pulse (eq. 3.2.4):

$$v_{\max} = \frac{1}{C} \int_0^{d/w} \frac{ewN}{d} \exp(-\bar{v}_a t)\, dt = \frac{eN}{C} \frac{1 - \exp(-\eta d)}{\eta d} \qquad (3.3.30)$$

In Fischer-Treuenfeld's method the current $i_p(\tau)$ will also show a superimposed decay as $\exp(-\bar{v}_a t)$ over most of its range [Hessenauer (1967)]. Figure 3.3.6 shows the behavior in a mixture of nitrogen and oxygen; the initial slope of the curve yields $\bar{v}_a$.

When the various shutter methods, as well as methods d, e, and f of Section 3.2, are used, the problem of attachment can be handled in exactly the same way as reported for pure diffusion, except that now proper attention must be given to the presence of the factor $\exp(-\bar{v}_a t)$ in the density distribution expressions. In particular Pack and Phelps (1966) have established that, in the first-order approximation, attachment makes an additional contribution to the fractional error (3.3.20) equal to:

$$r = \frac{2D_L}{w^2} \bar{v}_a = 2 \frac{D_L}{w} \eta \qquad (3.3.31)$$

Since error (3.3.31) is independent of d, the delay difference for two drift spacing is also affected by the same fractional error (3.3.31). This was not the case, however, for diffusion error (3.3.20), which is inversely proportional to d and thus disappears from the delay difference expression (3.3.21).

In the present analysis, as stated at the beginning, we disregard the cases of E/p values so large that ionization becomes of importance. However, it

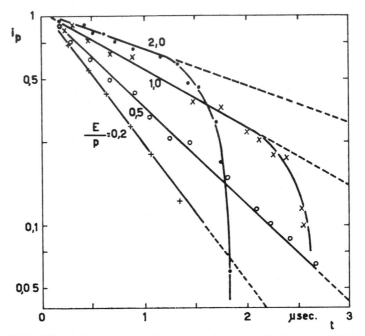

Fig. 3.3.6. Effect of electron attachment on the normalized photomultiplier current, measured at different E/p values (V cm^{-1} torr^{-1}) using the method of Fischer-Treuenfeld [mixture of nitrogen and oxygen (10.5% O_2) at $p = 200$ torr, $T_g = 293°$K]. At $E/p = 0.2$ and 0.5 the slope is so large that the electron transit time cannot be measured [after Hessenauer (1967)].

is easy to see that, when some ionization is present, it can be taken into account simply by replacing the frequency $\bar{\nu}_a$ in the preceding formulas with the quantity $(\bar{\nu}_a - \bar{\nu}_i)$, $\bar{\nu}_i$ being the electron ionization frequency. In this case, when $\bar{\nu}_i > \bar{\nu}_a$, the induced current will increase exponentially with time, and the electron swarm will convert into a growing avalanche. Thus, at very high ionization levels, the electron transit time will be approximately given by the time of the collector current peak (end of the pulse). Brambring (1964) and Schlumbohm (1965), who performed extensive measurements in various gases using the induced current method at E/p values ranging from some hundreds to several thousands of volts per centimeter per torr under dominant ionization conditions, have worked out different formulas assuming that the additional term for ionization on the right-hand side of eq. (3.3.1) is not $\bar{\nu}_i n$, but is a quantity proportional to the electron flow Γ. Appropriate criticisms of this approach, which has no valid justification, have been published by Crompton (1967) and by Burch and Huxley (1967).

The foregoing discussion on attachment and ionization assumes no appreciable detachment and no significant positive-ion current during the electron transit time; this is usually the case, since the negative ions have long enough lives and positive-ion drift velocities are much smaller than electron ones. The interested reader may, however, find in a paper by Ryżko (1966) a discussion of more general cases, where attachment, detachment, ionization, and positive-ion current are taken into consideration in evaluating the behavior of the integrated induced current in the region of importance for electron drift velocity determinations.

C. The assumption of a perfectly plane geometry and of a uniform electric field implies the existence of infinite plane electrodes and of an electron source which generates a transversally uniform density, that is, a density independent of x and y (see Fig. 3.2.1). In actual experiments, however, electrodes of finite size and transversally nonuniform electron distributions are used.

When plane parallel electrodes of finite size are used, the field may deviate significantly from uniformity in the drift region; the situation is easily improved, however, by the use of a Rogowski profile for the electrode shape, or by the addition of guard rings, properly positioned in the drift space and charged to appropriate potentials. Lowke (1963) investigated in a Bradbury-Nielsen tube the effect of electric field nonuniformity on drift velocity determinations, introducing substantial changes in the potentials of some of the guard rings. In this way he found, both experimentally and theoretically, a remarkably small influence of pronounced field distortions, so that in any well-designed experiment errors due to field distortions can probably be considered negligible.

The same result applies to field distortions due to the voltage between the shutter wires. It should be noted that the largest voltages are used in the Bradbury-Nielsen method, but that fortunately this method uses ac voltages, so that the effect of field distortions on any electron pulse is largely compensated on the successive pulse by the inversion of the grid voltage polarity.

The influence on the electron swarm motion of the finite size of the chamber and of the transverse nonuniformity of a narrow electron packet injected into the drift region can be determined by solving the equation of motion (3.3.1') with a nonuniform initial distribution of electrons in the $z = 0$ plane and introducing a cylindrical wall boundary. Let $F(\rho)$ be the initial electron distribution, where ρ is the perpendicular distance from the drift tube axis; we write the distribution as a series in the form:

$$F(\rho) = \sum_{i=1}^{\infty} C_i J_0\left(\frac{\sigma_i \rho}{R}\right) \qquad (3.3.32)$$

where R is the radius of the cylindrical boundary and σ_i are the roots of the J_0 Bessel function. The solution of eq. (3.3.1') equivalent to (3.3.3') is then:

$$n = \frac{1}{2\sqrt{\pi D_L t}} \sum_{i=1}^{\infty} C_i J_0\left(\sigma_i \frac{\rho}{R}\right) \exp\left[-\frac{(z - wt)^2}{4 D_L t} - \frac{\sigma_i^2 D}{R^2} t\right] \quad (3.3.33)$$

Since (1) the coefficient C_1 is usually larger than the other coefficients, the electron swarm being denser near the tube axis, and (2) higher-order terms decay faster with time, the σ_i values becoming larger with increasing i, we can restrict this discussion to the simplest case $C_1 \neq 0$, $C_i = 0$ for $i > 1$. In this case eq. (3.3.33) shows that the electron distribution is changing with axial position and time, as it was in the presence of attachment; this result is very simple to understand, since lateral diffusion introduces a linear loss process for electrons, like the one due to attachment. Therefore the results previously obtained for the attachment case can be applied here, substituting $\sigma_1^2 D/R^2$ for $\bar{\nu}_a$. In particular, according to (3.3.31), the contribution to the fractional error (3.3.20) will be:

$$r = 2\sigma_1^2 \frac{D D_L}{(wR)^2} = 11.6 \frac{D D_L}{(wR)^2} \quad (3.3.34)$$

D. The last problem to be considered concerns the energy distribution of the electrons in the drift region, for which we have assumed equilibrium values. In general this is not correct near the generation region of the electrons, however, since electrons will have to travel some distance along the field direction before steady state energy conditions are attained. We can obtain a significant estimate of this distance, if we evaluate the distance required for the electron swarm mean energy to reach its steady state value. Of course, the error due to this effect will be most important when electrons are generated inside the drift region or on one of its boundaries; however, in these cases the error may be basically avoided any time the chamber allows differential measurements to be performed, using various drift distances.

Let us integrate energy equation (2.5.12), which in the present case can be written in the simple form:

$$\frac{d\bar{u}}{dt} = ewE - (\bar{u} - \tfrac{3}{2}kT_g)\nu_u$$

The following assumptions are made:

a. The energy dependence of the quantities w and ν_u is neglected, as for these quantities we take the equilibrium values corresponding to the applied E/p conditions.

b. At time $t = 0$ the electrons are released by the source with thermal energy $3kT_g/2$.

The solution is:

$$\bar{u} = \tfrac{3}{2}kT_g + \frac{ewE}{\nu_u}\,[1 - \exp{(-\nu_u t)}] \tag{3.3.35}$$

After time t the electron excess energy will have reached the fraction:

$$\xi = 1 - \exp{(-\nu_u t)} \tag{3.3.36}$$

of its steady state value. Therefore the distance traveled before attaining the fraction ξ of the steady state value of the excess energy has to be:

$$s = wt = \frac{w}{\nu_u}\ln{(1 - \xi)^{-1}} \tag{3.3.37}$$

Substituting for ν_u the approximate effective value given by (2.6.7) and (2.6.8):

$$\nu_u = \tfrac{2}{3}\nu_u^* = \frac{2}{3}\frac{wE}{(D/\mu) - (kT_g/e)}$$

and multiplying both sides by p, we obtain from eq. (3.3.37):

$$ps = \frac{3}{2(E/p)}\left(\frac{D}{\mu} - \frac{kT_g}{e}\right)\ln{(1 - \xi)^{-1}} \tag{3.3.38}$$

Hence, to attain $\xi > 0.99$ in a case where the characteristic energy D/μ is 1 eV and $E/p = 10$ V cm^{-1} torr^{-1}, a typical situation previously considered also for the diffusion errors, the electrons must travel a distance s such that $ps > 0.69$ cm torr. For correct measurements, then, the quantity pd must be many times larger than this value; we recall that, when considering the induced collector current, neglecting diffusion errors under the same conditions was found to require $pd > 10$ cm torr.

For more accurate determinations of the preceding distances the interested reader is referred to a group of papers by Braglia and Ferrari (1970, 1971). These authors have solved the transport equations and given formulas for the determination of the times and distances required to attain steady-state energy conditions when all the electrons have initially the same speed.

As we have seen, for a given tube and a given E/p ratio, errors due to diffusion and to energy transients become smaller as the gas pressure increases. This has been verified experimentally by various authors; a typical case is shown in Fig. 3.3.7, where Lowke's (1962) data on the drift velocity in helium, measured at $E/p = 0.4$ V cm^{-1} torr^{-1} using the method of Bradbury and Nielsen, are shown as a function of the gas pressure. Experimentally, it is thus possible to determine the value of the drift velocity, free of the diffusion

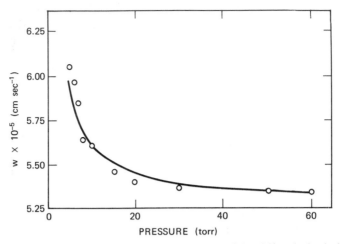

Fig. 3.3.7. Pressure dependence of the measured value of the drift velocity in helium, as measured using the method of Bradbury and Nielsen ($E/p = 0.4$ Vc m^{-1} torr^{-1}, $T_g = 293°$K) [after Lowke (1962)].

and energy transient errors, by measuring this velocity as a function of pressure and assuming as the correct value the asymptotic limit at high pressures.

3.4   TUBES AND EXPERIMENTAL TECHNIQUES USED FOR THE MEASUREMENT OF ELECTRON TRANSIT TIMES

In this section we shall describe briefly some of the most modern versions of the tubes used to measure electron transit times and some related equipment and techniques.

A. As a typical example of a modern version of a chamber for measurements of the induced collector current we can consider the one used by Ryzko (1966) for determinations in water vapor and dry air. The chamber, made from a vertical glass cylinder 50 cm in diameter and 75 cm long, covered at both ends with steel plates, houses two copper electrodes with a Rogowski profile ($\psi = 120°$) and an overall diameter of 27 cm. The upper electrode is the cathode and can be moved vertically over a distance of 10 cm. For this purpose the cathode bracket is movable within a guide tube, affixed directly to the upper plate of the chamber, and is connected to a micrometer screw enabling the gap to be set to within 10 μ from the outside of the chamber. The lower electrode, the anode, can be placed parallel to the cathode by means of three nylon brackets on which the electrode is mounted.

Illumination of the cathode takes place through a quartz window, 5 cm in diameter, in the lower plate and then through a set of holes, each 0.75 mm in diameter, which occupy a 4-cm-diameter area at the center of the anode. Ultraviolet light photopulses of about 30 and 60 nsec are produced in a separate chamber by a spark in compressed nitrogen.

Worthy of mention also are two other tubes: one used by Allen and Prew (1970) for measurements at very high pressures, and another by Chanin and Steen (1964) for measurements at high temperatures in cesium (Fig. 3.4.1). The first tube consists of a conventional electrode system mounted in

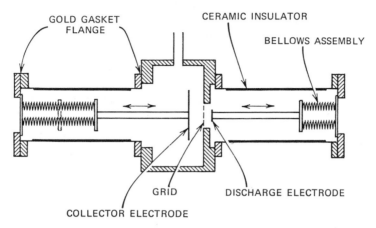

Fig. 3.4.1. Schematic drawing of the drift tube used by Chanin and Steen for measurements in cesium [after Chanin and Steen (1964)].

a pressure vessel, capable of withstanding pressures up to 1000 atm. The second tube is much smaller than Ryżko's one, so that it can be placed conveniently in an oven. Two interesting features are that electrons are generated by a very short discharge (~30 nsec) and that both drift and discharge distances can be varied from the outside of the oven, out to 5.0 and 2.5 cm, respectively, by means of micrometer-driven bellow assemblies, with an accuracy of 10 μ. The special construction of the tube has made it possible to perform measurements at temperatures up to 725°K.

B. As already indicated in the preceding sections, the most recent use of a discharge technique for detecting the presence of electrons in the drift region is due to Fischer-Treuenfeld (1965). A schematic of the tube used by this investigator is shown in Fig. 3.2.4.

The alpha-particle source is a well-screened and collimated polonium sample. The alpha-particle detector consists of a ZnS screen, a light guide, and a photomultiplier. The high voltage (60 kV maximum) pulse generator must have a very short rise time; for switching it, two triggerable sparks in compressed nitrogen have been used, and a 10-nsec rise time has been obtained. The minimum delay between the passage of the alpha-particle across the chamber ($d = 2.7$ cm) and the application of the high-voltage pulse to the electrodes is, in Fischer-Treuenfeld's experiment, 85 nsec. The pulse duration was chosen short enough to include the first current maximum, but no more than one of the successive discharge bursts, which could impair the life of the tube without providing any useful data for the experiment.

C. The method of Bradbury and Nielsen is presently used in all drift velocity determinations performed at Canberra by Crompton and his co-workers. A diagram of one of their recent drift tubes [Crompton, Elford, and Jory (1967)] is shown in Fig. 3.4.2.

Electrons are emitted by a heated platinum filament, properly outgassed by heating to dull red in an atmosphere of hydrogen for 1 hour; during operation the filament current is highly stabilized. But in subsequent measurements performed in parahydrogen [Crompton and McIntosh (1968)] it proved impossible to use a heated filament, since this rapidly converted parahydrogen into normal hydrogen; hence the filament was replaced by a silver-coated foil of americium 241, which provides electrons by volume ionization of the gas without conversion effects. In order to avoid at 77°K the small temperature gradients due to the heated filament, which may lead to significant errors in calculating the molecular number density from the gas pressure, the same source has also been used at Canberra in all low-temperature measurements performed since its introduction in 1968.

Thick cylindrical guard electrodes are used. This structure enables a highly uniform electric field to be generated, while at the same time it ensures a high degree of mechanical stability and a good screening of the drift space from stray electric fields exterior to the apparatus. If ΔV is the voltage difference between two adjacent electrodes, the potential ϕ of the point P (Fig. 3.4.3), at distances ρ from the tube axis and z from a reference plane passing through the center of an interelectrode gap, except for an inessential arbitrary constant, is:

$$\phi = \left[1 + 2 \sum_{i=1}^{\infty} \frac{\sin(\pi i g/l)}{\pi i g/l} \frac{\sin(2\pi i z/l)}{2\pi i z/l} \frac{I_0(2\pi i \rho/l)}{I_0(2\pi i R/l)} \right] \frac{z}{l} \Delta V \qquad (3.4.1)$$

where g is the gap spacing, l the distance between the midplanes of two adjacent electrodes, R the inner radius of the electrodes, and I_0 the modified Bessel function of zero order.

Using eq. (3.4.1), one finds that with $g/l \ll 1$ and $R/l = 3$ the potential is within 0.05% of the value corresponding to a uniform field at any point inside an axial cylindrical volume of radius $0.6R$. Therefore it is possible to

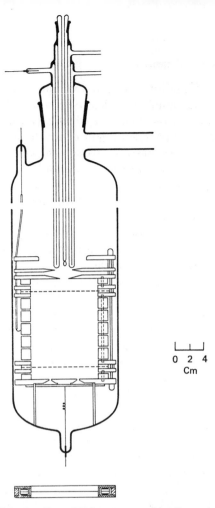

Fig. 3.4.2. Drift tube of the Bradbury-Nielsen type used by Crompton's group in their most recent measurements [after Crompton, Elford, and Jory (1967)].

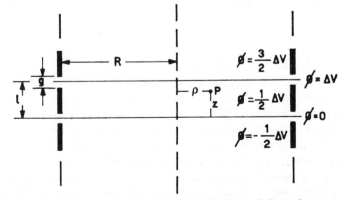

Fig. 3.4.3. Geometry of the thick cylindrical guard electrodes.

165

obtain a high degree of uniformity within a volume of adequate dimensions, without the use of a large number of electrodes or of a large diameter.

According to eq. (3.4.1), the planes at which z is a half-integer times the distance l are equipotential (ϕ independent of ρ); these are the midplanes of the electrodes and the planes passing through the centers of the gaps. The drift region can be terminated, therefore, without disturbing the field uniformity by replacing, as shown in Fig. 3.4.2, two of the standard electrodes by composite electrodes with the source and the collector shutters accurately positioned at their midplanes.

Each shutter, shown in the lower part of Fig. 3.4.2, consists of about 200 nichrome wires of 80-μ diameter with their centers accurately spaced 0.4 mm apart and mounted between two soda-glass annular rings. The electrodes are made of copper and are separated by glass spacers 0.5 mm thick. All surfaces exposed to the electron stream are gold coated, by electroplating (guard electrodes) or by vacuum deposition (shutter wires). Because of the possibility of such effects as stray contact potential differences, usually no measurement is taken at values of E less than $2 - 3$ V cm^{-1}.

The voltages applied to the shutters have amplitudes from a few tenths of a volt up to a few tens of volts; frequencies are in the range 1 kHz to 1 MHz. Precautions must be taken to ensure that the voltages do not change with frequency and that the phase difference between the voltages applied to the two shutters is less than 1°. Electron currents range from 10^{-11} to 10^{-13} A. Frequencies are usually read with a counter, and for better accuracy the positions of current maxima are determined from several frequency measurements of equal current points on both sides of each maximum.

The tube is usually immersed in a water bath, the temperature of which remains stable at about 293°K to within 0.1°C per hour. Similar tubes, made as small as possible, have also been operated in a liquid-nitrogen bath for measurements at about 77°K [Crompton, Elford, and McIntosh (1968)].

D. A schematic of the Westinghouse tube according to the original design by Phelps, Pack, and Frost (1960) is shown in Fig. 3.4.4. The latest tubes of this type have a ceramic terminal assembly, and a second evacuated chamber around the tube end, in order to maintain high leakage resistance between the terminals and to reduce noise when the tube is operated in liquid baths at temperatures different from room temperature. All of the electrodes and shields, except the grids, are made of gold-plated Advance metal; the grids, on the other hand, are 75-μ gold-plated molybdenum wires, whose spacing is chosen on the basis of a compromise between the requirement of a large electron transmission and that of efficient shutter action (3.5 mm is the preferred spacing).

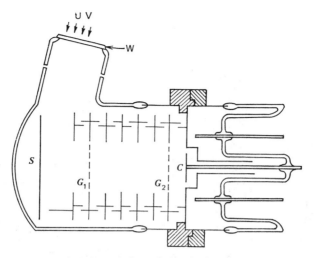

Fig. 3.4.4. Original Westinghouse design of the double-shutter drift tube (S = photocathode source, C = collector electrode, G_1 and G_2 = grids of the electron shutters) [after Phelps, Pack, and Frost (1960)].

Except in the earliest experiments, Phelps and his coworkers have used a hot-cathode hydrogen lamp as a pulsed light source. Light pulses and grid voltage gates $1 - 2 \, \mu$sec long, repetition rates of few thousand hertz, and drift lengths of 2.54 and 5.08 cm are typical of the experiments performed with this technique. Shorter pulses (0.1 μsec wide) have more recently been used by Prasad and Smeaton (1967) to improve resolution at high E/p; these authors have adopted a technique similar to the original one by Phelps, Pack, and Frost, but with a heated filament instead of a photocathode source.

E. In Fig. 3.4.5 the tube used by Christophorou, Hurst, and Hendrick (1966), who adopted the TOF method for drift velocity measurements in ethylene, is shown. The distance between the two electrodes is 9 cm; the alpha-particle source and the pulse height output are used normally for electron-capture studies and here for monitoring gas purity during the experiment.

For drift velocity measurements, however, electrons are ejected from the photosurface of the E.M. plate by ultraviolet radiation (down to 2000 Å) from a pulsed xenon flash tube, located outside the chamber. The ultraviolet pulse (rise time of approximately 0.35 μsec) is detected by a photodiode, which triggers an oscilloscope; swarm transit times are estimated by displaying on this oscilloscope the gas-discharge tube G.M. output. A guard ring G.R. around the earthed collector plate C ensures field uniformity. The chamber

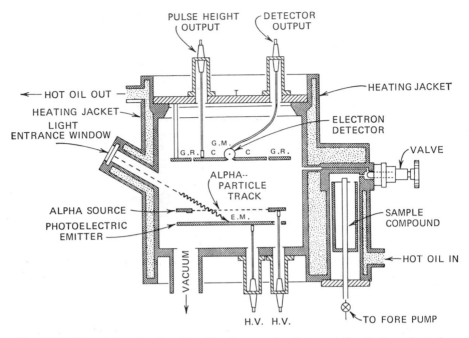

Fig. 3.4.5. Tube designed and used by Hurst's group for electron drift velocity and attachment coefficient measurements [after Christophorou, Hurst, and Hendrick (1966)].

is gold plated, and its temperature can be controlled by means of a heating bath to within 0.5°C over the range from 298 to 473°K.

Hamilton and Stockdale (1966), using a very similar tube and light pulses, with a rise time of $<0.2\ \mu$sec and a half-width of 1 μsec, measured transit times in nonattaching gases with an accuracy of $\pm 0.1\ \mu$sec in the range 2–50 μsec.

Similar tubes, but of larger size [e.g., $d = 27$ cm in the Hurst and Parks (1966) experiment], have been employed in cases in which the complete determination of the electron transit time distribution was desired; in fact, wider distributions correspond to larger transit times, as shown in Fig. 3.3.1. In these tubes field uniformity is maintained by a set of field rings, brought to the appropriate potentials through a voltage divider. Furthermore, Hurst and Parks (1966) used as detector a specially designed Geiger-Müller counter, with 10 equally spaced 0.8-mm-diameter holes and with an appropriate mechanism for selectively covering the holes, in order to control the probability of an electron entering the counter (we recall that this probability per swarm must be much less than 1). In other experiments Wagner, Davis, and Hurst (1967) used as detector in a differentially pumped region (swarm region 1–200 torr, detection region 1 × 10^{-6} torr) an RCA C7185J, 14-stage copper-beryllium dynode, electron multiplier; electrons pass from the drift chamber into a transition region (∼1 × 10^{-3} torr) through a 1.3-mm aperture and then are drawn by means of a 350-V potential into the detection region through a 4-mm aperture. The authors claim to have obtained a resolution far superior to that attainable with the Geiger-Müller tube.

In these experiments the electron transit time distribution can be determined accurately only if the instrumental noise is properly taken into account. This noise is due to the light source and to all the delays and fluctuations of the electrons inside the detector tube. When the Geiger-Müller counter is used, the instrumental noise distribution may thus be identified with the distribution of count rates as a function of time, which is observed in the presence only of photoelectrons emitted by scattered light inside the Geiger-Müller tube alone; in order to perform this type of measurement a reverse field is then applied to the swarm region to prevent electrons from entering from outside the tube. When the electron multiplier is used, the noise distribution is simply obtained by measuring the detector count rate versus time without gas in the drift chamber. If $T_0(t)$ is the unit-normalized instrument noise distribution, the actual distribution of the electron transit times $E(t_d)$ and the measured distribution $E'(t_d)$ are related by a convolution integral:

$$E'(t_d) = \int_0^{t_d} E(t)T_0(t_d - t)\, dt \qquad (3.4.2)$$

By means of unfolding numerical techniques or of appropriate first-order approximations, $E(t_d)$ can then be computed.

F. A cross section of the drift tube and a block diagram of the associated circuitry used by Comunetti and Huber (1960) are shown in Figs. 3.4.6 and 3.4.7. The alpha-particles are generated by a polonium source and are detected by a scintillator-photomultiplier combination with a very fast response (rise time ∼8–10 nsec). The anode-collector structure is formed by a grid of phosphorous bronze wire and by the quartz window of the 51-UVP photomultiplier, coated with a conductive layer; the distance between the grid and the coated window is 1 mm, and the applied voltage between them is over 1000 V, as required for significant light emission by electron impact at gas pressures of a few hundreds torr.

The delay between the alpha-particle emission and the maximum of the light in the anode structure can be measured by displaying this light on an oscilloscope, triggered by the scintillator-photomultiplier output. Better accuracy is obtained, however, with the scheme of Fig. 3.4.7. The outputs of the scintillator and of the anode light photomultipliers enter separate threshold detectors, where they are converted into very short pulses (see

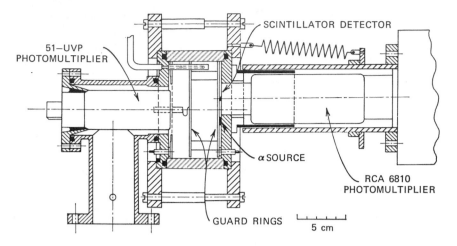

Fig. 3.4.6. Cross section of the Comunetti-Huber drift tube [after Comunetti and Huber (1960)].

Fig. 3.4.8); the time interval t_y between these pulses is measured by step delaying (maximum 9 μsec in 120 steps of 75 nsec each) the scintillator pulse, applying this delayed pulse and the anode pulse to a coincidence detector, where both pulses are first brought to a common 32-nsec length, and then finding which delay gives the maximum reading of a counter fed by the detector output. The transit time is obtained by adding to the measured delay t_y the correction:

$$\tau_y = 0.8\tau\sqrt{\ln (1/\nu)} \qquad (3.4.3)$$

where τ is the rise time (between 0.1 and 0.9 of the peak) of the anode light pulse, observed on an oscilloscope, and ν is the fractional threshold level adopted for this pulse. The time resolution in t_y measurements with the above system appears to be 75 nsec. The accuracy, however, is much better, since

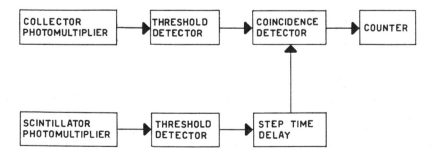

Fig. 3.4.7. Block diagram of the delay measuring system adopted by Comunetti and Huber.

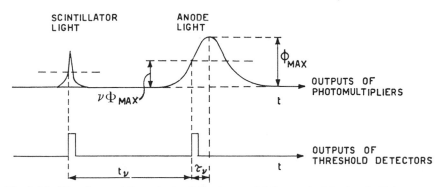

Fig. 3.4.8. Waveforms at the outputs of the photomultipliers and of the threshold detectors, with reference to the scheme of Fig. 3.4.7.

the applied voltage across the drift region can be adjusted so that t_ν is exactly a multiple of 75 nsec, in which case the counter reading will always be zero, except for one position only of the step-delay circuit (the condition of equal readings at two successive positions can also be considered, and this enlarges the choice to all multiples of 37.5 nsec).

G. The drift tube used by Nolan and Phelps (1965) for measurements in cesium with the alternating field method is rather conventional; the electrode structure is a parallel plate condenser, with a guard ring around the anode, and the distance between the electrodes can be varied from 0.05 to 1 cm by means of a micrometer-driven bellow assembly. At the tube temperature, which is kept around 250°C, both electrodes act as thermionic sources, but the cathode temperature is slightly larger than the anode one and thus is the available emission current. The square wave being perfectly symmetrical, the tube behaves in this case as another tube, whose cathode available current is equal to the difference between the available currents from the two electrodes.

More important is the circuit design (see Fig. 3.4.9), since the applied square wave must have good time symmetry and be free from any significant distortion. For this purpose the square wave is obtained by conversion from a sine wave at twice its frequency, and it is fed into the cathode of the drift tube through a properly designed cathode follower. Nolan and Phelps have thus been able to apply to the cathode of the tube a square wave with an amplitude up to 50 V, a rise time of about 15 nsec, and a time symmetry within 3% for all frequencies (typically in the kilohertz range). In order to minimize on the anode, which must be at ground potential, the ac signal due to coupling from the cathode across the tube, a compensating square wave of opposite polarity is applied to the anode through a variable impedance

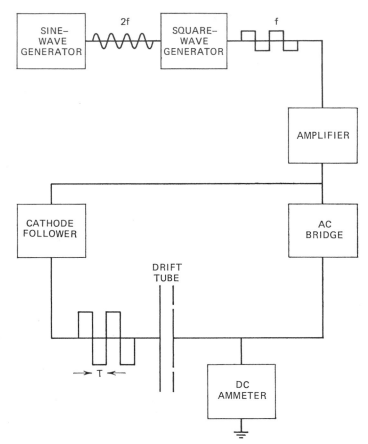

Fig. 3.4.9. Block diagram of the circuitry associated with the drift tube used by Nolan and Phelps for measurements in cesium with the alternating field method [after Nolan and Phelps (1965)].

(designated as "ac bridge" on Fig. 3.4.9). The average anode currents, in the range between 10^{-8} and 10^{-10} A, are measured by a suitable dc ammeter, consisting of a 10^4-Ω resistor, much below the anode-cathode leakage resistance, and of a microvoltmeter.

3.5 CHARACTERISTIC ENERGY: GENERAL DESCRIPTION OF THE BASIC MEASUREMENT METHODS

In Section 3.3 we saw that, when the electron burst drifts from the source to the collector, it spreads and the fractional distribution width of the electron swarm at the collector is proportional to the square root of the characteristic energy D/μ [see eq. 3.3.4)]. It should be feasible, therefore, to measure the characteristic energy by means of the same setups used for drift velocity measurements.

Consideration of the detailed analysis performed in Section 3.3 indicates, however, that transit time distribution measurements of the type there described can actually provide values, not of D/μ, but of D_L/μ. Only recently Lowke and Parker (1969) fully demonstrated with appropriate computations that D_L may be rather different from D; this explains why in the past most authors have disregarded the results of their own transit time distribution measurements, which were apparently inconsistent if simply compared to the D/μ values obtained by other methods. The best available determinations of D_L/μ are those obtained using the most recent and sophisticated implementations of the Stevenson and Hurst method [Hurst and Parks (1966), Wagner, Davis, and Hurst (1967)], which make it possible to determine the transit time distributions with the highest accuracy, as explained in Section 3.4.

Townsend's method, the most conventional method of measuring D/μ, is also based on the determination of the diffusive spreading of a drifting electron swarm, but it considers the lateral diffusion instead of the longitudinal one. For this purpose the role of time and space are interchanged, with respect to the previous case of drift velocity measurements: now a point source or a line source emits electrons continuously in the drift chamber and the lateral spread while drifting is determined; in the previous case, on the other hand, an electron pulse of infinite transverse extension was considered and its longitudinal time development was measured.

The principle of the method is due to Townsend (1915). A schematic diagram of the apparatus is shown in Fig. 3.5.1. Electrons generated by a heated filament or by a photoemissive cathode enter the main drift chamber through a small hole or a narrow slit, after having drifted through a region where an electric field equal to that in the main chamber has been applied, in order to bring electrons to the appropriate steady-state energy distribution. In the main chamber the electrons drift under the action of the applied uniform field E; to measure the swarm spread, the collector plate is divided into insulated areas. In one method of division (Fig. 3.5.1b) there is a central disk and concentric annuli; in another (Fig. 3.5.1c) a central strip is flanked by other strips or by the remaining portions of the electrodes. The ratio of the currents to any two insulated areas of the collector provides a measure of the electron swarm divergence and then of the D/μ ratio.

The differential equation governing the electron flow is the stationary form of (3.3.1'):

$$DV_t^2 n + D_L \frac{\partial^2 n}{\partial z^2} - w \frac{\partial n}{\partial z} = 0 \qquad (3.5.1)$$

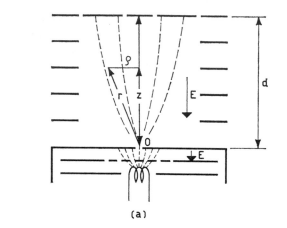

(a)

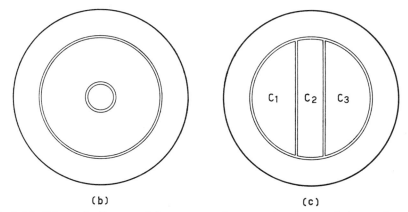

(b)   (c)

Fig. 3.5.1. Schematic diagram of the apparatus for measuring characteristic energies with Townsend's method and alternative modes of division of the collector.

If it is assumed that the source aperture is a small circular hole located at the origin of the cylindrical coordinate system (ρ, φ, z) and that no electron-absorbing boundary is present, the proper solution of (3.5.1) is:

$$n(\rho, z) = \frac{A}{r} \exp\left[-\frac{w}{2D_L}(r - z)\right] \qquad (3.5.2)$$

where $r = [(D_L/D)\rho^2 + z^2]^{1/2}$ is the distance from the origin in a coordinate system, in which the radial coordinate ρ is scaled by the factor $\sqrt{D_L/D}$ [Lowke (1971)].

A different solution holds when the entire aperture plate is considered to be electron absorbing, except for the point area of the hole; the following conditions must then be considered: $n = 0$ for $\rho \neq 0$ on the $z = 0$ plate, and

$n > 0$ at any finite point of the $z > 0$ region. These conditions yield the dipole-like solution [Huxley (1940)]:

$$n(\rho, z) = A \frac{z}{r^3}\left(1 + \frac{w}{2D_L}r\right)\exp\left[-\frac{w}{2D_L}(r - z)\right] \qquad (3.5.3)$$

If the collector plate is present and is considered to be also electron absorbing, a solution must be obtained that is zero both on the collector plate $z = d$ and on the aperture plate, except at the origin. The collector plate boundary condition can in effect be satisfied by simply adding to the dipole solution (3.5.3) a second dipole solution, representing an image source of suitable strength located at $z = 2d$. Since, however, this second solution does not satisfy the boundary condition on the aperture plate, a third dipole term must be added, representing a source at $z = -2d$; this in turn requires a fourth dipole at $z = 4d$, and so on until an infinite series of terms is obtained. The solution [Warren and Parker (1962)] is:

$$n(\rho, z) = A \exp\left(\frac{wz}{2D_L}\right)\sum_{k=-\infty}^{+\infty}\frac{z - 2kd}{r_k^3}\left(1 + \frac{wr_k}{2D_L}\right)\exp\left(-\frac{wr_k}{2D_L}\right) \qquad (3.5.4)$$

where:

$$r_k = \left[\frac{D_L}{D}\rho^2 + (z - 2kd)^2\right]^{\frac{1}{2}} \qquad (3.5.5)$$

The electron flow at the collector is:

$$\Gamma(\rho, d) = \left[wn - D_L\frac{\partial n}{\partial z}\right]_{z=d} = -D_L\frac{\partial n}{\partial z}\bigg|_{z=d} \qquad (3.5.6)$$

and the current collected by any annulus of inner radius a and outer radius b is:

$$I(a, b) = 2\pi e\int_a^b\Gamma\rho\, d\rho = -2\pi eD_L\int_a^b\frac{\partial n}{\partial z}\bigg|_{z=d}\rho\, d\rho \qquad (3.5.7)$$

When differentiation and integration are carried out, the following result is obtained:

$$I(a, b) = 4\pi eDA\exp\left(\frac{wd}{2D_L}\right)$$

$$\times \sum_{k=0}^{\infty}\left\{\frac{\exp(-wr_k/2D_L)}{r_k^3}\left[\frac{wr_k}{2D_L}(1 - 2k)^2 d^2 - \frac{D_L}{D}\rho^2\right]\right\}\bigg|_{\rho=a,z=d}^{\rho=b,z=d} \qquad (3.5.8)$$

The dependence on w/D and on D_L/D of the ratios of the currents to any two of the insulated areas of the collector plate can then be derived using this expression.

A large number of experiments are based on measurements of the ratio R of the current falling on a central disk ($a = 0$) of radius b to the total current arriving at the collecting electrode ($a = 0$, $b = \infty$). Tube dimensions are usually such that the terms after the first one in the summation of eq. (3.5.8) are small enough to be neglected. In this case the ratio R computed from eq. (3.5.8) becomes [Huxley and Crompton (1955), Crompton and Jory (1962)]:

$$R = \frac{I(0, b)}{I(0, \infty)} = 1 - \frac{I(b, \infty)}{I(0, \infty)}$$

$$= 1 - \left\{ \frac{1}{1 + (b'/d)^2} - \frac{2 D_L}{wd} \frac{(b'/d)^2}{[1 + (b'/d)^2]^{3/2}} \right\}$$

$$\times \exp \left[-\frac{wd}{2 D_L} (\sqrt{1 + (b'/d)^2} - 1) \right] \tag{3.5.9}$$

where $b' = \sqrt{D_L/D}\, b$ is the scaled radius of the central disk.

Under typical experimental conditions, $(b/d)^2 \ll 1$ and $D/wd (= D/\mu V$, electron characteristic energy/applied voltage) $\ll 1$, so that eq. (3.5.9) becomes in the first approximation:

$$R = 1 - \exp \left(-\frac{wb^2}{4 D d} \right) \tag{3.5.9'}$$

It is then apparent that Townsend's method is appropriate for measuring the characteristic energy D/μ.

Only recently [Parker and Lowke (1969), Lowke (1971)], however, has the necessity of distinguishing D_L from D in the application of Townsend's method been realized; hence in all the experiments reported in the literature so far the measured data have always been analyzed using formulas derived under the assumption that $D_L = D$ (and therefore $b' = b$). It is not surprising that eq. (3.5.9) with $D_L = D$ gave an excessively large and unjustified spread of results when applied in the derivation of D/μ from the experimental data. Since no explanation of these discrepancies was then known, various empirical schemes were worked out for the reduction of the experimental data.

1. Warren and Parker (1962) and later Cottrell and Walker (1967) derived from experiment empirical calibration curves of R versus wd/D for their own setups. These curves were chosen so as to reduce the scattering of the final results to within acceptable limits and to satisfy two conditions: for a given geometry R is a function of $wd/D = \mu V/D$ alone; for a given gas D/μ is a function of E/p alone and attains the Einstein thermal value kT_g/e as E/p becomes sufficiently small.

2. Huxley and Crompton (1955) and later Crompton and Jory (1962) found that in their experiments consistent results over a wide range of values of the physical parameters were obtained using the formula:

$$R = 1 - \frac{1}{\sqrt{1 + (b/d)^2}} \exp\left[-\frac{wd}{2D}(\sqrt{1 + (b/d)^2} - 1)\right] \quad (3.5.10)$$

instead of (3.5.9). This eq. (3.5.10) follows ($D_L = D$ always) either from (3.5.7) when the distribution is (3.5.2) with the inclusion only of an image term to make $n = 0$ at $z = d$, or from the mobility current density (wn) at the collector when the distribution is (3.5.3) without the inclusion of any collector boundary condition [Huxley (1940)].

Formula (3.5.10) has been largely used in the last two decades for the reduction of experimental data. Actually most of the recent determinations of the characteristic energy have been performed under experimental conditions such that a negligible difference exists between the w/D values computed from the two different equations (3.5.9) with $D_L = D$ and (3.5.10) [Crompton, Elford, and Gascoigne (1965)]; this was considered a safe criterion for using the theoretical expressions, instead of deriving empirical calibration curves. In order to satisfy this condition, however, b/d must be quite small; in particular, calculations have shown that to have a discrepancy never larger than 1%, one must have b/d ratios smaller than ~ 0.05. The choice of such low b/d values has been quite fortunate, since Lowke (1971) has found that for b/d ratios as small as ~ 0.05 the current ratio R given by (3.5.9) is insensitive to values of D_L/D; hence using eq. (3.5.10) in place of (3.5.9) is justified.

The ratio b/d must be chosen, however, with proper attention also to other parameters, and in particular to the fact that for adequate precision the current ratio R cannot be smaller than 0.1 or larger than 0.9. In order to have R values not less than 0.1, we obtain from (3.5.10) the condition:

$$\frac{wd}{D}\left(\frac{b}{d}\right)^2 \leqslant 9.2$$

provided $(b/d)^2 \ll 1$. This can be rewritten in the form:

$$V\left(\frac{b}{d}\right)^2 \leqslant 9.2 \frac{D}{\mu} \quad (3.5.11)$$

The minimum D/μ value to be measured is the Einstein value $kT_g/e = 0.0067$ eV at 78°K. The voltage must be large enough to avoid excessive influence of contact potential differences within the diffusion chamber: we take $V_{min} = 30$ V according to the findings of Crompton and Jory (1962).

Then, in order to satisfy (3.5.11), the ratio b/d must be smaller than ~ 0.05 which is the same limit required for the use of eq. (3.5.10).

Most of the recent determinations of the characteristic energy below 1 eV have actually been made by Crompton's group [Crompton, Elford, and Gascoigne (1965), Crompton, Elford, and Jory (1967)], using diffusion chambers with $b = 0.5$ cm and $d = 10$ cm, so that $b/d = 0.05$. Disk diameters smaller than 1 cm are undesirable, mainly because the width of the annular gap around the disk cannot be reduced in the same ratio, as required for good accuracy. Chamber lengths much larger than 10 cm are also undesirable, since the chamber has to be eventually cooled at low temperatures in a refrigerant container.

The above choice of dimensions makes it possible to perform determinations of the D/μ ratio down to the Einstein thermal value; on the other hand, the maximum D/μ which can be measured is given by the condition $R \leqslant 0.9$. In fact, this condition substituted into eq. (3.5.10) yields:

$$\frac{wd}{D}\left(\frac{b}{d}\right)^2 \geqslant 0.42$$

provided $(b/d)^2 \ll 1$, or, in a different form:

$$\frac{D}{\mu} \leqslant \frac{V}{0.42}\left(\frac{b}{d}\right)^2 \tag{3.5.12}$$

If $b/d = 0.05$ and $V_{\max} = 250$ V (a typical maximum value for no breakdown within the tube), the maximum D/μ which can be measured is 1.5 eV.

Measurements of larger D/μ values are possible using adequately larger b/d ratios. The electron collector of the chamber can be divided into a central disk and a few concentric annuli, so that tying one or more annuli to the central disk allows measurements to be performed at increasing b/d ratios without changing the chamber [Crompton and Jory (1962)]. At large b/d ratios formula (3.5.9) would have to be used. Although this was not done in the past, Lowke (1971) has found that expression (3.5.10), used by most authors, provides still acceptable results when $D_L/D \simeq 0.5$, which is fortunately a frequent case (as, e.g., in helium, hydrogen, nitrogen, and oxygen over the range $0.05 < E/p < 2$ V cm^{-1} torr^{-1}).

In this section we have specifically discussed the most common case, in which the ratio R of the electron current at the central disk to the total collected current is measured. However, it is a simple matter to modify the above results for other common cases. For instance, in terms of the previously defined $R(b)$ current ratio, we can say that the ratio of the current received by an annulus of radii a and b to the total received current is $R(b) - R(a)$, and the ratio of the current received by a central disk of radius b to the sum of the currents at the disk and a surrounding annulus of outer radius c is simply $R(b)/R(c)$. A similar analysis could be extended also to the case of a slit aperture and strip-like collecting electrodes.

3.6 CHARACTERISTIC ENERGY DETERMINATIONS: ERROR SOURCES AND SECONDARY EFFECTS

In Section 3.5 the theoretical relationships between the current ratio in the Townsend method and the characteristic energy D/μ were derived. We shall report now various theoretical and experimental analyses, which have been performed in order to evaluate the amount of deviations to be expected under actual experimental conditions, and the modifications to be introduced when secondary effects (such as electron attachment and ionization) take place. Following the original papers, which have never distinguished between D_L and D, we also assume, without further notice, that $D_L = D$ throughout this section; the formulas of Section 3.5, we shall refer to, have then to be considered as modified accordingly.

A. Warren and Parker (1962) worked out the extension of formula (3.5.8) to the case of a finite-size, uniform-density source hole. This formula and similar numerical evaluations by Crompton and Jory (1962) and by Naidu and Prasad (1968) indicate that this effect can be neglected for apertures of usual size, which are of the order of 1 mm diameter.

Before 1940, following the original work of Townsend, diffusion chambers with relatively large source apertures were employed. In all these experiments the calibration curves of R versus wd/D were determined for each individual geometry by solving first eq. (3.5.1) with the appropriate boundary condition on the source plane, and then computing the mobility current density (wn) at the collector without the inclusion of any specific boundary condition at this electrode.

B. Crompton and Jory (1962) performed extensive numerical computations on the errors arising from the misalignment of the source hole with respect to the collecting disk axis. Some of their results are shown in Fig. 3.6.1, where the percentage error in the measured w/D value is shown as a function of the ratio R for a number of combinations of the parameters b and d, and for the case of a point source off axis by 0.020 cm.

C. The diameter of the diffusion chamber is usually large enough to ensure that only a negligible portion of the electrons that enter from the source hole diffuse to the lateral walls. Therefore, a specific mathematical analysis of the role of lateral walls is of interest only for verification purposes, or in cases wherein secondary electrons produced by the bombardment of

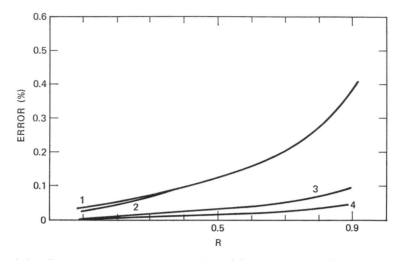

Fig. 3.6.1. Curves showing the variation with R of the error (expressed as a percentage) in the measured w/D values introduced when the point source is 0.020 cm from the axis. Curve 1: $d = 2$, $b = 0.5$; curve 2: $d = 10$, $b = 0.5$; curve 3: $d = 10$, $b = 1$; curve 4: $d = 10$, $b = 1.5$ (d and b are given in centimeters) [after Crompton and Jory (1962)].

the cathode by positive ions or photons, generated by the swarm movement, enter the chamber outside the main electron source aperture.

If zero electron density at the cylindrical wall ($\rho = \rho_w$) and at the anode ($z = d$) is assumed, the current passing through a disk of radius ρ at position z can be easily computed. The expression for this current, as derived by Lawson and Lucas (1965b) (the questionable term due to their second image charge has been dropped), is:

$$I(\rho, z) = 2\pi e D \rho_w \rho \sum_{i=1}^{\infty} \frac{A_i}{\sigma_i} J_1\left(\sigma_i \frac{\rho}{\rho_w}\right)$$
$$\times \left\{ \left(\frac{w}{2D} + u_i\right) \exp\left[\left(\frac{w}{2D} - u_i\right)z\right] \right.$$
$$\left. - \left(\frac{w}{2D} - u_i\right) \exp\left[\left(\frac{w}{2D} + u_i\right)z - 2u_i d\right] \right\} \tag{3.6.1}$$

where σ_i are the roots of the J_0 Bessel function, and:

$$u_i^2 = \left(\frac{w}{2D}\right)^2 + \left(\frac{\sigma_i}{\rho_w}\right)^2 \tag{3.6.2}$$

The values for A_i are obtained by setting $I(\rho, 0)$ equal to the corresponding radial distribution of the current entering the chamber from the cathode electrode. The collected current ratio R is then given by:

$$R = \frac{I(b, d)}{I(\rho_w, d)} \qquad (3.6.3)$$

D. Crompton, Elford, and Gascoigne (1965). thoroughly investigated the errors arising from field distortions. They identified two major sources of field distortions that can introduce serious errors: the radial fields, which are most likely to occur near the gap between the collecting central disk and the external electrode, and the contact potential differences, either over any single electrode or guard-ring surface or between different surfaces.

The radial fields near the annular gap of the collecting electrode have maximum effect since they may cause a spurious distribution of current between the disk and the external electrode. Their effect and all other effects due to inaccuracy of the chamber geometry or of the guard-ring voltage ratios are independent of the field magnitude E, so that the same D/μ result is obtained when measurements are performed by changing E at one fixed value of E/p.

On the contrary, the error on D/μ arising from the field distortions introduced by contact potential differences is recognized by the fact that its magnitude is approximately inversely proportional to E.

E. When an experiment is performed in electronegative gases, the electrons in the diffusion chamber generate negative ions by electron attachment. In this case it is a simple matter to evaluate the electron density distribution in the chamber by solving eq. (3.5.1) modified by the addition of the term $-\nu_a n = -\eta w n$ to the right side. In this case the simple pole solution, corresponding to (3.5.2), is:

$$n(\rho, z) = \frac{A}{r} \exp\left[-\frac{w}{2D}(\gamma r - z)\right] \qquad (3.6.4)$$

where:

$$\gamma = \left(1 + \frac{4D}{w}\eta\right)^{1/2} \qquad (3.6.5)$$

Since eq. (3.5.10), which was found to provide a very good fit with the experimental data, can be obtained from (3.5.2), we are interested in developing from (3.6.4) the equivalent formula for the current ratios in the presence of attachment. This has been done by Huxley (1959b). If an image source is introduced at $z = 2d$ to make $n = 0$ at $z = d$, the distribution becomes:

$$n(\rho, z) = A \exp\left(\frac{wz}{2D}\right)\left[\frac{1}{r_0}\exp\left(-\frac{w\gamma r_0}{2D}\right) - \frac{1}{r_1}\exp\left(-\frac{w\gamma r_1}{2D}\right)\right] \qquad (3.6.6)$$

where r_0 and r_1 are given by (3.5.5).

The current collected by any annulus of radii a and b is the sum of an electron current, whose general expression is still (3.5.7), and of a negative-ion current resulting from the attachment process. Most measurements are performed under conditions in which the electron characteristic energy is much larger than the negative-ion energy, which is always very close to the thermal value. This implies that the diffusion spread of negative ions in the chamber is much smaller than the spread of electrons; since the experiments we are considering are designed for adequate measurements of the electron spread, it follows that ion diffusion can be neglected and that the current due to ions formed in the chamber can be simply found by integration of the ion generation rate over the cylinder volume corresponding to the collecting annulus (the detachment is assumed negligible throughout). Moreover we assume that no negative ion enters the diffusion chamber through the hole of the electron source, or that the inner radius a is large enough for the annulus not to collect any negative ion entering from this hole. When these assumptions are satisfied, the current can be written as:

$$I(a, b) = 2\pi e \int_a^b \left[-D \left. \frac{\partial n}{\partial z} \right|_{z=d} + \bar{v}_a \int_0^d n \, dz \right] \rho \, d\rho \qquad (3.6.7)$$

In the 1920s the presence or absence of a significant number of negative ions in attaching gases was checked by applying a large transverse magnetic field, so as to deflect all the electrons, but not the ions, out of the central strip of the collector.

In experiments where attachment is expected to be present, it is common practice today to use Huxley's technique: we divide the anode into a central disk and two annuli and measure the ratio R of the current received by the inner annulus to the total current received by both annuli, so as to exclude the negative ions, which arrive directly from the hole of the electron source and are all collected by the central disk [Huxley (1959b)]. Using eq. (3.6.7), where the distribution is given by (3.6.6), we obtain the current ratio R (the outer radius of the external annulus is set equal to ∞, since we assume that an inappreciable current falls outside this annulus):

$$
\left.
\begin{aligned}
R &= \frac{I(a, b)}{I(a, \infty)} \\[2mm]
&= 1 - \frac{[1 + (b/d)^2]^{-\frac{1}{2}} \exp\left[-(wd/2D)(\gamma\sqrt{1 + (b/d)^2} - 1)\right] + (\eta d/\gamma)\phi(b/d)}{[1 + (a/d)^2]^{-\frac{1}{2}} \exp\left[-(wd/2D)(\gamma\sqrt{1 + (a/d)^2} - 1)\right] + (\eta d/\gamma)\phi(a/d)} \\[2mm]
\phi(\zeta) &= \int_0^1 \left\{ \exp\left[-\frac{wd}{2D}(\gamma\sqrt{s^2 + \zeta^2} - s)\right] \right. \\[2mm]
&\qquad\qquad \left. - \exp\left[-\frac{wd}{2D}(\gamma\sqrt{(2-s)^2 + \zeta^2} - s)\right] \right\} ds
\end{aligned}
\right\} \qquad (3.6.8)
$$

When η is small, eq. (3.6.8) reduces to:

$$R = 1 - \left[\frac{1 + (a/d)^2}{1 + (b/d)^2}\right]^{\frac{1}{2}} \exp\left[\frac{wd}{2D} \gamma(\sqrt{1 + (a/d)^2} - \sqrt{1 + (b/d)^2})\right] \quad (3.6.9)$$

Huxley's method of deriving both the characteristic energy D/μ and the attachment probability η with Townsend's setup and the use of eq. (3.6.8) is based on measuring the current ratio R as a function of d, keeping constant the applied field, the gas pressure, and other geometrical quantities. A pair of R measurements is actually sufficient, but a better result is obtained when more chamber lengths can be used. As a check on the reliability and self-consistency of the method we must find that D/μ is a function of E/p only, when the gas pressure is changed, and that the same result holds for η/p or η/p^2, depending on whether attachment is a two-body or a three-body collision process.

The radius a of the central disk, which ensures that the fraction ε of the negative ions admitted through the aperture of the electron source reaches the disk, can be computed with the formula [Huxley (1959b)]:

$$\frac{a}{d} \simeq \left\{\frac{\ln\left[1/(1 - \varepsilon)\right]}{10V}\right\}^{\frac{1}{2}} \quad (3.6.10)$$

where V (in volts) is the applied voltage across the chamber. For $\varepsilon = 0.99$ we must have $a/d \simeq 0.68/\sqrt{V}$.

As for drift velocity determinations, here too the presence of ionization can be taken into account, replacing in the density distribution expressions $\bar{v}_a$ by $(\bar{v}_a - \bar{v}_i)$ or η by $(\eta - \alpha_i)$, α_i being the ionization coefficient $(= \bar{v}_i/w)$. It follows that the current ratio is still given by (3.6.8), where now γ is [Huxley (1959b)]:

$$\gamma = \left[1 + \frac{4D}{w}(\eta - \alpha_i)\right]^{\frac{1}{2}} \quad (3.6.11)$$

At the highest E/p values, when ionization is fairly large, the production of secondary electrons caused by bombardment of the electrodes by both positive ions and photons may also be of importance. The problem was treated by Lawson and Lucas (1965b), who computed electron, ion, and photon flow distributions and from them the coefficients A_i of formula (3.6.1) appropriate for the corresponding cathode production of secondary electrons; coefficients u_i in formula (3.6.1) are also different and become:

$$u_i^2 = \left(\frac{w\gamma}{2D}\right)^2 + \left(\frac{\sigma_i}{\rho_w}\right)^2 \quad (3.6.12)$$

Hurst and Liley (1965), on the other hand, treated the problem differently, neglecting the role of lateral walls, but including photon emission of secondary electrons at the anode. The interested reader is referred to the original paper for the complete formulas and for the details. Here we shall report only an expression for the current ratio, which is equivalent to (3.5.10) but includes volume ionization and secondary electron emission by photon impact at the cathode. This expression can be written as:

$$
\begin{aligned}
R = \frac{wb}{2D} & \left[\exp\left\{ -\frac{wd}{2D}(1-\gamma) \right\} - \frac{\delta}{\alpha_i}\left\{ 1 - \exp\left[-\frac{wd}{2D}(1-\gamma) \right] \right. \right. \\
& \left. \left. - \frac{1-\gamma}{1+\gamma}\left[1 - \exp\left\{ -\frac{wd}{2D}(1+\gamma) \right\} \right] \right\} \right] \\
& \times \int_0^\infty \frac{J_1[(wb/2D)\xi]\, \exp\,[(wd/2D)(1-\sqrt{\gamma^2+\xi^2})]}{1+(\delta/\alpha_i)M(\xi)}\, d\xi \\
M(\xi) = & -\frac{\gamma(1-\gamma)}{\sqrt{\gamma^2+\xi^2}}\left\{ \frac{\exp\,[(wd/2D)(1-\xi-\sqrt{\gamma^2+\xi^2})]-1}{1-\xi-\sqrt{\gamma^2+\xi^2}} \right. \\
& - \frac{\exp\,[(wd/2D)(1-\xi+\sqrt{\gamma^2+\xi^2})]-1}{1-\xi+\sqrt{\gamma^2+\xi^2}} \\
& \times \exp\left[\frac{wd}{D}\sqrt{\gamma^2+\xi^2} \right] \right\}
\end{aligned} \right\} \quad (3.6.13)
$$

where γ is defined according to (3.6.11) [$= (1-4D\alpha_i/w)^{1/2}$], and δ is the Townsend secondary ionization coefficient for photons, namely, δ/α_i represents the number of secondary electrons generated by photons at the cathode per ionizing collision in the gas, when we consider a one-dimensional flow geometry between infinite parallel plates. The importance of formula (3.6.13) rests on the fact that it has been successfully and extensively applied by Crompton, Liley, McIntosh, and Hurst (1966) to the discussion of their measurements in hydrogen at E/p values near electrical breakdown.

Equation (3.6.13) cannot be evaluated explicitly, but R can be calculated numerically; for a chamber of known geometry (given b and d values) and using published data for α_i/p versus E/p, R is computed and plotted as a function of p and E/p for sets of values of w/D and of δ/α_i. Comparison, at a given value of E/p, of the measured R versus p curve with the corresponding calculated ones leads to unique values of w/D and δ/α_i.

Before Huxley's modification of Townsend's technique, different methods devised many years before by Bailey for measuring the attachment coefficient in electron swarms were used [Healey and Reed (1941)]. In Section 3.9 we shall discuss those of these methods which make use of a static magnetic field; here, however, we describe the first Bailey method, which, like Huxley's, affords the determination of both the electron characteristic energy and the attachment coefficient [Bailey (1925)].

A stream of electrons, having already acquired a steady energy distribution in a uniform field E, enters the first diffusion chamber of the apparatus (Fig. 3.6.2) through a narrow slit cut into plate 0 and moves under the same field E to plate 1. This plate has an aperture slit, which acts as a source for a

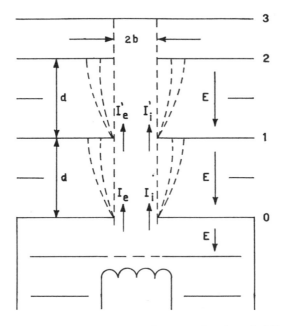

Fig. 3.6.2. Schematic diagram of the apparatus for measuring characteristic energies with Bailey's method.

second diffusion chamber identical to the first one. Also, anode plate 2 of the second chamber has a slit, and nearby plate 3 collects all the charged particles that have passed through. All the slits are of equal width and are accurately aligned, one above the other.

At the entrance of the first chamber we actually have both electrons and negative ions, since ions are already formed in the region between the cathode and plate 0. Let I_e and I_i be, respectively, the electron and ion currents at the entrance slit. The current passing through the slit in plate 1 and entering into the second chamber will be the sum of an electron current $I_e' = R_e I_e$ and of an ion current $I_i' = R_{ie} I_e + R_i I_i$, due partially ($R_{ie} I_e$) to the new ions formed in the chamber and partially ($R_i I_i$) to those of the input ion stream I_i. Coefficients R_e and R_{ie}, in addition to being dependent on the geometry, will be functions of both the electron characteristic energy and the attachment coefficient; coefficient R_i, however, will be dependent only on geometry, since negative ions of the I_i stream have thermal equilibrium energies. Here too detachment is assumed to be negligible. The ratio of the current passing through the slit to that arriving at the slit plane is then for plate 1:

$$S_1 = \frac{I_e' + I_i'}{I_e + I_i} = \frac{R_e I_e + R_{ie} I_e + R_i I_i}{I_e + I_i} \tag{3.6.14}$$

Likewise we can find the corresponding ratio for plate 2:

$$S_2 = \frac{R_e I_e' + R_{ie} I_e' + R_i I_i'}{I_e' + I_i'} = \frac{R_e^2 I_e + (R_e + R_i) R_{ie} I_e + R_i^2 I_i}{S_1 (I_e + I_i)} \tag{3.6.15}$$

Solving eq. (3.6.14) for $R_{ie} I_e$ and substituting into (3.6.15), we obtain after straightforward mathematics:

$$(S_1 S_2 - R_e S_1 - R_i S_1 + R_e R_i)(I_e + I_i) = 0$$

This implies that the first factor is zero; hence:

$$R_e = S_1 \frac{R_i - S_2}{R_i - S_1} \tag{3.6.16}$$

Current ratios S_1 and S_2 can be easily determined by appropriate measurements. In fact, if we measure the ratio ξ of the currents received by plates 2 and 1, respectively, and ζ, the corresponding ratio for plates 3 and 2:

$$S_1 = \frac{\xi(1 + \zeta)}{1 + \xi(1 + \zeta)}$$

$$S_2 = \frac{\zeta}{1 + \zeta} \tag{3.6.17}$$

Since coefficient R_i can be easily evaluated for the actual geometry, relationship (3.6.16) yields the value of R_e. If we can write an analytical expression for R_e as a function of D/μ and η, these quantities are determined from an appropriate set of ξ and ζ measurements.

In order to evaluate coefficient R_e the electron density is first determined by solving eq. (3.5.1) with the additional term $-\tilde{v}_a n$ for the case of a line source along axis y. When the usual image term is added to make $n = 0$ at $z = d$, the electron distribution becomes:

$$n(x, z) = A \exp\left(\frac{wz}{2D}\right)$$

$$\times \left[K_0\left(\frac{w\gamma}{2D} \sqrt{x^2 + z^2}\right) - K_0\left(\frac{w\gamma}{2D} \sqrt{x^2 + (2d - z)^2}\right) \right] \tag{3.6.18}$$

where K_i is the modified Bessel function of the second kind and of ith order. Substituting this expression for the density in the formula for the electron current through a central slit of width $2b$ at $z = d$:

$$I = -eD \int_{-b}^{b} \frac{\partial n}{\partial z}\bigg|_{z=d} dx \qquad (3.6.19)$$

we obtain the following expression for R_e:

$$R_e = \frac{w\gamma d}{\pi D} \exp\left(\frac{wd}{2D}\right) \int_0^b \frac{K_1[(w\gamma/2D)\sqrt{d^2 + x^2}]}{\sqrt{d^2 + x^2}} dx \qquad (3.6.20)$$

In general, $x^2 \ll d^2$ since $b^2 \ll d^2$; moreover, the applied voltage over characteristic energy ratio wd/D is much larger than unity under typical experimental conditions, so that the asymptotic approximation:

$$K_1(\xi) = \sqrt{\pi/2\xi} \exp(-\xi) \qquad (3.6.21)$$

can be used. Then eq. (3.6.20) yields:

$$R_e = \text{erf}\left(\frac{b}{2}\sqrt{w\gamma/Dd}\right) \exp\left[-\frac{wd}{2D}(\gamma - 1)\right] \qquad (3.6.22)$$

Since usually ηd is much less than the ratio wd/D, γ can be approximated by:

$$\gamma = 1 + \frac{2D}{w}\eta \qquad (3.6.23)$$

and an adequate expression for R_e becomes:

$$R_e = \text{erf}\left(\frac{b}{2}\sqrt{w/Dd}\right) \exp(-\eta d) \qquad (3.6.24)$$

The hypothesis of a two-dimensional geometry with infinitely long line sources and slits, on which the previous equation is based, is sometimes too crude an approximation. In these cases the general Bailey formula [Bailey (1925)]:

$$R_e = \psi\left(\frac{w}{Dd}\right) \exp(-\eta d) \qquad (3.6.25)$$

is used, where the function ψ is either theoretically evaluated for the actual geometry or experimentally derived from measurements in gases for which D/μ and η values are known. The function ψ may be slightly different for the two diffusion chambers of the same apparatus.

The values of D/μ and η can be found by changing E and p by the same factor q ($\eta \propto p$ is assumed), so that E/p remains constant and the second set of measurements yields:

$$R'_e = \psi\left(q\,\frac{w}{Dd}\right)\exp\left(-q\eta d\right) \tag{3.6.26}$$

Or, if d can also be varied, it is better to change $E, p,$ and d by the same factor q, yielding:

$$R'_e = \psi\left(\frac{w}{Dd}\right)\exp\left(-q^2\eta d\right) \tag{3.6.27}$$

In this second case calculations are clearly simpler and more reliable: η is derived by the ratio R_e/R'_e independently of the function ψ [Bailey and McGee (1928)].

A final remark concerns coefficient R_i. It can be computed from R_e expressions by setting $\eta = 0$ (ion detachment is negligible) and $D/\mu = kT_g/e$ (the ions are in thermal equilibrium with the gas molecules).

F. In spite of the low swarm currents which are typical of these experiments (10^{-12} to 10^{-13} A), space-charge effects may be significant at the lowest E/p and gas temperatures. In fact, it is the ratio of the space-charge potentials to the electron characteristic energy that determines the magnitude of these effects, and the electron characteristic energy is a decreasing function of both E/p and the gas temperature.

In general, the presence of a significant amount of space charge is recognized by the apparent current dependence of the electron characteristic energy, as computed from experimental data; the limiting value of this energy as the current tends to zero, that is, its true value, may in these cases be readily determined. Liley (1967), however, gives two good reasons for the development of an appropriate, quantitative analysis of space-charge phenomena: the first is to guarantee that any experimentally observed current dependence can be adequately explained in terms of these effects, and the second is to use the results of such a treatment to determine characteristic energy values and attachment coefficients also at the lowest E/p and gas temperatures.

The reader is referred to Liley's original paper for the complete analytical treatment. Here we shall report only his main result, that is, appropriate expressions for the ratio $R(\rho, z)$ of the current passing through a disk of radius ρ in the plane z to the total swarm current I_t.

If $4D/wz \ll 1$, then for a swarm consisting of equally charged electrons and ions, the ions possibly being produced by attachment, $R(\rho, z)$ satisfies the differential equation:

$$\rho\,\frac{\partial}{\partial\rho}\left(\frac{1}{\rho}\,\frac{\partial R}{\partial\rho}\right) - \frac{w}{D}\,\frac{\partial R}{\partial z}$$

$$= 2\delta\left\{f_{e0}\exp\left(-\eta z\right) + \frac{w}{w_i}\left[1 - f_{e0}\exp\left(-\eta z\right)\right]\right\}\frac{R}{\rho}\,\frac{\partial R}{\partial\rho} \tag{3.6.28}$$

where f_{e0} is the fraction of the total current carried by the electrons at $z = 0$ [$\simeq \exp(-\eta L)$, if L is the distance between the electron-emitting source and the cathode], w_i is the ion drift velocity, and:

$$\delta = \frac{I_t}{4\pi\varepsilon_0 DE} \tag{3.6.29}$$

For small δ, a solution for $R(\rho, z)$ to first order in δ is:

$$R = [1 - \exp(-\nu)]$$

$$- \delta\left\{\left[f_{e0} + \frac{w}{w_i}(1 - f_{e0})\right]a_{10} + \frac{20}{7}\frac{w}{w_i}f_{e0}[1 - \exp(-0.7\eta z)]a_{11}\right\} \tag{3.6.30}$$

where:

$$\nu = \frac{w\rho^2}{4Dz} \tag{3.6.31}$$

and:

$$a_{10} = [1 - \exp(-\nu)]\ln 2 + E(0,\nu)[1 + \exp(-\nu)] - E(0,2\nu)$$
$$a_{11} = \nu[\ln 2 + E(0,\nu) - E(0,2\nu)] - \tfrac{1}{2}\exp(-\nu)[1 - \exp(-\nu)] \tag{3.6.32}$$

with:

$$E(0, \xi) = \int_0^{\xi}[1 - \exp(-u)]u^{-1}\,du \tag{3.6.33}$$

being a tabulated function. Numerical values of coefficients a_{10} and a_{11} are given in Liley's paper. Solution (3.6.30) is inexact in the term $\exp(-0.7\eta z)$; the inaccuracy involved, however, is only of the order of 1%, provided $\eta z \leqslant 1$, $\nu \leqslant 2$.

When negative ions are absent ($\eta = 0$, $f_{e0} = 1$), the swarm consists entirely of electrons and the above solutions are valid no matter what the values of the electron characteristic energy. In the presence of both ions and electrons the results are valid only when the characteristic energies of ions and electrons are about equal, as happens in thermal swarms, where, furthermore, space-charge effects are expected to be most important. Factor w/w_i, which appears in eq. (3.6.30) in front of the terms due to the ions, is large, so that these terms may provide significant contributions also when ion currents are only small fractions of the total current. Space-charge effects are then likely to be of special importance in experiments where the swarm moves through a gas in which attachment is possible.

G. All previous theoretical developments assume that electrons in the drift region have an energy distribution independent of position, so that the electron motion can be characterized by constant μ and D coefficients. Parker (1963) reconsidered this problem in terms of his more exact theory, which provides the position-dependent distribution function (2.3.66).

Equation (2.3.68) gives Parker's result for the electron density at the anode, when electrode boundary effects are not taken into account and a constant collision frequency is assumed. If we normalize this density distribution with respect to the value on the axis ($\rho = 0$) and introduce the ratio D/μ in place of $2\mathcal{U}/3e$ [see eq. (2.6.2)], we find:

$$n = \left[1 + \left(\frac{\rho}{2d}\right)^2\right]^{-3/2} \exp\left(-\frac{w}{4D}\frac{\rho^2}{d}\right) \qquad (3.6.34)$$

This expression has to be contrasted with the equivalent distribution (3.5.2), which assumes position-independent coefficients; when the same normalization is used, (3.5.2) reads:

$$n = \left[1 + \left(\frac{\rho}{d}\right)^2\right]^{-1/2} \exp\left(-\frac{wd}{2D}(\sqrt{1 + (\rho/d)^2} - 1)\right) \qquad (3.6.35)$$

What is important here is how different the D/w values predicted by the two formulas are for a given n. Expanding (3.6.34) and (3.6.35) in powers of $(\rho/d)^2$, Parker obtained the following expression for the first-order approximation of the fractional difference in D/μ:

$$\frac{\delta(D/\mu)}{D/\mu} \simeq \frac{1}{2}\frac{D}{wd} - \frac{1}{4}\left(\frac{\rho}{d}\right)^2 \qquad (3.6.36)$$

If the quantities involved are those discussed as typical in Section 3.5, the above difference is negligibly small. In fact, assuming $(D/\mu) \leqslant 1$ eV, $V_{min} = E_{min}d = 30$ V, $\rho/d = b/d = 0.05$, difference (3.6.36) will never attain a value larger than 2%.

3.7 CHARACTERISTIC ENERGY: TUBES AND EXPERIMENTAL TECHNIQUES

In this section three tubes will be described: the most modern one with fixed elements, for use with Townsend's method in nonattaching gases down to the lowest near-thermal electron energies; an apparatus of variable length for measurements over a wide range of characteristic energies and for experiments in attaching gases, according to Huxley's modification of Townsend's method; and one of the old tubes used for measurements in attaching gases according to Bailey's method. The first tube will be the most fully described since it includes the best technological solutions.

A. A tube typical of those used by Crompton and his coworkers in their most recent measurements of near-thermal characteristic energies of electron

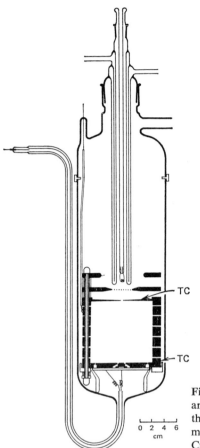

Fig. 3.7.1. Tube designed and used by Crompton and his coworkers for measurements of near-thermal characteristic energies with Townsend's method; TC denotes a thermocouple [after Crompton, Elford, and Gascoigne (1965)].

swarms is shown in Fig. 3.7.1 [Crompton, Elford, and Gascoigne (1965)]. The most significant dimensions are as follows:

Height d of the chamber	10 ± 0.005 cm
Diameter of central collector disk	0.9955 ± 0.0005 cm
Internal diameter of annulus of collector	1.0050 ± 0.0005 cm
External diameter of annulus of collector	8.5 cm
Internal diameter of guard electrodes	10 cm
Thickness of full guard electrodes	1.616 ± 0.003 cm
Thickness of glass spacers	0.051 ± 0.002 cm
Diameter of source hole	0.1 ± 0.005 cm

The design of the tube and the above values of the parameters satisfy the conditions discussed in Section 3.5 for accurate measurements in low-energy electron swarms and in Section 3.4 for the production of a uniform axial electric field. Considerable care has also been paid to the procedure of assembling the apparatus to obtain accurate alignment of all the electrodes.

According to the above dimensions, the width of the gap between disk and annulus is nominally 0.005 cm; this choice is a compromise between the desire to have the gap as small as possible, since we do not know really how the electrons arriving at the gap divide between disk and annulus, and the necessity to avoid too high an interelectrode capacity, which may add to the difficulty of determining the current ratio R. Moreover, a very small gap may seriously limit the lowest attainable temperature for the use of the chamber, since differential contraction of the components of the tube can cause the disk and annulus to touch. [In the above design this temperature was $90°K$; a different structure of the collecting electrode, suitable for use at $77°K$, is described by Crompton, Elford, and McIntosh (1968).]

The effective diameter of the central disk $2b$ is generally taken as the actual diameter plus the gap width, on the assumption that electrons at the gap divide equally between disk and annulus. Then in this case $b = 0.500$ cm.

The electrode structure is made of copper coated with gold for reducing contact potentials. Since contact potential differences across the surface of the anode have the largest effects, the anode and the cathode have been coated by vacuum deposition. By careful attention to the conditions the most uniform and stable gold layer can in fact be obtained with the vacuum deposition technique; a maximum contact potential difference between any two areas of the anode of 8 mV has been reported, with differences of only 3 mV occurring in the central region.

Guard rings have electroplated gold surfaces. Large potential differences of up to 100 mV may exist between the anode or cathode surfaces and the guard-ring surfaces. A compensating voltage of the same order has thus been placed between the set of guard electrodes and the anode and cathode; it has actually been proved that within reasonable limits this voltage compensates potential differences regardless of where they exist. The applied voltage may be set to the correct value using positive ions instead of electrons; in this case the same method can equally well be used, but now the characteristic energy of the ions is known, since at low E/p they are virtually in thermal equilibrium with the gas molecules. The compensating voltage may also be set at the correct value by adjusting it until, for any given value of E/p, the same value of the electron characteristic energy is obtained over a wide range of electric field strengths; this procedure has been preferred at very low temperatures, where the positive-ion method has failed [Crompton, Elford, and McIntosh (1968)].

The electron source consists of two filaments: a platinum one for electrons and a heated potassium aluminosilicate bead for positive (potassium) ions; a water cooling system removes heat generated by the filaments. As a current control, the inner surface of the cooling water jacket in the filament region is coated with a layer of platinum, the potential of which can be varied with respect to the filament. Electrons are made to cross a region 2 cm in length, where they attain the equilibrium energy [see eq. (3.3.38) for the evaluation of the required distance]. The gas temperature is monitored by means of two copper-constantan thermocouples. For measurements at E/p values near electrical breakdown, Crompton, Liley, McIntosh, and Hurst (1966) modified the chamber, using electrodes with a Rogowski profile to minimize the possibility of sparking, other than across the gap; the nominal dimensions of the tube are $d = 2$ cm, $2b = 0.7$ cm.

B. Figure 3.7.2 shows another tube used at Canberra, which was designed to satisfy the following major requirements [Crompton and Jory (1962)]:

a. Length of diffusion chamber to be variable from 1 to 10 cm with length setting accurate to 20 μ.
b. Above variation to be achieved without recourse to an external drive in order to eliminate sources of contamination such as shaft seals.
c. Anode to be divided to enable the diameter of the central disk to be 1, 2, 3, or 4 cm.

The mechanism used to vary the chamber length consists of a copper rotor driven by an external rotating magnetic field, a 25:1 reduction gear to increase the torque, a stainless steel lead screw with a 1-mm pitch, a nut that engages with the screw and is made of a copper-tin alloy with an expansion coefficient matching that of the stainless steel, and a graduated drum for accurate measurements of the anode position.

The other features of this tube have no special interest, since the more modern design of Fig. 3.7.1 provides much better technological solutions. We shall mention only the difficulty of aligning the movable electrode center on the axis of the chamber in all positions; in the present case this error was always kept within 0.020 cm.

This tube, like the one previously described, makes use of a platinum filament as electron source. Before leaving this subject, however, it is worth mentioning also different solutions used for similar tubes by other authors. Warren and Parker (1962), Cottrell and Walker (1967), and Naidu and Prasad (1968) employ photoelectrons ejected from a gold cathode by ultraviolet light; Huxley and Zaazou (1949) and Lawson and Lucas (1965a) use electrons extracted, respectively, from a brush discharge on a thin tungsten wire and from a glow discharge between a spherical copper electrode and a

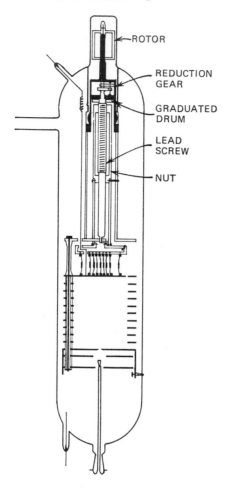

ROTOR

REDUCTION GEAR

GRADUATED DRUM

LEAD SCREW

NUT

Fig. 3.7.2. Variable-length tube designed and used by Crompton and his coworkers for measurements of characteristic energies with Townsend's method [after Crompton and Jory (1962)].

brass plate anode. In the case of photoelectrons the cathode current is controlled by acting on the ultraviolet lamp; moreover, Warren and Parker coated all the electrodes except the cathode with colloidal graphite, which provides low reflectivity and hence reduces stray currents due to scattered light, and low contact potentials. In the case of the glow discharge the current has been regulated by a series of control grids, which are traversed by the electrons as soon as they leave the anode of the glow discharge.

More recently, in order to avoid conversion of parahydrogen into normal hydrogen when in contact with a heated filament, Crompton and McIntosh (1968) replaced, in the case of this gas, the filament with a silver-coated foil of americium 241, which provides electrons by volume ionization of the gas. The source for this application, however, has to be carefully designed, so as to prevent penetration of alpha-particles through the hole into the diffusion chamber; furthermore, compensation techniques are required insofar as background electrons generated by γ-rays are present in the chamber.

C. Figure 3.7.3 shows a schematic of the original tube with fixed electrodes, used at Sydney for measurements in attaching gases according to Bailey's method [Bailey (1925), Healey and Reed (1941)]. The distance between successive guard rings R_0, R_1, R_2, R_3, and R_4 is 2 cm; the height of the two diffusion chambers is 4 cm. The electrodes E_1, E_3, and E_5 consist of rings over which silver foil with a 4-mm-wide slit is tightly stretched; the slits in

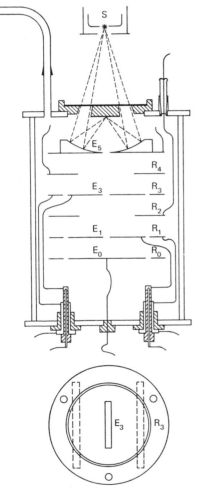

Fig. 3.7.3. Tube for measurements of characteristic energies and attachment coefficients with Bailey's method [after Bailey (1925)].

E_1 and E_3 are considerably longer than the electron source slit in E_5, so that their length can be regarded as infinite in the computation of current ratios. Ultraviolet light from the spark S is the source of the photoelectrons of the diffusing swarm. A similar tube, in which the height of diffusion chambers can be changed from 2 to 4 cm, is described by Bailey and McGee (1928).

The usual techniques for measuring very small ratios of currents in all these experiments are mostly modern elaborations of the original Townsend induction-balance method, which has the important characteristic that during each observation the electrostatic conditions within the drift chamber remain undisturbed [Townsend (1915)].

When the Townsend-Huxley method is used, the anode is at earth potential and this condition must be kept also in the course of measurements; we shall describe here the induction-balance technique as used for this purpose by Crompton and his coworkers in their most recent experiments of this type [Crompton, Elford, and Gascoigne (1965)]. The circuit diagram is shown in Fig. 3.7.4. A voltage V (1–10 V) is applied to the highly linear potentiometer P_1 (linearity within $\pm 0.05\%$), which is driven at a constant

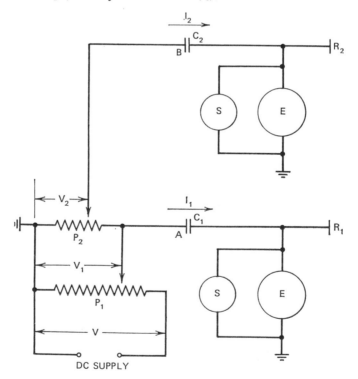

Fig. 3.7.4. Circuit diagram of the double induction balance [after Crompton, Elford, and Gascoigne (1965)].

rate of 1 revolution per minute by a synchronous motor. Thus two highly linear sweep voltages are generated at A and B, their relative amplitudes being determined by the setting of potentiometer P_2 (resolution and accuracy 1 part in 1000). Two equal (within 0.01 %) condensers C_1 and C_2 are fed by these voltages on one side and are connected to the two anode electrodes R_1 and R_2, whose current ratios we have to measure, on the other side. Two constant displacement currents I_1 and I_2 will flow through the condensers and will be equal to the respective electrode currents, when electrometers E indicate zero currents. This is achieved by adjusting voltage V and the setting of P_2, which thus provides the value of the current ratio. In this way electrodes R_1 and R_2 are maintained at earth potential (within 0.2 mV) as desired. For maximum accuracy switches S unearth electrometers when the slider of P_1 passes through the zero position, so that the measuring process is actually an integrating one. Condensers C_1 and C_2 are of special design to eliminate leakage currents [Crompton and Sutton (1952)]. Highly stable electrometers, capable of detecting currents of the order of 10^{-15} to 10^{-16} A, are employed.

When Bailey's method is used, one must measure the ratio of currents collected by electrodes at different potentials. In this case also, the induction-balance technique can be employed. However, if one wants to keep electrometers at earth potential, two double condensers must be used for C_1 and C_2, as shown in the schematic diagram of Fig. 3.7.5 [Bailey (1925), Healey and Reed (1941)].

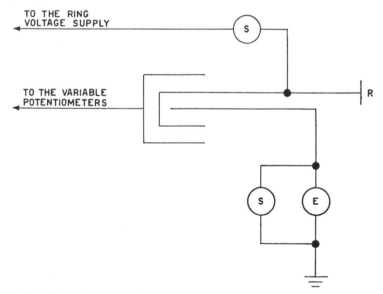

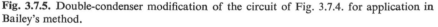

Fig. 3.7.5. Double-condenser modification of the circuit of Fig. 3.7.4. for application in Bailey's method.

3.8   DIRECT CURRENT CONDUCTIVITY

In this section we consider the techniques for measuring the dc conductivity of isothermal ionized gases using very low fields, so as not to perturb electron energies. The ionized gases that we shall here consider are characterized by approximately equal electron and positive-ion densities, so that quasi neutrality of the electron charge is typical of our cases; ionized gases satisfying this condition are customarily called plasmas. Whereas electrons and positive ions have about equal densities, their drift velocities under the action of the same applied electric field are quite different, the electrons being much faster; therefore, the total plasma conductivity, which is here the observable parameter, can be identified with the conductivity due to the electrons alone. It follows also that, if the electron density is known, from these measurements we can derive the electron drift velocity.

In general, dc conductivity measurements are much less accurate than the corresponding high-frequency conductivity measurements, which we shall consider in Section 3.11. This is mainly due to the presence of voltage drops across the sheaths on the electrodes and on the probes, and to deviations of the actual plasma and field configurations from those of the unidimensional case, to which simple theories apply; therefore, electric fields and corresponding current densities in the plasma cannot be determined with high accuracy. For this reason dc conductivity measurements are seldom considered a source of reliable data for the determination of electron collision parameters. The only interesting exceptions are those in which the above difficulties can be avoided or reduced, such as the case of electrodeless measurements in a shock tube, where a transversally homogeneous plug of moving ionized gas interacts and modifies a magnetic field (induction probe, external to the tube); the case of thermally ionized gases in high-temperature chambers, where the conductivity is measured between electrodes, whose thermionic emission strongly reduces any voltage drop across the electrode sheaths; and the case of measurements using an ac field of low frequency, so that the conductivity maintains the dc value, but external exciting structures can be used.

In spite of the above limitations on the measuring accuracy of this method, its usefulness, in comparison to the methods discussed so far, is beyond any doubt. It is the only practical dc method that can be used for determining electron collision parameters in encounters with positive ions or with temperature dissociation products (atomic species, e.g.) of molecular gases.

A. Induction Probe

This probe, devised by Lin, Resler, and Kantrowitz (1955) [see also Fuhs (1965)] for use in pressure-driven shock tubes, is based on the interaction between the radial component of an axially symmetric magnetic field, applied to a section of the shock tube, and the electrically conducting slug of the test gas, which forms between the shock front and the interface with the driver gas. Following Pain and Smy (1960), the appropriate magnetic field is generated using two equal opposite coils fed by equal direct currents (Fig. 3.8.1); when the conductive plug of ionized gas passes through the coils, the radial component of the magnetic field induces azimuthal currents in the plug and in turn these currents induce an emf voltage in a search coil placed between the field coils. This voltage is a function of the plug conductivity as follows.

When the shock front is at the distance s from the search coil, the electrical conductivity* may be expressed as $\sigma(s - x)$, since the thermodynamic state of the gas is a function of the distance from the shock front only and transverse nonuniformities are usually negligible. The total induced magnetic flux passing through the search coil can be written as:

$$\Phi(s) = U_g \int_{-\infty}^{s} h(x)\sigma(s - x)\,dx \tag{3.8.1}$$

since the induced current in the gas is proportional to the gas flow speed U_g and to its local conductivity; $h(x)\,dx$ is a quantity proportional to the field coil current and dependent on the geometry of the coil system: it represents the induced flux generated by an ionized slab, having unity conductivity and moving at unity velocity, which is located at x and has the thickness dx. The induced voltage in the search coil will be:

$$V = A\frac{d\Phi}{dt} = AU_s\frac{d\Phi}{ds} \tag{3.8.2}$$

where $U_s\,(= ds/dt)$ is the speed of the shock front (in general $\neq U_g$), and A is a constant proportional to the number of turns of the search coil.

If σ is constant behind the shock front, the induced voltage becomes:

$$V(s) = AU_sU_g\sigma h(s) \tag{3.8.3}$$

The function $h(x)$ has a maximum at the position of tightest coupling (at $x = 0$ for the geometry depicted in Fig. 3.8.1); correspondingly the induced voltage peak will be:

$$V_p = AU_sU_g\sigma h_{\max} \tag{3.8.4}$$

The factor $Ah_{\max}$ can be computed if the coil geometry is known, but it is more convenient to calibrate the apparatus experimentally by shooting along the tube a metallic rod of known conductivity σ_r and of known velocity

---

* Throughout this section σ will mean σ_{dc} only.

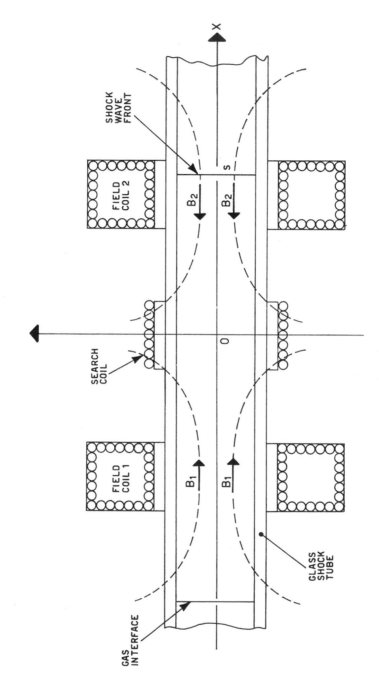

X

SHOCK
WAVE
FRONT

FIELD
COIL 2

B_2

B_2

s

SEARCH
COIL

0

FIELD
COIL 1

B_1

B_1

GLASS
SHOCK
TUBE

GAS
INTERFACE

Fig. 3.8.1. Schematic diagram of the induction probe for measurements of plasma conductivities in a shock tube.

U_r, and by observing the induced voltage peak V_{rp} ($= AU_r{}^2\sigma_r h_{max}$). The ionized gas conductivity is then obtained from the formula:

$$\sigma = \frac{U_r{}^2}{U_s U_g} \frac{V_p}{V_{rp}} \sigma_r \qquad (3.8.5)$$

If the conductivity of the ionized gas plug is not constant along the tube axis, we should derive its profile by unfolding the integral equation (3.8.1), where $\Phi(s)$ is obtained by integration of V according to (3.8.2) and the instrumentation function $Ah(x)$ is known from the calibration experiment. However, we usually confine our interest to the value of the maximum conductivity, since at this position the thermal and ionization conditions of the ionized gas are most close to equilibrium and its temperature can be evaluated with the largest degree of confidence. On the basis of formulas already given we may express the maximum conductivity either as a function of V_p by means of the same equation (3.8.5), where we introduce only a multiplying factor ψ_1 to take into account the conductivity profile, or as a function of the peak value of the induced voltage integral Φ_p, according to a similar formula:

$$\sigma = \frac{U_r{}^2}{U_s U_g} \frac{\Phi_p^i}{\Phi_{rp}} \psi_2 \sigma_r \qquad (3.8.6)$$

Functions ψ_1 and ψ_2, determined by Lin, Resler, and Kantrowitz for some of the most common conductivity profiles, are shown in Fig. 3.8.2. Since ψ_2 in the cases shown in the figure is unity or close to it, provided conductivity does not change too fast, it will usually be more convenient to determine the induced voltage integral and compute the maximum conductivity by means of (3.8.6), instead of measuring the peak voltage and using (3.8.5).

Radial conductivity variations, which are not taken into account by the above theory, are a possible source of error in the application of this method. Most important from this point of view is the presence of a boundary layer in the shock tube. Actually, in cases where the decay of the voltage integral immediately beyond the maximum can be interpreted in terms of a dominant change of the thickness of the boundary layer, which develops from the shock front, the true diameter of the plasma slug can be properly estimated from the measured decay slope and the conductivity can thus be corrected for the boundary effect [Enomoto, Goda, and Hashiguchi (1970)].

Furthermore it should be noted that the conductivity σ must not be altered by the presence of the applied magnetic field, and this condition holds if $\nu_m \gg \omega_b$. For a typical field of the order of 100 G this requires sufficiently high pressures to ensure that $\nu_m \gg 3 \times 10^8$ sec^{-1}.

Figure 3.8.3 shows schematically the design of the search coil used by Lin, Resler, and Kantrowitz (1955). It consists of two 25-turn coils wound side by side, but connected in a push-pull arrangement, so that any electrostatic

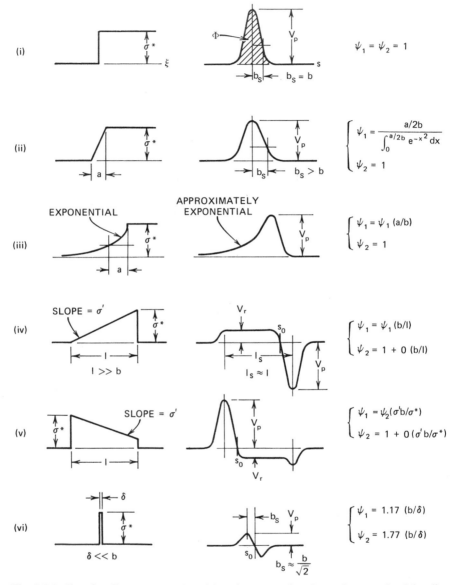

Fig. 3.8.2. Search coil response $V(s)$ and functions ψ_1 and ψ_2 for various conductivity distributions $\sigma(\xi)$; b is a measure of the longitudinal width of the magnetic field. Note: $O(\varepsilon)$ denotes a term of the order of ε [after Lin, Resler, and Kantrowitz (1955)].

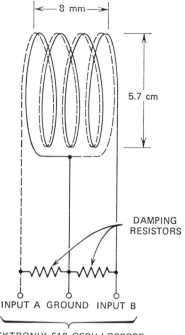

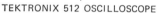

Fig. 3.8.3, Schematic diagram of the search coil of an induction probe [after Lin, Resler, and Kantrowitz (1955)].

pickup due to the capacity between the coil and the flowing ionized gas in the shock tube is minimized, while the magnetic pickup is unaffected. The induced voltage is photographically recorded on an oscilloscope, whose sweep is triggered by a photomultiplier that detects the luminous shock front, upstream of the search coil. The shock speed measurements are accomplished by taking rotating-drum camera pictures of the propagating shock wave.

On the assumption that thermal and ionization equilibrium has been reached, the gas temperature appropriate to the conductivity maximum can be computed. To this end one must use the mass, momentum, and energy conservation equations, which relate the pressure and enthalpy of the gas to its density and to the shock speed, and tabulations, available for most gases of interest, of pressure and enthalpy as functions of density and temperature [Teare (1963)]. The related electron density is obtained from Saha's equation, which for a low degree of ionization and for one ionizable species only, of density n_g, can be written as:

$$n_e = 6.95 \cdot 10^7 g^{0.5} n_g^{0.5} T_g^{0.75} \exp\left(-\frac{eE_i}{2kT_g}\right) \tag{3.8.7}$$

where densities are in units of centimeters^{-3}, g is the ratio of the ion to the neutral partition functions (statistical weight) [see, e.g., Sutton and Sherman (1965)], and E_i is the ionization energy in electron volts.

However, the presence of small amounts of easily ionizable impurities may considerably increase the electron density beyond the value predicted by (3.8.7). For this reason, and because the time required for ionization to reach equilibrium may be longer than the available observation time, it is strongly recommended that electron densities be measured experimentally; the use of microwaves is valuable for this purpose (see Section 3.11).

B. Voltage-Current Characteristic in High-Temperature Cells

This method applies to the case of easily ionizable gases, such as alkali-metal vapors or mixtures containing these vapors. By means of appropriate ovens the gases are maintained near thermal equilibrium at temperatures up to 2000°K in a diode test section. At these temperatures electrons are supplied by thermionic emission of the electrodes, so that if this emission is large enough any sheath drop should vanish in the zero current limit [Harris (1963)], and a linear V-I characteristic has to be found at the low voltages required by the condition that electron energies be unchanged.

In the typical setup used by Harris (1963, 1964) for studying the conductivities of cesium- and potassium-seeded rare gases, the diode test section is cylindrical in shape, bounded on the side by an alumina tube and on the top and bottom by plane molybdenum electrodes (sizes 1.27 cm diam. × 2.54 cm long, and 1.90 cm diam. × 3.81 cm long). The seeded gas flows slowly in through a small hole in the center of the top electrode, and out again, after 5–20 sec, through several baffled holes in the bottom electrode. An external oven provides a uniform temperature in the test section; the temperature is calibrated for the oven input electric power by means of pyrometric measurements with an unseeded gas, so as to avoid pyrometer errors due to optical absorption by cool vapors of the seeding gas in the optical path. From the initial linear portion of the V-I characteristic, the gas conductivity is computed with the Ohm formula:

$$\sigma = \frac{I}{V}\frac{L}{A} \tag{3.8.8}$$

where L and A are the length and the cross-sectional area of the test section.

Seeded combustion gases are very important for application in magnetohydrodynamic generators. When the conductivity is measured in these gases, it is not necessary to use an independent furnace to heat the test cell. The high-temperature combustion gas itself, flowing at low speed, heats the electrodes nearly to the gas temperature, provided the electrodes are well insulated thermally from the outside. Concentric electrodes, instead of plane ones, have been used for these measurements [Mullaney, Kydd, and Dibelius (1961)].

At thermal equilibrium with zero current, sheath drops vanish when there is enough thermionic emission from the electrodes, but a sheath resistance appears as soon as a current flows. Since a current must flow to determine the gas conductivity, it is important to determine the magnitude of this resistance, so that experimental conditions may be chosen properly to make its presence insignificant. Following Rynn (1964), we express the cathode current density j_c and the anode current density j_a at the boundaries of the two sheaths as:

$$j_c = j_s \exp\left(-\frac{e\phi_c}{kT_g}\right) - j_r$$
$$j_a = -j_s \exp\left(-\frac{e\phi_a}{kT_g}\right) + j_r \tag{3.8.9}$$

where j_s is the saturation electron current from each electrode, given by the Richardson formula and much greater than any other current; ϕ_c and ϕ_a are the two sheaths potentials; and:

$$j_r = \tfrac{1}{4}ne\bar{v} \simeq \tfrac{1}{4}ne\sqrt{3kT_g/m} \tag{3.8.10}$$

is the random current at the plasma ends, given by well-known kinetic theory expressions [Kennard (1938)]. Continuity requires that both j_c and j_a be equal to the total current density j. Solving (3.8.9) for ϕ_c and ϕ_a, and taking their difference V_d, which is the overall voltage drop across the sheaths, yields:

$$V_d = \phi_a - \phi_c = \frac{kT_g}{e}\ln\left(\frac{1 + j/j_r}{1 - j/j_r}\right)$$

$$= \frac{kT_g}{e}\frac{2j}{j_r} \tag{3.8.11}$$

insofar as $j/j_r \ll 1$. Thus the total sheath resistance per unit area becomes:

$$\frac{V_d}{j} = \frac{2kT_g}{ej_r} = \frac{8}{e^2 n}\sqrt{mkT_g/3} \tag{3.8.12}$$

This resistance can be neglected if it is small compared to $V/(I/A) = L/\sigma$.

If we consider a constant collision frequency, this condition can be written as:

$$\frac{m v_m}{e^2 n} L \gg \frac{8}{e^2 n} \sqrt{m k T_g / 3}$$

which becomes:

$$L \gg \frac{8}{v_m} \sqrt{k T_g / 3m} \tag{3.8.13}$$

In other words, we can say that the length of the plasma must be very much greater than the average electron mean free path. It is worth recalling that this condition is also one of those required for the validity of all our developments based on the kinetic theory.

In order to avoid the effects of any residual presence of unknown sheath resistance, Roehling (1963) used a cell with two parallel plates whose separation distance could be changed from outside; by measuring the total resistance as a function of the electrode distance, the plasma resistance could obviously be separated from the sheath resistance. More recently Japanese investigators [Yano, Matsunaga, Hiramoto, and Shirakata (1964), Sakao and Sato (1969)] used for the same purpose the four-electrode arrangement shown in Fig. 3.8.4. The voltage V to be used in eq. (3.8.8) is the one between the two floating grids, whereas the supply voltage is applied between the two most external electrodes.

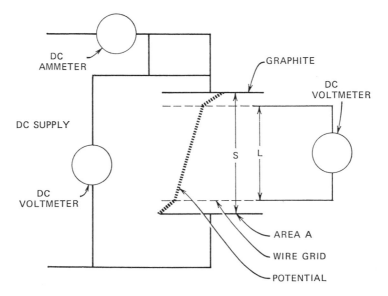

Fig. 3.8.4. Four-electrode arrangement to measure electrical conductivity [after Fuhs (1965)].

C. Voltage-Current Characteristic in Q-Machines

The Q-machine, originally developed by D'Angelo and Rynn (1960), is an appropriate device for producing quiescent plasmas by surface ionization on hot plates without applying any external ionizing electric field. A schematic is shown in Fig. 3.8.5. Two hot plates, usually made of a refractory metal, such as tungsten or tantalum, are placed at 45° with the axis of the machine

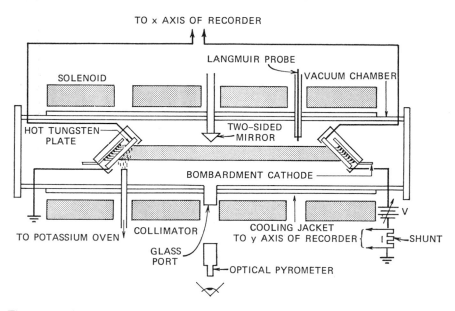

Fig. 3.8.5. Schematic diagram of a Q-machine used for measurements of the electrical and thermal conductivities of an alkali-metal plasma [after Rynn (1964)].

and heated to electron emission temperatures. A collimated neutral beam of alkali metal (or barium) illuminates one of the plates, which thus ionizes most of the impinging atoms of the neutral beam. The ions, together with the electrons emitted by thermionic emission, stream along the tube axis toward the second plate under the confining action of a strong, axial magnetic field; thus a cylindrical plasma is formed in the central region between the two plates. The vacuum chamber walls are cooled so as to condense out any neutral atom present in the background gas or resulting from recombination.

When a voltage is applied between the plates and a current flows, a sheath will form on each plate. The expressions derived in the previous case of high-temperature cells apply here too for computing the plasma conductivity and the sheath resistance, since the magnetic field has no influence on the axial motion of the charged particles. The above formulas are correct, however, only in the limit case of very small currents; thermal effects at larger currents will be considered later in Section 3.13. Rynn (1964) measured in this way the initial conductivity of a nearly fully ionized potassium plasma and observed the sheath effect when condition (3.8.13) is not satisfied.

D. Alternating Current Low-Frequency Measurements

At high pressures and low frequencies, such as to have $v_m{}^2 \gg \omega^2$, the real part of the plasma conductivity is dominant and can be identified with the dc value. In this case, if the plasma is contained in a nonconducting cylindrical tube, the dc conductivity can be determined by measuring the change in the Q of a solenoid, concentric to the tube but external to the plasma.

The problem is simplified if the fields generated by the solenoid are not disturbed by the presence of the plasma. This requires that the plasma skin depth δ_s, given by the usual formula for conductors:

$$\delta_s{}^2 = \frac{2}{\omega \mu_0 \sigma_r} = \frac{2mv_m}{e^2 \omega \mu_0 n} \tag{3.8.14}$$

(v_m is taken here as velocity independent), be sufficiently larger than the radial plasma dimensions, thus establishing the maximum usable frequency. Under these conditions the changes in the equivalent parameters of the solenoid can easily be related to the plasma conductivity.

The method was thoroughly discussed by Persson (1961), who used a bridge structure with two identical solenoids, one of which interacts with the plasma. He determined the plasma conductivity through measurements of the perturbation voltage induced in the solenoid by the plasma currents generated by the applied fields. The system can be calibrated using a medium of known conductivity, which replaces the plasma. Persson has used a single-layer solenoid with a special shield, provided by a tin oxide coating, which prevents the axial electric field, associated with a plain solenoid, from penetrating into the plasma; capacitive effects that would complicate the interaction picture are thus avoided. Persson considered applying to the bridge either a sinusoidally time-varying signal or a pulse; he found that, using a Gaussian-like pulse with a width of approximately 1 μsec and critically damped solenoids, a time resolution of less than $\frac{1}{3}$ μsec and a conductivity measuring capability over 4–6 decades are obtained.

Instead of using a bridge, Donskoi, Dunaev, and Prokof'ev (1962) determined the change in the solenoid Q by measuring the voltage change at resonance (1–10 MHz) of a resonator formed by coupling the solenoid in parallel with a capacitor. This technique has been used for measurements in shock tubes [Vasil'eva et al. (1970)].

In addition to the change in Q, the inductance of the solenoid also changes because of the presence of the plasma; if appropriate conditions are chosen, the inductance change ΔL depends chiefly on the real part of the conductivity (practically equal to the dc value) and not on the imaginary part. This method is due to Savic and Boult (1962), who determined the ΔL change from the frequency shift, given by a discriminator circuit, of a resonant circuit of which the solenoid is one of the basic elements. Curves of $\Delta L/a$ versus $\omega \sigma b^2$ for different values of the a/b ratio (a = radius of solenoid coil; $b(<a)$ radius of the plasma core) have been numerically computed, but here too calibration performed by means of a medium of known conductivity is preferred for actual experimental use (since ΔL is a function of the $\omega \sigma$ product, media of different conductivities can be used for the calibration in a given $\omega \sigma$ range simply by choosing appropriate operating frequencies). Savic and Boult and later Lau (1964) used this technique for measuring the conductivity in shock tubes; these authors adopted an operating frequency of 10.7 MHz and found for the considered application adequate time resolution and very low hydromagnetic effects. Use of a probe coil immersed in the plasma ($b > a$) is also feasible; Stubbe (1968) has determined the theoretical expressions for the change in coil impedance and has verified experimentally their applicability to conductivity measurements in electrolytes and in shock tube plasmas. No significant use of this method has however been reported so far.

3.9 DIRECT CURRENT TRANSPORT PARAMETERS IN STATIC MAGNETIC FIELDS

The methods customarily used for the determination of the dc transport coefficients in magnetic fields can be divided into four major categories: magnetic deflection, diffusion in parallel fields, diffusion in crossed fields, and conductivity in crossed fields.

A. Magnetic Deflection

These methods are based on observation of the deflection of an electron swarm moving in a magnetic field. The first measurements of this type were made by Townsend and Tizard (1913).

The original Townsend method makes use of a diffusion chamber with a slit aperture for the electron source and, parallel to it, strip-like collecting anode electrodes. This is exactly the same diffusion apparatus shown in Figs. 3.5.1a and 3.5.1c and used for the measurement of the ratio D/μ, but now the chamber is placed in a magnetic field, parallel to the source slit, and the electrodes C_1 and C_2 are connected to form a single electrode, with C_3 behaving as a separate one. The electron stream is deflected as shown schematically in Fig. 3.9.1.

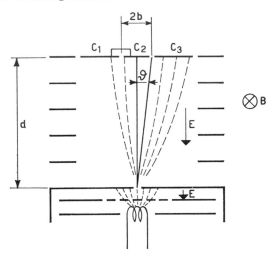

Fig. 3.9.1. Diagram of the electron stream deflection in crossed electric and magnetic fields, as used for measurements of the magnetic drift velocity.

In this experiment the motion of the center of the stream is given approximately by the composition of the perpendicular and transverse drift velocities; therefore the angular deflection ϑ will be determined by the relationship:

$$\tan \vartheta = \frac{w_\perp}{w_T} = \frac{B}{E} w_M \tag{3.9.1}$$

where eq. (2.6.77) has been used. In the experiment the magnetic field is adjusted until equal currents are received by the two electrodes $C_1 + C_2$ and C_3; this represents the condition for which the center of the stream falls in the narrow gap between $C_1 + C_2$ and C_3. If $2b$ is the width of C_2 and d is the distance between the slit and collector planes, $\tan \vartheta$ is equal to b/d and the magnetic drift velocity w_M turns out to be:

$$w_M = \frac{E}{B} \tan \vartheta = \frac{E}{B} \frac{b}{d} \tag{3.9.2}$$

Measurements of the type described are usually performed under conditions appropriate to the LMF limit, so that w_M data for each gas and temperature appear to be functions of the proper variable E/p only.

A slight modification of the above method was introduced by Bernstein (1962), who employed a coaxial cylindrical geometry (Fig. 3.9.2) instead of Townsend's parallel plane geometry, the purpose being to eliminate edge effects in the $\mathbf{E} \times \mathbf{B}$ direction and to allow better alignment of the electrodes

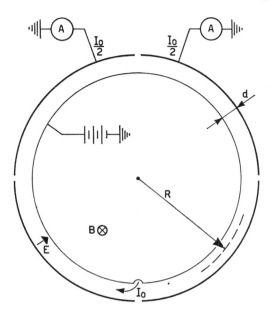

Fig. 3.9.2. Schematic diagram of Bernstein's apparatus for measuring magnetic drift velocities using a coaxial cylindrical geometry.

with the magnetic field. In Bernstein's experiment high magnetic fields are used; the anode is divided into four quadrants. As in Townsend's method the magnetic field is adjusted so that the electron current is divided equally between two adjacent quadrants; if Q is the number of quadrants traversed, d the diffusion gap separation, and R the median radius of the gap, the magnetic mobility is approximately:

$$\mu_M p = \frac{w_M}{E/p} = \frac{p}{B} \frac{\pi R Q}{2d} \tag{3.9.3}$$

in accordance with the equivalent (3.9.2).

In Bernstein's experiment the SMF limit conditions are satisfied, and $\mu_M p$ is, for each gas and temperature, a function of E/B only. Experimental details and dimensions of the apparatus are shown in Fig. 3.9.3.

A more substantial modification of Townsend's method is due to Huxley and Zaazou (1949). They consider a typical diffusion chamber for D/μ measurements, with a source hole and a collector which now not only is divided into a central disk and concentric annuli, but is also bisected diametrically by a narrow slit (Fig. 3.9.4). In the absence of deflecting magnetic fields the currents received by each semicircular half of the bisected collector are equal; when a magnetic field parallel to the slit is applied the ratio of the

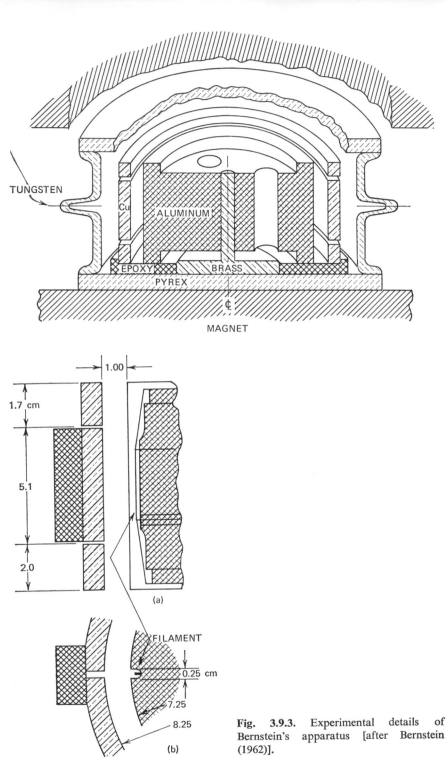

TUNGSTEN

Cu

ALUMINUM

EPOXY   BRASS

PYREX

¢

MAGNET

1.00

1.7 cm

5.1

2.0

(a)

FILAMENT

0.25 cm

7.25

8.25

(b)

Fig. 3.9.3. Experimental details of Bernstein's apparatus [after Bernstein (1962)].

212

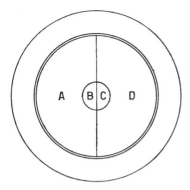

Fig. 3.9.4. Form of the collector electrode used by Huxley and Zaazou for measuring magnetic drift velocities.

two currents becomes [Huxley and Zaazou (1949)]:

$$R = \frac{1 - A}{1 + A} \tag{3.9.4}$$

where:

$$A = \frac{2}{\pi} W U^{1/2} \exp\left[W(1 - U)^{1/2}\right] \sum_{n=1}^{\infty} \frac{\Gamma(n)}{\Gamma(2n)} (2UW)^{n-1} K_{(n-1)}(W) \tag{3.9.5}$$

Γ denotes the gamma function, K_n is the modified Bessel function of the second kind and order n, and:

$$W = \frac{w_T}{2D_T} d\left[1 + \left(\frac{w_\perp}{w_T}\right)^2\right]^{1/2} = \frac{w_T}{2D_T} d\left[1 + \left(\frac{B}{E}\right)^2 w_M^2\right]^{1/2}$$

$$U = \frac{(w_\perp/w_T)^2}{1 + (w_\perp/w_T)^2} = \frac{w_M^2}{(E/B)^2 + w_M^2} \tag{3.9.6}$$

The above result has been derived from the solution of the differential equation governing the steady-state electron flow in crossed fields [eq. (2.5.1) with $\partial n/\partial t = 0$ and Γ given by (2.4.14)]:

$$\mathbf{\nabla \cdot \Gamma} = D_T\left(\frac{\partial^2 n}{\partial x^2} + \frac{\partial^2 n}{\partial z^2}\right) + D_\parallel \frac{\partial^2 n}{\partial y^2} - w_\perp \frac{\partial n}{\partial x} - w_T \frac{\partial n}{\partial z} = 0 \tag{3.9.7}$$

where the axis x is perpendicular to both fields, and y is parallel to $\mathbf{B}$ and z to $\mathbf{E}$. The same mathematical steps required for the derivation of eq. (3.5.10) from (3.5.1) (see Section 3.5) yield here the reported result; the appropriate assumptions are either a pole source at the cathode and zero electron density on the collector [Huxley (1959b)], or a dipole source at the cathode and a purely geometrical plane for the collector [Huxley and Zaazou (1949)]. The current ratio R is plotted in Fig. 3.9.5 as a function of $w_T d/D_T$ for a range of values of $w_\perp/w_T$.

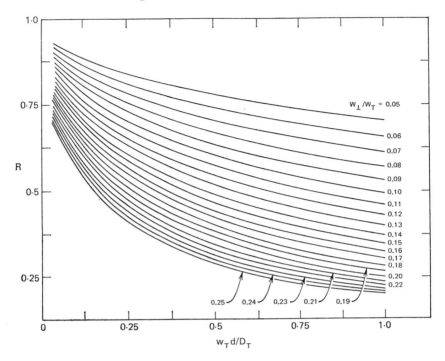

Fig. 3.9.5. Ratio R of the currents to the two semicircles of Fig. 3.9.4 as a function of $w_T d/D_T$ for a range of values of $w_\perp/w_T$ [after Huxley and Zaazou (1949)].

Since experiments are here performed at magnetic field values that satisfy the LMF limit conditions, the ratio w_T/D_T in (3.9.6) can be replaced by w/D for the same E/p. The value of w/D can be independently determined using the same apparatus in the usual way with no magnetic field; from the ratio R the magnetic drift velocity can then be calculated. The velocity w_M is here a function of E/p only, but since the method no longer relies on the apparatus geometry it is possible to determine w_M for a given E/p using a large number of combinations of electric and magnetic field strengths. This improves the reliability of the method as compared to the original Townsend version.

A diagram of a modern apparatus designed and used by Jory (1965) according to the Huxley-Zaazou modification of Townsend's method is shown in Fig. 3.9.6. The structures and dimensions of the cathode and of the guard electrodes are about the same as those of the diffusion tube shown in Fig. 3.7.1. The segments of the collecting electrode are mounted on a glass plate, which provides adequate electrical isolation and mechanical stability; the effective radius of the central disk is 6.00 ± 0.01 mm, the width of the

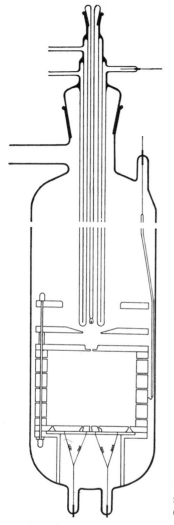

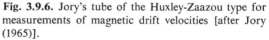

Fig. 3.9.6. Jory's tube of the Huxley-Zaazou type for measurements of magnetic drift velocities [after Jory (1965)].

gap around the disk is about 0.3 mm, and the width of the transverse cut is 0.25 ± 0.05 mm. The height d of the chamber is 83.3 mm. Magnetic fields are generated by a pair of Helmholtz coils; the presence of the magnetic field of the earth is taken into account.

The correct geometrical alignment and the proper behavior of the apparatus are easily checked by verifying that at zero magnetic field the current ratio is unity. Creaser (1967) found that the main cause of deviation from equality of the currents is the fact that the effective center of the electron source may not coincide with that of the source hole; a small electrode is then placed above this hole and adjacent to it, in order to produce a small transverse field capable of shifting the effective source center. Other causes of deviations are contact potential differences, which can be compensated as described in Section 3.7, and misalignment of the anode slit; the latter can be simply accounted for by repeating the measurements after reversal of the magnetic field and averaging the values.

B. Diffusion to Mobility Ratio $D_T/\mu_{\parallel}$ in Parallel Fields

At first the importance of the difference between w and w_M was not properly appreciated, so that measurements of w_M with Townsend's method were considered practically equivalent to measurements in drift tubes. Since, however, Townsend's method as previously described could not be used when an appreciable number of negative ions were present in the stream because of attachment processes, Bailey (1930) devised a new method for measuring drift velocities in magnetic fields, also in attaching gases. In the light of our modern understanding, however, Bailey's method actually provides from measurements only the ratio $D_T/\mu_{\parallel}$, and the drift velocity can then be derived only when very special assumptions are made.

In fact this method consists of using the same Bailey apparatus and technique employed for D/μ and η measurements, according to the scheme of Fig. 3.6.2, but with the addition of a magnetic field parallel to the electric one. If diffusion in the field direction can be neglected compared to the drift motion, it is a simple matter to verify that the current ratio R_e [to be computed from (3.6.16) using measured S_1 and S_2 data] will be given by the formula:

$$R_e = \psi\left(\frac{w_{\parallel}}{D_T d}\right) \exp\left(-\eta d\right) \tag{3.9.8}$$

derived from (3.6.25) with the appropriate substitutions of the parallel and transverse components of the transport coefficients.

Bailey's original procedure is based on first measuring D/μ and η without the magnetic field, and then adjusting the magnetic field until the divergence of the stream is the same as that of a stream composed entirely of thermal negative ions, that is, until S_1 and S_2 are equal to R_i. In this case we have:

$$\psi\left(\frac{\mu_{\parallel} E}{D_T d}\right) = \psi\left(\frac{eE}{kT_g d}\right)$$

Then, for this value of the magnetic field:

$$\frac{D_T}{\mu_{\parallel}} = \frac{kT_g}{e} \tag{3.9.9}$$

Introducing Townsend's energy factor k_1, according to definition (2.6.4), we obtain from this equation:

$$\frac{D_T}{D_\parallel} = \frac{1}{k_1} \tag{3.9.10}$$

If the collision frequency is velocity independent, the ratio of diffusion components, as given by (2.4.18) and (2.4.21), is:

$$\frac{D_\parallel}{D_T} = 1 + \left(\frac{\omega_b}{\nu_m}\right)^2 = 1 + (\mu B)^2 \tag{3.9.11}$$

where the mobility has replaced the collision frequency according to expression (2.4.20). The corresponding drift velocity is then:

$$w = \mu E = \frac{E}{B}\left(\frac{D_\parallel}{D_T} - 1\right)^{\frac{1}{2}} = \frac{E}{B}(k_1 - 1)^{\frac{1}{2}} \tag{3.9.12}$$

The data obtained with this technique by Bailey and his colleagues at Sydney [Healey and Reed (1941)] are all given as drift velocity values computed from the experimental results by using (3.9.12), where k_1 is obtained in each case from the value of D/μ determined, at the given E/p, by means of the same apparatus. In modern terms, however, we must interpret these old data more properly, by stating that they provide, as a function of E/p, only the B/p values at which in a certain gas $D_T/\mu_\parallel$ is equal to the fixed value kT_g/e; in accordance with (3.9.12), these values of the ratio B/p can be derived from the reported data by means of the relation:

$$\frac{B}{p} = \frac{E}{p}\frac{(k_1 - 1)^{\frac{1}{2}}}{w} \tag{3.9.13}$$

Modifications of the original Bailey method have also been employed. Thus Bailey and Rudd (1932), after having determined the ratio R_e in the absence of the magnetic field [eq. (3.6.25)]:

$$R_e = \psi\left(\frac{w}{Dd}\right)\exp(-\eta d) = \psi_1\left(\frac{E}{k_1 d}\right)\exp(-\eta d) \tag{3.9.14}$$

and of the field B, for which $S_1 = S_2 = R_i$, determined the ratio R_e for a magnetic field equal to $B/2$. Under the assumption of a velocity-independent collision frequency, this ratio will have the value:

$$R'_e = \psi\left\{\frac{w_\parallel}{D_\parallel d}\left[1 + \left(\frac{\mu B}{2}\right)^2\right]^{\frac{1}{2}}\right\}\exp(-\eta d)$$

$$= \psi_1\left\{\frac{E}{k_1 d}\left[1 + \left(\frac{\mu B}{2}\right)^2\right]^{\frac{1}{2}}\right\}\exp(-\eta d) \tag{3.9.15}$$

Substituting for μB its expression $\sqrt{k_1 - 1}$ according to (3.9.12), we obtain for the ratio:

$$R'_e = \psi_1 \left[\frac{(k_1 + 3)E}{4k_1 d} \right] \exp(-\eta d) \tag{3.9.16}$$

and k_1 can thus be calculated from the measured R_e and R'_e values by solving the equation:

$$\frac{\psi_1[(k_1 + 3)E/4k_1 d]}{\psi_1(E/k_1 d)} = \frac{R'_e}{R_e} \tag{3.9.17}$$

Thereafter η is determined from (3.9.14), and w from (3.9.12).

No recent measurement with Bailey's method has been reported in the literature. A similar approach, but using a Townsend type of diffusion chamber (see Section 3.5), was employed by Hall (1955a) for measuring $D_T/\mu_{\parallel}$ in hydrogen as a function of the applied fields.

C. Diffusion to Mobility Ratio $D_{\parallel}/\mu_T$ in Crossed Fields

Let us consider again a diffusion apparatus similar to the one shown in Figs. 3.5.1a and 3.5.1c, but with the magnetic field perpendicular both to the electric field and to the source slit. It is a simple matter to see that, if diffusion in the electric field direction is neglected in comparison to the drift motion, measurements of the ratio between the current falling on the central strip electrode C_2 and the total collector current yield the ratio

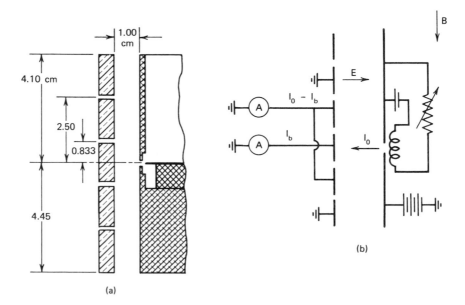

(a)

(b)

Fig. 3.9.7. Experimental details of the electrodes, and electrical schematic of the setup used by Bernstein for measuring parallel diffusion in crossed fields [after Bernstein (1962)].

$D_\parallel/\mu_T$. In fact, for a cathode-anode spacing d and a central strip of width $2b$, this current ratio will be:

$$R = \text{erf}\left(\frac{b}{2}\sqrt{w_T/D_\parallel d}\right) \qquad (3.9.18)$$

as for eq. (3.6.24) with $\eta = 0$ and substitution of the appropriate components for w and D.

This method has been used only by Bernstein (1962), who chose a cylindrical geometry, like the one employed for w_M measurements. Details of the diffusion chamber and an electrical schematic of the measuring setup are shown in Fig. 3.9.7. Experimental difficulties in strong magnetic fields were experienced, with the release of impurities, the related formation of negative ions, and unwanted electron diffusion in the cathode region.

D. Conductivity in Crossed Fields

Among the very few experiments designed for measuring σ_T under conditions similar to those of Section 3.8 (high-temperature gases, applied fields below the electron heating threshold) we mention only two.

a. A setup consisting of four cylindrical concentric electrodes, designed by Cottereau and Valentin (1967, 1968) for measuring currents and fields in the ionized gas plug of a shock wave. The magnetic field is longitudinal, namely, parallel to the axes of the shock tube and of the electrode setup; the voltage is applied between the outer electrodes, whereas the inner electrodes, designed to be small enough not to perturb plasma conditions, allow proper measurement of the local electric field in a way similar to that shown in Fig. 3.8.4 for a plane configuration. It should be possible, then, to obtain σ_T in high-temperature gases. Up to now, however, results have been rather discouraging, since the observed values disagree significantly from the theoretical predictions based on the most generally accepted collision parameter values [Cottereau, Guilly, and Valentin (1970)].

b. A high-temperature diode cell with two cylindrical concentric electrodes and the magnetic field applied parallel to the axis of the electrode configuration. The general remarks made in Section 3.8B apply here too. This technique was used by Mullaney and Dibelius (1961) for measuring the conductivity of cesium, both in the presence and in the absence of a magnetic field, and deriving from these data the electron collision frequency by means of eq. (2.6.93).

3.10 FREE ELECTRON DIFFUSION

A. Diffusion when no Magnetic Field is Present

In the first sections of this Chapter we considered the experimental methods for measuring the transport coefficients μ and D/μ. The product of these two coefficients at the same E/p yields the corresponding value of the diffusion coefficient D. Analogously, if the electron swarm is in thermal equilibrium with the gas and the mobility is measured without significantly changing this condition, the diffusion coefficient is given by $(kT_g/e)\mu$ according to Einstein relation (2.6.2). These are all indirect methods for determining D.

Now we shall consider, instead, how D can be measured directly; this approach is most important in the case of an electron swarm in thermal equilibrium with a gas, since in some gases (neon, e.g.) it is very difficult to measure μ and D/μ at E/p low enough not to perturb the electron energy distribution (the presence of contact potentials in the tube forbids the use of very low applied fields, and large pressures may require a special pressurized apparatus).

Let us consider, then, an electron swarm diffusing freely through a gas to the walls of a container; the swarm is in thermal equilibrium with the gas, and any existing electric or magnetic field is assumed to be weak enough not to modify the equilibrium regime of the diffusion process. Accordingly, electrons are considered to have everywhere a Maxwell distribution at the gas temperature T_g, and their rate of loss due to diffusion is assumed to exceed by far attachment and recombination losses. In this case eqs. (2.5.3) and (2.4.16) or (2.4.17) yield:

$$\frac{\partial n}{\partial t} - \nabla^2(Dn) = \frac{\partial n}{\partial t} - D\,\nabla^2 n = 0 \qquad (3.10.1)$$

where D is independent of time and position.

The solutions of eq. (3.10.1) for containers of various geometries and for the usual boundary condition of zero density at the walls have been discussed thoroughly by many authors; a review of the most significant results can be found in Chapter 10 of McDaniel's (1964) book. In general the solution for the density distribution can be written as the sum of an infinite number of diffusion modes, each decaying exponentially in time, when all electron sources have been turned off. If τ is the decay time constant of a mode, the appropriate density distribution is given by the corresponding eigenfunction solution of the equation, derived from (3.10.1):

$$\Lambda^2\,\nabla^2 n + n = 0 \qquad (3.10.2)$$

where we have introduced the characteristic diffusion length of the mode:

$$\Lambda = \sqrt{D\tau} \qquad (3.10.3)$$

This length is particularly useful in describing the shape and size of the container from the point of view of the diffusion process.

The mode with the largest Λ value (the fundamental mode) is the one that decays at the lowest rate and hence dominates at the latest times, although many other modes can initially be present when electron sources are turned off. In order to determine the diffusion coefficient the experimental conditions are chosen accordingly: the electron density decay after an ionization period is observed at late enough times so that the time constant of the fundamental mode alone is obtained from the measurements; the container has a simple shape, for which the value of the diffusion length for the fundamental mode is known; the diffusion coefficient is then derived from (3.10.3). The usual container shapes and the corresponding diffusion lengths for the fundamental modes are as follows:

a. A rectangular parallelepiped of sides a, b, c:

$$\frac{1}{\Lambda^2} = \pi^2\left(\frac{1}{a^2} + \frac{1}{b^2} + \frac{1}{c^2}\right) \tag{3.10.4}$$

b. A right circular cylinder of radius R and height L:

$$\frac{1}{\Lambda^2} = \left(\frac{2.405}{R}\right)^2 + \left(\frac{\pi}{L}\right)^2 \tag{3.10.5}$$

c. A sphere of radius R:

$$\Lambda = \frac{R}{\pi} \tag{3.10.6}$$

However, in these experiments the electron swarm is typically obtained by volume ionization, so that positive ions also are present. Since positive ions are slower, a space-charge field is created by the unequal electron and ion free diffusion rates, and this space charge may build up to the point where, because of its field action, both ions and electrons diffuse at the same rate. At this limit the density distribution is still given by (3.10.2), but the diffusion coefficient will now have so-called ambipolar value D_a, which for thermal equilibrium conditions is twice the free diffusion coefficient for the positive ions. The ambipolar limit is attained at sufficiently high electron densities to allow for the formation of an adequately large space charge; at the other end, at very low densities, free diffusion conditions prevail. Experiments, as described for the determination of the free electron diffusion coefficient D, must then be performed at very low densities; we shall now evaluate how low they have to be.

For this purpose it is convenient to consider the concept of the Debye shielding length [see Section 1.5 and details in most textbooks, such as Spitzer (1956), Longmire (1963), Shkarofsky, Johnston, and Bachynski (1966)]:

$$\lambda_D = \sqrt{\varepsilon_0 k T_g / e^2 n} \tag{3.10.7}$$

which is a measure of the size of the maximum volume over which an ionized gas may, on the average, be nonneutral. Free electron diffusion conditions can then exist only when the container size, considered from the point of view of diffusion, is much smaller than the Debye length, that is, when $\Lambda \ll \lambda_D$. After substitution of the numerical values into (3.10.7), this condition, for $T_g = 300°\mathrm{K}$ and Λ in centimeters, becomes:

$$n \ll 1.4 \times 10^4/\Lambda^2 \text{ electrons cm}^{-3} \tag{3.10.8}$$

The theoretical treatments of Allis and Rose (1954) and Cohen and Kruskal (1965), who have all been concerned with the transition region between ambipolar and free diffusion, indicate that condition (3.10.8) has to be satisfied by 3 or 4 orders of magnitude, before the decay time constant attains within a reasonable approximation the free electron diffusion value. Therefore densities have to be below $\sim 10/\Lambda^2$ or $\sim 10^3$ electrons cm^{-3} for a typical small container with $\Lambda = 0.1$ cm.

Except in the case of very weak ionization, these extremely low densities are obtained very late after the ionization period, so that both requirements—having electrons in thermal equilibrium with the gas and decaying according to the fundamental diffusion mode alone—are very well satisfied at the same time. The difficulty of this procedure for determining the free diffusion coefficient lies instead in measuring these low densities with adequate accuracy, using existing techniques. We shall now describe three methods that can be considered for this purpose.

a. The first method is due to Cavalleri, Gatti, and Principi (1964). At the instant when density has to be measured, a well-defined, damped radiofrequency oscillation is applied, strong enough to produce a proportional avalanche multiplication of the original electrons. The intensity of the light pulse emitted by the avalanche is proportional to the electron density, and its amplitude distribution can be measured by means of a photomultiplier tube connected, through an appropriate amplifier, to a multichannel pulse-height analyzer. By measuring several of these distributions at different delays after the ionization period and by plotting on a semilogarithmic paper the relative amplitudes of the centroids of the distributions as a function of time, one obtains the time constant of the electron density decay directly from the slope of this plot.

The latest experiments with this method were performed by Cavalleri (1969a), using a collimated burst of monoenergetic X-rays as the primary source of diffused low-density ionization. A schematic of the measuring chamber is shown in Fig. 3.10.1; it consists of a cylindrical glass envelope (radius 3.735 cm) containing an X-ray window D and two plane stainless steel electrodes (A and B, separated by $L = 2.952$ cm) with a diameter very

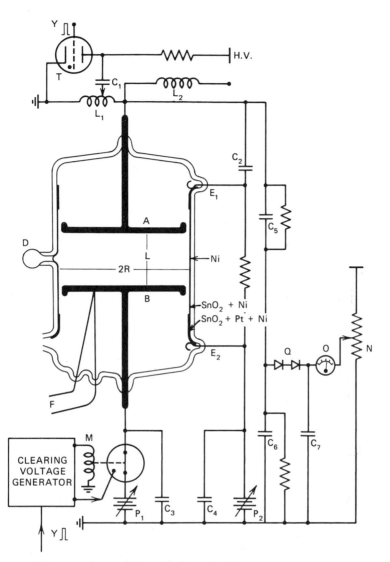

Fig. 3.10.1. Diagram of Cavalleri's diffusion chamber and electrical schematic of the measuring setup [after Cavalleri (1969a)].

close to the inside diameter of the chamber. All the inside walls of the useful volume between the electrodes are covered with a semitransparent vacuum-deposited layer of nickel, whose purpose is to make negligible any contact potential difference and to shield the gas volume from any external stray field; residual contact potential differences are balanced by the applied variable polarizations P_1 between the electrodes and P_2 between the electrodes and the lateral nickel layer. The damped oscillating pulse is applied via L_1 to the upper electrode and to the high-resistivity nickel metalization of the chamber walls. For the last connection high-conductivity SnO_2 layers are deposited over the lateral walls on the two sides of the nickel central film and are connected to the feed-through contacts E_1 and E_2 by means of appropriate platinum strips.

   After each radio-frequency pulse a long dc clearing pulse is applied between the electrodes, so that positive ions are swept out of the useful volume and the primary ionization process can be repeated (the time interval between two successive pulses can be varied in the range 20 msec to 2.56 sec). To avoid the effects of instrumental drifts, two different delays between the ionization and the measuring instants are alternated from one ionization cycle to the next one; the amplitudes of the photomultiplier output are converted into corresponding numbers of short pulses, which are sent alternately to two counters, which thus accumulate and record over a long time the sums of the amplitudes for the two delays. The ratio of the counter indications is equal to the ratio of the centroids of the density distributions after the two delays.' In order to avoid accumulation of impurity atoms on one electrode only, the clearing voltage is changed from positive to negative or vice versa after each complete cycle of two delays. A precision of $\pm 0.5 \%$ in the determination of the diffusion coefficient is claimed for this method.

   In addition to measuring the electron diffusion coefficient in a field-free environment, Cavalleri (1969a) in his last experiment measured this coefficient also in the presence of a superimposed electric field. If this field is dc, the swarm energy conditions are the same as exist for equal E/p values in the drift velocity experiments discussed in the preceding sections of this chapter. In the actual implementation of the method, Cavalleri applied between the electrodes a sinusoidal voltage of constant amplitude but selected the voltage frequency (16.26 MHz) and the gas pressure to satisfy two conditions: (1) the electron distribution component f^0 must have negligible alternating variations at the frequency ω (see Section 2.3), and (2) $\omega^2 \ll \nu_m^2$, so that according to eq. (2.3.20) the sinusoidal applied voltage is equivalent to a dc field with an amplitude equal to the rms value of the ac field. An appropriate modification has to be introduced into the expression of the diffusion length, since because of the field each electrode captures all electrons within an adjacent layer of thickness $2^{3/2}w/\omega$, the electron mean drift length during a semiwave of the ac field; therefore L in eq. (3.10.5) has to be replaced by the effective height $(L - 2^{5/2}w/\omega)$. The ac field is applied via the inductance L_2 (Fig. 3.10.1), and its amplitude (0–400 V) is measured with an accuracy of $\pm 0.5\%$ by means of the capacitive voltage divider C_5-C_6 and of the "peak sensing" potentiometric voltmeter Q-C_7-O-N.

b. The second method of determining D from low electron density measurements is based on the fact, mentioned in Section 3.2, that the sparking efficiency of a nuclear spark chamber is approximately $1 - \exp(-N)$, N being the number of electrons at the time of application to the chamber electrodes of the high-voltage pulse for breakdown. Therefore measuring sparking efficiency against the time delay of this pulse provides information on the electron density decay and thus makes it possible to determine the diffusion coefficient. Lloyd (1960) developed the pertinent theory for this method; his results are given as a set of curves of the sparking efficiency η as a function of Dt/a^2 for different values of the parameter af_1Q_1, where a is the internal radius of the cylindrical spark chamber, f_1 is the probability that an avalanche will be started when in the gas chamber there is one free electron only, and Q_1 is the number of electrons produced by the ionizing particle per unit path length (delayed ionization by metastables being disregarded). By comparing an experimental η versus t curve with Lloyd's theoretical set, one can determine the quantity f_1Q_1 from the best fitting of the curve shape and the diffusion coefficient D from the abscissa axis scale.

Measurements of η versus t decay curves have been performed in neon-filled chambers by various authors [Gardener, Kisdnasamy, Rössle, and Wolfendale (1957), Coxell and Wolfendale (1960)], who used for improved accuracy not a tube alone, but a stack of flash tubes, properly arranged into layers with interposed electrodes. The passage of an ionizing particle through all layers of the stack is detected by means of an appropriate arrangement of ordinary ionization gauges, placed at the stack boundary. It has been found experimentally that the efficiency, for a given tube size, is a function of t/p only; this result is obviously in agreement with Lloyd's theory and confirms it, under the assumption that the quantity f_1Q_1 is pressure independent.

c. The third method is an extension to very low densities of the conventional microwave method of measuring electron densities in resonant cavities containing an ionized medium. This standard technique is quite general and most suitable for accurate measurements of electron densities in repetitive discharge afterglows. The basic elements of this technique will first be reviewed in general and then applied to the specific case of interest.

Let $E_0 \exp(j\omega_0 t)$ and $H_0 \exp(j\omega_0 t)$ be the natural free oscillation fields inside a nondegenerate microwave resonant cavity in the absence of the ionized medium, and $E \exp(j\omega_1 t)$ and $H \exp(j\omega_1 t)$ be the corresponding fields when this medium is present. The pulsations ω_0 and ω_1 are then the complex free resonance angular frequencies of the cavity without the ionized medium and with it; by analogy to the free oscillations of a simple L-C circuit, it is convenient to write the complex resonant frequency ω_0 (the same applies to ω_1 as well) in the form:

$$\omega_0 = 2\pi f_0 \sqrt{1 - (1/2Q)^2} + j\pi \frac{f_0}{Q} \qquad (3.10.9)$$

or:

$$\omega_0 \simeq 2\pi f_0 + j\pi \frac{f_0}{Q} \qquad (3.10.10)$$

the latter expression being accurate enough for our purposes, since all the measurements of concern here are performed in very-high-Q cavities. From the above we see that field amplitudes decrease in a shock-excited cavity according to $\exp\left[-\pi(f_0/Q)t\right]$, so that the fractional decrease in energy per unit cycle ($f_0 t = 1$) is $2\pi/Q$, and we may assume this as a convenient general definition of Q in resonant cavities:

$$Q = 2\pi \frac{\text{energy stored}}{\text{energy dissipated per cycle}} \qquad (3.10.11)$$

We recall also that, according to the universally known theory of a simple L-C circuit, f_0/Q, which appears in the imaginary part of ω_0, is the half-power band width δf_0 of the resonant circuit.

We consider now Maxwell equations, both in the cavity without the ionized medium and in the cavity containing the ionized medium, whose microwave conductivity is σ:

$$\left.\begin{array}{l} \nabla \times \mathbf{H}_0 = j\omega_0 \varepsilon_0 \mathbf{E}_0 \\[4pt] \nabla \times \mathbf{E}_0 = -j\omega_0 \mu_0 \mathbf{H}_0 \\[4pt] \nabla \times \mathbf{H} = j\omega_1 \varepsilon_0 \mathbf{E} + \sigma \cdot \mathbf{E} \\[4pt] \nabla \times \mathbf{E} = -j\omega_1 \mu_0 \mathbf{H} \end{array}\right\} \qquad (3.10.12)$$

We multiply the first equation by $\mathbf{E}$, the second by $\mathbf{H}$, the third by $\mathbf{E}_0$, and the last by $\mathbf{H}_0$. Then we add the first two and subtract the last two, and finally we integrate all over the volume V of the cavity. We obtain, S being the surface-enclosing and $d\mathbf{S}$ the axial vector of its area elements:

$$\int_V \nabla \cdot (\mathbf{H}_0 \times \mathbf{E} + \mathbf{E}_0 \times \mathbf{H})\, dV$$

$$= \oint_S (\mathbf{H}_0 \times \mathbf{E} + \mathbf{E}_0 \times \mathbf{H}) \cdot d\mathbf{S}$$

$$= j(\omega_0 - \omega_1) \int_V (\varepsilon_0 \mathbf{E} \cdot \mathbf{E}_0 - \mu_0 \mathbf{H} \cdot \mathbf{H}_0)\, dV - \int_V (\sigma \cdot \mathbf{E}) \cdot \mathbf{E}_0\, dV \qquad (3.10.13)$$

The left-hand side vanishes, because of the usual boundary conditions of a perfectly conducting wall, and we obtain:

$$\Delta\omega = \omega_1 - \omega_0 = \frac{j\int_V(\boldsymbol{\sigma} \cdot \mathbf{E}) \cdot \mathbf{E}_0 \, dV}{\int_V(\varepsilon_0\mathbf{E} \cdot \mathbf{E}_0 - \mu_0\mathbf{H} \cdot \mathbf{H}_0) \, dV} \tag{3.10.14}$$

which gives the shift in the complex resonant pulsation of the cavity when the ionized medium is present.

At low electron densities, we can assume that the field configuration is not changed by the plasma, and hence set $\mathbf{E} = \mathbf{E}_0$, $\mathbf{H} = \mathbf{H}_0$. Within this approximation and because of the well-known fact that in a microwave cavity the energy stored in the electric field is equal to that stored in the magnetic field:

$$\varepsilon_0 \int_V E_0^2 \, dV = -\mu_0 \int_V H_0^2 \, dV \tag{3.10.15}$$

we obtain:

$$\Delta\omega = \frac{j\int_V(\boldsymbol{\sigma} \cdot \mathbf{E}_0) \cdot \mathbf{E}_0 \, dV}{2\varepsilon_0\int_V E_0^2 \, dV} \tag{3.10.16}$$

In the absence of an applied magnetic field the conductivity is a scalar quantity given by eq. (2.4.19); at low pressures, where $v_m^2 \ll \omega_0^2$, and at frequencies near the cavity resonance it can be written as (in the following formulas ω_0 stands for $2\pi f_0$ only):

$$\sigma = \frac{e^2 n}{m\omega_0^2}(\langle v_m \rangle - j\omega_0) \tag{3.10.17}$$

Substitution of (3.10.17) into (3.10.16) and separation of the real from the imaginary part gives the shift Δf_0 of the resonant cyclic frequency f_0 of the cavity as follows:

$$\frac{\Delta f_0}{f_0} = \frac{\mathrm{Re}\,\Delta\omega}{\omega_0} = \frac{e^2}{2m\varepsilon_0\omega_0^2}\frac{\int_V nE_0^2 \, dV}{\int_V E_0^2 \, dV} = \frac{e^2\bar{n}}{2m\varepsilon_0\omega_0^2} \tag{3.10.18}$$

Then the frequency shift is a measure of the weighted mean electron density $\bar{n}$, defined by the equation:

$$\bar{n} = \frac{\int_V nE_0^2 \, dV}{\int_V E_0^2 \, dV} \tag{3.10.19}$$

At very low electron densities, as here required by the goal of determining the free electron diffusion coefficient, very small frequency shifts have to be measured. We shall describe now the techniques used by Weaver and Freiberg (1968) for extending these measurements down to the lowest electron densities.

With a constant input the field in the cavity as a function of frequency will

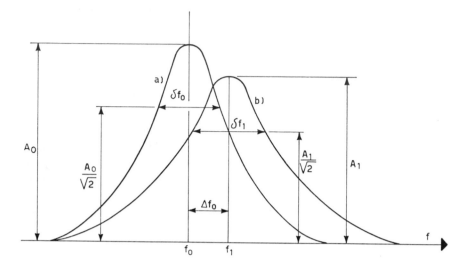

Fig. 3.10.2. Strength of cavity fields as a function of frequency, through resonance, for the cases of an empty cavity (a) and of a cavity containing the ionized medium (b).

show the standard resonant behaviors sketched in Fig. 3.10.2. In our present case the change in band width $\delta f_1 - \delta f_0$ is much less than the resonant frequency shift Δf_0; in fact, from the above formulas we have:

$$\delta f_1 - \delta f_0 = \frac{1}{\pi} \operatorname{Im} \Delta\omega = \frac{e^2 \langle \nu_m \rangle \bar{n}}{2\pi m \varepsilon_0 \omega_0^2} \qquad (3.10.20)$$

$$= \frac{2\langle \nu_m \rangle}{\omega_0} \Delta f_0 \qquad (3.10.21)$$

which proves our statement insofar as the gas pressure is low enough to ensure that $\langle \nu_m \rangle \ll \omega_0$. Therefore any variation in the strength of the cavity fields for a constant probing frequency is a measure of the Δf_0 shift alone; if we detect this variation when the probing frequency is adjusted on either side of the resonance curve at the points of maximum slope, the largest strength variations will be observed.

   In the experimental arrangement described by Weaver and Freiberg, the discharge is repetitively pulsed for a few microseconds and the electron density is measured at different delays in the afterglow by means of a cylindrical cavity excited to the H_{010} mode and located along the axis of the discharge tube. If the electron density takes time t_d to decay to an unmeasurable level after the ionization period, the repetition rate f_r of the discharge has to be less than $1/2t_d$, so that the cavity fields may be sampled twice during each afterglow period: once at the instant of the desired density measurement, and once half a period later when the cavity looks empty. Thus, for a given delay, the sampled signal is a train of very short pulses of alternating amplitudes, whose difference in successive heights is a measure of the frequency shift Δf_0.

As shown in Fig. 3.10.3, the electronic circuit not only views the reflected cavity fields at rate $2f_r$, but also stretches the sampled pulses into a square

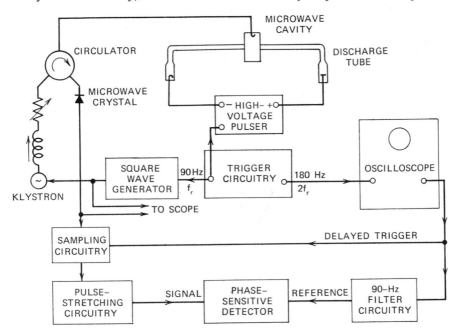

Fig. 3.10.3. Schematic diagram of the cavity apparatus used for measuring very low electron densities in late discharge afterglows [after Freiberg and Weaver (1968)].

wave and provides synchronous detection of the resulting signal. In order to further improve the sensitivity of this scheme, shift Δf_0 is measured by synchronously upshifting the microwave probing frequency by the same amount as the upshift of the cavity resonant frequency, as indicated by a null in the phase-sensitive detector reading. The klystron is frequency modulated by applying a small-amplitude square wave of frequency f_r to the repeller, care being taken that no amplitude modulation of the microwave level is produced. Calibration of the repeller modulation amplitude against the frequency shift of the klystron output thus provides a direct measurement of the resonant cavity shift Δf_0 due to the afterglow electron density being probed.

The minimum measurable density is determined by the lower level at which the microwave probing signal begins to heat the electrons appreciably above their thermal equilibrium value. Thus Weaver and Freiberg, using an 8-mm-I.D., 35-mm-long quartz tube for the discharge, a C-band cavity with a Q of $\sim$5500, a microwave signal power no larger than $10 \ \mu W$, and a sampling pulse length of $0.5 \ \mu sec$ for adequate resolution, found a minimum measurable density of about $\sim 2 \times 10^4 \ cm^{-3}$. We must recognize, therefore, that, in spite of the considerable sophistication of the scheme of Fig. 3.10.3, this minimum density appears to be not yet as low as we would like to obtain for a determination of the free electron diffusion coefficient ($\sim 10^3 \ cm^{-3}$). Hence different experimental conditions or further improvements of the techniques will be required for this purpose.

We would like to add two remarks of general interest in concluding our discussion on the determination of the free electron diffusion from measurements of the electron density decays. In our analysis we have neglected both recombination and attachment losses. The first ones are never of importance here, since they are proportional to n^2, and we must operate at very low n values. The second losses may be present in certain gases and never become negligible, since they are proportional to n, like the diffusion losses. However, for this reason they can easily be taken into account; if the term $\bar{\nu}_a n$ is added into (3.10.1), according to (2.5.2), relationship (3.10.3) becomes modified as follows:

$$\Lambda = \left(\frac{D\tau}{1 - \bar{\nu}_a \tau} \right)^{\frac{1}{2}} \qquad (3.10.22)$$

so that D can still be derived from measurements of τ if the attachment coefficient $\bar{\nu}_a$ is known, or if it can be obtained from the same measurements, taking into account the different dependences on gas pressure of D and $\bar{\nu}_a$. A similar solution holds for Cavalleri's (1969a) case of a superimposed electric field so large as to produce also ionization; in fact, for this case it is sufficient to replace $\bar{\nu}_a$ by ($\bar{\nu}_a - \bar{\nu}_i$) in eq. (3.10.22).

Another point of importance to be considered is that the condition of zero density at the container walls can be correctly used only in so far as the electron mean free path λ_m is much shorter than the container dimensions [see McDaniel (1964), pp. 496–497]. The condition $\lambda_m \ll \Lambda$ must then always be satisfied in the type of experiments discussed.

Whereas all methods discussed so far are based on measuring electron density decays, a different approach has been proposed and used by Nelson and Davis (1969). It is a modification of the TOF method of measuring μ and D_L/μ quantities (see Sections 3.2 and 3.5); the method is characterized by bringing the electric field to zero after the electron swarm has drifted to the center of the diffusion chamber, and then reapplying the field after varying periods of dwell time. The increasing width in the electron transit time distribution with dwell time provides the field-free diffusion coefficient D.

It is easy to see that in the Nelson-Davis experiment eq. (3.3.3') has to be changed into:

$$n = \frac{N}{2\sqrt{\pi(DT + D_L t')}} \exp\left[-\frac{(z - wt')^2}{4(DT + D_L t')}\right] \qquad (3.10.23)$$

where T is the dwell time duration, and $t' = t - T$. Equation (3.3.17) will change correspondingly, and the width of the transit time distribution (δt_d, measured between the points at $1/e$ of the peak value) will be given in the first approximation by:

$$\left(w\frac{\delta t_d}{2}\right)^2 = 4(DT + D_L t_d) \qquad (3.10.24)$$

where t_d is the drift time corresponding to the distribution peak (which, in the same approximation, is not changed by the dwell period). Therefore, if $(\delta t_d)^2$ is plotted versus the dwell time T, a straight line must result; if M is the measured slope of this straight line, the coefficient D is obtained from:

$$D = \left(\frac{w}{4}\right)^2 M \qquad (3.10.25)$$

Obviously, the above derivation implies that the energy relaxation times between the field-free and the applied-field electron velocity distributions are always negligible compared to the drift and dwell times. For this reason the shortest dwell times used by Nelson and Davis to determine M are 5 or 10 μsec.

The apparatus used by these authors is similar to the Wagner, Davis, and Hurst TOF systems, described in Section 3.4. The diffusion chamber has been modified by replacing the guard-ring assembly for uniform field by a continuous cylinder of 1400 $\Omega/\square$ resistive paper, thus avoiding the presence during the dwell time of unwanted fields produced by charged dielectric surfaces. The drift voltage is composed of two square pulses, separated by a variable, accurately known delay (T).

B. Diffusion in a Static Magnetic Field

When a magnetic field is present, we should be able to measure the transverse diffusion coefficient D_T. In principle, this case appears more manageable than that without a magnetic field, since a strong magnetic field reduces the positive-ion transverse diffusion much less than the electron one, so that also at large plasma densities the total transverse diffusion can be controlled by electrons.

In fact, let us consider an isothermal plasma, namely, a swarm of about equal densities of electrons and of positive ions with Maxwellian distributions at the gas temperature, and let us assume that the plasma is cylindrical, so that densities change only radially, the magnetic field being directed along the plasma axis. The classical picture of so-called ambipolar diffusion assumes that a slight deviation from charge neutrality creates a radial field E_r, which forces the radial electron flow Γ_{er} and the radial positive-ion flow Γ_{ir} to be equal, so as to maintain the quasi-neutrality condition. According to eq. (2.4.13), we have:

$$\Gamma_{er} = -\mu_T n E_r - \frac{\partial(D_T n)}{\partial r}$$

$$= -D_T \frac{en}{\hat{u}} E_r - \frac{\partial(D_T n)}{\partial r} \tag{3.10.26}$$

where the latter equation is obtained by using Einstein relation (2.6.2′). For the ions we may write a similar equation:

$$\Gamma_{ir} = D_{iT} \frac{en}{\hat{u}} E_r - \frac{\partial(D_{iT} n)}{\partial r} \tag{3.10.27}$$

where D_{iT} is the transverse diffusion coefficient for the ions. The equal-flow condition provides:

$$E_r = \frac{\hat{u}}{en} \frac{1}{D_{iT} + D_T} \left[\frac{\partial(D_{iT} n)}{\partial r} - \frac{\partial(D_T n)}{\partial r} \right] \tag{3.10.28}$$

which, substituted into (3.10.26), yields:

$$\Gamma_{er} = -\frac{1}{D_{iT} + D_T} \left[D_T \frac{\partial(D_{iT} n)}{\partial r} + D_{iT} \frac{\partial(D_T n)}{\partial r} \right] \tag{3.10.29}$$

Usually electron and ion diffusion coefficients have the same dependence on position, so that we may assume:

$$\frac{1}{D_T} \frac{\partial(D_T n)}{\partial r} = \frac{1}{D_{iT}} \frac{\partial(D_{iT} n)}{\partial r} \tag{3.10.30}$$

which, substituted into (3.10.29), yields:

$$\Gamma_{er} = -\frac{2 D_{iT}}{D_{iT} + D_T} \frac{\partial(D_T n)}{\partial r} \tag{3.10.31}$$

The electron cyclotron frequency $\omega_b = eB/m$ is always much larger than the ion cyclotron frequency eB/M, since $M \gg m$; therefore, on the basis of eq. (2.6.89), we can predict in general for strong fields $D_T \ll D_{iT}$. In this case eq. (3.10.31) becomes:

$$\Gamma_{er} \simeq -2 \frac{\partial(D_T n)}{\partial r} \tag{3.10.32}$$

which may be interpreted by saying that electrons diffuse radially according to an effective diffusion coefficient, called transverse ambipolar and given by:

$$D_{aT} \simeq 2D_T \tag{3.10.33}$$

Therefore measuring the diffusion transverse to a strong magnetic field also in a quasi-neutral plasma provides the value of the electron transverse diffusion coefficient.

However, the experimental and theoretical works of Bohm, Burhop, and Massey (1949), of Simon (1955), and of many others [see, e.g., Boeschoten (1964), and Geissler (1968)] indicate greater diffusion values and suggest a much more complicated picture than the one predicted by the simple ambipolar theory. Only in a few cases have results in accordance with the simple theory been obtained. D'Angelo and Rynn (1961), measuring by means of a probe the radial density decays of nearly fully ionized cesium and potassium plasmas in a Q-machine, similar to the one described in Section 3.8, obtained with the application of the ambipolar theory D_T values having the correct order of magnitude, appropriate for the case of plasmas dominated by electron-ion collision processes. Ganichev et al. (1964) measured in afterglow helium plasmas D_T values in accordance with ambipolar theory, but only when the diameter of the cylindrical discharge tube was greater than 4 cm. However, Geissler (1970) has criticized this result and has proposed different experimental evidence suggesting that the assumption of ambipolarity is in general inappropriate for the treatment of actual problems of plasma diffusion across a magnetic field. The uncertainties in this area are such that no diffusion experiment in a magnetic field will be considered hereafter for the derivation of electron collision parameters.

3.11   HIGH-FREQUENCY CONDUCTIVITY RATIO

The high-frequency conductivity ratio ρ, defined in Chapter 2 by (2.6.33), can be determined from measurements of the change of the characteristics of an electromagnetic wave which interacts with an ionized gas medium, whose state properties can be regarded as practically stationary during each measurement interval. We shall deal first with the various types of ionized media that are usually considered for this kind of measurements, and then with the different types of electromagnetic experiments that provide applicable results. A comprehensive review by Golant (1960) covers the theoretical and experimental studies made up to about 1960.

Experimental p data are suitable for the determination of electron collision parameters according to any of the procedures described in Section 2.6, provided they have been taken in an ionized gas medium that satisfies the following conditions:

a. Electrons and gas particles in the absence of applied fields have Maxwell distributions and equal average kinetic energies (thermal equilibrium condition).

b. Any dimension of the container is significantly larger than the average electron mean free path, so as to make the collision theory of Chapter 2 applicable.

c. The amplitude of oscillation of the average electron, under the action of the applied field, is smaller than the dimensions of the container, so that electrons do not travel completely across and collide with the walls in every half-cycle.

The most common ionized medium that satisfies the thermal equilibrium condition is provided by the late afterglow of a repetitively pulsed discharge; the use of a high-frequency discharge for this purpose was introduced by Brown's group at Massachusetts Institute of Technology [Phelps, Fundingsland, and Brown (1951)], whereas the similar use of a dc discharge was conceived by Goldstein's group at the University of Illinois [Goldstein, Anderson, and Clark (1953), Anderson and Goldstein (1955)]. Likewise, the afterglow following a burst of an ionizing beam of particles or of radiation through a gas can also provide appropriate conditions for these experiments, as demonstrated by Van Lint (1959), who ionized the gas by means of a short pulse of MeV electrons generated from a linear accelerator. Independently of how an afterglow electron swarm is created, and provided that gas pressure is not so low to make diffusion cooling effects (see Section 2.3) significant, we can properly assume that the electrons cool down, from the high average energies that they have during the ionization period, to the temperature of the gas, and that this happens with a time constant of the order of the electron energy relaxation time τ_r, given by (2.5.9) or (2.5.10). An upper limit for τ_r is obtained when we take for G its minimum value $2m/M$, which is the value for elastic collisions only; this limit is then $M/2m\langle v_m \rangle$. As an example of the times involved, for typical values like $G = 10^{-4}$ and $\langle v_m \rangle / p = 10^9$ (sec · torr)$^{-1}$, the time constant becomes $(10^{-5}/p)$ sec and the equilibrium is practically well satisfied at all postionization times larger than $(10^{-4}/p)$ sec; in general these times are quite suitable for accurate propagation measurements.

However, measurements of the electron radiation temperature during the afterglow have indicated in some cases energy decays substantially slower than those expected from the above considerations [Formato and Gilardini

(1960)]. Therefore the possible presence of secondary sources of electron heating during the afterglow has to be carefully considered in any experiment designed to determine electron collision frequencies from afterglow studies. A known case of a secondary heating source is provided by the presence of appreciable concentrations of metastable states of the neutral gas particles, which may feed kinetic energy to the electrons, either through de-excitation collisions or through metastable-metastable ionizing collisions. The decay of the electron radiation temperature in a helium afterglow under these conditions was investigated both theoretically and experimentally by Ingraham and Brown (1965), who found that pronounced heating effects occur only for gas pressures from 0.1 to 1 torr, and that in this range de-excitation collisions dominate the decay at low pressures and metastable-metastable ionizing collisions at high pressures. Noon, Blaszuk, and Holt (1968) measured radiation temperatures in the afterglow of a nitrogen discharge immersed in a magnetic field, their purpose being to gain information concerning, and possibly to explain, the slow energy decays found by Formato and Gilardini (1960), using 10-μsec pulse widths and currents of the order of 1 A, in opposition to the fast decays predicted by the calculated relaxation times of the electron energy and observed by Mentzoni and Row (1963), using what they referred to as a mildly driven discharge. With the method described in Section 3.12, Noon et al. derived data on the average electron energy and on the distribution function at different times in the afterglow (till a maximum postdischarge time of 50 μsec) and for different discharge conditions. Their results, which show the strong influence of the initial conditions of the discharge, are consistent with the hypothesis of an electron heating process due to a strong coupling between metastable vibrational levels of the nitrogen molecules and the electrons.

All the above results indicate how important, in any afterglow experiment, is measuring the electron radiation temperatures, as can be done using the techniques to be described in Section 3.12: when the radiation temperature is found equal to the gas temperature, it may be reasonably assumed that electrons have actually cooled down to the same kinetic energies as the gas particles. Today it is common practice to perform these temperature measurements in all experiments designed to collect significant data on electron collision parameters. Quite recently Blue and Stanko (1969) demonstrated the possibility of using also pulsed Langmuir probes for this purpose of measuring afterglow temperatures.

In addition to afterglows, there are other relevant cases in which propagation measurements have been performed in electron swarms in thermal equilibrium with the gas. These cases, characterized by the presence of a contemporary source of ionization, include: (a) the shock wave plasma in a shock tube that allows operating up to a few thousand degrees [this technique was introduced by Lin and Kivel (1959) and has since been used by many investigators], (*b*) a gas weakly ionized by a continuous X-ray source [Carruthers (1962)], and (*c*) a gas ionized by the radiation field within the core of a pulsed water-cooled nuclear reactor [Bhattacharya, Verdeyen, Adler, and Goldstein (1967)]. For all these cases it is always highly desirable to perform a direct check on the isothermicity condition between electrons and gas molecules. For instance, it has been found in case *c* by means of radiation temperature measurements that, if the reactor is pulsed from a 10-W steady-state power level to several hundred megawatts in a short interval of time ($\sim$30 msec), electrons formed in relatively high-pressure light noble gases ($\gtrsim$ 100 torr) are essentially at ambient temperature, whereas at lower pressures peak electron temperatures over 1000°K can be attained. It is also worth mentioning that in chemically active gases, such as nitric oxide, employing photoionization by ultraviolet photons could be particularly advantageous, since this process produces much less dissociation and excitation of neutral molecules than any other plasma-generating method mentioned [Mahan (1960), Gunton and Shaw (1965)].

All the processes we have considered imply generation of the ionized medium by volume ionization; equal numbers of electrons and positive ions are then produced. Moreover, electron densities and temperatures are such that, with the exception of the case of the X-ray ionized gas in Carruthers' experiment, the Debye length λ_D [eq. (3.10.7)], representing a measure of the maximum size of nonneutral regions, is always much smaller than any significant container dimension. The ionized medium thus satisfies the property of approximately equal electron and ion densities, or of charge quasi neutrality, and therefore we shall call it currently a plasma (see Section 3.8).

Two characteristics of electromagnetic waves whose changes due to the presence of an ionized medium have actually been used for the purpose of determining electron collision parameters are (*a*) propagation of a plane wave and (*b*) resonance of a microwave cavity or of an appropriate resonator. In both cases the analysis of the experimental data is simpler, and the final result more reliable, if the conditions for the experiment are chosen so that any observed change in the characteristics of the electromagnetic wave is small enough to justify treating the problem as a perturbation one. We shall now examine the two cases separately, assuming that they may actually be treated by small perturbation techniques. Thereafter we shall discuss how to handle the case of a dense plasma, whose electron density is so large as to forbid considering the problem as a small perturbation one.

A. Propagation of a Plane Wave, Slightly Perturbed by an Ionized Gas

We discuss the case of a probing plane wave, which may be either a free space TEM wave or a guided wave in an open or closed structure; appropriate observable features of this propagation problem are the changes in the characteristics of the transmitted wave and of the reflected and scattered waves. Under typical experimental conditions (plasma electron densities below $\sim 10^{11}$ cm^{-3}, and microwave frequencies below 10 GHz), for accurate measurements of the conductivity ratio it is best to adopt metal-bounded cylindrical structures (coaxial lines, rectangular and circular waveguides), a section of which is filled completely or partially with the ionized medium. In this case, no radiation loss is present, and, provided the dimensions of the guiding structure are chosen so that only the dominant wave mode of the structure can propagate, the entire problem is reduced to a consideration of the transmitted and reflected waves of this mode alone. Thus the problem lends itself to a manageable analytical treatment: the observed features can be easily interpreted, and the parameters of the ionized medium correctly derived. In what follows we shall discuss first and thoroughly the case of propagation in a nonradiating transmission line loaded by a plasma, and thereafter we shall give only a brief account of the similar use of free space beam waves.

When a section of a transmission line is filled with a dielectric, such as our ionized medium, two effects on wave propagation can be recognized: (1) the line propagation constant is changed in the region occupied by the dielectric, and (2) a reflection takes place at each discontinuity of the line characteristic impedance, as at the end faces of the dielectric and at any position along the line axis, where eventually the dielectric parameters would change. If the propagation problem has to be handled by perturbation, according to one of the requirements previously stated, the ionized medium must be such that any change in the propagation constant and in the characteristic impedance of the line will be small. In this case, reflection effects will be correspondingly small and unsuitable for accurate measurements of the properties of the medium; on the other hand, a change in the propagation constant, also if small, can provide large effects on the wave transmission, since an adequately long section of the line can be filled with the ionized medium. For this reason plasma conductivities, and hence electron collision parameters, are actually derived from the changes of the propagation constant of a transmission line, as obtained from measurements of the changes of the line transmission characteristics, proper attention being paid in these experiments to avoid the presence of any appreciable wave reflection at the line discontinuities. For the last purpose the shapes of the ionized gas container and of the waveguide are generally such as to provide smooth and tapered transitions at the two section ends (Fig. 3.11.1).

Let us evaluate now the propagation constant for waves propagating along a section of a cylindrical guiding structure filled with the ionized medium, under the assumption that the structure and the medium are axially uniform, which implies that the quantities of importance (like the electron density and the electron kinetic energy distribution) are functions only of the

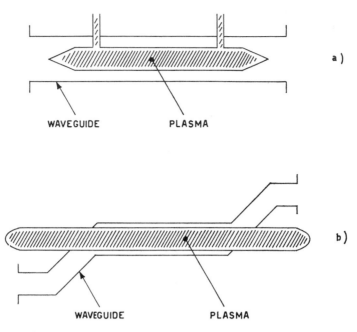

WAVEGUIDE   PLASMA

WAVEGUIDE   PLASMA

Fig. 3.11.1. Schematic diagram of two typical arrangements of plasma containers in waveguides, designed to provide low reflectivity of the propagating wave.

position in the structure cross section, and not of the position along the axis. In this case the wave propagates as $\exp(-\gamma z)$, the axis z being the axis of the guide and γ the complex propagation constant. The fields in the structure will be $\mathbf{E} \exp(j\omega t - \gamma z)$ and $\mathbf{H} \exp(j\omega t - \gamma z)$, where $\mathbf{E}$ and $\mathbf{H}$ are functions only of the position in the cross section, and can be derived from the solution of the appropriate Maxwell equations. These equations, for the structure with the ionized medium and without it (the quantities for the structure without the ionized medium being characterized by the subscript 0), are:

$$\left.\begin{aligned}
\mathbf{V}_t \times \mathbf{H}_0 - \gamma_0 \mathbf{a}_z \times \mathbf{H}_0 &= j\omega\varepsilon_0\mathbf{E}_0 \\
\mathbf{V}_t \times \mathbf{E}_0 - \gamma_0 \mathbf{a}_z \times \mathbf{E}_0 &= -j\omega\mu_0\mathbf{H}_0 \\
\mathbf{V}_t \times \mathbf{H} - \gamma \mathbf{a}_z \times \mathbf{H} &= j\omega\varepsilon_0\mathbf{E} + \boldsymbol{\sigma} \cdot \mathbf{E} \\
\mathbf{V}_t \times \mathbf{E} - \gamma \mathbf{a}_z \times \mathbf{E} &= -j\omega\mu_0\mathbf{H}
\end{aligned}\right\} \qquad (3.11.1)$$

Here $\mathbf{V}_t$ is the transverse part of the $\mathbf{V}$ operator, $\mathbf{a}_z$ is the unit vector along the positive z axis, and the medium conductivity $\boldsymbol{\sigma}$ is written in the tensor form to include the most general case of the presence of an external dc magnetic field, which will be of interest later.

Let us multiply the first equation by $\mathbf{E}$, the second by $\mathbf{H}$, the third by $\mathbf{E}_0$, and the last by $\mathbf{H}_0$; then we add up the first two equations, subtract the last two, and integrate over the cross section S of the structure. Applying the two-dimensional form of the divergence theorem to the first term, we obtain:

$$\oint_C (\mathbf{H}_0 \times \mathbf{E} - \mathbf{H} \times \mathbf{E}_0) \cdot ds$$

$$- (\gamma + \gamma_0) \int_S (\mathbf{E} \cdot \mathbf{a}_z \times \mathbf{H}_0 - \mathbf{E}_0 \cdot \mathbf{a}_z \times \mathbf{H}) \, dS$$

$$= - \int_S (\boldsymbol{\sigma} \cdot \mathbf{E}) \cdot \mathbf{E}_0 \, dS \quad (3.11.2)$$

where the first term is now a circulation integral around the cross-section line boundary C (ds is the line element vector, directed along the normal to the structure wall). This circulation integral vanishes since $\mathbf{E} \times ds = \mathbf{E}_0 \times ds = 0$ on C, if perfectly conducting walls are assumed.

Solving the second and fourth equations of (3.11.1) for $\mathbf{H}_0$ and $\mathbf{H}$ and substituting these in (3.11.2) gives:

$$(\gamma + \gamma_0) \int_S [\mathbf{E}_0 \cdot \nabla_t E_z - \mathbf{E} \cdot \nabla_t E_{0z} + (\gamma - \gamma_0)\mathbf{E} \cdot \mathbf{E}_0] \, dS$$

$$= j\omega\mu_0 \int_S (\boldsymbol{\sigma} \cdot \mathbf{E}) \cdot \mathbf{E}_0 \, dS \quad (3.11.3)$$

Within the limits of the small perturbation assumption, and provided that only the dominant mode can propagate along the guiding structure (higher-order modes being evanescent), the field $\mathbf{E}$ can be set equal to $\mathbf{E}_0$. Then eq. (3.11.3) yields:

$$\gamma^2 - \gamma_0^2 = j\omega\mu_0 \frac{\int_S (\boldsymbol{\sigma} \cdot \mathbf{E}_0) \cdot \mathbf{E}_0 \, dS}{\int_S E_0^2 \, dS} \quad (3.11.4)$$

This formula provides the propagation constant γ for the waveguide containing the ionized medium.

Let us introduce the attenuation and the phase constants of the propagating wave in the guide without the ionized medium (α_0, β_0) and their changes $(\Delta\alpha, \Delta\beta)$ due to the presence of the ionized medium:

$$\gamma_0 = \alpha_0 + j\beta_0 = (k_c^2 - k_0^2)^{\frac{1}{2}} \quad (3.11.5)$$

$$\gamma = \gamma_0 + \Delta\alpha + j\,\Delta\beta \quad (3.11.5')$$

In eq. (3.11.5) k_0 and k_c are, respectively, the free-space wave number and the cutoff complex wave number of the guiding structure [see, e.g., Marcuvitz (1951)]. Usual structures and transparent plasmas are such that α_0 and $\Delta\alpha$

can be considered as small quantities compared to β_0; in this case, neglecting their products and squares, we have:

$$\gamma^2 - \gamma_0^2 = -(2\beta_0 + \Delta\beta)\,\Delta\beta + 2j[(\beta_0 + \Delta\beta)\,\Delta\alpha + \alpha_0\,\Delta\beta]$$

$$(3.11.6)$$

If we substitute this expression into (3.11.4), separate the real and imaginary parts, and take their ratio, the following result is obtained:

$$\frac{\text{Re}\int_S (\boldsymbol{\sigma} \cdot \mathbf{E_0}) \cdot \mathbf{E_0}\, dS}{\text{Im}\int_S (\boldsymbol{\sigma} \cdot \mathbf{E_0}) \cdot \mathbf{E_0}\, dS} = \frac{(\beta_0 + \Delta\beta)\,\Delta\alpha + \alpha_0\,\Delta\beta}{(\beta_0 + \Delta\beta/2)\,\Delta\beta} \qquad (3.11.7)$$

This is a very important result. The left-hand side gives, under appropriate conditions, an average value of the conductivity ratio in the ionized medium. The right-hand side is an expression which can be easily written in terms of the observable propagation parameters: attenuation $A = l\alpha_0$ of the probing signal in nepers and its change $\Delta A = l\,\Delta\alpha$, l being the length of the guide section containing the ionized medium; and phase shift $\Delta\phi = -l\,\Delta\beta$ in radians (the minus sign is introduced to make $\Delta\phi$ positive, since $\Delta\beta$ is in these cases always negative).

Restricting the discussion from now on to the case of no external static magnetic field, we obtain for the left-hand side of (3.11.7):

$$\frac{\int_S \sigma_r E_0^2\, dS}{\int_S \sigma_i E_0^2\, dS} = -p\,\frac{\int_S \rho\sigma_i E_0^2\, dS}{\int_S \sigma_i E_0^2\, dS}$$

and we can write:

$$\frac{\int_S \rho\sigma_i E_0^2\, dS}{\int_S \sigma_i E_0^2\, dS} = \rho_{\text{av}} = \frac{(1 - \varphi)\,\Delta A - \varphi A}{(1 - \varphi/2)p\,\Delta\phi} \qquad (3.11.8)$$

where ρ_{av} represents the conductivity ratio, averaged over the cross section as indicated, and:

$$\varphi = -\frac{\Delta\beta}{\beta_0} = \frac{\Delta\phi}{l\beta_0} \qquad (3.11.9)$$

has been introduced as a convenient parameter. Small perturbation conditions imply that $\varphi \ll 1$; in this case we can use the simplified formula:

$$\rho_{\text{av}} = \frac{\Delta A}{p\,\Delta\phi} \qquad (3.11.10)$$

It seems appropriate at this point to identify some convenient way of predicting whether or not the small perturbation assumption is actually applicable in a specific experiment. Equation (3.11.4) provides this way, since it appears clearly from the equation that the perturbation assumption may be accepted any time the absolute value of the right-hand side is much less than β_0^2; this condition, assuming the simplest case of no magnetic field and of space uniform conductivity, becomes:

$$\omega\mu_0\,|\sigma| \ll \beta_0^2 \tag{3.11.11}$$

In the case of a velocity independent collision frequency ν_m, eq. (2.4.19) yields:

$$|\sigma| = \frac{e^2 n}{m\sqrt{\omega^2 + \nu_m^{\,2}}} = \varepsilon_0 \frac{\omega_p^{\,2}}{\sqrt{\omega^2 + \nu_m^{\,2}}} \tag{3.11.12}$$

where the radian plasma frequency ω_p has been introduced as the most appropriate parameter to represent the electron density:

$$\omega_p^{\,2} = \frac{e^2 n}{m\varepsilon_0} \tag{3.11.13}$$

Substitution of (3.11.12) into (3.11.11) yields:

$$\frac{\omega_p^{\,2}}{\sqrt{1 + (\nu_m/\omega)^2}} \ll \frac{\beta_0^{\,2}}{\mu_0\varepsilon_0} = \left(\frac{\beta_0}{k_0}\right)^2 \omega^2 \tag{3.11.14}$$

In typical experiments $\nu_m \lesssim \omega$, $(\beta_0/k_0)^2 \simeq 0.5$; therefore the above condition states that the small perturbation approximation requires an ionized medium with a low density, such that its plasma frequency will be much less than the applied signal frequency. The cyclic value of the plasma frequency in hertz is (n in centimeters^{-3}):

$$f_p = 8979\sqrt{n} \tag{3.11.13'}$$

An important assumption that we introduced originally in the derivation of our formulas was that of axial uniformity of the ionized medium characteristics. The presence of axial nonuniformities can then be handled, if important, by considering successive layers of practically uniform characteristics, whose individual attenuation and phase shift contributions are added up to provide the total observable changes of the propagation parameters. In particular, when there is no magnetic field, if φ may be neglected compared to unity, integrating along the guide axis z the elementary $\Delta\beta$ and $\Delta\alpha$ values, as derived from (3.11.4) and (3.11.6), yields:

$$\Delta\phi = -\frac{\omega\mu_0 \int_V \sigma_i E_0^{\,2}\, dV}{2\beta_0 \int_S E_0^{\,2}\, dS} = \frac{e^2\mu_0}{2m\beta_0}\,\bar{n}l \tag{3.11.15}$$

$$\Delta A = \frac{\omega\mu_0 \int_V \sigma_r E_0^{\,2}\, dV}{2\beta_0 \int_S E_0^{\,2}\, dS} \tag{3.11.16}$$

where V is the total volume of the ionized medium, and $\bar{n}$ is the average electron density, weighted according to $E_0^{\,2}$; the second expression of (3.11.15) holds only for the case $\nu_m^{\,2} \ll \omega^2$. It must be noted that by definition E_0

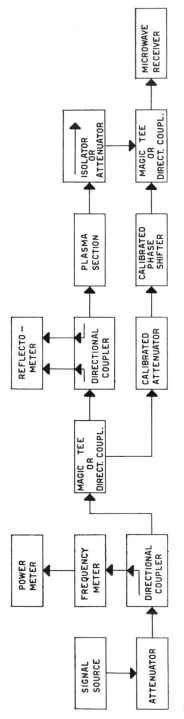

Fig. 3.11.2. Block diagram of the microwave bridge interferometer.

depends, not on z, but only on the position in the cross section; actually, E_0 can be regarded as the field configuration of the fundamental propagation mode and as such may be conveniently normalized, for instance, to unity at its maximum. From (3.11.15) and (3.11.16) we obtain:

$$\frac{\Delta A}{p\,\Delta\phi} = \frac{\int_V \rho\sigma_i E_0^2 \, dV}{\int_V \sigma_i E_0^2 \, dV} = \rho_{\text{av}} \qquad (3.11.17)$$

where now the averaging process has to be performed over the entire volume of the ionized medium.

Of paramount importance in our discussion is the case in which ρ is constant throughout the ionized gas volume. In fact, eq. (3.11.17) shows that in this case the constant ρ value is simply given by the ratio $\Delta A/p\,\Delta\phi$ also if the other medium parameters (e.g., the electron density) change with position over the cross section or along the axis. This constant ρ condition is very convenient from the point of view of simplicity in interpreting measurements; and, insofar as it provides not an average, but a single value of ρ for any set of experimental conditions, it becomes most desirable when electron collision data are to be determined by the procedures described in Section 2.6. Having established the importance of performing experiments when ρ is position independent, we will now specify how and when this condition is satisfied.

A general requirement for its validity is that only electron-neutral collisions be relevant, since electron-electron and electron-ion scattering is dependent on electron and ion densities, which are typically nonuniform quantities in any bounded plasma. A second requirement could be that the electron collision frequency be velocity independent in the relevant range of energy; this condition, however, is seldom met in actual gases. Hence, when ν_m is dependent on velocity we consider another requirement, which can be easily satisfied if the experiment is properly conceived: that the electron energy distribution be constant over the entire plasma. On the assumption that electrons and gas molecules are in thermal equilibrium in the absence of applied fields, two cases of major practical interest can easily be identified, in which the electron energy distribution can be regarded as constant: (1) the low-field case, where the power of the probing signal is so low that no appreciable change of electron kinetic energies takes place; and (2) the uniform-field case, where the electric field is large, but the plasma is confined to the region in which this field is nearly uniform. The low-field, no-electron-heating condition was used in most of the experiments whose results have been published.

As we have seen, amplitude and phase changes have to be measured; the most appropriate experimental setup, therefore, is a microwave bridge interferometer, like the one shown schematically in Fig. 3.11.2. The basic concept of this method, developed at the University of Illinois by Goldstein's group [Goldstein, Lampert, and Geiger (1952), Anderson and Goldstein (1955)], has been applied ever since in most of the experiments with only minor modifications. The signal source is usually a klystron, whose output is properly attenuated and accurately monitored, so that the desired level of field intensity is obtained in the test section containing the ionized medium; as already noted, this level is chosen in most experiments according to the requirement of no electron heating (in such cases typical values of the micro-wave power are in the microwatt range). Most experiments are performed in waveguides at X-band frequencies, since in this case moderate volumes of plasma are sufficient to fill the entire guide cross section. Frequencies and gas pressures are generally such that the condition $\omega^2 \gg \nu_m{}^2$ is fulfilled, and the convenient procedures described in Chapter 2 can be used for determining from these measurements the electron collision parameters.

In the following discussion details will be given only for the basic experi-mental arrangements, since extensions to all other cases can easily be con-ceived. The situation that is by far the most common and typical and hence has been chosen for a more detailed discussion is that of a discharge-generated afterglow plasma; the discharge is usually produced in a thin-walled Pyrex or quartz tube, housed in the waveguide, by application of a short (1–10 μsec) high-voltage dc pulse between a cathode and an anode. For experiments in rare gases the electrodes of the discharge tube are con-veniently made out of high-purity titanium because of its good gettering property for atmospheric gases [Chen, Leiby, and Goldstein (1961)]. Typical geometries and preferred mode configurations are shown in Fig. 3.11.3; in some experiments the discharge tube also performs the waveguide function, its outer surface being properly coated by a conductive layer (e.g., gold). In order to attain lower radiation temperatures during the afterglow decay in atmospheric gases, on the basis of a result obtained by Anderson (1960) some authors have preferred using a negative discharge, generated by appropriate internal electrodes, in contrast to the more conventional positive column plasma, generated by external electrodes at the opposite ends of the discharge tube. In particular, Formato and Gilardini (1962) produced a negative glow by means of two parallel axial wire electrodes, and Taft, Stotz, and Holt (1963) by means of anode wires extending through holes in the waveguide, whereas the waveguide itself acts as a hollow cathode. The plasma columns are many decimeters and in some cases over 1 meter in length. Since afterglows last in these experiments a few milliseconds at most, it is convenient to repeat periodically the gas breakdown and the measuring process in the afterglow at an appropriate rate, usually in the range from 10 to a few hundreds hertz.

Plasmas produced by shock waves in a shock tube have been considered

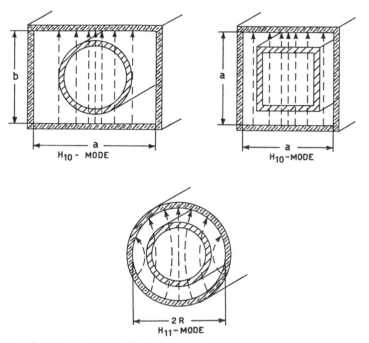

Fig. 3.11.3. Three typical waveguide and container geometries, and commonly used field modes. The dashed lines depict the electric field direction and distribution.

mainly for the purpose of determining electron collision parameters in atomic species other than rare gases, since these species result from the dissociation of molecules in the high-temperature shock waves. In the experiment of Daiber and Waldron (1966), 3000-MHz microwaves were propagated through the shock tube (3.81 cm × 6.35 cm), as they would be in an ordinary waveguide, in the direction opposite to the shock wave (Fig. 3.11.4). The operating

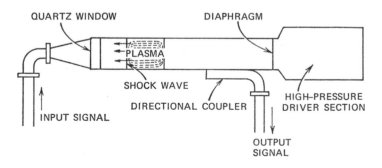

Fig. 3.11.4. Schematic diagram of a shock tube, acting also as an ordinary waveguide [after Daiber and Waldron (1966)].

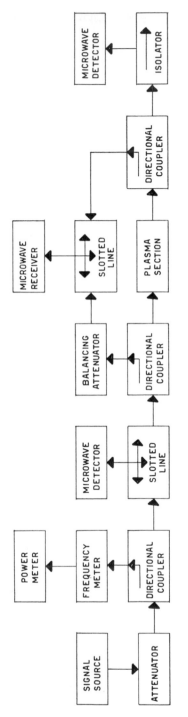

Fig. 3.11.5. Block diagram of the Anderson and Goldstein slotted-line bridge.

conditions were selected so that the species distribution and the gas and electron temperatures behind the shock front could be calculated from the equilibrium normal shock-wave relations, the time required to relax to equilibrium being very short. Makios (1965, 1966, 1967) verified the possibility of using a Lecher wires system (0.1-mm-diameter copper wires, separated by 2 mm), placed transversely to the propagation direction of the shock wave, and matched to a waveguide bridge interferometer. When the system is operated at 4-mm wavelength, a space resolution of 2 mm is obtained, since Lecher wires support a truly guided wave; on the other hand, because of the very thin wires no perturbation of the plasma is observed.

In all well-designed experiments either a directional coupler, as in Fig. 3.11.2, or a circulator or a slotted line is located along the probing signal path just before the plasma-filled section. This is for the purpose of verifying that reflections due to the ionized medium are negligible, and hence that the measured attenuations and phase shifts of the transmitted signal are due entirely to plasma volume effects.

In the preferred interferometer scheme of Fig. 3.11.2 a precision attenuator and a phase shifter in the bridge reference path (or in the probing signal path as well) are successively adjusted for zero output at the receiver in the absence and in the presence of the plasma, so that their reading changes are ΔA and $\Delta \phi$, respectively. It is convenient to keep the reference and the probing signal paths of about the same electrical length, to avoid differential phase changes if the signal frequency drifts. The splitting of the signal between the two paths and the later sum of the two signals in the waveguide network are obtained by means either of magic tees or of directional couplers.

A different bridge scheme was used originally by Anderson and Goldstein (1955) [see also Takeda and Holt (1959)]. The phase bridge of their type (Fig. 3.11.5) does not require the use of a calibrated attenuator and of a similar phase shifter, but rather employs a slotted line, in which the reference and the probing signals are sent along opposite directions to produce a standing wave pattern. In the absence of the plasma the wave amplitudes are adjusted to produce the maximum achievable standing wave ratio. Then the phase and attenuation changes due to the plasma are determined, respectively, from the shift of the position of the minimum in the standing wave pattern and from the change in the standing wave ratio (to be determined here by measuring the separation of the twice-minimum points of the pattern [Purcell (1947)]).

The receiver for the interference comparison is either a square-law detector, such as a silicon crystal diode, followed by a video amplifier, or a linear superheterodyne receiver of traditional type. Since electron densities in afterglow plasmas change with time, transient types of measurements have to be performed. In repetitive afterglows this is done alternatively by (a)

248

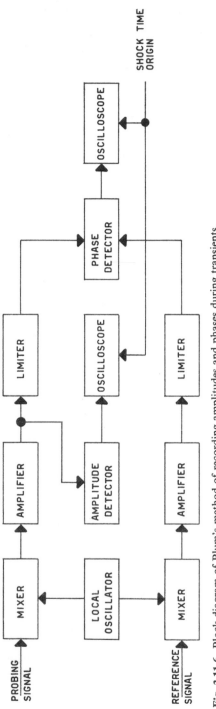

Fig. 3.11.6. Block diagram of Blum's method of recording amplitudes and phases during transients.

pulse-modulating the signal klystron at the same rate as the discharge but with an appropriate delay and pulse length (usually a few microseconds), (b) pulse-gating the receiver in a similar fashion, or (c) displaying on a scope the receiver output as a function of the postdischarge time, so that instant-by-instant data can be obtained by rebalancing on the scope the bridge for zero output at the desired time.

For nonrepetitive short-duration plasmas, such as those produced in shock tubes, manual rebalancing at each instant during the run is obviously impossible. In this case, following a method due to Blum (1965), the amplitude of the probing signal and its phase relative to the reference signal are recorded separately on oscilloscopes during the run, by means of the system shown in Fig. 3.11.6. The reference and probing signals are sent to separate mixers fed by a common local oscillator, so that their amplitude and relative phase information is preserved at the i.f. frequency. In order to obtain an amplitude-insensitive phase output, the two i.f. signals are first amplified and limited separately, and then combined in a conventional phase detector. The probing signal amplitude, on the other hand, is obtained by detecting this signal, at i.f. frequency, before limiting. A precision attenuator and a phase shifter incorporated in the probing signal path are used to calibrate the scales of the oscilloscopes. In an alternative solution, offered by Blackburn and Tevelow (1965) [see also Tevelow (1967)], the reference and probing signals are combined in a waveguide magic tee, so that the amplitude and phase of the probing signal relative to the reference one can be derived from the amplitudes of the two signals in the output arms of the tee, which are detected by crystal diodes and displayed on an oscilloscope. In a more rudimentary solution, applicable only when $\Delta\phi \gg 2\pi$, the reference and probing signals are simply combined on a mixer; the resulting detected envelope is an interferogram, which for a low-loss growing or decaying plasma appears quasi-sinusoidal in time, so that the roles of the phase changes and of the amplitude changes can be estimated separately.

The attenuation and phase shift changes due to the plasma are doubled if the probing signal is forced to propagate twice through the plasma, by means of a short circuit at the end of the waveguide test section. The reflected probing signal can be separated from the incident signal by means of a circulator, and then processed as before in a phase bridge to determine the attenuation and phase shift changes. Alternatively, with a more simple arrangement, these changes can be determined from slotted-line measurements of the standing wave pattern in front of the plasma: shift in the minimum position and change in the standing wave ratio. This reflection method is seldom employed, since any reflection from discontinuities in the test section adds up to the desired reflected signal to degrade the measuring accuracy. However, it is well known in microwave practice how to separate the discontinuity contributions from the main reflected wave, by using a movable short circuit, instead of the fixed one, to produce a variable phase reflection; this technique could be applied also in the present case if one is willing to undertake the extra experimental and computational work required.

We have already remarked that in most experiments the signal power sent through the plasma is chosen to be low enough not to perturb the thermal equilibrium condition between electrons and gas molecules. A quantitative evaluation of this maximum power is based on condition (2.3.28), which we assume to be well satisfied when $E_e \leqslant E_{et}/10$; using relation (2.3.22) between the actual and the effective fields, we obtain for the maximum permissible field in the waveguide the value (rms):

$$E_m{}^2 = \frac{3mkT_g}{200e^2}(\omega^2 + \nu_m{}^2)G \qquad (3.11.18)$$

The quantities ν_m and G are actually functions of the electron velocity; in formula (3.11.18) it is advisable to take their minimum values over the appropriate range. For a given mode in an empty waveguide the power P_m, which corresponds to the maximum field E_m, can be written as [see, e.g., Marcuvitz (1951)]:

$$P_m = \eta\left[1 - \left(\frac{\lambda_0}{\lambda_c}\right)^2\right]^{\frac{1}{2}} SE_m{}^2 = YE_m{}^2 \qquad (3.11.19)$$

where λ_0 is the free-space wavelength, λ_c the mode cutoff wavelength, S the area of the waveguide cross section, and η a coefficient that depends on the mode and waveguide geometry alone. In particular, we have (η in ohms^{-1}):

$$\eta = 1.33 \times 10^{-3}, \qquad \lambda_c = 2a \qquad (3.11.20)$$

for the dominant H_{10} mode in a rectangular waveguide of inner side a in the direction perpendicular to the electric field, and:

$$\eta = 1.26 \times 10^{-3}, \qquad \lambda_c = 3.41R \qquad (3.11.21)$$

for the dominant H_{11} mode in a circular waveguide of radius R.

As a typical case we consider room temperature ($T_g = 300°\text{K}$), low-pressure ($\nu_m{}^2 \ll \omega^2$) neon ($G = 5.4 \times 10^{-5}$) in a square waveguide with $a = 0.6\lambda_0$; the resulting maximum power is 55 μW.

When the electric field intensity of the probing signal is increased beyond limit (3.11.18), the purpose being the determination of electron collision parameters by the methods discussed in Chapter 2, it is advisable to reduce the size of the plasma container until the condition of nearly uniform field (constant ρ) becomes satisfied. Actually, instead of using a large probing signal, a different technique is usually followed: the probing and electron heating functions of the electromagnetic field are separated, one field of large intensity being launched for the purpose of increasing electron kinetic energies, and another of slightly different frequency and low power being used for measuring ΔA and $\Delta \phi$ as functions of the heating field power. This approach is particularly advantageous since the heating field can be pulsed, so that the probing signal detects the propagation changes not only under steady-state thermal conditions, but also during transients, thus providing additional useful data, as we shall see later in this chapter. Furthermore, two heating fields of equal power, but different frequency, can be sent through the ionized medium from the opposite ends of the test section to provide a more uniform heating of electrons along the axis; in fact, in this manner the heating nonuniformity associated with the attenuation of a single wave as it moves along the axis can be compensated, when attenuation is modest, from the opposite behavior of the other wave. Figure 3.11.7 shows a block diagram of a typical setup for implementation of the described technique.

The reduction of the ionized medium size to values consistent with the requirement of a nearly uniform field over the medium volume, without violating the small perturbation assumption, makes it increasingly difficult to perform measurements with adequate accuracy [Gilardini (1957)]. Therefore most experiments have actually been conducted under conditions that do not satisfy the uniform field requirement. In these cases experiments provide only values of ρ_{av} as a function of the maximum intensity of the heating field, and when the collision parameters have to be determined from these data more sophisticated procedures are required. As a simple but relevant example of these procedures we illustrate the case of the determination of electron collision frequencies in rare gases.

The experimental conditions are chosen such that (a) $\omega^2 \gg \nu_m{}^2$; (b) at each point the electron distribution function is Maxwellian (see case 2, Table 2.6.1) with a temperature corresponding to the local intensity of the electric field, as it would be in a uniform equivalent case; and (c) two heating fields, with about the same power, launched along opposite directions make this electron temperature nearly uniform axially. Thus, on the basis of (2.6.60) with $G = 2m/M$, the constant electron temperature on the waveguide axis can be taken as:

$$T_{e0} = T_g + \frac{e^2 M}{3km^2}\left(\frac{P_1}{Y_1\omega_1{}^2} + \frac{P_2}{Y_2\omega_2{}^2}\right) \qquad (3.11.22)$$

where by means of (3.11.19) we have introduced P_1, the power flow of heating field 1 (ω_1, Y_1), and P_2, the power flow of heating field 2 (ω_2, Y_2), both averaged along the waveguide section containing the plasma. Correspondingly the electron temperature distribution as a function of the position $\mathbf{r}$ in the cross section will be:

$$(T_e - T_g) = (T_{e0} - T_g)E_h{}^2(\mathbf{r}) \qquad (3.11.23)$$

252

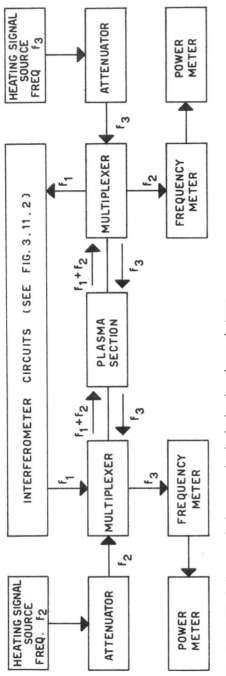

Fig. 3.11.7. Block diagram of microwave circuits for heating plasma electrons.

where $E_h(\mathbf{r})$ is the electric field configuration of the heating signal in the cross section, normalized to unity on the axis.

A relationship between the observable $\rho_{av} = \Delta A/p \, \Delta\phi$ and the electron temperature T_{e0}, computed by means of (3.11.22), is obtained from (3.11.17), if we use expression (2.6.48) for the conductivity ratio and note furthermore that $\sigma_i = -ne^2/m\omega$ under condition a above. The result is:

$$\omega\rho_{av} = \sum_j \left(\frac{2k}{m}\right)^{j/2} A_j \frac{\int_V [T_g + (T_{e0} - T_g)E_h^2]^{j/2} nE_0^2 \, dV}{\int_V nE_0^2 \, dV} \tag{3.11.24}$$

Quite often the electron density distribution can be written as the product of $n_z(z)$, the density on the plasma column axis z, multiplied by $n_r(\mathbf{r})$, the density distribution in the cross section, independent of the cross-section position and normalized to unity on the axis; result (3.11.24) becomes:

$$\omega\rho_{av} = \sum_j \left(\frac{2k}{m}\right)^{j/2} A_j \frac{\int_S [T_g + (T_{e0} - T_g)E_h^2]^{j/2} n_r E_0^2 \, dS}{\int_S n_r E_0^2 \, dS} \tag{3.11.25}$$

Since the field configurations are known and the density distribution can be assumed properly, it is possible to determine coefficients A_j by fitting the experimental points of ρ_{av} versus T_{e0} to the theoretical curve calculated from (3.11.25); the collision probability curve is then obtained from (2.6.49) after substitution of the A_j values.

Other analogous cases of nonuniform electron heating can be handled in similar ways; details can be found in the literature [Anderson and Goldstein (1956b), Gilardini and Brown (1957), Gilardini (1963)]. When a standing wave pattern is formed along the guide axis, the electron temperature becomes also a function of z; such temperature variations have been discussed by Cronson (1966).

We recall now that, when electron-ion collisions are important, the constant ρ condition cannot be satisfied in an isothermal plasma, since ρ is proportional to the electron density [see eq. (2.6.43)], and this is a function of the position in the waveguide. In this case the contribution of electron-ion collisions to ρ_{av} is obtained by substituting (2.6.43) into (3.11.17); if $\omega^2 \gg \nu_m^2$, this yields expression (2.6.43) again, provided, however, that one regards n as an average density n_{av}, according to the relation:

$$n_{av} = \frac{\int_V n^2 E_0^2 \, dV}{\int_V nE_0^2 \, dV} \tag{3.11.26}$$

In this way methods 3 and 4 described in Section 2.6B for the case of isothermal plasmas can both be properly used. We specify only that the $d\rho/dt = 0$ condition has been determined experimentally [Anderson and Goldstein (1955)] by observing on an oscilloscope the change of the transmission loss through an afterglow plasma, when a pulsed heating signal is applied (cross-modulation effect), and by finding the conditions for which the initial rate of this change is zero. In this experiment the transmission loss rate of change can properly be considered a measure of dp/dt, since electron density decays proceed at a much slower rate. Furthermore, we note that the electron density, usually known as $\bar{n}$ from $\Delta\phi$ measurements according to (3.11.15), has to be transformed into n_{av} by means of appropriate assumptions concerning the volume distribution of n.

Free-space microwave transmission arrangements have not been considered as yet, since, as we stated earlier, the use of a nonradiating transmission line makes it possible to attain the best accuracy. Only at millimeter and submillimeter wavelengths does the free-space beam technique become attractive, since these wavelengths are much smaller than common plasma dimensions, so that diffraction and radiation effects are sufficiently small. In this case the propagation problem can be treated by considering the model of a simple straight ray crossing a plane-slab plasma, and the conductivity ratio is provided by the same formula (3.11.10) as in the transmission line case. Since the relative amounts of perturbation ($\Delta\phi/\beta_0$ and $\Delta A/\beta_0$) vary as $(\omega_p/\omega)^2$, accurate measurements at millimeter and submillimeter wavelengths require electron densities correspondingly higher than at lower frequencies. In practice, this free-space beam technique has been frequently adopted for determining electron density distributions in dense plasmas of different size and geometry, but it has not been used to any significant extent for measuring conductivity ratios. Therefore, we shall not discuss this technique any further here. The interested reader can find more detailed information on the relevant problems in the reviews by Heald and Wharton (1965) and by Hermansdorfer (1968); simplified design charts for focused-beam interferometers are presented by Musal (1969).

B. Resonance of a Microwave Cavity, Slightly Perturbed by an Ionized Gas

Equation (3.10.16) gives the change in the complex free resonance angular frequency of a cavity due to the presence of an ionized gas inside the cavity. For our purpose we must consider the ratio of the change in the cavity band width $\delta f_1 - \delta f_0$, or of the change $\Delta(1/Q)$ of the cavity $1/Q$, to the resonant frequency shift Δf_0:

$$\frac{\delta f_1 - \delta f_0}{\Delta f_0} = \frac{\Delta(1/Q)}{\Delta f_0/f_0} = 2\frac{\mathrm{Im}\ \Delta\omega}{\mathrm{Re}\ \Delta\omega}$$

$$= -2\frac{\mathrm{Re}\int_V(\boldsymbol{\sigma}\cdot\mathbf{E_0})\cdot\mathbf{E_0}\,dV}{\mathrm{Im}\int_V(\boldsymbol{\sigma}\cdot\mathbf{E_0})\cdot\mathbf{E_0}\,dV} \qquad (3.11.27)$$

reference being made to the discussion of Section 3.10 [eqs. (3.10.18) and (3.10.20)]. We recall also that the above formulas were obtained in the hypothesis of the ionized gas introducing only a small perturbation, so that the field configuration is not appreciably different from the one of the cavity without the ionized medium; this means that the absolute magnitude of the electron current must be much smaller than the displacement current over the whole cavity and that coupling to different resonant modes must be negligible.

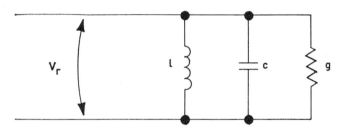

Fig. 3.11.8. Equivalent circuit of an isolated cavity.

Equation (3.11.27) is basically the three-dimensional version of (3.11.7), so that the same considerations apply to the cavity as we saw above for the guide. In the no-magnetic field case we obtain:

$$\rho_{av} = \frac{\int_V \rho\sigma_i E_0{}^2 \, dV}{\int_V \sigma_i E_0{}^2 \, dV} = \frac{\delta f_1 - \delta f_0}{2p \, \Delta f_0} = \frac{\Delta(1/Q)}{2p \, \Delta f_0/f_0} \qquad (3.11.28)$$

Here too the constant-ρ case is of paramount importance, experimental conditions being chosen in most cases so as to fulfill this requirement (low-field and uniform-field cases).

We shall now discuss how the parameters of a resonant cavity can be determined; for this purpose it is convenient to consider first the equivalent circuit of an isolated cavity, as shown in Fig. 3.11.8. In terms of these circuit parameters the resonant frequency and the Q are given by:

$$f_0 = \frac{1}{2\pi\sqrt{lc}} \qquad (3.11.29)$$

$$Q = 2\pi f_0 \frac{U}{P} = \frac{1}{\sqrt{lc}} \frac{cV_r^2}{gV_r^2} = \frac{1}{g}\sqrt{c/l} \qquad (3.11.30)$$

The formula for Q is derived from definition (3.10.11); U is the stored energy, P the energy dissipated per second, and V_r the rms voltage across the resonant circuit.

At frequencies of the order of 100 MHz or less the equivalent circuit parameters of the cavity can be directly measured by an impedance bridge;

these measurements are performed both in the absence and in the presence of the ionized gas in the cavity, so that from them the value of ρ_{av} can easily be computed. In fact, if Δg and Δb are the observed changes of the cavity conductance and susceptance due to the ionized gas, as long as this presence perturbs the resonant frequency only slightly, it can be derived from (3.11.29) and (3.11.30) that:

$$\frac{\Delta f_0}{f_0} = -\tfrac{1}{2}\sqrt{l/c}\,\Delta b \qquad (3.11.31)$$

$$\Delta\left(\frac{1}{Q}\right) = \sqrt{l/c}\,\Delta g \qquad (3.11.32)$$

Thus eq. (3.11.28) may be put in the simple form:

$$\rho_{av} = -\frac{\Delta g}{p\,\Delta b} \qquad (3.11.33)$$

This formula is easily understood if we regard Δg as due to σ_r and Δb to σ_i, properly integrated over the cavity volume, in accordance with the numerator and the denominator of (3.11.28).

This impedance-measuring method was used by Carruthers (1962), who investigated atmospheric gases ionized by 100-kV X-rays at about 4300 roentgens per hour. Figure 3.11.9 shows the block diagram of the experimental setup. The ionization chamber (outer cylinder 15.2 cm O.D. × 25.4 cm long, inner cylinder 7.6 cm O.D. × 17.8 cm long) and the wax-filled

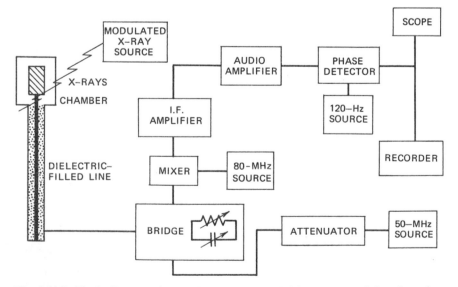

Fig. 3.11.9. Block diagram of Carruthers' impedance bridge apparatus [after Carruthers (1962), reproduced by permission of the National Research Council of Canada].

coaxial line (89 cm long) form a resonant circuit, connected to the bridge (a Wayne-Kerr type 801) by a quarter-wave line. Very high sensitivity is achieved by modulating the X-ray source at 120 Hz, so that a phase-sensitive detecting scheme can be used.

At higher frequencies, in the microwave region, the cavity is driven through an input line and its parameters are determined by power or field measurements on this same line or on an output line beyond the cavity. In this case

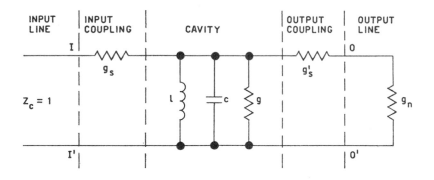

Fig. 3.11.10. Equivalent circuit of cavity and coupling lines.

we must consider the entire system formed by the cavity and the input and output lines. The appropriate equivalent circuit of this system is shown in Fig. 3.11.10: on the input line the cavity is viewed at plane II' and all admittances are reflected back on this line, whose characteristic admittance has been normalized to unity; g_s and g_s' represent the series losses of the input and output couplings; g_n, the output line admittance. Measurements yield the Q of the entire system, which is called Q_L, the loaded Q, and can be written like (3.11.30) as:

$$\frac{1}{Q_L} = \left(g + \frac{g_s}{1 + g_s} + \frac{g_s' g_n}{g_n + g_s'} \right) \sqrt{l/c} = \frac{1}{Q} + \frac{1}{Q_e} \qquad (3.11.34)$$

Here Q_e represents the Q of the reduced resonant circuit, derived from the above when g is removed, so that the conductance is provided by the external line and coupling elements alone.

In general, variations of Q_e due to the introduction of the ionized gas in the cavity are comparatively small and can be neglected. Thus measuring the change $\Delta(1/Q_L)$ of the inverse loaded Q yields directly the change $\Delta(1/Q)$ to be used in (3.11.28).

The resonant frequency and the loaded Q are obtained from measurements of the transmitted P_t over the incident P_i power, of the reflected P_r

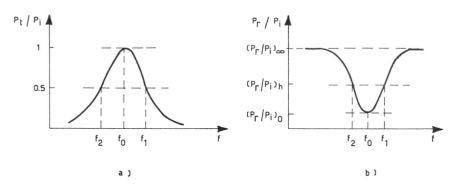

Fig. 3.11.11. Ratios of transmitted and reflected powers to the incident power as a function of frequency near the cavity resonance.

over the incident P_i power, or of the standing wave ratio R, that is, the ratio of the maximum to the minimum voltage on the input line. We shall now discuss separately the various possibilities that have actually been implemented in experiments for the determination of electron collision parameters.

a. Transmission method. From the analysis of Gould and Brown (1953) it can easily be derived that, in the case of a very small loss in the output coupling ($g'_s \simeq \infty$), the ratio P_t/P_i normalized to unity at resonance is given by:

$$\frac{P_t}{P_i} = \left[1 + \left(2Q_L \frac{f - f_0}{f_0}\right)^2\right]^{-1} \tag{3.11.35}$$

A convenient procedure consists of changing the frequency until we find (Fig. 3.11.11a) the two frequencies $f_1(>f_0)$ and $f_2(<f_0)$ on the two sides of the resonance, for which $P_t/P_i = 0.5$. From these we obtain:

$$f_0 = \tfrac{1}{2}(f_1 + f_2) \tag{3.11.36}$$

$$Q_L = \frac{f_0}{f_1 - f_2} = \frac{f_1 + f_2}{2(f_1 - f_2)} \tag{3.11.37}$$

b. Reflection method. In this case we have no output line ($g_n = 0$), and the following procedure can be performed. The ratio P_r/P_i is measured (Fig. 3.11.11b) at resonance $(P_r/P_i)_0$, where it attains its minimum value, and far from resonance $(P_r/P_i)_\infty$, where the maximum is obtained. The frequencies f_1 and f_2 at which the reflected to incident power ratio is:

$$\left(\frac{P_r}{P_i}\right)_h = \frac{1}{2}\left[\left(\frac{P_r}{P_i}\right)_0 + \left(\frac{P_r}{P_i}\right)_\infty\right] \tag{3.11.38}$$

are found, and f_0 and Q_L are determined by means of the same relationships (3.11.36) and (3.11.37).

The same procedure can be followed if the corresponding standing wave ratios R_0, R_∞, and R_h are considered. Brown and Rose (1952) give the appropriate curves of R_h versus R_0 for various values of R_∞.

In order to determine $\Delta(1/Q_L)$ the loaded Q is measured twice—in the absence and in the presence of the ionized medium; the second time, however, the procedure can be simplified since usually the role of the ionized gas is to change g, but not so much $\sqrt{c/l}$ and the coupling losses. For instance, in a reflection type of experiment, we readily obtain from (3.11.34) that, if g alone changes to a new value g', the loaded Q attains the new value Q'_L, given by:

$$\frac{Q_L}{Q'_L} = \frac{g' + g_s/(1 + g_s)}{g + g_s/(1 + g_s)} \tag{3.11.39}$$

Very far from resonance the signal from the input line sees the admittance g_s alone, so that measuring the ratio $(P_r/P_i)_\infty$ yields g_s; at resonance the admittance is given by $1/(1/g + 1/g_s)$, the series combination of g_s and g, so that measuring $(P_r/P_i)_0$, together with the previous knowledge of g_s, provides g. When these appropriate formulas are substituted for g_s and g into (3.11.39), the following result is obtained:

$$\frac{Q_L}{Q'_L} = \frac{\sqrt{(P_r/P_i)_\infty} \mp \sqrt{(P_r/P_i)_0}}{\sqrt{(P_r/P_i)_\infty} \mp \sqrt{(P_r/P_i)'_0}} \tag{3.11.40}$$

where $(P_r/P_i)'_0$ indicates the power ratio at resonance in the presence of the ionized gas, and the upper sign has to be chosen when the cavity system is undercoupled $[(1/g + 1/g_s) < 1]$, the lower one when overcoupled $[(1/g + 1/g_s) > 1]$. More conveniently (3.11.40) can also be written in the form:

$$\Delta\left(\frac{1}{Q_L}\right) = \frac{1}{Q'_L} - \frac{1}{Q_L} = \pm \frac{\sqrt{(P_r/P_i)'_0} - \sqrt{(P_r/P_i)_0}}{\sqrt{(P_r/P_i)_\infty} \mp \sqrt{(P_r/P_i)'_0}} \frac{1}{Q_L} \tag{3.11.41}$$

Thus we have demonstrated that, after Q_L has been measured in the absence of the ionized gas, to obtain ρ_{av} it is sufficient to determine the new resonant frequency and the corresponding ratio of reflected to incident power when the ionized gas is present. Measurement of the power ratio at resonance, rather than the width of the resonance curve, avoids the problem of changes in the electric field, and subsequently of electron heating, which may occur as the power absorbed from the cavity changes with the frequency.

Determinations of the conductivity ratio in resonant cavities are typically performed in the late afterglow of a repetitively pulsed microwave discharge.

For generation of the discharge a different resonant mode is employed, and the gas is broken down periodically by a pulsed signal from a magnetron. Probing is performed by a continuous wave signal, properly attenuated in order to obtain the desired field in the cavity. Since the resonant frequency and the loaded Q have to be determined simultaneously at a specific time in the late afterglow, all power and standing-wave-ratio measuring setups are made sensitive for only a short time interval (usually from few microseconds up to a maximum of about 100 μsec) at the desired instant in the afterglow.

Transient superheterodyne receivers, whose local oscillator (klystron) frequency is controlled by a fast sweep or by a short gate, properly delayed in time with respect to the breakdown pulse, are frequently used for this purpose. The receiver output is observed on an oscilloscope, whose horizontal sweep can be the same sweep that modulates the local oscillator frequency. When the ratio of the transmitted to the incident power is considered, samples of the two signals are separately received and then introduced to the push-pull input of the oscilloscope [Gould and Brown (1953)]. By adjusting the gain of the two receivers a null deflection is obtained on the oscilloscope at the resonant frequency [$P_t/P_i = 1$ according to (3.11.35)]; when the frequency is changed, deflection is rebalanced to zero by adjusting a calibrated wave attenuator preceding one of the receivers: the change of attenuation provides the new power ratio. A similar technique can be used also for measuring the ratio of the reflected to the incident power. When instead the standing wave ratio has to be measured, only a probe with a calibrated attenuator followed by a transient receiver is required; the attenuator is adjusted to maintain a constant deflection on the scope when the probe is moved along the slotted-line section [Phelps, Fundingsland, and Brown (1951)], and this provides the standing-wave-pattern information.

A different approach makes use of a very short swept-frequency probing signal, which is applied to the cavity after a specified delay, and of video detectors and amplifiers instead of superheterodyne receivers. In this case we can proceed to determine on an oscilloscope the ratios P_t/P_i and P_r/P_i, as described above for the superheterodyne receivers, the only difference being that now the oscilloscope sweep axis represents the signal frequency, so that the resonant response curves of the cavity appear on the scope. A cavity wavemeter is connected on the signal line in an absorption mode, so that a sharp absorption pip appears on the curves, providing the appropriate calibration of the frequency axis. The method is seldom used, however, since its sensitivity is less than that obtainable with superheterodyne receivers.

In all experiments it is important to know the electric field in the cavity, since electron heating is a function of this field. Under steady-state conditions the magnitude of this field can be determined by the following method [Brown and Rose (1952)]. We consider a cavity with a simple shape, for which field configuration is known, and we call E_m the rms value of the maximum electric field inside the cavity. The stored energy in an empty cavity of volume V can be expressed as:

$$U = \eta \varepsilon_0 E_m{}^2 V \qquad (3.11.42)$$

where $\eta < 1$ is a dimensionless parameter, which can easily be computed when the field configuration is known. Substituting (3.11.42) into (3.11.30), we obtain:

$$E_m = \sqrt{PQ/2\pi \eta \varepsilon_0 V f_0} \qquad (3.11.43)$$

The power P absorbed by the cavity and its Q can be determined from an appropriate set of measurements of incident, reflected, and transmitted powers, as can be easily inferred from the preceding discussion on the techniques for measuring the cavity loaded Q and the other equivalent circuit parameters. In fact, P is given by $(P_i - P_r - P_t)$ and Q is computed from Q_L according to (3.11.34) if we know the relevant conductivities, which again can be derived from measurements of the off-resonance ratio $(P_r/P_i)_\infty$ and of the resonance ratios $(P_r/P_i)_0$ and $(P_t/P_i)_0$.

The parameter η depends on the cavity shape and on the operation resonant mode. Thus, for a cylindrical E_{010}-mode cavity $\eta = 0.2695$, and for a rectangular parallelepiped E_{110}-mode cavity $\eta = 0.25$.

When E_m in the cavity is below value (3.11.18), the constant-ρ condition holds, since the ionized gas remains isothermal. When E_m is larger, however, the constant-ρ condition requires that the ionized gas be confined within a small glass or quartz bottle, where the field is nearly uniform. Moreover, in order to avoid unnecessary complications, we have two additional requirements:

1. The Q of the cavity has to be sufficiently low so that the electric field remains nearly constant for a time long enough to ensure that electrons have reached equilibrium with the field. In fact, whereas the probing signal frequency is usually kept constant, the resonant frequency of the cavity changes continuously during the afterglow, as imposed by the varying characteristics of the decaying plasma.

2. The $\Delta(1/Q_L)$ has to be determined by means of power ratio measurements at resonance rather than by measurements of the width of the resonance curve, since the latter technique requires changing the signal frequency and thus the electric field inside the cavity. The method based on eq. (3.11.41) is a good example of this technique.

The low-Q requirement may be in opposition to the fact that accurate measurements are better performed in a high-Q cavity; for this reason the method has been changed so as to separate the measuring field, kept to very low values not to perturb the plasma, from the field applied to heat the

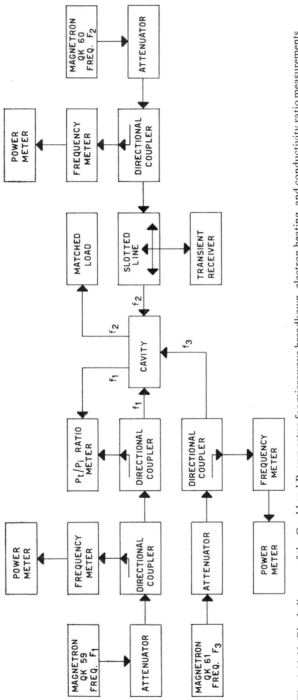

Fig. 3.11.12. Block diagram of the Gould and Brown setup for microwave breakdown, electron heating, and conductivity ratio measurements.

electrons, and this is implemented by using two different resonant modes of the cavity. Figure 3.11.12 shows the general block diagram of this experimental setup designed by Gould and Brown (1954) and used thereafter by Gilardini and Brown (1957) and Bekefi and Brown (1958) for different gases.

The microwave cavity used by these authors is a rectangular parallelepiped, designed to resonate in its three fundamental modes at wavelengths of 9.5, 10.0, and 10.5 cm. A tunable pulsed magnetron (QK 61), driven by a well-regulated power supply, provides 100-W peak power to the 9.5-cm mode for the gas breakdown; the magnetron is pulsed for a duration varying between 0.1 and 5 msec and at a repetition rate varying from 20 to 120 Hz. A continuous-wave tunable magnetron (QK 59) is used for sending a highly attenuated measuring signal (a few microwatts) to the 10.5-cm mode. The resonant frequency and the loaded Q are determined by measurements of P_t/P_i versus frequency, employing transient superheterodyne receivers located in the power-measuring section, and operative for periods of 20–100 μsec at any desired delay in the afterglow. A continuous-wave tunable magnetron (QK 60) is used for producing the 10.0-cm mode electric field, which increases the average electron energy; this field is determined according to eq. (3.11.43), with an accuracy of $\pm 3\%$, from measurements of the power incident on the cavity (bolometer and bridge) and from standing-wave-ratio measurements (slotted section and transient receiver). The Q of the 10.0-cm heating mode is properly decreased to values as low as 200 by means of an external load connected to this mode alone; in all the experiments mentioned, these Q values have allowed the electric field to remain constant for a time sufficient for the electrons to attain steady-state energy values. The conductivity ratio is measured during this interval of time.

A small quartz bottle concentric with the cavity would be required to contain the plasma in a region of nearly uniform field; however, if the volume becomes very small, whereas the small perturbation assumption is still to be satisfied, the measuring accuracy may decrease significantly. To avoid this difficulty a larger bottle is generally used, and the field nonuniformity is properly taken into account in the derivation of the collision parameters from the measurements of ρ_{av} as a function of the maximum intensity of the heating field. Gould and Brown (1953) and Gilardini and Brown (1957) have carefully analyzed this problem for the case of electron-neutral collisions is rare gases, when $\omega^2 \gg \nu_m^2$. When local thermal equilibrium is assumed, a formula like (3.11.24) is again obtained; the appropriate formulas for the more general case, in which energy flows are of importance, are given by Gilardini and Brown (1957), who also indicate how one can determine suitable conditions for a pressure-independent conductivity ratio ρ_{av}.

To avoid some of the inherent difficulties and uncertainties connected with the procedures described above, Hoffmann and Skarsgard (1969) have

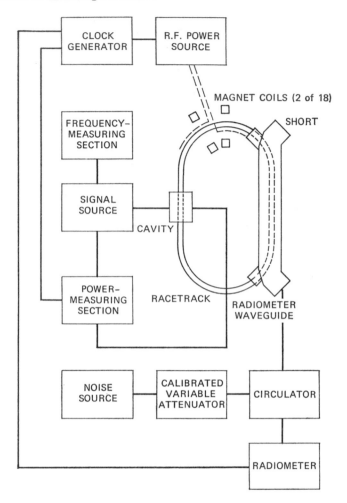

Fig. 3.11.13. Block diagram of the Hoffmann and Skarsgard setup for breakdown and electron heating at 3.5 MHz, conductivity ratio measurements at S-band, and radiation temperature measurements at C-band [after Hoffmann and Skarsgard (1969)].

preferred the following technique. The discharge tube is made from a quartz tube (2-cm diameter) bent in the shape of a racetrack with 20-cm straight sections and 15-cm U-bend radii (Fig. 3.11.13). On one of the straight sections there is a cylindrical cavity for measuring ρ at S-band; on the other one, a waveguide for measuring the plasma radiation temperature with a C-band radiometer, using the techniques to be described in Section 3.12. The gas is broken down at a repetition rate of 30 Hz by r.f. pulses (frequency

3.5 MHz; duration 100–500 μsec) applied to three turns of wire wound around the racetrack; these turns are re-energized to heat the electrons with a short burst of r.f. power (duration 10 μsec) at a late time in the afterglow. Measurements of ρ and T_r are made at times later than 40 μsec after the heating pulse. Hoffmann and Skarsgard state that at these times ρ can be taken as uniform over the entire tube volume; furthermore, the electron distribution function can be recognized as Maxwellian experimentally, by applying the method of Fields, Bekefi, and Brown (1963), described in Section 2.6D. These measurements provide ρ versus T_e, so that the appropriate formulas (2.6.48) and (2.6.49), illustrated in Section 2.6B, can be applied for the derivation of electron collision probabilities.

Another approach based on the use of a resonant cavity is the one of Fundingsland, Faire, and Penico (1954), who operate with a pulsed heating signal, so as to measure by means of an adequately fast transient receiver the conductivity ratio, not only at equilibrium, but also as a function of time during the application of the heating pulse. Since the method of these authors for the evaluation of the electron collision parameters is the one based on eq. (2.6.67), a determination of the strength of the heating field is not required in their experiments; this represents a significant simplification of the measuring process.

At millimeter and submillimeter wavelengths the free-space beam technique can be used. In this case the resonator is of the focused Fabry-Perot type, and its open structure allows measurements without restricting the plasma dimensions. So far, this arrangement has usually been applied for measuring electron densities; information on this application can be found, for example, in Hermansdorfer (1968). The possibility to perform also ρ measurements, however, has been successfully demonstrated by Chaffin and Beyer (1968), who used, for this purpose, a 35-GHz Fabry-Perot resonator with confocal spherical mirrors.

C. Measurements in the Case of a Dense Plasma

When the problem cannot be handled by perturbation methods, a complete electromagnetic solution must be considered. In general this undertaking becomes quite complicated, and the derivation of the plasma parameters from the observable propagation or resonance features involves approximated analytical procedures and *a priori* schematizations of the geometrical and physical configurations of the plasma, which may introduce significant errors into the final results. For these reasons, as we said earlier in this section, it is generally preferred to operate with tenuous plasmas, which can be handled with more straightforward procedures by the small perturbation techniques. However, this becomes increasingly difficult as soon as we try to perform measurements of the electron collision parameters in plasmas containing large fractions of temperature dissociation products (atomic species, e.g.) of molecular gases, since at the desired high temperatures the electron densities of the equilibrium plasmas may be so large that, if we operate in the usual microwave range, the small perturbation condition $\omega_p{}^2 \ll \omega^2$ cannot be satisfied. On the other hand, the lack of data on the electron collision parameters in hot gases justifies using methods that are compatible with the experimental conditions, even if they are not so accurate as other methods.

We consider here the most common and important case, that of a shock tube operating at large Mach numbers ($\geqslant 10$); typically the cross sections of such tubes have dimensions of the order of a few centimeters, and the initial gas pressures are a few torr. The resulting electron densities are $\geqslant 10^{12}$ cm^{-3}, to which correspond plasma frequencies of $\geqslant 10$ GHz. The following approaches appear to be the most promising ones and have actually been adopted in some cases:

a. Use of millimeter and submillimeter wavelengths, which still satisfy the small perturbation condition. In this case the use of the shock tube for a waveguide or cavity function is ruled out by the excessive size of the tube cross section; on the other hand, conditions are favorable for applying a free-space microwave beam technique, since the plasma dimensions are much larger than the wavelength of the probing signal. This solution has been discussed in previous paragraphs; since known results are rather inaccurate [see, e.g., Kelly (1966)], this approach will be eliminated from any further consideration.

b. Use of the shock tube as a waveguide for a lower frequency signal ($\omega \leqslant \omega_p$), so that the shock-wave plasma plays the role of a mismatched dielectric load of the waveguide [Daiber and Glick (1961), Brandewie and Williams (1964), Krylov and Tusnov (1966), Glick and Jones (1969)]. Insofar as the plasma slab can be considered homogeneous, measuring the reflected wave amplitude and phase makes it possible to determine the plasma conductivity; in the real case, however, the axial uniformity of the shock-produced plasma is not very good (see Fig. 3.11.14) and assumptions must be made concerning the electron density and temperature profiles if we desire to derive correct information on the conductivity. This technique is rather unsatisfactory, however, since final results depend to a significant extent on the mathematical models of the plasma profile which have been used; actually, this difficulty could be greatly reduced by performing simultaneous measurements at sufficiently different microwave frequencies, since the propagation of the electromagnetic field in the plasma depends on the ratio ω/ω_p, so that the profiles could also be derived from these observations.

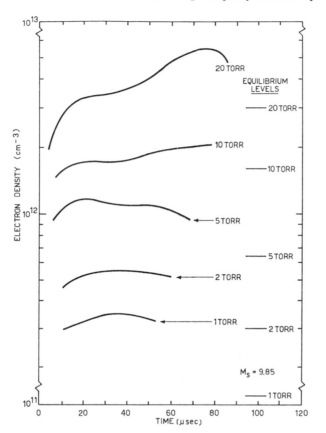

Fig. 3.11.14. Electron density, measured as a function of time, behind a shock in air at Mach number 9.85 and for different initial pressures [after Tevelow (1972)].

We recall that an attempt by Bethke and Ruess (1964) to introduce the axial density distribution, determined independently by means of transverse propagation measurements, has not provided consistent results. Furthermore, data can be altered by the presence of precursor electrons [see, e.g., Weymann (1969) and Holmes and Weymann (1969)], ahead of the shock, or of some preionization from the driving discharge in the case of electric and electromagnetic shock tubes.

c. Use of a microwave reflection probe, whose dimensions are considerably less than the size of the shock tube cross section. This is inserted into the shock tube perpendicular to its axis or placed at the tube end parallel to the axis [Takeda, Tsukishima, and Funahashi (1966), Funahashi and Takeda (1968)]. The probe consists of a piece of rectangular waveguide, facing at its end a plasma characterized by a sharp boundary and nearly uniform density and temperature distributions; this situation is provided either by the transverse uniformity of the shock wave (except for the thin boundary layers), when the probe axis is orthogonal to the direction of propagation of the shock front, or by the nearly uniform plasma, which is produced at the tube end when the shock wave is reflected there. For a very dense plasma ($\omega_p^2 \gg \omega^2$), the penetration depth of the microwave signal is small compared to the probe dimensions, so that fringing field effects at the probe end can be neglected; in this case the reflected wave in the probe waveguide is practically the same as the one we should observe if a uniform infinite plasma terminated the waveguide (the presence of the sheath, which is inevitably formed at the boundary between the plasma and the probe end, is neglected under the assumption that the sheath thickness is much thinner than the penetration depth of the microwave signal).

The uniform plasma, like any homogeneous medium of complex conductivity σ, is equivalent to a dielectric of relative permittivity [see, e.g., Marcuvitz (1951)]:

$$\varepsilon_r = 1 - j\frac{\sigma}{\varepsilon_0 \omega} \tag{3.11.44}$$

and the reflection coefficient in the waveguide is [see, e.g., Collin (1960)]:

$$\Gamma_\infty = \frac{1 - \sqrt{\varepsilon_r}}{1 + \sqrt{\varepsilon_r}} \tag{3.11.45}$$

Therefore measuring Γ_∞ with transient receivers and by any one of the many conventional microwave procedures provides, after the appropriate mathematics, the complex conductivity σ, from which the conductivity ratio ρ can be computed. Since measurements must be performed on a one-shock basis, Γ_∞ can be conveniently determined by computing the standing wave ratio from the simultaneous transient observation of the signal amplitudes at two different positions along the probe waveguide; calibration and normalization of both signal amplitudes and waveguide positions can be performed by short-circuiting the probe end [Tsukishima and Takeda (1962)].

d. Modification of method *c* so as to measure transmission instead of reflection for a microwave signal propagating in a direction normal to the shock tube axis. For this purpose a rectangular waveguide runs along a diameter of the tube; the waveguide has slots in the narrow walls and a corresponding affixed nozzle, properly designed to section out a portion of the flowing plasma with minimal disturbance to the plasma itself, while maintaining the condition of an essentially closed waveguide [Fig. 3.11.15 shows the nozzle designed for this purpose by Tevelow (1967)].

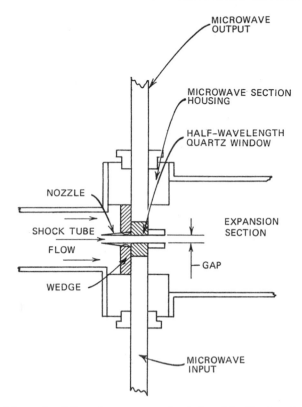

Fig. 3.11.15. Schematic of the microwave nozzle, designed by Tevelow for transmission measurements in an X-band waveguide, a section of which is filled by a flowing portion of the shocked gas [after Tevelow (1967)].

The complex transmission coefficient for a slab of thickness L is [Stratton (1941), Heald and Wharton (1965)]:

$$T = \frac{(1 - \Gamma_\infty^2) \exp\left[-(\gamma - \gamma_0)L\right]}{1 - \Gamma_\infty^2 \exp(-2\gamma L)} \qquad (3.11.46)$$

where Γ_∞ is given by (3.11.45). In the first approximation the empty waveguide propagation constant γ_0 is given by (3.11.5), and the propagation constant of the plasma-loaded waveguide γ is [see, e.g., Marcuvitz (1951)]:

$$\gamma = (k_c^2 - \varepsilon_r k_0^2)^{1/2} \qquad (3.11.47)$$

When $L > 0.08\lambda_0$ we must take into account the effect of the large sidewall slots on the propagation characteristics of the waveguide; this case is discussed by Tevelow and Tischer (1971). The transmission coefficient T can be measured by inserting the waveguide section into a bridge interferometer scheme, equipped with transient receivers for experiments of short duration, thus adopting the techniques already described for ΔA and $\Delta \phi$ measurements.

e. Insertion of the shock tube, when its diameter is sufficiently small, across a rectangular waveguide with the tube axis parallel to the microwave electric field (dominant H_{10} mode). Thus the flowing plasma takes the appearance of a lossy dielectric post located in a certain position of the waveguide. The impedance of this post can be measured by various conventional microwave methods, modified only insofar as transient observations must be performed. For instance, a short circuit is placed to terminate the waveguide, and the change in the standing wave pattern due to the plasma post is observed [Sekiguchi and Herndon (1958), Takeda and Roux (1961)]. Alternatively, two partial shorts are symmetrically erected in the waveguide, before and after the traversing tube, to create a resonant cavity, and its resonant frequency shift and the change in Q due to the plasma post are determined [Anderson (1961)]. From these measurements we can derive the plasma relative permittivity, using Marcuvitz's (1951) accurate expressions for the equivalent circuit parameters of a lossy dielectric post in a rectangular waveguide. An important source of error for this method, however, is the perturbation effect on the microwave field (surface polarization) due to the presence of holes in the waveguide walls to permit the shock tube to pass through [Coffey (1965)].

3.12   RADIATION TEMPERATURE MEASUREMENTS

The importance of radiation temperature measurements to the subject of our concern is basically twofold: (*a*) to verify that in an ionized gas electrons and gas particles are, at a certain instant of time, in thermal equilibrium in the absence of applied fields, as discussed in Section 3.11, and (*b*) to know the values of this quantity, which must often be used together with the values of the conductivity ratio for the determination of collision parameters, when electrons are heated by an applied electric field (see Section 2.6). In both cases the relevant experimental conditions are those described in Section 3.11 as the most typical ones for the measurements of the conductivity ratio: a plasma is generated by volume ionization in a waveguide or in a resonant cavity, where it perturbs only slightly the electromagnetic behavior of the supporting structure.

The problem of plasma radiation in microwave structures has been thoroughly reviewed in Bekefi's (1966) book, and we shall consider here only the aspects that are relevant to specific cases related to the determination of electron collision parameters [for a more general theory see also Wright and Bekefi (1971)]. Here too we shall discuss the waveguide case first.

Let us consider a unit length dx of a waveguide with perfectly conducting walls, containing a tenuous plasma; for each propagating mode the presence of the plasma introduces an attenuation, whose coefficient α can be computed with the appropriate formulas of Section 3.11 [see eqs. (3.11.4) and (3.11.6), where it was called $\Delta\alpha$]. Correspondingly the plasma slab emits incoherent radiation; its power, which in one mode, per unit radian frequency interval $d\omega$ centered around ω, flows down the waveguide in one direction, is:

$$p_\omega = 2\alpha\lambda_0^2 B_\omega \qquad (3.12.1)$$

This equation represents the waveguide analogue of (2.4.39). Substituting the Rayleigh-Jeans limit (2.4.40) for B_ω yields:

$$p_\omega = \frac{\alpha k T_r}{\pi} \qquad (3.12.2)$$

The above equations apply to the case of a plasma whose electron energy distribution is constant over the entire cross section, so that T_r is a constant too. In the more general case we shall have a local radiation temperature, in accordance with (2.6.36), and each cross-section element will contribute to the total incoherent radiation according to Kirchhoff's law, that is, proportionally to the product of the local T_r value by the local absorption $\sigma_{er} E_0^2$. Then eq. (3.12.2) may still be used, provided we take for the radiation temperature the following appropriate average value:

$$T_{av} = \frac{\int_S T_r \sigma_{er} E_0^2 \, dS}{\int_S \sigma_{er} E_0^2 \, dS} \qquad (3.12.3)$$

When we consider in the waveguide a plasma of axial length L, extending from $x = -L$ to $x = 0$, the power p_ω emitted by an element of unit length dx at position x will contribute to the total incoherent radiation, flowing out of the plasma in the positive x direction, the amount:

$$p_\omega \exp\left(-2\int_x^0 \alpha \, dx\right) = p_\omega \exp\left(-\tau\right)$$

where $\tau(x)$, the so-called optical depth, has been introduced for convenience of discussion. Thus, assuming an axially constant radiation temperature, we obtain for the total emitted power per unit radian frequency interval $d\omega$

and per mode:

$$P_\omega = \frac{kT_r}{\pi} \int_{-L}^{0} \alpha \exp(-\tau)\, dx$$

$$= \frac{kT_r}{2\pi} \int_{\tau(L)}^{0} \exp(-\tau)\, d(-\tau)$$

$$= \frac{kT_r}{2\pi} \{1 - \exp[-\tau(L)]\}$$

$$= \frac{kT_r}{2\pi} [1 - \exp(-2\,\Delta A)] = \frac{kT_r}{2\pi} A_p \qquad (3.12.4)$$

In the last expressions ΔA is the attenuation change due to plasma, discussed in Section 3.11, and A_p is the plasma absorption coefficient, defined as the fraction of incident power of a test wave that is absorbed by the plasma.

When the optical thickness $\tau(L)$ of the plasma is large, $A_p = 1$ and the power P_ω is simply $kT_r/2\pi$. However, in most of our experiments conditions are such that this assumption cannot be made. Usually, when dealing with semitransparent plasmas, we can recognize (Fig. 3.12.1a) behind the plasma some kind of noise source, which radiates as a black body at temperature T_0; in some cases this source may simply be a load with an absorption coefficient $A = 1$, at temperature T_0. The total incoherent power flowing

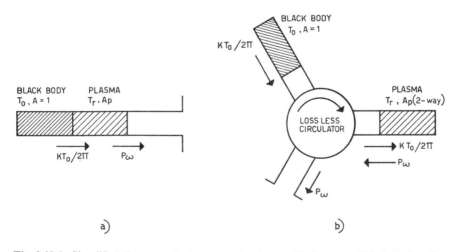

Fig. 3.12.1. Simplified diagrams of microwave structures with plasma and black-body noise sources, and indication of radiation power flows.

down the waveguide past the plasma is the sum of the radiation emitted by the plasma, plus that part of the radiation emitted by the source behind which is not absorbed in traversing the plasma (reflection is assumed to be negligible, in accordance with the hypothesis of small perturbation):

$$P_\omega = \frac{kT_r}{2\pi} A_p + \frac{kT_0}{2\pi}(1 - A_p)$$

$$= \frac{k}{2\pi} [T_0 + (T_r - T_0)A_p] \tag{3.12.5}$$

It is immediately apparent that eq. (3.12.5) applies also to the less frequent, but quite important, configuration shown in Fig. 3.12.1b. In this case, there is a short circuit behind the plasma section and A_p represents the two-way absorption. The circulator feeds the noise source power $kT_0/2\pi$ to the plasma and sends to the output the reflected part of it, plus the plasma radiation contribution; the reflected power is obtained by subtracting the absorbed power from the incident one. This configuration is important, since now eq. (3.12.5) is derived without the assumption that reflections at the plasma interface are negligible.

Our goal is measuring T_r regardless of whether A_p is known. If a stationary plasma is considered, this goal is achieved by comparing P_ω to the power $kT_0/2\pi$ from the source and changing T_0 until the two powers are equal: at this point T_0 is equal to T_r [see eq. (3.12.5)], independently of the value of plasma absorption.

A typical experimental setup [Bekefi and Brown (1961)], to be used when configuration a of Fig. 3.12.1 is chosen, is shown in Fig. 3.12.2. The source, which radiates as a black body at the variable temperature T_0, is actually implemented by the series combination of (1) a white-noise source of temperature T_s, which here is obtained by placing a standard noise gas discharge tube in the waveguide; (2) a calibrated attenuator, and (3) a 3-dB power divider, here provided by a magic tee. According to (3.12.5) the combination of the source and of the attenuator behaves as a new source having the temperature:

$$T'_s = T_s - (T_s - T_g)A \tag{3.12.6}$$

where A is the attenuator absorption coefficient, and T_g its ambient temperature. The magic tee provides on one arm the noise power at the temperature T_0, which illuminates the plasma, and on the opposite arm the same power for reference; since on the fourth arm of the tee there is a load at ambient temperature, T_0 will be:

$$T_0 = \tfrac{1}{2}(T'_s + T_g) = \tfrac{1}{2}[(1 - T_s)A + (1 + A)T_g] \tag{3.12.7}$$

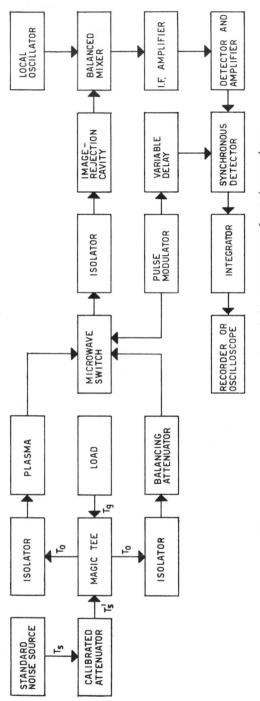

Fig. 3.12.2. Block diagram of a radiometer appropriate for measuring the radiation temperature of a stationary plasma.

274

By means of a ferrite switch, like a pulsed circulator [Mayer (1956)] (20–1000 Hz have been used as convenient switching rates), noise powers P_ω and $kT_0/2\pi$ are alternately sent to a radiometer, where the resulting signal is properly amplified and detected in a low-noise superheterodyne receiver, and then passed into a synchronous detector, whose reference phase is provided by the ferrite switch. After a final integration (for a time between 0.1 and 10 sec), the signal is displayed, metered, or recorded. The two noise powers are equal when zero output is observed (balance condition). A properly designed diode switch may replace the ferrite switch if desired.

Four additional features of the arrangement described are a balancing attenuator to equalize the power flows in the reference and in the plasma arms, when the plasma is off; ferrite isolators to cancel any possible wave reflection [the insertion loss of the isolators on the two arms of the magic tee can easily be taken into account by properly modifying eq. (3.12.7) according to general relation (3.12.6)]; an image-rejection cavity preceding the r.f.-to-i.f. conversion mixer, when detection of only one of the two side bands is desired; and a balanced mixer to eliminate the noise contribution of the local oscillator.

When configuration b of Fig. 3.12.1 is chosen, either for the possibility of placing a short circuit at one end of the plasma or for the presence of substantial reflections at the plasma interface, the r.f. part of the scheme of Fig. 3.12.2 must be modified as shown in Fig. 3.12.3 [Bekefi (1966)]. A switched three-port circulator could be appropriate, but a four-port circulator is preferred because it offers the possibility of equalizing, by means of a balancing attenuator in the path of the reference signal, power flows along the alternative paths when the plasma is off. The rest of the circuit does not require any modification.

Both of the arrangements described can actually be considered as particular forms of the well-known comparison radiometer devised by Dicke (1946); as such, at balance they are insensitive to any fluctuation of the receiver gain and noise factor, if this occurs at a frequency low compared to the switching rate. Therefore the system sensitivity is here determined only by the statistical fluctuations of the noise-like signal.

In this case, if B_R is the 3-dB band width of the system before detection and τ is the time constant of the final integration, $B_R\tau$ represents approximately the number of rectified i.f. noise pulses upon which the measurement is performed. The fractional error in regard to temperature is thus expected to be of the order of $1/\sqrt{B_R\tau}$. Actually, a more accurate analysis [Strum (1958)] shows that ΔT_0, the minimum detectable change in the radiation temperature, is given by:

$$\frac{\Delta T_0}{T_0 + T_R} = \frac{C}{\sqrt{B_R\tau}} \tag{3.12.8}$$

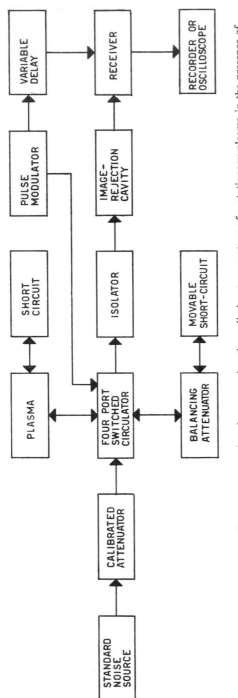

Fig. 3.12.3. Block diagram of a radiometer appropriate for measuring the radiation temperature of a stationary plasma in the presence of substantial reflections at the plasma interface.

where T_R is the receiver noise temperature, and C is a constant associated with the type of detector, the switching waveform, and the band shape of the receiver elements. Considering that $C \simeq 2$ is a typical value for this constant, it appears from (3.12.8) that adequate accuracy in plasma experiments will usually be obtained by taking $\sqrt{B_R \tau} \simeq 10^3$. For instance, appropriate

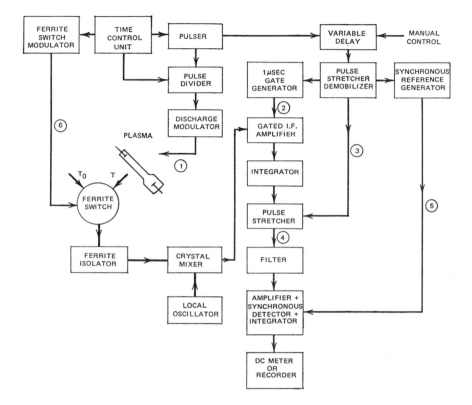

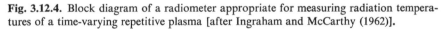

Fig. 3.12.4. Block diagram of a radiometer appropriate for measuring radiation temperatures of a time-varying repetitive plasma [after Ingraham and McCarthy (1962)].

values which can be adopted without difficulty are $B_R = 1$ MHz and $\tau = 1$ sec.

A few appropriate modifications (Fig. 3.12.4) allow the use of the same basic setup for radiation temperature measurements in time-varying repetitive plasmas [Ingraham and McCarthy (1962), Bekefi (1966)]. The plasma repetition frequency must be the same as the switching rate of the ferrite switch. The receiver is gated, usually in the i.f. or in the video stage, with a very narrow pulse (typically a few microseconds), so that it samples radiation for a time interval short enough to justify the assumption of constant plasma characteristics during the gate pulse. This gating is performed twice during the basic switching period: one gate samples the plasma radiation, and the following one the source radiation (see Fig. 3.12.5). Then the video train of alternate signal pulses is stretched, so as to obtain a square waveform, which is the same waveform that we have in the Dicke radiometer; thus the rest of the circuit does not have to be modified. The gate can be positioned in

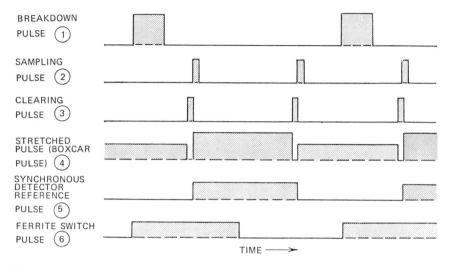

Fig. 3.12.5. Signal waveforms at typical points (marked 1–6) of the radiometer of Fig. 3.12.4 [after Ingraham and McCarthy (1962)].

time as desired for the purpose of determining the radiation temperature at different instants; the reference phase for the synchronous detector is also provided by the gate position.

Whereas in the steady-state Dicke radiometer $B_R\tau$ is the approximate number of noise pulses upon which the measurement is performed, the gating process reduces this number to $B_R\tau\delta$, where δ is the gate duty cycle, that is, the ratio of the gate width to one half of the switching period [Taft, Stotz, and Holt (1963)]. This causes an increase $1/\sqrt{\delta}$ in the minimum detectable change ΔT_0 given by eq. (3.12.8). Since relevant values of δ can be as low as 10^{-3} to 10^{-4}, whereas maximum practical magnitudes for B_R and τ are $\sim$10 MHz and $\sim$10 sec, respectively, we have typically an order-of-magnitude loss of sensitivity when a well-designed gated radiometer is compared with a normal steady-state radiometer, both used for measurements of plasma radiation temperatures.

The discussed technique can be applied to radiation temperature measurements in the afterglow of a repetitively pulsed discharge. In this case, however, advantage may be taken of the practical disappearance of the plasma in the very late afterglow, in such a way as to eliminate the bridge and the ferrite switch arrangement and to receive the reference noise radiation $kT_0/2\pi$ through the same waveguide that contains the plasma, during the time interval when this is completely transparent [Formato and Gilardini (1960), Taft, Stotz, and Holt (1963)].

Let us consider the setup shown in Fig. 3.12.6; both configurations of Fig. 3.12.1 can be used as inputs. The temperature T_i of the noise power arriving at the receiver is $T_0 + (T_r - T_0)A_p$, where T_0 is $T_s - (T_s - T_g)A$, according to eqs. (3.12.5) and (3.12.6); both T_r and A_p change with time during the afterglow. In Fig. 3.12.7 we show as a function of time the separate contributions to the total received noise from the source $T_0(1 - A_p)$ and from the plasma T_rA_p; the behavior is due mainly to the fact that A_p drops from 1 during the discharge period to 0 in the very late afterglow. By gating we sample the received noise twice every afterglow, the first time at the instant when we wish to measure T_r, and the second time half a period later, in the very late afterglow, when $T_i = T_0$ since $A_p \simeq 0$; the two noise pulses will be equal only when we have set T_0, by adjusting A, equal to T_r. The equal-pulse condition is shown by the zero output of a selective amplifier, tuned at the discharge repetition frequency, or of a synchronous detector following a pulse stretcher.

The techniques we have described allow measuring T_r regardless of whether A_p is known and hence are strongly recommended. However, we must note that some of the data we shall report in Chapter 4 have been obtained using more elementary techniques: the noise power and the absorption and reflection coefficients of the plasma are all measured, so that the radiation temperature can be derived from them.

Let us now consider the different case of a plasma in a resonant cavity. If the cavity is assumed to be lossless and to be the terminal load of a waveguide, we may handle this configuration exactly like that of a plasma backed up by a reflecting short in a waveguide. Null methods can be used in the same way; attention has only to be paid to the fact that, not to lose sensitivity, the band width B_R of the receiving system has to be consistent with the resonant cavity band width.

In the analysis performed so far, corrections must be introduced, however, to take into account the wall and container losses, both in the waveguide and in the resonant cavity cases. Let us call α_g the attenuation constant of the waveguide with the empty plasma container, and α_p the contribution of the plasma alone; both values are assumed to be constant along the guide axis.

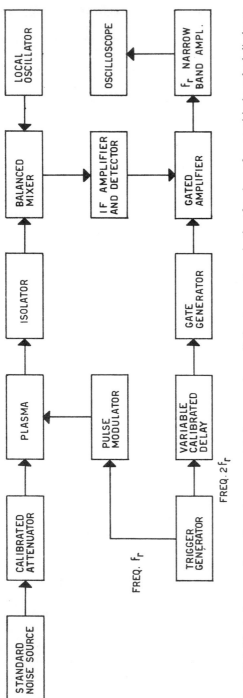

Fig. 3.12.6. Block diagram of a radiometer appropriate for measuring radiation temperatures in the afterglow of a repetitively pulsed discharge.

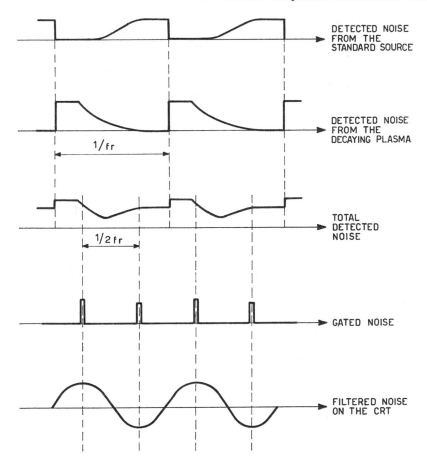

Fig. 3.12.7. Waveforms of the noise received when the radiometer of Fig. 3.12.6 is used.

The power P_ω emitted by the plasma will now be [see derivation of eq. (3.12.4)]:

$$
\begin{aligned}
P_\omega &= \frac{kT_r}{\pi} \int_{-L}^{0} \alpha_p \exp\left[2(\alpha_p + \alpha_g)x\right] dx \\
&= \frac{kT_r}{2\pi} \frac{\alpha_p}{\alpha_p + \alpha_g} \{1 - \exp\left[-2(\alpha_p + \alpha_g)L\right]\} \\
&= \frac{kT_r}{2\pi} \frac{\alpha_p}{\alpha_p + \alpha_g} A_t
\end{aligned}
\tag{3.12.9}
$$

where A_t is the absorption coefficient of the entire plasma-filled section of the waveguide. Thus, when the wall and container losses are considered, eq. (3.12.5) becomes:

$$P_\omega = \frac{kT_r}{2\pi} \frac{\alpha_p}{\alpha_p + \alpha_g} A_t + \frac{kT_g}{2\pi} \frac{\alpha_g}{\alpha_p + \alpha_g} A_t + \frac{kT_0}{2\pi}(1 - A_t) \quad (3.12.10)$$

where the second term at the right-hand side represents the noise power generated by the waveguide walls and by the container (at room temperature T_g). At balance, whichever method is used, this power has to be equal to the reference, or no-plasma noise power:

$$\frac{kT_g}{2\pi} A_g + \frac{kT_0}{2\pi}(1 - A_g)$$

where A_g is the absorption coefficient of the plasma section, when the plasma is off. Then the unknown temperature of the plasma is not given exactly by T_0, but has to be computed from the expression [Formato and Gilardini (1960)]:

$$T_r = \left(1 + \frac{\alpha_g}{\alpha_p}\right)\left(1 - \frac{A_g}{A_t}\right)(T_0 - T_g) + T_g \quad (3.12.11)$$

which thus requires knowledge of the various contributions to the signal attenuation, except when the attenuation due to the plasma dominates.

In the foregoing discussion, we have neglected the presence of reflections at the plasma interface. If we assume the plasma container section of the waveguide to be perfectly matched in the plasma absence and call Γ the plasma reflection coefficient, it is a simple matter to show that we must add the term $(k/2\pi) |\Gamma|^2 (T_g - T_0)$ to the right-hand side of eq. (3.12.10) to provide P_ω, and modify (3.12.11) into:

$$T_r = \left(1 + \frac{\alpha_g}{\alpha_p}\right)\left(1 - \frac{A_g}{A_t} + \frac{|\Gamma|^2}{A_t}\right)(T_0 - T_g) + T_g \quad (3.12.11')$$

A modification of the method has been proposed by Delpech and Gauthier (1971) in order to avoid having to know $|\Gamma|$. The precision of this approach seems definitely lower, however, since the method requires two separate, successive measurements of noise power unbalance, instead of a null determination alone. On the other hand, measuring $|\Gamma|$ is not cumbersome, since A_t has to be measured anyhow during the afterglow decay.

In the case of a resonant cavity, the appropriate A_p and A_t coefficients depend strongly on frequency, unless we choose to operate using a small B_R near resonance only; within this limitation A_g is given by $1 - (P_r/P_i)_0$ and A_t by $1 - (P_r/P_i)_0'$, where the notations of Section 3.11 have been adopted. Furthermore the ratio α_g/α_p can here be recognized as $(1/Q)/\Delta(1/Q)$, so that the expression for computing T_r becomes:

$$T_r = \left[1 + \frac{1/Q}{\Delta(1/Q)}\right]\frac{(P_r/P_i)_0 - (P_r/P_i)'_0}{1 - (P_r/P_i)'_0}(T_0 - T_g) + T_g \quad (3.12.12)$$

The same remarks apply here as for the waveguide case.

When heating microwave fields are used for the purpose of changing the electron energies in an afterglow and of determining from appropriate microwave measurements the electron collision parameters by the methods discussed in Chapter 2, radiation temperatures can be measured as previously discussed, provided heating is performed at different microwave frequencies and an appropriate multiplexing system is used to separate the heating and noise signals (Fig. 3.12.8). For the reasons discussed in Section 3.11 these experiments must usually be performed applying nonuniform fields, so that measurements provide only values of T_{av} [see eq. (3.12.3)] as a function of the applied microwave heating power.

Formato and Gilardini (1962) discussed one of these cases and showed that, if G is velocity independent, its value can still be computed by means of eq. (2.6.61), where now (2.6.63) reads:

$$\Delta T = (T''_{av} - T_g) - \frac{(\Delta A/\Delta\phi)'}{(\Delta A/\Delta\phi)''}(T'_{av} - T_g) \quad (3.12.13)$$

one apex being the values in the absence of the heating field and two apices those in the presence of this field, and E^2 is replaced by an appropriate average value. Thus, if we have two heating signals, flowing along opposite directions, for the purpose of making the radiation temperature nearly uniform axially, with the notations of the previous section we have either:

$$\frac{E^2}{\omega^2} = \left(\frac{P_1}{Y_1\omega_1{}^2} + \frac{P_2}{Y_2\omega_2{}^2}\right)\frac{\int_S n_r E_h{}^2\, dS}{\int_S n_r\, dS} \quad (3.12.14)$$

in the limit case of an infinite thermal conductivity of the plasma, or:

$$\frac{E^2}{\omega^2} = \left(\frac{P_1}{Y_1\omega_1{}^2} + \frac{P_2}{Y_2\omega_2{}^2}\right)\frac{\int_S n_r E_h{}^2 E_0{}^2\, dS}{\int_S n_r E_0{}^2\, dS} \quad (3.12.15)$$

in the limit case of zero thermal conductivity (local equilibrium case) and of small energy gain ($\Delta T \ll T'_{av}$). A simple criterion may be given for the choice between (3.12.14) and (3.12.15): the first one will be used when for the average electron the energy relaxation time ($\simeq 1/\langle G\nu_m\rangle$) is much longer than the time required by the electron to diffuse over a distance of the order of the heating field nonuniformities ($\simeq l^2/2D$, where l is the distance cited), and the second one will be used in the opposite case.

284

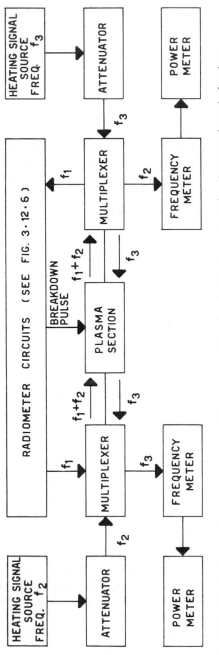

Fig. 3.12.8. Block diagram of a microwave circuit for measurements of radiation temperatures in microwave-heated afterglows.

Furthermore, in cases where we have a very large thermal conductivity, the distribution function tends to become independent of the position, since an extremely fast redistribution of the absorbed microwave heating energy takes place in the cross section; in this case the plasma behaves as a plasma heated by a uniform field, given by (3.12.14), and all methods and formulas applicable to the uniform-field situations can be used here as well.

3.13 THERMAL TRANSPORT PARAMETERS

A. General Description of the Basic Experimental Methods

Application of a microwave power pulse to an isothermal afterglow plasma heats the electrons. Direct or indirect measurements of the electron temperature decay after removal of the heating pulse allow determination of the electron energy relaxation time. In a similar fashion measurements of the electron temperature distribution in space or in time, when only a small volume of plasma is selectively heated, yield values for the energy transport coefficients.

Two conditions, $\omega^2 \gg \nu_m{}^2$ and f^0 Maxwellian, are considered to be satisfied in all the experiments we shall now discuss. The Maxwellian hypothesis is based on the results of Chapter 2, applied to these experiments: either weak microwave heating provides only a small energy disturbance over the isothermal regime, or strong heating is used in the case of rare-gas plasmas only (velocity-independent G). Furthermore all experiments are performed using microwave propagation in waveguides; this, however, is by no way essential. The experiments are described by eq. (2.5.8); other sources of energy flow, like mass diffusion and heat losses by radiation and absorption of photons, can be considered negligible under actual experimental conditions. In particular the following basic methods have been implemented.

a. Energy decay method. After the end of a microwave heating pulse the electron average energy $\bar{u}$ and the time derivative of its decay are simultaneously observed. Since the experimental conditions are arranged so that electron heating over the entire plasma is uniform and the electron density decay is much slower than the energy decay, eq. (2.5.8) provides the following formula for computing $\tau_r(\bar{u})$:

$$\frac{1}{\tau_r(\bar{u})} = -\frac{1}{\Delta \bar{u}} \frac{\partial \Delta \bar{u}}{\partial t} = -\frac{\partial (\ln \Delta \bar{u})}{\partial t} \tag{3.13.1}$$

where:

$$\Delta \bar{u} = \bar{u} - \bar{u}_g = \tfrac{3}{2} k (T_e - T_g) \tag{3.13.2}$$

b. Transient energy flow method. At time $t = 0$ a microwave pulse suddenly raises the electron average energy in a localized region of the plasma. The corresponding increase $\Delta \bar{u}$ of the electron energy over the thermal value, which takes place in the nonheated region of the plasma at time $t > 0$, is obtained by solving the equation:

$$\frac{\partial(n \, \Delta \bar{u})}{\partial t} = -\nabla_r \cdot \mathbf{q} - \frac{n \, \Delta \bar{u}}{\tau_r} \tag{3.13.3}$$

When the presence of a static magnetic field is included (a case of interest in Section 3.14), the energy flow $\mathbf{q}$, according to (2.4.30) and (2.6.72) for a Maxwellian distribution, is given by:

$$\mathbf{q} = -\frac{\hat{u}}{e} \, \mathfrak{M} \cdot \nabla_r n - n \mathfrak{D} \cdot \nabla_r \bar{u} - n \mathfrak{M} \cdot \mathbf{E} \tag{3.13.4}$$

The dc field $\mathbf{E}$ is due to electrostatic space charges and can be evaluated as follows. The electron flow $\mathbf{\Gamma}_e$ in the plasma is much less than the free diffusion value $-\nabla_r \cdot (\mathbf{D}n)$, since the slower diffusion rate of positive ions allows the formation of the space-charge field $\mathbf{E}$ which opposes this flow. Hence we can set $\mathbf{\Gamma}_e \simeq 0$ in eq. (2.4.13) and derive from it the value of $\mathbf{E}$:

$$\mathbf{E} = -(\mu n)^{-1} \cdot \nabla_r \cdot (\mathbf{D}n)$$

$$= -\mu^{-1} \cdot \mathbf{D} \cdot \frac{\nabla_r n}{n} - \mu^{-1} \cdot \frac{\partial \mathbf{D}}{\partial \bar{u}} \cdot \nabla_r \bar{u}$$

$$= -\frac{\hat{u}}{e} \frac{\nabla_r n}{n} - \mu^{-1} \cdot \frac{\partial \mathbf{D}}{\partial \bar{u}} \cdot \nabla_r \bar{u} \tag{3.13.5}$$

The last expression is obtained by substituting for $\mathbf{D}$ its Einstein value (2.6.2') for a Maxwellian distribution. When value (3.13.5) is introduced, eq. (3.13.4) reduces to:

$$\mathbf{q} = -n \mathfrak{D}' \cdot \nabla_r \bar{u} \tag{3.13.6}$$

where:

$$\mathfrak{D}' = \mathfrak{D} - \mathfrak{M} \cdot \mu^{-1} \cdot \frac{\partial \mathbf{D}}{\partial \bar{u}} \tag{3.13.7}$$

represents an equivalent thermal diffusivity tensor. Now we assume further that the electron density does not change appreciably with time during the heat flow, and that both $n\mathfrak{D}'$, which is $2/3k$ times a corresponding equivalent conductivity $\mathscr{K}'$, and τ_r are practically constant, the electron energy deviations from thermal values being sufficiently small. Then substitution of (3.13.6) into (3.13.3) yields:

$$\frac{\partial \, \Delta \bar{u}}{\partial t} = \mathfrak{D}'_T \, \nabla_T{}^2 (\Delta \bar{u}) + \mathfrak{D}'_{\parallel} \, \nabla_{\parallel}{}^2 (\Delta \bar{u}) - \frac{\Delta \bar{u}}{\tau_r} \tag{3.13.8}$$

where ∇_T^2 and $\nabla_{||}^2$ indicate ∇^2 space operators acting, respectively, in the plane orthogonal to the static magnetic field and along this field direction.

We consider now the no-magnetic-field, one-dimensional-flow case, in which heating takes place suddenly at $t = 0$ in the plane $z = 0$. The energy flow equation becomes:

$$\frac{\partial \, \Delta \bar{u}}{\partial t} = \mathfrak{D}' \frac{\partial^2 \, \Delta \bar{u}}{\partial z^2} - \frac{\Delta \bar{u}}{\tau_r} \qquad (3.13.9)$$

where $\mathfrak{D}'$ is the no-magnetic-field scalar value of $\mathfrak{D}'$:

$$\mathfrak{D}' = \mathfrak{D} - \frac{\mathfrak{M}}{\mu} \frac{\partial D}{\partial \bar{u}} \qquad (3.13.7')$$

The Green function solution of this equation for a δ pulse at $t = 0$ in the plane $z = 0$ is:

$$\Delta \bar{u} \propto t^{-\frac{1}{2}} \exp \left(-\frac{z^2}{4 \mathfrak{D}' t} - \frac{t}{\tau_r} \right) \qquad (3.13.10)$$

Nygaard (1967a) has determined $\mathfrak{D}'$ by measuring the energy increases $\Delta \bar{u}_1$ and $\Delta \bar{u}_2$ in two positions z_1 and z_2 at the same instant of time; from (3.13.10) $\mathfrak{D}'$ is obtained as:

$$\mathfrak{D}' = \frac{z_2^2 - z_1^2}{4t \ln (\Delta \bar{u}_1/\Delta \bar{u}_2)} \qquad (3.13.11)$$

It must be noted that this equation does not include τ_r.

In previous experiments by Goldstein and Sekiguchi (1958) and by Sekiguchi and Herndon (1958), on the other hand, experimental conditions were arranged so as to have τ_r large enough to justify neglecting the energy decay term $\Delta \bar{u}/\tau_r$ in comparison to the thermal diffusion one. Furthermore, microwave heating took place for times longer than those describing the energy diffusivity in the nonheated region: this implies assuming as a more appropriate boundary condition that of a constant $\Delta \bar{u}_0$ in the position $z = 0$ at all times $t \geqslant 0$. In this case the solution of eq. (3.13.9) is:

$$\Delta \bar{u} = \Delta \bar{u}_0 \left[1 - \text{erf} \left(\frac{z}{2\sqrt{\mathfrak{D}'t}} \right) \right] \qquad (3.13.12)$$

For any fixed value of z this equation describes a curve of $\Delta \bar{u}$ versus t; this curve can also be observed experimentally, and the time t_0 of its maximum slope, designated here as the delay time, determined. According to (3.13.12) this time is given by:

$$t_0 = \frac{z^2}{6\mathfrak{D}'} \qquad (3.13.13)$$

so that $\mathfrak{D}'$ can be easily derived from measurements. It is also worth noting that the value of $\Delta \bar{u}$ at $t = t_0$ is only about 8% of $\Delta \bar{u}_0$, the value to be attained as $t \to \infty$.

   c. Steady-state energy flow method. A limited volume of plasma is heated by a microwave signal pulse, whose duration is long enough to provide for a steady-state distribution of electron energy both in the heated and in the nonheated regions. In the latter region the distribution of the excess electron energy $\Delta \bar{u}$ for a one-dimensional flow is obtained by solving eq. (3.13.9) with $\partial \Delta \bar{u}/\partial t = 0$, which yields:

$$\Delta \bar{u} \propto \exp\left(-\frac{z}{\sqrt{\mathfrak{D}'\tau_r}}\right) \tag{3.13.14}$$

If the $\Delta \bar{u}$ distribution is measured and plotted as $\ln \Delta \bar{u}$ versus z, a straight line must result; its slope is negative and can be set equal to the inverse of the "relaxation distance" $\sqrt{\mathfrak{D}'\tau_r}$, according to (3.13.14). Thus this method, which is also due to Goldstein and Sekiguchi (1958), determines the product $\mathfrak{D}'\tau_r$ alone. Then $\mathfrak{D}'$ can be obtained if τ_r is independently determined for the same experimental conditions; this is easily done by using the energy decay method or, in rare gases, by measuring the conductivity ratio and applying eq. (2.6.69).

   Instead of heating only a small volume of plasma, Delpech and Gauthier (1969) imposed on a rare-gas afterglow plasma an electromagnetic standing wave of known amplitude $E_m |\cos(\beta_0 z)|$; in this case eq. (2.5.8) yields:

$$\frac{e^2 M E_m^2}{2m^2\omega^2}\cos^2(\beta_0 z) + \mathfrak{D}'\tau_r \frac{\partial^2 \Delta \bar{u}}{\partial z^2} - \Delta \bar{u} = 0 \tag{3.13.15}$$

Here, measuring the energy profile $\Delta \bar{u}(z)$ makes it possible to determine $\mathfrak{D}'\tau_r$ by means of a best fitting procedure based on eq. (3.13.15), where all other quantities are known.

   Before discussing how electron energy determinations are performed, a few words are in order on the $\mathfrak{D}'$ coefficient, which was introduced above and is the quantity directly derived from measurements. Recalling Einstein's relation (2.6.2), we can write definition (3.13.7') of $\mathfrak{D}'$ as:

$$\mathfrak{D}' = \mathfrak{D} - \frac{2}{3e}\left(1 + \bar{u}\frac{\partial \ln \mu}{\partial \bar{u}}\right)\mathfrak{M} \tag{3.13.16}$$

When a power-law dependence $\nu_m \propto v^h$ is assumed, we have $\mu \propto \bar{u}^{-h/2}$, and $\mathfrak{D}'$ becomes:

$$\mathfrak{D}' = \mathfrak{D} - \frac{2}{3e}\left(1 - \frac{h}{2}\right)\mathfrak{M} \tag{3.13.17}$$

Substitution of expressions (2.6.73) and (2.6.74) finally yields:

$$\mathfrak{D}'p = \tfrac{5}{3}\gamma_7^{-h}\frac{\hat{u}}{m}\frac{p}{v_m(\hat{u})} = \frac{2}{4-h}\mathfrak{D}p \tag{3.13.18}$$

which in particular, for dominating electron-ion collisions, gives:

$$\mathfrak{D}' = \tfrac{2}{7}\mathfrak{D} \tag{3.13.19}$$

According to the results of Spitzer and Härm (1953), the coefficient $\tfrac{2}{7}$ becomes 0.2293 when electron-electron collision effects are taken into account.

B. Experimental Implementation of the Basic Methods

The various methods will now be reconsidered for the purpose of specifying which observable parameters have actually been used for measuring the average electron energy, and for adding details concerning the experimental arrangements.

a. Energy decay method. Let the observable parameter be the quantity ζ, whose relation to $\bar{u}$ for Maxwellian distributions is known: $\bar{u}(\zeta)$. At the instant of the heating pulse end, one measures the value ζ_0 of the observable parameter and the value $\dot{\zeta}_0$ of its initial time derivative. Equation (3.13.1) yields:

$$\frac{1}{\tau_r} = -\left.\frac{\partial\bar{u}}{\partial\zeta}\right|_{\zeta=\zeta_0}\cdot\frac{\dot{\zeta}_0}{\bar{u}(\zeta_0)-\bar{u}_g} \tag{3.13.20}$$

Frequently the relationship between $\bar{u}$ and ζ can be written in the form $\zeta \propto \bar{u}^l$; in this case (3.13.20) becomes:

$$\frac{1}{\tau_r} = -\frac{\dot{\zeta}_0}{l\zeta_0[1-(\zeta_\infty/\zeta_0)^{1/l}]} \tag{3.13.21}$$

where ζ_∞ is the value of ζ that corresponds to $\bar{u}_g$, namely, the value of ζ at the same afterglow time, in the absence of the heating pulse.

Anderson and Goldstein (1956b) applied this method, taking as an appropriate observable parameter the plasma attenuation ΔA suffered by a weak microwave probing signal, which goes through an afterglow rare-gas plasma contained in a waveguide. The experimental arrangement is the same as the one these authors used for conductivity ratio measurements. If $v_m^2 \ll \omega^2$ and if the average electron energy due to microwave heating is uniform over the entire plasma, $\Delta A \propto \langle v_m \rangle \propto \bar{u}^{h/2}$, the power law $v_m \propto v^h$ being used in deriving this relationship. Then we can apply (3.13.21), where now ζ is ΔA and $l = h/2$.

In most experiments operation is restricted to electron energies near the thermal value, so that $(\zeta - \zeta_\infty) \ll \zeta_\infty$. In this case we have:

$$\bar{u}(\zeta) \simeq \bar{u}_g + \left.\frac{\partial \bar{u}}{\partial \zeta}\right|_{\zeta=\zeta_\infty} (\zeta - \zeta_\infty) \qquad (3.13.22)$$

and (3.13.1) becomes:

$$\frac{1}{\tau_r} = -\frac{\dot{\zeta}}{\zeta - \zeta_\infty} \qquad (3.13.23)$$

If τ_r is constant over the pertinent energy range, this equation provides for an exponential decay with time constant τ_r; the value of this constant can easily be determined by observing the ζ decay on an oscilloscope. If τ_r is not constant, its value at the energy attained at the end of the heating pulse can be measured anyhow, according to (3.13.23), from the values of ζ_0 and $\dot{\zeta}_0$ at the beginning of the energy decay, as shown in Fig. 3.13.1.

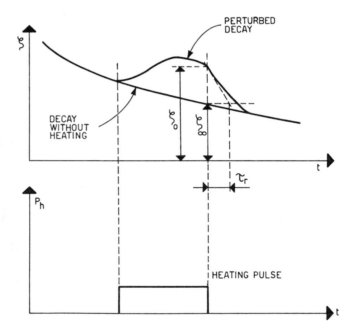

Fig. 3.13.1. Waveform of the observable ζ during an afterglow decay, perturbed or not by a microwave heating pulse.

Common observable parameters have been the transmitted power or the power loss of a weak microwave probing signal [Anderson and Goldstein (1955), Dougal and Goldstein (1958), Mentzoni and Row (1963), Mentzoni and Narasinga Rao (1965), Mentzoni and Donohoe (1968)] and the luminous intensity associated with the electron-ion recombination process in the decaying plasma [Dougal and Goldstein (1958)]. The latter technique exploits the phenomenon of "afterglow quenching" [Kenty (1928), Goldstein, Anderson, and Clark (1953)], that is, the reduction of recombination light intensity in the presence of r.f. heating, due to the inverse dependence of the recombination coefficient on the electron temperature. As in conductivity ratio measurements, the afterglow plasma is typically generated by a high-voltage dc pulse with an appropriate repetition rate in a long glass tube, housed in a rectangular or square waveguide. In the case of light intensity measurements, a viewing port is cut in the waveguide walls at the maximum E field position, so as not to perturb significantly the microwave field; the light is filtered, and then detected by photomultipliers, amplified, and displayed on an oscilloscope.

b. Transient energy flow method. Sekiguchi's and Herndon's (1958) experimental implementation of this method is shown in Fig. 3.13.2. Here too the observable parameter is the recombination light, generated within a small volume of the afterglow plasma by a repetitive discharge, and detected from a photomultiplier, which can be positioned along the discharge tube. For the case of small deviations from the no-heating condition, the instant of maximum slope of the light radiated as a function of time at any given position should give t_0 [see eq. (3.13.13)]. Actually, however, since the inflection point was not always clear enough in the experiment, Sekiguchi and Herndon took as t_0 the instant at which the light trace on the oscilloscope begins to diverge from that obtained in the absence of heating pulse. Plotting t_0 versus z^2 for each power level of the microwave heating signal, they obtained straight lines (Fig. 3.13.3); their slopes determine the corresponding $\mathcal{D}'$ values: the limit $\mathcal{D}'$ value, as heating power approaches zero, is the value appropriate to the plasma thermal conditions in the absence of heating.

The same Fig. 3.13.2 conveys information on the dimensions of the discharge tube and of the heating signal waveguide; this waveguide is terminated by a short circuit and, at a distance of one-quarter of a guide wavelength from the short, is traversed in the center, where the electric field is largest, perpendicularly by the discharge tube. This configuration makes it possible (1) to enhance heating for any given microwave input power, (2) to obtain high resolution in position, and (3) to accommodate the discharge without excessive disturbance of the microwave field; in this way it is also possible to determine the plasma conductivity by means of standing-wave-ratio measurements (see Section 3.11). The microwave conductivity, in turn, allows us to find, when $\nu_m{}^2 \ll \omega^2$, the electron density, the average collision frequency $\langle \nu_m \rangle$, and, when the $\langle \nu_m \rangle$ versus $\bar{u}$ relation is known, the average electron energy $\bar{u}$ of the plasma in the waveguide; all these data, in spite of

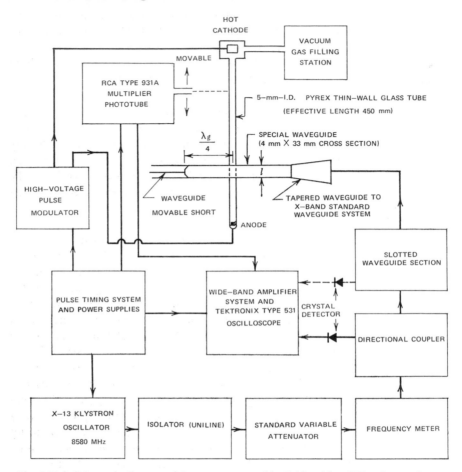

Fig. 3.13.2. Schematic diagram of the apparatus used by Sekiguchi and Herndon to observe in an afterglow plasma the electron energy diffusion from a microwave-heated region; the energy observable is here the radiation recombination light from the plasma [after Sekiguchi and Herndon (1958)].

being actually only average values over the cross section, are usually sufficient to characterize appropriately the plasma in which $\mathfrak{D}'$ is measured.

Alternatively $\langle \nu_m \rangle$ in rare gases can be derived from τ_r [eq. (2.6.69)], which is measured using the energy decay method, the observable parameter being the radiated recombination light from the plasma volume in the waveguide.

Earlier, Goldstein and Sekiguchi (1958) had also tried using a second waveguide, equal to the first one, in order to determine the change in the electron temperature along the discharge tube, the observable parameter being in

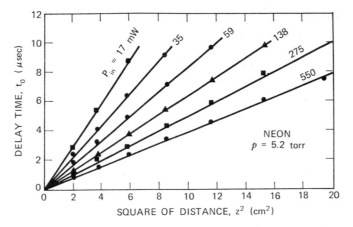

Fig. 3.13.3. Curves of the delay time versus the square of the distance from the heated volume, for various incident power levels (P_{in}) of the microwave heating signal [after Sekiguchi and Herndon (1958)].

this case the transmitted power of a probing weak signal in the second waveguide. The method was later abandoned in favor of the optical one described above, which was found to yield much better accuracy.

Nygaard (1967a) determined the quantities $\Delta \bar{u}_1$ and $\Delta \bar{u}_2$, to be used in eq. (3.13.11) for the calculation of $\mathfrak{D}'$, by choosing as the appropriate energy observable parameter either the occurrence in the afterglow of a Tonks-Dattner resonance at a given frequency or the radiated recombination light from the plasma. Let us discuss first the Tonks-Dattner resonance case, for which a short account of the theory underlying these resonances will be given. [This subject is extensively reviewed in a book by Vandenplas (1968), to which the reader is referred for details and bibliography.]

We consider a collisionless cylindrical plasma column of radius a, placed transversally in a waveguide or in a strip line, as shown in Fig. 3.13.4; for

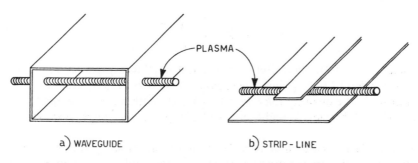

Fig. 3.13.4. Two typical microwave structures, in which Tonks-Dattner resonances can be observed.

particular values of the electron density and temperature a microwave signal propagating along the waveguide is strongly attenuated by the plasma (Tonks-Dattner resonances). This phenomenon is interpreted in terms of the formation of standing waves in the outer region of the plasma column, where the electron density is such that the corresponding plasma frequency:

$$\omega_p = \frac{e\sqrt{n}}{\sqrt{m\varepsilon_0}} \tag{3.11.13}$$

is below the signal frequency, and longitudinal waves may propagate in the plasma, according to the relation of Bohm and Gross (1949) for the propagation phase constant β:

$$\beta^2 \simeq \frac{\omega^2 - \omega_p{}^2}{3kT_e/m} \tag{3.13.24}$$

Equation (3.13.24) follows from the fact that purely longitudinal waves in a plasma require the total current $\mathbf{J}$ to be identically zero, and then:

$$j\omega\varepsilon_0 + \sigma = 0 \tag{3.13.25}$$

The appropriate expression for σ to be substituted into (3.13.25), however, is not (2.4.19), which would yield $\omega = \omega_p$ only, but:

$$\sigma \simeq -j\frac{e^2 n}{m\omega}\left(1 + \frac{3kT_e}{m}\frac{\beta^2}{\omega^2}\right) \tag{3.13.26}$$

which is appropriate for a collisionless Maxwellian plasma, in which both the electromagnetic field and the electron distribution component $\mathbf{f}^1$ are treated as wavelike quantities, $\exp[j(\omega t - \boldsymbol{\beta} \cdot \mathbf{r})]$. In fact, when the above quantities are written in this form and substituted in the component Boltzmann equation (2.3.1), one obtains (ν_m and $\boldsymbol{\omega}_b$ being zero):

$$j(\omega - \mathbf{v} \cdot \boldsymbol{\beta})\mathbf{f}^1 = \frac{e\mathbf{E}_p}{m}\frac{\partial f^0}{\partial v} \tag{3.13.27}$$

in place of (2.3.7); thereafter, substitution of this expression for $\mathbf{f}^1$ into (2.4.5) yields (3.13.26) [Allis, Buchsbaum, and Bers (1963)] accurate to the first order in the quantity $(3kT_e/m)(\beta^2/\omega^2) \ll 1$. Within the same approximation, (3.13.26) under condition (3.13.25) gives the dispersion relation (3.13.24).

In the plasma column β is a function of the radial position, and the eigenvalue of the propagation equation for a longitudinal wave in the radial direction can be obtained through the WKB method, which yields a typical quantization condition for the total phase:

$$\int_{r_0}^{a} \beta \, dr = \left(\frac{m}{3kT_e}\right)^{\frac{1}{2}} \int_{r_0}^{a} [\omega^2 - \omega_p^{2}(r)]^{\frac{1}{2}} \, dr = (i + \alpha)\pi \qquad (3.13.28)$$

where r_0 is the radius at which $\omega_p = \omega$ (the "turning point" since, for $r < r_0$, β becomes imaginary and the wave evanescent), i is an integer, and α is an appropriate constant (<1). When (3.13.28) is integrated, assuming a spatial density distribution given by J_0 (2.405r/a), and convenient approximations are made, one finds that each resonance is characterized by a value of the product $\omega_p^{4}(0)T_e$ or $n^2(0)T_e$ [Schmitt, Meltz, and Freyheit (1965)]. Thus, if the electron temperature changes slightly by ΔT from the isothermal value T_g, the resonance will appear at a different density $(n_i + \Delta n)$, n_i being the previous values of the central density for resonance. The relationship between ΔT and Δn is easily derived by considering the constancy of the product n^2T_e:

$$2\frac{\Delta n}{n_i} + \frac{\Delta T}{T_g} = 0 \qquad (3.13.29)$$

Nygaard's problem was that of measuring ΔT in an afterglow plasma. Since the electron density is decaying in the afterglow, the Δn shift in the density required for resonance will cause a Δt shift in the postdischarge time of appearance of the resonance absorption curves. If the density decay is exponential and τ_d is the time constant, we will have:

$$\frac{\Delta t}{\tau_d} = -\frac{\Delta n}{n_i} = \frac{\Delta T}{2T_g} \qquad (3.13.30)$$

provided τ_d and the density are not significantly altered by the electron temperature change. Nygaard's experimental setup is shown in Fig. 3.13.5. The electrons are selectively heated by a $\sim$1-μsec pulse in the waveguide, and the Tonks-Dattner resonances are observed by displaying the reflected power versus time in the strip lines, in the absence and in the presence of the heating pulse; the time shift Δt is measured as a function of the postheating time, and the curve of the relative electron temperature transient is obtained from (3.13.30). The external strip line can be positioned at various distances from the waveguide, so that all necessary data for determining $\mathfrak{D}'$ from (3.13.11) are obtained. Nygaard claims that with special care temperature changes of the order of 1% can be detected with this method.

Likewise, when the recombination light method is applied, $\mathfrak{D}'$ is derived from the curves of the electron temperature change ΔT versus the postheating time t, for different distances from the waveguide. In this case the curves are obtained by recording for each distance $L_h(t)$ and $L_0(t)$, the radiated intensities with and without application of the microwave heating

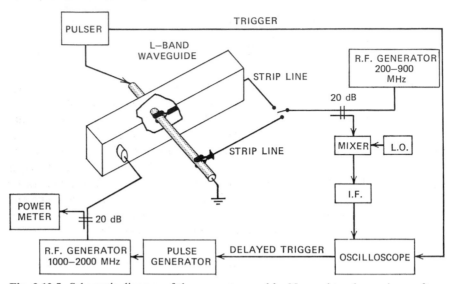

Fig. 3.13.5. Schematic diagram of the apparatus used by Nygaard to observe in an afterglow plasma the electron energy diffusion from a microwave-heated region; the energy observable is here the occurrence of a Tonks-Dattner resonance [after Nygaard (1967a)].

pulse. If we assume that (1) the dependence of the recombination coefficient α on the electron temperature can be approximated by:

$$\alpha \propto T^{-s} \tag{3.13.31}$$

(2) the recombination light is described by:

$$L(t) \propto \alpha(T)n^2(t) \tag{3.13.32}$$

and (3) the electron density is not significantly changed by the temperature variation, then $\Delta T(t)$ can be computed from the relation:

$$1 + \frac{\Delta T(t)}{T_g} = \left[\frac{L_0(t)}{L_h(t)}\right]^{1/s} \tag{3.13.33}$$

The results obtained by Nygaard with the two methods agree within the experimental uncertainty. However, it should be noted that the increase in the electron temperature as determined by the optical method represents an average over the plasma cross section, whereas that determined by the resonance method takes into account the outer region only of the plasma column.

c. Steady-state energy flow method. The experimental setups that implement the transient energy flow method can be used also for the steady-state method. Attention need be paid only to the fact that the duration of the microwave heating pulse must be long enough to ensure the attainment of steady-state energy conditions within the pulse duration itself.

In particular, Goldstein and Sekiguchi (1958) and Sekiguchi and Herndon (1958) determined the relaxation distance $\sqrt{\mathfrak{D}'\tau_r}$ by measuring, at various points along the plasma, L_h, the radiated light intensity near the end of the heating pulse, and L_0, the radiated light intensity at the same postdischarge time, but in the absence of heating. The appropriate equations are (3.13.14) and (3.13.33). For a significant check of the correctness of the $\Delta\bar{u}$ determinations, the $\Delta\bar{u}$ value extrapolated at $z = 0$ from the $\ln\Delta\bar{u}$ versus z straight line is compared to the value derived from $\langle\nu_m\rangle$, this quantity being obtained from either microwave conductivity or energy decay measurements in the waveguide.

Delpech and Gauthier (1969) implemented their method by inserting a tube containing the afterglow plasma in an X-band waveguide, terminated with a sliding short circuit. The standing wave was moved by means of this short, and the corresponding variations in side light intensity were observed, thus allowing the authors to determine the electron energy variations along the plasma.

C. Thermal Diffusion Effects in Q-Machines

It seems appropriate to discuss in this Section also a completely different approach that has been used for determining experimentally the thermal diffusion coefficient $\mathfrak{D}'$ [Rynn (1964)]. We reconsider the voltage-current characteristic of the quiescent plasmas generated by surface ionization on hot plates in Q-machines; in Section 3.8 this problem was examined only in the extreme case of very small currents, and no attention was paid to thermal effects at larger currents. These effects will now be included.

We shall apply eq. (2.5.11) to the present case, assuming, as appropriate for a Q-machine operating at low currents, steady-state conditions, one-dimensional dependence on the position along the Q-machine axis (z axis) only, and nearly thermal equilibrium between electrons and gas particles. This yields:

$$\frac{dq}{dz} = en\mu E^2 = -j\frac{d\phi}{dz} \tag{3.13.34}$$

where:

$$j = en\mu E = -en\mu\frac{d\phi}{dz} \tag{3.13.35}$$

is the current density, which flux continuity requires to be constant (independently of the z position), and $\phi(z)$ is the potential of the applied field

$(E = -d\phi/dz)$. Integrating (3.13.34) and equating the result to (3.13.6) yields:

$$q = -j\phi = -n\mathfrak{D}'\frac{d\bar{u}}{dz}$$

which, using (3.13.35), becomes:

$$e\mu\phi\frac{d\phi}{dz} = -\mathfrak{D}'\frac{d\bar{u}}{dz} \qquad (3.13.36)$$

Equations (2.6.23′) and (3.13.18) indicate that, for a Maxwellian distribution and a law $v_m \propto v^h$, we have:

$$\mathfrak{D}' = \frac{5}{3}\frac{\gamma_7^{-h}}{\gamma_5^{-h}}\frac{\hat{u}}{e}\mu = \frac{5-h}{3}\frac{\hat{u}}{e}\mu \qquad (3.13.37)$$

Using this relationship and the electron-volt equivalent ϑ of the electron energy $\hat{u}(\vartheta = \hat{u}/e)$, eq. (3.13.36) becomes after integration:

$$\phi^2 = \varepsilon(\vartheta_0^2 - \vartheta^2) \qquad (3.13.38)$$

where:

$$\varepsilon = \frac{3e\mathfrak{D}'}{2\hat{u}\mu} = \frac{5-h}{2} \qquad (3.13.39)$$

and ϑ_0 is the maximum of the energy, located at the zero potential point. Symmetry about this point requires that on the hot plates $(\vartheta = \vartheta_c)$ the potential be half of the applied voltage V, that is:

$$\left(\frac{V}{\vartheta_c}\right)^2 = 4\varepsilon(y_c^{-2} - 1) \qquad (3.13.40)$$

having introduced the normalized variable $y = \vartheta/\vartheta_0$.

Substituting (3.13.38) into (3.13.35) and invoking (2.6.23′), which indicates that $\mu \propto \hat{u}^{-h/2} \propto \vartheta^{-h/2}$, yields after integration:*

$$j|z| = e\sqrt{\varepsilon}\, n\mu\vartheta^{h/2}\int_{\vartheta}^{\vartheta_0}\frac{\vartheta^{1-(h/2)}\,d\vartheta}{(\vartheta_0^2 - \vartheta^2)^{\frac{1}{2}}} \qquad (3.13.41)$$

This equation implies symmetry about $z = 0$, so that on the plates $(|z| = L/2)$ it becomes:

$$\frac{jL}{en\mu_c\vartheta_c} = \frac{IR_0}{\vartheta_c} = 2\sqrt{\varepsilon}\,y_c^{-1+(h/2)}\int_{y_c}^{1}\frac{y^{1-(h/2)}}{(1-y^2)^{\frac{1}{2}}}\,dy \qquad (3.13.42)$$

where L is the length of the plasma, μ_c the electron mobility at the plate temperature, I the total current $(= Aj$, A being the area of the plasma cross

---

* In Rynn's (1964) published formula, ε appears erroneously instead of $\sqrt{\varepsilon}$.

section), and R_0 the zero-current slope of the voltage-current charac-teristic ($= L/en\mu_c A$).

Eliminating y_c between (3.13.40) and (3.13.42) and performing the appro-priate numerical computations, we can plot a set of IR_0/ϑ_c versus V/ϑ_c curves, one for each value of parameter ε. Comparing an experimental curve with such theoretical ones allows us to determine the appropriate value of ε, and then of $\mathfrak{D}'$ if μ is known, as is actually the case since the value of μ at the plate temperature can be obtained from the observable R_0 (for details see Section 3.8).

In Q-machines the plasma is typically dominated by electron-ion collisions, for which $h = -3$. Hence ε should be 4 on the basis of (3.13.39), but con-sideration of the electron-electron collision effects, and then of the Spitzer-Härm correction factors, as specified previously with reference to expressions (2.6.29), (2.6.76), and (3.13.19), brings this value to 1.622.

3.14 HIGH-FREQUENCY AND THERMAL TRANSPORT PARAMETERS IN MAGNETIC FIELDS

A. High Frequency Measurements

If an appropriate static magnetic field is applied to the electron swarm in any one of the microwave experiments described in the previous sections, more favorable conditions for accurate measurements can be attained [Gilardini (1957)]. We shall examine a typical case in which this approach is easily implemented.

The ionized medium is contained in a rectangular waveguide; the con-ductivity ratio is determined by measuring the ratio $\Delta A/\Delta\phi$ for propagation of the dominant H_{10} mode; a static uniform magnetic field is applied in the plane perpendicular to the direction of the microwave electric field (either along the axis or transversally). In this case the formulas derived in Section 3.11, which relate the observable parameter $\Delta A/\Delta\phi$ to the conductivity ratio, are still applicable, the only difference being the replacement every-where of σ_r and σ_i by σ_{Tr} and σ_{Ti}. If we operate at fairly low pressures and not too near resonance, we may satisfy the condition $v_m^2 \ll (\omega - \omega_b)^2$ and obtain from (2.4.23) the appropriate expression for the conductivity ratio:

$$\rho_T = -\frac{\sigma_{Tr}}{\rho\sigma_{Ti}} = \frac{\omega^2 + \omega_b^2}{\omega^2 - \omega_b^2}\frac{\langle v_m\rangle}{\rho\omega} = \frac{\omega^2 + \omega_b^2}{\omega^2 - \omega_b^2}\rho \qquad (3.14.1)$$

The value of the observable parameter $\Delta A/\Delta\phi$ may thus be a few times larger than the value in the absence of the magnetic field (the factor will be exactly $(\omega^2 + \omega_b^2)/(\omega^2 - \omega_b^2)$ in the case of a space-independent ρ), and this signifies improved accuracy when the $\Delta A/\Delta\phi$ value without a magnetic field happens

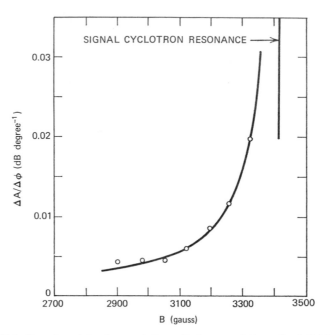

Fig. 3.14.1. Experimental values and theoretical curve of the ratio $\Delta A/\Delta \phi$ as a function of the applied magnetic field, for the case of an H_{11} mode propagating in a circular waveguide through an oxygen isothermal afterglow plasma [after Gilardini (1964)].

to be very small. Figure 3.14.1 shows the experimental values and the theoretical curve of the ratio $\Delta A/\Delta \phi$ as a function of the applied magnetic field for a similar case: propagation of a dominant H_{11} mode in a circular waveguide with the magnetic field perpendicular to both the waveguide axis and the maximum electric field direction.

Likewise the magnetic field plays a useful role for the purpose of heating electrons selectively in the plasma, since more power is absorbed when cyclotron resonance is approached, and larger heating is thus obtained with the same flowing power. For the magnetic field cases of actual experimental interest, the derivation of the appropriate formulas for the electron energy increase and for the radiation temperature, which correspond to the no-field formulas of the foregoing sections, is straightforward, and hence this subject will not be pursued further here. We note only that the problem can also be solved formally by replacing the microwave electric field in the waveguide with a fictitious equivalent field, which is a function of position and of the ω_b/ω ratio, so that all pertinent formulas reduce to those for the no-magnetic-field case [Gilardini (1964)].

Whereas previous considerations refer to the use of a static magnetic field, whose cyclotron frequency never becomes too close to the operating one, we shall examine now the cases that are based on observation of the entire cyclotron resonance behavior. Only methods that have actually been implemented experimentally will be described.

In a first group of experiments the attenuation ΔA due to a tenuous ionized medium filling a rectangular waveguide (dominant H_{10} mode) is measured as a function of the intensity of the static magnetic field, which lies in the plane perpendicular to the direction of the microwave electric field. In this case, with the same procedure and within the same limitations that yielded expression (3.11.16) from (3.11.4) and (3.11.6), we obtain from these two equations:

$$\Delta A = \frac{\omega \mu_0}{2\beta_0} \frac{\int_V \sigma_{Tr} E_0^2 \, dV}{\int_S E_0^2 \, dS} \qquad (3.14.2)$$

Near cyclotron resonance $\sigma_{Tr} \simeq \frac{1}{2}\sigma_{C_r}$ [see eq. (2.6.94)], so that for an iso-thermal plasma we can write:

$$\Delta A = \frac{\omega \mu_0}{4\beta_0} \frac{e^2}{m} \left\langle \frac{\nu_m}{\nu_m^2 + (\omega - \omega_b)^2} \right\rangle \frac{\int_V n E_0^2 \, dV}{\int_S E_0^2 \, dS} \qquad (3.14.3)$$

and the ΔA resonance line shape, normalized to unity at the peak, coincides with $\sigma_{C_r}^*$, defined accordingly by (2.6.95). Therefore, eq. (2.6.97) or (2.6.99) can be used for the derivation of electron collision frequency values from the experimental data.

Figure 3.14.2 shows the configurations of the discharge tube, solenoid

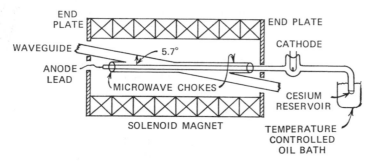

Fig. 3.14.2. Simplified diagram of Ingraham's configuration for discharge tube, solenoid magnet, and waveguide [after Ingraham (1966)].

magnet, and waveguide used by Ingraham (1966) for measurements in cesium afterglows. In this particular case the length of the solenoid magnet was 85 cm, and the field uniformity over the center 40 cm of the solenoid axis was ±0.15%; the solenoid current was controlled within ±0.10%. In order to measure ΔA Ingraham used a microwave bridge, substantially equal to that of the radiometer of Fig. 3.12.2; half of the attenuated signal from a klystron ($\sim 10^{-8}$ W), which here replaces the standard noise source, passes into the arm containing the plasma and half into the reference arm, which has a calibrated balancing attenuator. A ferrite (or diode) switch sends the signals from the two arms alternately into a gated microwave receiver, which is similar to that of Fig. 3.12.4, so that their power difference is detected. Adjusting the calibrated attenuator each time for a null reading at the receiver output, when the magnetic field is changed across cyclotron resonance, makes it possible to determine the absorption line shape at any given afterglow time. The arrangement includes a radiometer for measuring the electron temperature, which also operates in accordance with the schemes of Figs. 3.12.2 and 3.12.4.

In these experiments a simple procedure can be used to verify whether electron-ion collisions are contributing significantly to the results: the band width $\Delta\omega_{0.5}$ is plotted as a function of the ΔA value at resonance, which at a given gas pressure and electron temperature is linearly related to the electron density [eq. (3.14.3)]. If $\Delta\omega_{0.5}$ is independent of this maximum ΔA value, the electron-ion collisions will not be important.

Meyerand and Flavin (1964), who also performed measurements in cesium, employed the arrangement shown in Fig. 3.14.3. Electrons are supplied by thermal ionization of the cesium vapor for temperatures above 750°K, and by r.f. (600-MHz) breakdown of the gas external to the waveguide cell for temperatures below 750°K; in both cases electrons in the waveguide are in thermal equilibrium with the gas molecules. To increase the sensitivity, the magnetic field is modulated at 500 Hz by means of two modulation coils on the cell sides, and the microwave signal is phase detected with an integration time of 10 sec.

The cyclotron resonance investigations by Ingraham and by Flavin and Meyerand cited above were performed adopting a constant probing frequency and a variable magnetic field. Recently Coffey (1970) switched the roles of these two quantities and observed resonant absorption curves using short-duration, frequency-modulated microwave pulses and a constant magnetic field. In this way no danger exists of altering the breakdown characteristics and afterglow properties while sweeping; in addition, shortening the experiment duration (each microwave pulse provides the complete resonance curve) may improve the reliability of the results.

Instead of using waveguide arrangements Bayes, Kivelson, and Wong (1962), Fehsenfeld (1963), and others adopted a microwave cavity for measurements of the cyclotron resonance. For this purpose it is convenient to lock the frequency of the probing signal to the resonant frequency of the acvity, to sweep the magnetic field through resonance, and to add a small,

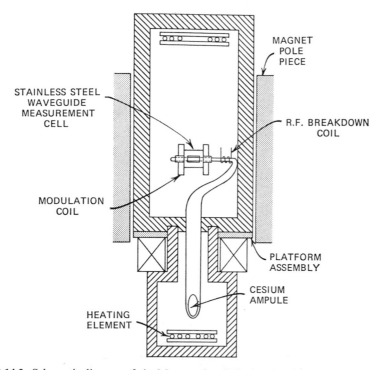

STAINLESS STEEL
WAVEGUIDE
MEASUREMENT
CELL

MODULATION
COIL

HEATING
ELEMENT

MAGNET
POLE
PIECE

R.F. BREAKDOWN
COIL

PLATFORM
ASSEMBLY

CESIUM
AMPULE

Fig. 3.14.3. Schematic diagram of the Meyerand and Flavin setup for waveguide measurements of the cyclotron-resonance absorption of a cesium plasma, with a double oven for control of vapor pressure and electron energy [after Meyerand and Flavin (1964)].

low-frequency modulation to this field; in this case the signal reflected by the cavity, preferably detected by phase-sensitive techniques so as to recover only the in-phase component of modulation, is proportional to the field derivative of the microwave power absorbed by the plasma. Bayes, Kivelson, and Wong (1962), and Tice and Kivelson (1967) performed measurements by sending through the cavity an electron-containing active nitrogen stream, independently generated by means of an electrodeless discharge; small amounts of oxygen must be added, however, in order to achieve the required electron densities. Gases other than nitrogen are studied by introducing them downstream from the discharge, but the reactivity of the active nitrogen in the gas stream limits the technique to a few stable substances. Fehsenfeld has used instead a flowing afterglow plasma, generated in a nearby suitable cavity; adequate times for thermalization before measurements are thus attainable.

The method of Hirshfield and Brown (1958), described in Section 2.6, was implemented by these authors using a microwave cavity containing an isothermal afterglow plasma. For this experiment they chose a cylindrical cavity, a H_{011} mode of oscillation, and a magnetic field along the axis of the cavity, so that the microwave electric field and the static magnetic field would be orthogonal. In this case the condition $\sigma_{Ti} = 0$ is satisfied when the resonant frequency of the cavity containing the magnetized plasma has the same value as in the case of the empty cavity.

The strength of the magnetic field at which this happens is independent of the electron density for dominating electron-neutral collisions [see eq. (2.6.102)], and therefore it must also be independent of the postdischarge time at which the frequency shift is measured. By using a probing signal, which is saw-tooth frequency modulated at a high rate compared with the breakdown rate, successive transmission curves of the cavity during the late afterglow can be displayed superimposed on an oscilloscope. The zero-shift situation can be detected by observing either these transmission curves directly (Fig. 3.14.4) or, with higher accuracy, the difference between these

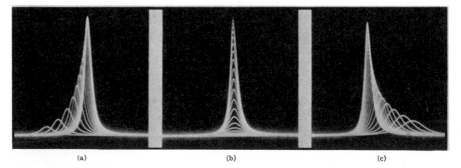

(a) (b) (c)

Fig. 3.14.4. Amplitude of microwave signal transmitted through a cavity as a function of frequency, for successive afterglow times (different traces) and for increasing magnetic fields (different photographs); the zero-shift ($\sigma_{Ti} = 0$) situation is satisfied by the field of (b) [after Hirshfield and Brown (1958)].

transmission curves and the transmission curve of a high-Q reference cavity. Implementing the last technique according to the diagram shown in Fig. 3.14.5, and using a probe cavity with a resonant frequency of 4200 MHz and a reference cavity with a Q of approximately 30,000, Hirschfield and Brown have been able to observe resonant frequency changes as low as 10 kHz. This precision is more than adequate, since the magnetic field, measured by means of a flip-coil method [Strandberg, Tinkham, Solt, and Davis (1956)], was determined with a precision of 2 parts in 10^4. Other features of the experiment are similar to those of cavity experiments discussed in Section 3.11; the magnetic field was designed to be homogeneous within 1 % over the plasma region.

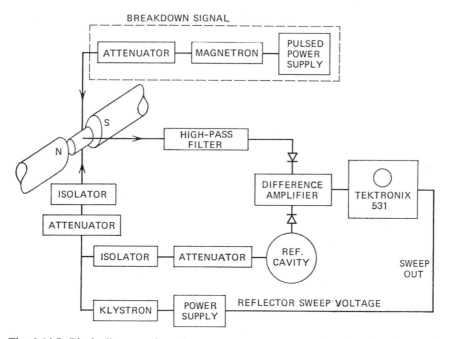

Fig. 3.14.5. Block diagram of a microwave apparatus appropriate for observing small resonant frequency shifts, and for determining the value of the magnetic field at which $\sigma_{Ti} = 0$ (zero-shift condition) [after Hirshfield and Brown (1958)].

Implementation of the method of Narasinga Rao, Verdeyen, and Goldstein (1961), described in Section 2.6, requires determining the magnetic field at which $\partial\sigma_{Tr}/\partial\nu_m$ vanishes. For this purpose a short pulse of microwave power is applied to an afterglow plasma, filling a rectangular waveguide, and axially magnetized. This pulse is of sufficient power to heat the electrons slightly, and then to change the attenuation ΔA of a microwave probing signal (cross-modulation effect). The attenuation being given by (3.14.2), its initial rate of change, neglecting the much slower electron density variations, will be:

$$\frac{\partial\,\Delta A}{\partial t} \propto \frac{\partial\sigma_{Tr}}{\partial t} \propto \frac{\partial\sigma_{Tr}}{\partial\nu_m}\frac{\partial\nu_m}{\partial\hat{u}}\frac{\partial\hat{u}}{\partial t}$$

and $\partial\sigma_{Tr}/\partial\nu_m$ will vanish when ΔA does not change in the presence of the heating pulse. The experiment consists, then, in observing on an oscilloscope the cross-modulation effect and determining the value of the magnetic field, on either side of the cyclotron resonance, for which the effect disappears.

Coffey (1970) has slightly modified this method. Instead of sweeping the magnetic field, he probes the plasma with a frequency-modulated microwave pulse, lasting a few tens of microseconds; then, in order to heat selectively electrons and to find the ω value at which cross modulation vanishes, he applies a much shorter pulse of microwave power, lasting no longer than a few tenths of a microsecond, at a variable time position within the main probing pulse.

Bruce, Crawford, and Harp (1968) implemented their method, described in Section 2.6, according to the scheme of Fig. 3.14.6. Measurements are

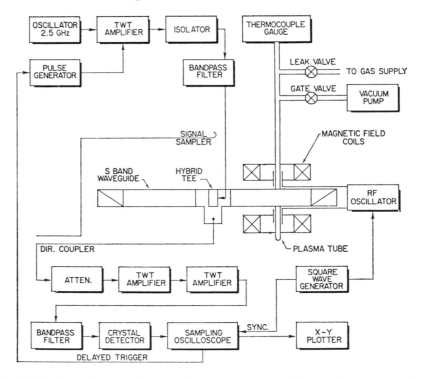

Fig. 3.14.6. Block diagram of a setup for the observation of the transient response of a magnetoplasma, subjected to a short microwave pulse at the frequency of the electron cyclotron resonance [after Bruce, Crawford, and Harp (1968)].

performed in the late afterglow of a repetitively pulsed r.f. discharge, when the plasma is isothermal and the electron density sufficiently low, so as to avoid the possible presence of Tonks-Dattner resonances. A pulse-modulated TWT amplifier is used to generate a 2.5-GHz signal with a 10-nsec duration; this is fed into a rectangular S-band waveguide, where it excites at the cyclotron frequency the plasma, which is contained in a cylindrical tube, oriented transversely to the waveguide axis and to the electric field of the dominant H_{10} mode. The current response of the plasma induces in the waveguide an H_{10} field of corresponding strength; this, plus an attenuated sample of the input signal, is amplified, detected, and displayed on an oscilloscope. The static magnetic field is directed along the discharge axis and is furnished by two coils, which provide a field homogeneous within 0.1 % over the plasma volume inside the waveguide.

The main difference between the basic theory of this method, as developed in Section 2.6, and its actual implementation in the scheme of Fig. 3.14.6 is the electric field nonuniformity. In fact, in the case of the H_{10} mode in a rectangular waveguide, there is a cosinusoidal electric field variation across the section, and the initial velocity $v(0)$ thus becomes a function of position. Bruce, Crawford, and Harp have worked out appropriate formulas for this case and have shown how, by solving an integral equation, $v_m(v)$ can be obtained from the initial decay rate of the plasma response, measured as a function of the strength of the applied field. The same authors, however, suggest using in future experiments different waveguide and plasma structures to attain uniform field conditions [Harp and Moser (1967)] and thus avoid the loss of accuracy inherent in all methods requiring the unfolding of experimental data by means of integral equations. With this modification the method would offer the peculiar opportunity of making collision measurements with low-energy monoenergetic electrons, with the same advantages offered by beam techniques at higher energies.

B. Measurements of Thermal Transport Parameters

We shall consider now which of the energy relaxation and flow experiments described in Section 3.13 has been performed also in magnetoplasmas and shall analyze the required modifications. A little consideration should be given to the determinations of the electron energy relaxation times by Narasinga Rao and Taylor (1967), since the magnetic field is used only for selectively heating the electrons at cyclotron resonance, and the modifications that the presence of the magnetic field introduces into the energy decay measurements (performed by means of a weak probing signal outside resonance) can easily be conceived.

The application of the steady-state energy flow method to magnetoplasmas, as done by Rostas, Bhattacharya, and Cahn (1963), is less straightforward. Here too the electrons of an afterglow plasma column are heated selectively in a localized volume by a long pulse of X-band microwave power, and the steady-state distribution of the electron energy along and across the column is determined by observing the quenching of the afterglow recombination light. The magnetic field direction (x axis) is perpendicular to the plasma column axis (z axis); two geometries have been used, as shown in

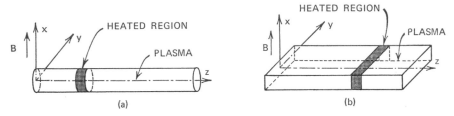

Fig. 3.14.7. Geometries used in the study of electron energy flows in magnetoplasmas [after Rostas, Bhattacharya, and Cahn (1963)].

Fig. 3.14.7. The plasma is heated uniformly in the x direction, and thus the distribution of the excess electron energy $\Delta \bar{u}$ becomes a function of the longitudinal z position and of the transverse y position only.

This distribution must be a solution of the steady-state form of the energy flow equation (3.13.8), which here becomes:

$$\frac{\partial^2 \Delta \bar{u}}{\partial y^2} + \frac{\partial^2 \Delta \bar{u}}{\partial z^2} = \frac{\Delta \bar{u}}{\mathfrak{D}_T' \tau_r} \tag{3.14.4}$$

By analogy to (3.13.14) and in accordance with the experimental results, a solution of the form:

$$\Delta \bar{u} \propto \exp\left(-\frac{y}{\rho_y} - \frac{z}{\rho_z}\right) \tag{3.14.5}$$

is considered; this solution satisfies eq. (3.14.4) if the relaxation distances ρ_y and ρ_z are such that:

$$\frac{1}{\rho_y^{\,2}} + \frac{1}{\rho_z^{\,2}} = \frac{1}{\mathfrak{D}_T' \tau_r} \tag{3.14.6}$$

This solution will not, in general, be consistent with the $\Delta \bar{u}$ distribution in the heated volume, so that it will be applicable only at some distance from this region, where all other terms of the complete solution, which decay more rapidly with distance, have become negligible compared to the slower decay (3.14.5).

Equation (3.14.6) provides a relationship between ρ_y and ρ_z, but does not permit a separate evaluation of these two quantities. For this purpose the boundary condition $q_y = 0$ at the transverse walls is tentatively introduced (the consistency of the final experimental results proves the validity of this condition). Hence we have from (3.13.6):

$$q_y = -n\left(\mathfrak{D}_T' \frac{\partial \Delta \bar{u}}{\partial y} + \mathfrak{D}_\perp' \frac{\partial \Delta \bar{u}}{\partial z}\right)$$

$$= n\left(\frac{\mathfrak{D}_T'}{\rho_y} + \frac{\mathfrak{D}_\perp'}{\rho_z}\right) \Delta \bar{u} = 0 \tag{3.14.7}$$

which yields:

$$\frac{\rho_y}{\rho_z} = -\frac{\mathfrak{D}'_T}{\mathfrak{D}'_\perp} \tag{3.14.8}$$

We saw in Section 2.6D that the ratios $\mathfrak{D}_T/\mathfrak{D}$ and $\mathfrak{D}_T/\mathfrak{D}_\perp$ are known functions of h and of $\omega_b/\nu_m(\hat{u})$ only; the same is true for the ratios $\mathfrak{D}'_T/\mathfrak{D}'$ and $\mathfrak{D}'_T/\mathfrak{D}'_\perp$. Therefore, if we measure experimentally ρ_y, ρ_z, and τ_r in the same plasma, both in the presence of a magnetic field and in its absence, we can determine the ratios $\mathfrak{D}'_T/\mathfrak{D}'$ and $\mathfrak{D}_T/\mathfrak{D}'_\perp$, and then derive from these data appropriate values for h and $\nu_m(\hat{u})$. To ensure equal plasma conditions from the point of view of these parameters, data are taken in the late isothermal afterglow; furthermore, when electron-ion collisions dominate, either electron densities are the same, or the thermal diffusivity value $\mathfrak{D}'$ has to be scaled accordingly [$\propto 1/n$ according to (2.6.76), if we neglect the n dependence of the $\mathscr{L}$ term].

The experiments of Rostas, Bhattacharya, and Cahn were actually performed in an afterglow plasma dominated by electron-ion collisions. In the case of the rectangular tube geometry they determined ρ_y, ρ_z, and τ_r by means of recombination light measurements, using a light pipe affording an observation area of 3 mm²; the relative electron density values for scaling $\mathfrak{D}'$ are derived from the τ_r measurements [$\propto 1/\langle\nu_m\rangle \propto 1/n$]. In the case of the cylindrical tube geometry ρ_y could not be measured. The authors have shown, however, that in their experiment ρ_z was approximately equal to $\sqrt{\mathfrak{D}'_T\tau_r}$, so that by measuring ρ_z and τ_r as before, $\mathfrak{D}'_T$ could be determined; relative electron density values are here derived from measurements of the microwave conductivity with a probing signal sent through the same waveguide used for electron heating. Independent determinations of $\nu_m(\hat{u})$ have also been derived by the same authors from the measured τ_r values and, in the case of the cylindrical geometry, from additional conductivity ratio measurements, performed using the microwave probing signal. In the last case ρ measurements are also used for the purpose of scaling to equal values the electron temperatures, when they are believed to be higher than the isothermal value: $T_e \propto (n/\rho)^{2/3}$, in accordance with (2.6.43).

Chapter 4

Experimental Determinations of Electron Collision Parameters

In this chapter we collect in a comprehensive way all significant information that is available on the experimental determinations of electron collision parameters in gases; both the original experimental data and the analyses performed for deriving the collision parameters are reviewed for each gas.

In Section 4.1 we illustrate all the basic criteria that have been adopted for the selection, presentation, and discussion of the data and for the classification of the methods of analyses. Rare gases are treated in Sections 4.2–4.6, the major molecular gases in Sections 4.7–4.13, and the alkali-metal vapors in Section 4.14; Section 4.15 is devoted to all other remaining gases for which data are available. Data on transport parameters in mixtures have been considered in some cases for the derivation of the collision parameter values appropriate to the constituent gases; these cases are discussed under the headings of each individual gas to which the analyses apply. A separate section (Section 4.16) is devoted to air, which is the mixture of most practical importance, and to all other gaseous mixtures, the goal being in this case to provide only selected bibliographical references for the interested reader. Data on transport parameter measurements under conditions of dominating elastic collisions between electrons and positive ions are discussed in Section 4.17, the last section of the chapter.

4.1 GENERAL REMARKS ON THE PRESENTATION OF MEASURED TRANSPORT DATA AND OF COLLISION PARAMETER DETERMINATIONS FOR THE VARIOUS GASES

In the following sections we shall present for each gas first the experimental data of measured transport quantities and then the numerical or analytical determinations of the electron collision parameters. All significant papers published before December 1970 have been considered for the purpose of our discussion. Results of papers published during 1971 have also been included, but only when regarded as outstanding contributions.

The experimental data are presented separately for each transport parameter. Experiments are listed in chronological order; for each of them basic information is provided as available on the method used, on the ranges of the pertinent variables,* on the measurement accuracy, and on the gas purity. When people of the same group or school have performed repeated measurements using the same basic method, only the most accurate and reliable experiment is included in the chronological list, the others being at most mentioned in the same paragraph for completeness; in general, the best data are also the latest ones, since these have been obtained with improved instrumentation and experimental techniques. Because of the common features the diffusion to mobility ratio (D/μ, D_L/μ, and $D_T/\mu_\parallel$) experiments are all discussed under the same heading; however: (1) except when otherwise noted, the experiments refer to the most common case only, that of D/μ measurements; (2) the experiments concerning measurements of D/μ alone or of D/μ and $D_T/\mu_\parallel$ together are listed first, and are followed by the very few cases of D_L/μ measurements.

In accordance with the aims of this book, we consider only measurements of transport parameters that are appropriate for the purpose of deriving elastic and inelastic, nontransformation electron collision parameters at energies below about 1 eV. Therefore, except for the case of alkali-metal vapors, we shall disregard all measurements performed in the presence of applied electric fields large enough to produce significant electronic excitation and ionization of molecules. This implies that for each gas we shall consider dc transport parameters up to a maximum E/p value only; this value and the corresponding characteristic energy value are given at the beginning of each section.

With a few exceptions all the old data, obtained as far back as five decades ago, have also been included in the lists; unfortunately, however, the accuracy of these data is rarely ascertainable. A major group of rejected experiments is represented by the set of drift velocity measurements performed by Loeb and Wahlin in the 1920s, using the alternating electric field method (see Section 3.2C); these have been excluded since the method, according to Loeb himself, left much to be desired, and because the assumption of an E/p independent mobility was too crude. Also, magnetic drift velocities and characteristic energies, measured in electronegative gases by the original Townsend methods, have been disregarded any time that the effect of attachment on the results does not appear to have been properly established. Thus D/μ data are considered only when the absence of any significant

---

* Since range specifications are only indicative, the inclusion or exclusion of the lower and upper limit values has no real significance here; accordingly the simple notations $>$ and $<$ are used throughout.

number of negative ions has been proved by the magnetic deflection technique, or by the consistency of current ratios for different dimensions of the collecting disks and rings.

In the experiments of the last two decades commercial reagent or research-grade gases of very high purity have been used, except when otherwise specifically noted. In the older experiments less pure gases have, in general, been used; their purity is given whenever this information is available in the literature.

For each transport parameter, after having listed all the pertinent experiments, the results are discussed in a comprehensive way. Whenever possible, particularly when the results of many experiments are known, a set of data regarded as the best one for accuracy, range coverage, and overall consistency is selected; all others are reported only by specifying any major deviation from the reference best set. Tables are preferred to graphs when available; in fact, in many cases measured values span over a few orders of magnitude, and also logarithmic graphs cannot be read with adequate precision.

To reduce reading and redrawing inaccuracies, graphs, when necessary, are reproduced as they appeared in the original papers, even if this has meant some loss of uniformity in the style of presentation. For the same reason gas densities are specified, according to the various authors, not only as pressures in torr at a temperature of 0°C (p_0), but also as pressures in torr at 293°K (p_{293}) and at 300°K (p_{300}), and as number densities in molecules per unit volume (N). The relation between these quantities is:

$$N = 3.54 \times 10^{16} p_0 = 3.30 \times 10^{16} p_{293} = 3.22 \times 10^{16} p_{300} \quad \text{(molecules cm}^{-3}\text{)}$$

Furthermore, in this chapter we shall use p for pressure without any subscript every time a definite reference to a precise value of ambient temperature is not given in the original paper or appears unnecessary because only ranges of measurement values are of concern. Throughout the chapter E/p is intended in units of volts per centimeter per torr. A new unit named the townsend, 1 townsend being defined as 10^{-17} V cm^2, has been proposed by Huxley, Crompton, and Elford (1966) for the ratio E/N [symbol: Td; E/N (Td) $= 2.83 E/p_0 = 3.03 E/p_{293}$].

For the purpose of discussing the determinations of electron collision parameters, we have found it convenient to classify the calculation and evaluation procedures into four groups:

A. Numerical iterative procedure, which provides the best fitting of the experimental data over a large electron energy range, by introducing a set of tentative values of elastic and inelastic collision cross sections into the appropriate equations.

B. Verification of agreement between the experimental data and the corresponding curves, computed numerically by introducing into the appropriate equations a set of collision cross-section values, which resulted from a type A analysis of other experiments.

C. Derivation of collision parameter curves in analytical form, as they result from the applicable specific equations, when the transport parameters, measured over a sufficiently large range of the controlling state quantities (for the electron energy at least $\bar{u}_{max} > \sim 3\bar{u}_{min}$), are introduced into these equations in the form of appropriate analytical expressions representing the experimental results. This procedure is intended to be such as to provide, over the appropriate electron energy range, a substantially unique solution within the experimental error uncertainties.

D. Derivation of collision parameter curves as with procedure C, but introducing *a priori* significant assumptions regarding the shapes of these curves, since without these assumptions unique solutions are not obtained, because of the insufficient range over which experimental data are known or because of the large scattering of measurement results.

For each gas we shall report and discuss all the appropriate available analyses of types A, B, and C that have appeared in the literature so far. Analyses of type D, on the other hand, are presented only when applicable for the cases of transport parameters not covered by the more accurate types of analyses, and for the interpretation of experimental results that disagree significantly from those for which analyses of types A, B, and C are reported.

Curves representing the whole set of discussed determinations of the collision cross sections Q_m as a function of the electron energy are given for each gas. When more appropriate estimates are lacking but the pertinent data are given, we have considered it reasonable, if also arbitrary, to set, for each curve obtained using type A, B, or C analyses, the lower energy limit at half the minimum average energy of the electrons in the swarm under discussion, and the upper energy limit at twice the maximum average energy of the same electrons. Dashed lines are used to represent curves resulting from analyses of type D; their limits are set restrictively at the corresponding minimum and maximum average energies (in the case of cesium, according to the most common practice, the role of the average energy $\bar{u}$ is taken by the energy $\hat{u}$). A heavy solid line shows, for each gas, the curve believed to be the most accurate and reliable one. When available, curves of the collision cross section Q_m, derived from a known cross section Q_t by adopting appropriate $\sigma_9(\vartheta)$ distributions, are also plotted as point-dash lines on the same graphs. On the basis of Section 1.5 results, Q_m is always considered independent of the gas temperature.

4.2   HELIUM

Experimental Data [dc parameters for $E/p < 2(D/\mu \simeq 2.5$ eV)]

a. Drift velocity

(1) Nielsen (1936) performed measurements at $293°$K for $E/p > 0.1$, using the original double-shutter method he and Bradbury had developed ($d = 5.93$ cm). The estimated accuracy of the measurements in rare gases, based on the scatter of the experimental points, is of the order of 4%. Helium purity was better than 99.8%.

(2) Unpublished data of Errett (1951) were reported by Frost and Phelps (1964) for $E/p > 0.2$ and $T_g = 300°$K.

(3) Pack and Phelps (1961) performed measurements at $77°$K, $195°$K, and $300 \pm 3°$K over the range $1 \times 10^{-4} < E/p < 0.4$, using their modernized version of the double-shutter method [see Section 3.4D]. Drift velocities measured at $300°$K with drift distances of 5.08 and 10.16 cm agree with those obtained with distances of 2.54 and 5.08 cm. The data can be extended up to $E/p = 1$ by adopting the previous results obtained at $300°$K by Phelps, Pack, and Frost (1960) using the same method; in fact, over the common range of measurements, the two sets of data agree fairly well (attention must be paid to the fact that the older results at $E/p < 10^{-2}$ must be increased by about 10% because of an overestimate of the end-effect correction). The impurity content of the reagent-grade helium was less than 0.005 mole $\%$.

(4) Bowe and Langs (1966) determined drift velocities at room temperature over the range $0.1 < E/p < 1$ from measurements of the pulse length of the induced collector current following a very short flash of radiation on a photocathode. These results modify significantly and supersede earlier measurements performed by Hornbeck (1951) and by Bowe (1960), who used the same method but inferior measuring techniques and instrumentation. Worthy of mention, however, is the fact that in the earlier work Bowe found no change in the measurements after purification of research-grade tank helium by forced circulation over calcium turnings kept at $450°$C. For this reason all recent measurements of drift velocities in helium have been made without attempting any specific gas purification, apart from that obtained by passing the gas through cold traps on admittance to the drift chambers.

(5) Crompton, Elford, and Jory (1967) performed measurements at $293°$K over the range $6 \times 10^{-3} < E/p < 1.2$, using the double-shutter drift tube of Fig. 3.4.2; Crompton, Elford, and Robertson (1970) subsequently extended these measurements to $77°$K in the range $2.6 \times 10^{-3} < E/p < 6.6 \times 10^{-1}$, using a smaller tube ($d = 5$ cm typically) and a radioactive source [see Section 3.4C]. The data of these authors, which were obtained using improved techniques and instrumentation, supersede previous results obtained similarly by the same Canberra group [in particular, those at

293°K of Crompton and Jory (1965), which, however, differ only slightly (by less than 1%) from the latest values]. The measurements have been corrected for diffusion errors according to eq. (3.3.19), where the authors have set:

$$r = 1.5 \frac{D}{wd} \qquad (4.2.1)$$

the coefficient 1.5 being chosen empirically to provide the best agreement between measurements performed at different pressures. A 1% error limit can be claimed for the final w data.

(6) Wagner, Davis, and Hurst (1967) performed measurements at room temperature over the range $0.1 < E/p < 1$, using the electron multiplier arrangement described in Section 3.4E. The impurity content of helium was less than 0.002 mole %.

(7) Grünberg (1968, 1969) performed measurements at room temperature for $E/p > 3 \times 10^{-2}$ in high-purity helium (99.999%), using the induced current method and very high pressures, from 1 to 43 atm. The author estimates that his results are accurate within 1% below 23.5 atm and within 1.5% above this pressure.

The room-temperature results in the above experiments are all within $\sim 15\%$ of each other. The results of (5), which are the most significant because of accuracy, extent of the E/p range, and their midway position, are summarized in Table 4.2.1; below $E/p = 5 \times 10^{-3}$ the data are from (3). The results of (4), which can be described by the relation:

$$w = (0.800 \pm 0.009) \times 10^6 \sqrt{E/p_0} \quad (\text{cm sec}^{-1})$$

are only $\sim 2\%$ below those of (5), whereas the results of (2) are significantly lower. Slightly (a few per cent only) above (5) are the results of (3), (6), and (7) at low E/p, whereas larger upward deviations and scattering characterize those of (1). The only existing data on drift velocities in helium at 77°K are those of (3) and (5), which agree well within the combined experimental errors; the results of (3) for $E/p \leqslant 2 \times 10^{-3}$ and of (5) for $E/p \geqslant 3 \times 10^{-3}$ are also summarized in Table 4.2.1.

It is worth noting that frequently in the drift velocity tables or plots one finds a typical linear dependence of w on E/p below a certain E/p value (this feature can be demonstrated by a straight line with a 45° slope on a log-log plot). This dependence provides a constant μp value, which indicates that except in very special cases (velocity-independent ν_m) a thermal equilibrium regime between electrons and gas molecules ($E^2 \ll E_t^2$) has been attained. In the case of helium at 77°K this linear dependence takes place for $E/p < 3 \times 10^{-4}$; in this region the results of (3) yield $\mu p_0 = 14.55, 9.10, 7.29 \times 10^6$ cm^2 V^{-1} sec^{-1} torr, respectively, at 77, 195, and 300°K.

Table 4.2.1 Experimental Values of the Drift Velocity, of the Characteristic Energy, and of the Magnetic Deflection Coefficient in Helium

Drift velocities are from Crompton and his coworkers [a(5)], except at low E/p values ($E/p_{293} \leqslant 2 \times 10^{-3}$ and 4×10^{-3} at 77 and 293°K, respectively), where they are from Pack and Phelps [a(3)]; characteristic energy values at room temperature are from Crompton, Elford, and Jory [b(3)], except at low E/p values ($E/p_{293} \leqslant 6 \times 10^{-3}$), where they are from Warren and Parker [b(2)]. The data attributed to a(3) and b(2) at 77°K are taken from Laborie, Rocard, and Rees (1968).

$10^{17}E/N$ (V cm²)	E/p_{293} (V cm⁻¹ torr⁻¹)	$10^{-5}w$ (cm sec⁻¹) [$T_g = 293°K$; a(3) and a(5)]	$10^{-5}w$ (cm sec⁻¹) [$T_g = 77°K$; a(3) and a(5)]	D/μ (eV) [$T_g = 293°K$; b(2) and b(3)]	D/μ (eV) [$T_g = 77°K$; b(2)]	Ψ [$T_g = 293°K$; c(2)]
3.03×10^{-4}	1.0×10^{-4}		1.6×10^{-2}		6.5×10^{-3}	
6.07	2.0		3.2		6.6	
1.214×10^{-3}	4.0		6.2		6.7	
1.820	6.0		9.3		6.8	
2.43	8.0		1.23×10^{-1}		7.0	
3.03	1.0×10^{-3}	7.8×10^{-2}	1.53	2.6×10^{-2}	8.2	
6.07	2.0	1.55×10^{-1}	2.80	2.6	9.6	
9.10	3.0	2.25	3.82	2.7	1.1×10^{-2}	1.15
1.214×10^{-2}	4.0	2.95	4.72	2.8	1.4	1.13
1.820	6.0	4.18	6.23	3.0	1.7	1.12
2.43	8.0	5.33	7.45	3.13	2.0	1.10
3.03	1.0×10^{-2}	6.37	8.50	3.37	2.75	1.09
4.55	1.5	8.63	1.07×10^{0}	4.00	3.5	1.08
6.07	2.0	1.05×10^{0}	1.25	4.68	5.0	1.09
9.10	3.0	1.36	1.54	6.04	6.4	1.08
1.214×10^{-1}	4.0	1.62	1.78	7.41	9.1	1.09
1.820	6.0	2.04	2.17	1.01×10^{-1}	1.15×10^{-1}	1.08
2.43	8.0	2.37	2.50	1.27	1.41	1.07
3.03	1.0×10^{-1}	2.67	2.78	1.54	2.00	1.06
4.55	1.5	3.28	3.37	2.17	2.60	1.06
6.07	2.0	3.79	3.87	2.79	3.80	1.06
9.10	3.0	4.63	4.70	3.99		1.05
1.214×10^{0}	4.0	5.33	5.39	5.20		1.05
1.820	6.0	6.55	6.58	7.55		1.04
2.43	8.0	7.57		9.96		1.05
3.03	1.0×10^{0}	8.57		1.24×10^{0}		1.04

Grünberg finds at the lowest E/p a small decrease of the drift velocity with increasing pressure above a few atmospheres ($N > 3 \times 10^{20}$ cm^{-3}). This behavior may recall similar results of Levine and Sanders (1967), who measured the mobility of photoelectrons in helium over the temperature range 2.6–4.2°K and found very anomalous behavior. At the highest gas densities ($N > 2 \times 10^{21}$ cm^{-3}; near the normal boiling point) the mobility is lower than the value expected according to kinetic theory by 4 orders of magnitude; at intermediate densities a transition region occurs; and at the lowest densities ($N < 5 \times 10^{20}$ cm^{-3}) the mobility approaches the kinetic theory limit. Since in both experiments the effect takes place over the same range of atom densities (10^{20}–10^{22} cm^{-3}), even if the orders of magnitude of the drift velocity variations are quite different, we should first look for some possible common cause. This common cause may be the fact that at these densities the mean free path of the electron ($l_m = 1/NQ_m$) becomes comparable with its de Broglie wavelength λ_e [eq. (1.5.2)]; in fact, the helium atom densities at which $l_m = \lambda_e$ are 1.2×10^{22}, 2.9×10^{21}, and 5×10^{20} cm^{-3} at 1, 4×10^{-2}, and 1×10^{-3} eV, respectively (see Table 4.2.5 for Q_m values). When l_m approaches λ_e, the gas no longer behaves as an ensemble of individual localized scattering centers; it has been postulated that under these conditions the gas must be treated either as a continuous medium or as a mixture of localized uncorrelated ("free") and correlated ("bubble") states, the latter ones arising from the interaction of a slow electron with a collection of helium atoms. Recently Young (1970), introducing relaxation times to characterize the life duration of the free and bubble states and allowing these times to be functions of the helium density, has shown how the bubble theory may account quantitatively for the low-temperature experimental results of Levine and Sanders and has determined the values of the relaxation times and of the free electron mobilities necessary for a good fit to these data. On the other hand, Legler (1970) has advanced a preliminary theory for treating the gas as a continuous scattering medium, which seems compatible with Grünberg's high-pressure results.

b. Diffusion to mobility ratio

(1) Townsend and Bailey (1923) performed measurements at room temperature for $E/p > 1.3 \times 10^{-2}$, using Townsend's method with calibration curves computed as discussed in Section 3.6A. They used two tubes with different diffusion chamber lengths ($d = 2$ and 4 cm), but the same striplike configuration of the collecting electrodes [b (half-width of the central strip, measured to the center of the separation gap) $= 0.25$ cm]. Commercial helium was purified by liquid-air-cooled charcoal.

(2) Warren and Parker (1962) performed measurements at 77°K and at 300°K over the range $1.2 \times 10^{-4} < E/p < 0.4$. These authors used

Townsend's method but adopted empirical calibration curves [see Section 3.5], which afforded very good consistency of results after they had been developed independently for the various situations: inner and outer current ratios between a central disk and two surrounding concentric annuli (outer radii c_1 and c_2) in a long ($d = 8.9$ cm, $b/d = 0.069$, $c_1/d = 0.136$, $c_2/d = 0.201$) and in a short ($d = 1.27$ cm, $b/d = 0.484$, $c_1/d = 0.954$, $c_2/d = 1.41$) diffusion tube. Since these authors, in spite of using relatively low field strengths, made no compensation for contact potential differences, their results in helium, as well as in all other gases, show significant scatter, often in excess of 5 %.

(3) Crompton, Elford, and Jory (1967) used a tube similar to the one of Fig. 3.7.1 for measurements at 293°K over the range $7 \times 10^{-3} < E/p < 1.0$. Experimental conditions were chosen to be those for maximum accuracy, as discussed in Section 3.5. The measurements made at the smallest values of E/p at the highest pressures have been corrected for a slight space-charge effect by means of a simple linear extrapolation to zero current. The authors estimate that their final values are in error by less than 1 %.

(4) Wagner, Davis, and Hurst (1967), using the Stevenson-Hurst method of transit time measurements, modified so as to obtain the distributions of these times of flight [see Sections 3.4E and 3.5], determined the quantity D_L/μ at room temperature over the range $0.1 < E/p < 1.0$. The impurity content of the gas is given in $a(6)$ above.

(5) Unpublished data of D_L/μ, obtained by Crompton and Elford at room temperature over the range $3 \times 10^{-3} < E/p < 0.24$, were reported by Lowke and Parker (1969).

The most accurate measurements of D/μ at room temperature appear to be those of (3). The room-temperature results of (2) are 5–10 % below those of (3); they attain Einstein's value for 300°K when $E/p < 2 \times 10^{-3}$. The results of (3) for $E/p \geqslant 8 \times 10^{-3}$ and of (2) for $E/p \leqslant 6 \times 10^{-3}$ are given in Table 4.2.1. The results of (1) agree well with those of (2) when $E/p > 0.1$, whereas larger D/μ values are found when $E/p < 0.1$; these larger apparent values of D/μ (particularly at low E/p values) can be due to almost any geometric distortion of the diffusion tube that affects the current flow. The results of (2) at 77°K, which are the only existing data at this temperature, are also given in Table 4.2.1. The D_L/μ room-temperature results of (4) and (5) are plotted in Fig. 4.2.1.

c. Magnetic drift velocity (LMF case)

(1) Townsend and Bailey (1923) performed measurements using the original Townsend magnetic deflection method, and the tubes and conditions discussed under $b(1)$ above.

(2) Crompton, Elford, and Jory (1967) adopted Jory's tube (Fig. 3.9.6)

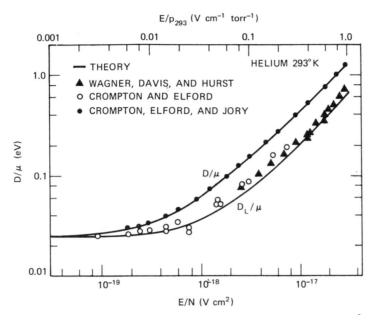

Fig. 4.2.1. Experimental values of D_L/μ in helium [b(4) and b(5); $T_g = 293°$K]. The D/μ values of Table 4.2.1 at the same temperature are also shown for comparison. Solid lines represent theoretical curves derived as indicated for No. 5 of Table 4.2.4 [after Lowke and Parker (1969)].

for measurements at 293°K over the range $6 \times 10^{-3} < E/p < 1.0$; two flux densities, 40 and 100 G, were used. These results supersede and correct previous data obtained by Crompton and Jory (1965); a reasonable estimate of their maximum experimental error is $\pm 2\%$.

The results of (1) and (2) can be divided by the drift velocity data of Table 4.2.1, so as to obtain the magnetic deflection coefficient Ψ. This quantity for the data of (2) is also given in Table 4.2.1. The results of (1) intersect those of (2) at $E/p \simeq 0.15$; above this value they are smaller and become inconsistent ($\Psi < 1$), whereas below they are significantly larger, particularly at the lowest E/p values ($\Psi = 1.44$ at $E/p = 1.3 \times 10^{-2}$).

d. Magnetic drift velocity (SMF case)

Bernstein (1962) performed measurements in strong magnetic fields with the coaxial tube of Fig. 3.9.3. His data, reduced to 293°K and given in the paper by a plot of the quantity B/p_{293} at which $w_\perp/w_T = 10$ as a function of E/B, have been converted into the more significant curve $e/m\mu_{MP_0}$ versus T_e (which corresponds to the $\omega\rho$ versus T_e curve of the microwave case, as discussed in Section 2.6), using the equivalence $1/\mu_{MP_0} = B/10p_0$, and eq.

(2.6.91) to transform E/B into T_e (in helium the electron distribution is Maxwellian, since G is velocity independent). If we restrict our range of interest to the maximum temperature of $\sim 25{,}000°K$, in order to avoid relevant gas ionization effects, Bernstein's experimental data are well represented by the relationship:

$$\frac{e}{m\mu_M p_0} = 2.08 \times 10^9 \left(\frac{T_e}{10{,}000}\right)^{0.4} \quad (\text{sec} \cdot \text{torr})^{-1}$$

These data can be directly compared with the microwave data of Table 4.2.3. Bernstein does not provide information concerning the accuracy of the results, but states that the data did not fall on a curve as smoothly for helium as for hydrogen and deuterium, except at the higher pressures; this suggests the presence of a transverse diffusion effect. Standard tanks of helium have been used for the gas supply.

e. Free electron diffusion

(1) Cavalleri (1969a) performed measurements at $300°K$ over the entire range of E/p values from zero up, using the r.f. avalanche technique and equipment described in Section 3.10A. Absolute errors of less than $\pm 0.5\%$ are claimed by the author.

Table 4.2.2 Cavalleri's [e(1); $T_g = 300°K$] Experimental Values of the Free Electron Diffusion Coefficient in Helium

$10^{17}E/N$ (V cm^2)	$10^{-21}DN$ (cm^{-1} sec^{-1}) = $3.54 \times 10^{-5}Dp_0$ (cm^2 sec^{-1} torr)	$10^{17}E/N$ (V cm^2)	$10^{-21}DN$ (cm^{-1} sec^{-1}) = $3.54 \times 10^{-5}Dp_0$ (cm^2 sec^{-1} torr)
0	6.41×10^0	2.5	1.25
6.0×10^{-3}	6.42	3.0	1.33
8.0	6.44	4.0	1.49
1.0×10^{-2}	6.46	5.0	1.62
1.5	6.58	6.0	1.73
2.0	6.69	8.0	1.93
2.5	6.81	1.0×10^0	2.09
3.0	7.03	1.5	2.44
4.0	7.4	2.0	2.77
5.0	7.78	2.5	3.08
6.0	8.09	3.0	3.37
8.0	8.65	4.0	3.95
1.0×10^{-1}	9.22	5.0	4.55
1.5	1.04×10^1	6.0	5.21
2.0	1.15		

(2) Nelson and Davis (1969) used their technique, described in Section 3.10A, for measurements at $300 \pm 2°K$ in helium with a minimum purity of 99.999%.

The most accurate and extensive data are those of (1), which are summarized in Table 4.2.2, whereas Fig. 4.2.2 shows the good agreement between these

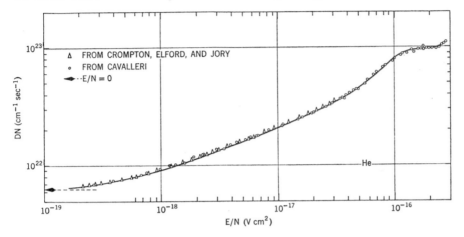

Fig. 4.2.2. Agreement between the *DN* values of Cavalleri (Table 4.2.2) and those of Crompton, Elford, and Jory (obtained from the third and fifth columns of Table 4.2.1) [after Cavalleri (1969a)].

and the *DN* values obtained by multiplying the D/μ data of Table 4.2.1 by the corresponding μN [$= w/(E/N)$] values of the same table. The value $Dp_0 = 1.95 \times 10^5 \, \text{cm}^2 \, \text{sec}^{-1} \, \text{torr}$ obtained by Nelson and Davis from measurements at 10 torr is about 8% higher than Cavalleri's value at $E/p = 0$; this difference, as well as the fact that the same authors have measured a larger value ($Dp_0 = 2.11 \times 10^5$) at 5 torr, suggests that 10 torr in helium is not a pressure large enough for accurate measurements (boundary effects). Therefore the results of (2) in helium will not be considered further.

It is also of interest that Freiberg and Weaver (1968) investigated, with the microwave technique described in Section 3.10A, the time decay of the electron density in pulsed helium afterglow discharges, down to very low electron densities. Actually, they did not attain the free diffusion limit well enough to obtain an accurate value of the coefficient; it appears, however, that the electron density decay approaches the asymptotic value corresponding to the Dp_0 value of Table 4.2.2 for $E/p = 0$.

f. Microwave conductivity ratio ($\omega^2 \gg \nu_m^2$)

(1) Gould and Brown (1954), using a rectangular, vacuum-tight copper cavity, designed to resonate in its three fundamental modes at the *S* band,

measured with good consistency the conductivity ratio in the late isothermal afterglow as a function of the gas temperature over the range 77–400°K. Furthermore, they extended the electron energy range up to 25,000°K, using a cubic quartz bottle as the gas container inside the microwave cavity, and applying a microwave heating field for increasing the electron energy; their setup was described in detail in Section 3.11B (see, in particular, Fig. 3.11.12). The gas for the experiments was of high purity, being obtained by liquid-helium evaporation. The data of Gould and Brown supersede preliminary measurements of Phelps, Fundingsland, and Brown (1951), obtained in the same laboratory using a similar resonant cavity, but less sensitive measuring techniques [see Kr $b(1)$].

(2) Fundingsland, Faire, and Penico (1954) measured, during the time interval of a microwave disturbing pulse, in accordance with the technique devised by them and described in Sections 2.6B and 3.11B, the conductivity ratio of a late-afterglow isothermal plasma produced by a microwave discharge and contained in a small pill-box cavity, resonant at 3 GHz. Since, however, the collision frequency data derived from these measurements alone agree only qualitatively with other data available for helium, it seems appropriate to disregard the less accurate measurements, namely, the transient behavior information, and to consider only the conductivity ratio in the undisturbed isothermal plasma at room temperature.

(3) Anderson and Goldstein (1955, 1956b) performed 9.4-GHz propagation measurements to determine the conductivity ratio in late isothermal afterglows of pulsed dc helium discharges; a slotted-line phase bridge (Fig. 3.11.5) was used. The authors employed a 1- to 100-μsec-duration square pulse of microwave energy at 8.6 GHz to increase the electron temperature from the isothermal 300°K value up to ~1700°K.

(4) Chen (1963) performed, over the gas temperature range 200–300°K, propagation measurements of the conductivity ratio through late isothermal afterglows of pulsed dc helium discharges, in an X-band square waveguide using a standard transient interferometer bridge. Because of the larger temperature range and of the better accuracy afforded by the interferometer bridge, these data supersede the results of previous propagation measurements in X-band waveguides performed by Chen, Leiby, and Goldstein (1961), using slotted-line measurements in a short-circuited waveguide. In these earlier experiments Chen et al. had used very-high-purity helium (impurity content less than 1×10^{-9}), obtained by the combined action of filtering through a Vycor glass membrane and of cataphoretic pumping of a commercial research-grade helium; the results show that here too very high purification does not seem significant for accurate ρ measurements. In Chen's $\omega\rho$ measurements there is a small contribution ($1-10\%$) due to electron-ion collisions; this is subtracted out, using for its evaluation expression (2.6.43).

(5) Bhattacharya, Verdeyen, Adler, and Goldstein (1967) measured the conductivity ratio of a plasma produced by the radiation field in the core of a water-cooled nuclear reactor (Triga Mark II), pulsed to peak powers of 250 and 500 MW during short intervals of time (tens of milliseconds). The gas is contained in a quartz tube, filling the interior of an aluminum cylindrical waveguide, which replaces one of the fuel elements in the reactor core. The waveguide (3.17-cm I.D.) is short circuited, and slotted-line measurements are performed at 6.3 GHz; a radiometer measures the radiation temperature at about the same frequency.

(6) Hoffmann and Skarsgard (1969) measured the conductivity ratio using an S-band cavity and an r.f. heating field, according to the method described in Section 3.11B (see Fig. 3.11.13). We have observed that the results, as plotted in Fig. 2 of their paper for the energy range 300–2000°K, can well be represented in a log-log scale by a linear relationship of $\omega\rho$ versus T_e.

Only the results obtained in isothermal plasmas or in the presence of uniform electron heating have general validity and can be directly compared among themselves; these results are given in Table 4.2.3. Microwave

Table 4.2.3 Experimental Values of the Microwave Conductivity Ratio in Helium

Authors	Temperature Range (°K)	$10^{-8}\,\omega\rho\ (\text{sec}\cdot\text{torr})^{-1}$
(1)	77–400	$2.63\left(\dfrac{T_e}{300}\right)^{0.5}$
(2)	300	2.55
(3)	300	3.43
(4)	200–300	$2.70\left(\dfrac{T_e}{300}\right)^{0.5}$
(5)	300	2.75
(6)	300–2000	$2.51\left(\dfrac{T_e}{300}\right)^{0.555}$

electron heating in the experiments of Gould and Brown and of Anderson and Goldstein was nonuniform in space. The conductivity ratio determined by the former authors, as a function of the electron temperature at the center of the quartz bottle, is plotted in Fig. 4.2.3; the conditions are such that local thermal equilibrium can be assumed. The data of Anderson and Goldstein, are not reported here, since they will not be considered further for the derivation of the electron collision parameters, inasmuch as they disagree significantly with all other results [see Table 4.2.3, (3)].

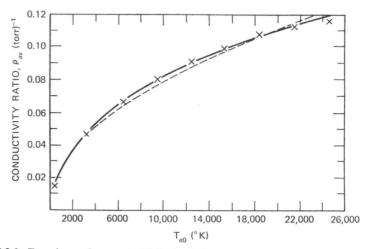

Fig. 4.2.3. Experimental curve (solid line) of the conductivity ratio measured in helium by Gould and Brown [$f(1)$], as a function of the electron temperature at the center of the quartz bottle. The dotted line derives from the relationship given in Table 4.2.3, (1). The crosses are calculated from the power series approximation of Table 4.2.4, No. 8 [after Gould and Brown (1954)].

Observing the data of Table 4.2.3, we note that, within experimental errors, the one-temperature values of (2) and (5) agree, respectively, with the values of (6) and (4) for the corresponding temperatures. Therefore we shall not further consider the data of (2) and (5) for an independent determination of the collision cross sections. In conclusion, for this purpose only the results of (1), (4), and (6) are worth analyzing.

Goldan and Goldstein (1965) performed measurements of the conductivity ratio in afterglow plasmas not far from thermal equilibrium with the gas, when this is kept at 4.2°K. Whereas measurements at pressures below 1 torr provide a pressure-independent $\omega\rho$ value [$\sim 1.4 \times 10^8$ (sec · torr)$^{-1}$], at higher pressures $\omega\rho$ becomes pressure dependent. This fact supports the hypothesis that in the low-temperature ionized helium a stable multiatom "bubble" state exists, like the one suggested by Levine and Sanders in order to interpret their mobility data at cryogenic temperatures; here, however, the effect appears at much lower atom densities ($N > 2 \times 10^{18}$ cm^{-3}).

Most of the experiments reported above were performed in the afterglow following a pulsed helium discharge. For an appropriate knowledge of the features of the electron energy decay in these discharges, the reader is referred to the radiometer study and theoretical analysis by Ingraham and Brown (1965).

g. Energy relaxation time. This quantity has been measured in the afterglow of a pulsed discharge, disturbed by a microwave heating pulse, the observable parameter being the transmitted power or the attenuation of a weak microwave probing signal. Anderson and Goldstein (1955) and Goldan and Goldstein (1965) performed measurements in helium kept at 300°K and at 4.2°K, respectively, for electron energies of the disturbed plasma near thermal values; Anderson and Goldstein (1956b) extended their earlier measurements up to electron energies of 0.23 eV. The spread of all these data, however, is comparatively much larger than the spread observed when measuring other dc and microwave parameters; this is a reflection of the difficulty of determining accurately the rate of change of the amplitude of the transmitted microwave probing signal. Therefore these data will not be further considered here, even though their consistency with the ν_m and G values derived more accurately from the other transport parameters has been verified.

h. Thermal diffusivity. Nygaard (1967b) and Delpech and Gauthier (1969) measured this parameter in an isothermal afterglow plasma, each one using his own method as described in Section 3.13. For electron densities below 10^{10} cm^{-3}, the measured diffusivity is practically independent of density, thus ensuring that electron-neutral collisions will be dominant over electron-ion collisions. Nygaard's value of $\mathcal{D}p_0$, extrapolated to zero electron density, is 4.6×10^5 cm^2 sec^{-1} torr; the data of Delpech and Gauthier show a larger spread but are also consistent with this value. Cronson's (1966) determinations of the spatial variations in electron temperature produced in an afterglow plasma by a microwave standing wave, as obtained from measurements of recombination light intensity, are not accurate enough to justify any serious attempt at deriving a diffusivity value.

i. Cyclotron resonance

(1) Hirshfield and Brown (1958) applied to an afterglow isothermal plasma their method based on determining the condition $\sigma_{Ti} = 0$. From the experimental data plotted in the paper it appears that the magnetic field changes necessary for returning the cavity to its empty resonant frequency (4200 MHz) can be expressed as a function of gas pressure by the relationship:

$$\delta B = 9 \times 10^{-2} p_0^2 \quad \text{(gauss)}$$

(2) Fehsenfeld (1963) determined the resonance line half-width $\Delta \omega_I$ at 300°K from the absorption of 9-GHz energy in TE_{012} rectangular and cylindrical cavities, loaded by a flowing gas afterglow [see Section 3.14A]. The result is:

$$\Delta \omega_I = (1.16 \pm 0.11) \times 10^8 p \quad \text{(sec}^{-1})$$

(3) Tice and Kivelson (1967) measured the resonance line shape at 9.5 GHz in a room-temperature helium-nitrogen mixture, obtained by titrating

Table 4.2.4 Summary of Q_m Determinations in Helium

Case No.	Source	Transport Parameters	Range of Swarm Mean Energies	Type of Analysis	Q_m (Å^2)
1	Frost and Phelps (1964)	$a(1), a(3), b(1), b(2)$ (77 and 300°K only)	0.010–3.6	A [eqs. (2.3.25), (2.4.20), (2.6.1)]	See Table 4.2.5(a)
2	Crompton, Elford, and Jory (1967)	$a(5)$ (293°K only), $b(3), c(2)$	0.039–1.6	A [as 1] plus eq. (2.6.83)]	See Table 4.2.5(b)
3	Crompton, Elford, and Robertson (1970)	$a(5), b(3)$	0.010–1.6	A [as 1]]	See Table 4.2.5(c)
4	Bowe and Langs (1966)	$a(4)$	0.2–1.6	C [eqs. (2.6.11), (2.3.47), (2.3.46'), $h = 1$]	7.3 ± 0.15
5	Lowke and Parker (1969)	$b(4), b(5)$	0.039–1.3	B [eqs. (2.3.25), (2.3.72) with Q_m from 2)]	Values of Table 4.2.5(b) provide good agreement
6	...	d	0.55–3.3	C [eq. (2.6.41)]	$7.6u^{-0.1}$
7	Cavalleri (1969)	$e(1)$	0.039–3.6	B [eqs. (2.3.25), (2.4.18) with Q_m from 2]	Values of Table 4.2.5(b) provide excellent agreement
8	Gould and Brown (1954)	$f(1)$	0.010–3.3	C [eqs. (2.6.49), (3.11.24)]	$5.12 + 0.49u^{0.5} - 0.21u$
9	Chen (1963)	$f(4)$	0.026–0.039	D [eq. (2.6.41), $h = 1$]	5.3
10	...	$f(6)$	0.039–0.26	C [eq. (2.6.41)]	$5.75u^{0.055}$
11	...	h	0.039	D [eq. (2.6.74), $h = 1$]	4.9
12	Hirshfield and Brown (1958)	$i(1)$	0.039	D [eq. (2.6.104), $h = 1$]	5.65
13	Fehsenfeld (1963)	$i(2)$	0.039	D [eq. (2.6.97), $h = 1$]	6.1 ± 0.6
14	Tice and Kivelson (1967)	$i(3)$	0.039	D [eq. (2.6.95), $h = 1$]	5.5 ± 0.5

helium through a long glass capillary into the active nitrogen stream, which has been used by these authors as explained in Section 3.14A [see, for the evaluation of the nitrogen contribution, N_2 h].

Determination of Collision Parameters

The most significant determinations of the collision cross section Q_m, derived from analyses of the experimental data reported above, are given in Tables 4.2.4 and 4.2.5 and are plotted in Fig. 4.2.4.

Table 4.2.5 The Momentum Transfer Cross Sections Q_m, Determined by Frost and Phelps (*a*), by Crompton, Elford, and Jory (*b*), and by Crompton, Elford, and Robertson (*c*), for Electrons in Helium

	Q_m (Å²)				Q_m (Å²)		
u (eV)	(*a*)	(*b*)	(*c*)	u (eV)	(*a*)	(*b*)	(*c*)
0.0	4.87			0.30	6.54	6.35	6.35
0.008	5.07		5.18	0.40	6.60	6.50	6.49
0.010	5.10		5.21	0.50	6.63	6.60	6.59
0.013	5.14		5.26	0.60	6.63	6.67	6.66
0.016	5.17		5.30	0.80	6.60	6.78	6.77
0.020	5.20	5.40	5.35	1.0	6.56	6.87	6.85
0.025	5.24	5.45	5.41	1.2	6.51	6.94	6.91
0.030	5.29	5.50	5.46	1.5	6.42	6.99	6.96
0.040	5.38	5.58	5.54	2.0	6.26	7.00	6.99
0.050	5.47	5.65	5.62	2.5	6.08	6.96	6.96
0.060	5.54	5.71	5.68	3.0	5.89	6.89	6.89
0.080	5.65	5.82	5.79	4.0	5.53	6.60	6.60
0.10	5.75	5.89	5.86	5.0	5.22	6.26	6.26
0.12	5.86	5.97	5.94	6.0	4.90	6.01	6.01
0.15	6.04	6.07	6.04	10.0	4.10		
0.20	6.31	6.19	6.16	15.0	3.48		
0.25	6.49	6.29	6.27	20.0	2.99		

The results of Crompton and his coworkers given in Table 4.2.5, (*b*) and (*c*) (the two sets are practically coincident), must be recognized as the most reliable ones, since (1) they are based on accurate measurements of different transport parameters, all performed by the same authors; (2) the numerical method of curve fitting has been used, thus avoiding the introduction of any *a priori* assumption regarding the $Q_m(v)$ behavior; (3) the agreement between calculated and experimental values of the transport parameters is always within the accuracy claimed for the measurements; and (4) the results are in good agreement also with different measurements by other authors. Crompton, Elford, and Jory have also analyzed the possibility of the existence

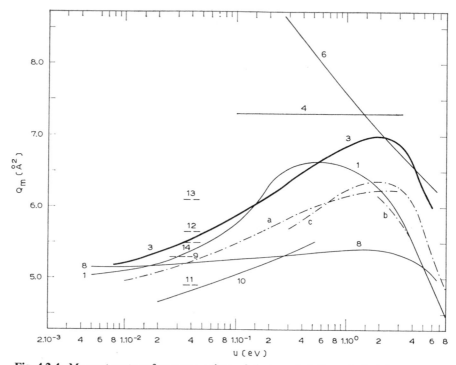

Fig. 4.2.4. Momentum transfer cross sections of electrons in helium, derived from swarm experiments (numbers as in Table 4.2.4) and deduced from single scattering experiments (letters as in the text). Curve 2 is not represented, being very near to curve 3 over the entire range.

of a fine structure in the $Q_m(v)$ curve at low energies, since such a structure was observed by Ramsauer and Kollath (1929) and Normand (1930), who measured the total collision cross section by beam techniques; their conclusion is that, if there is fine structure, it has to be far less significant than reported by the earlier investigators.

The curves plotted in Fig. 4.2.4 and derived from measurements of the total collision cross section are as follows:

(a) Golden's (1966) computations, based on the atomic effective range theory formulas (1.5.1), (1.5.3), and (1.5.5) and the experimental $Q_t \simeq q_t$ data of Golden and Bandel (1965) ($A = 1.15$, $B = 0.273$, $A_1 = 0.264$);

(b) the computations of Crompton, Elford, and Jory (1967), based on application of formulas (1.3.22′) and (1.3.22″) to the results of the measurements of P_t performed by Golden and Bandel (1965), and adopting the angular distribution $s_3(\vartheta)$ form observed by Ramsauer and Kollath (1932);

(c) like (b), but using La Bahn's and Callaway's (1966) theoretical prediction of the $s_3(\vartheta)$ form.

4.3 NEON

Experimental Data [dc parameters for $E/p < 0.4(D/\mu \simeq 2.8$ eV)]

a. Drift velocity

(1) Nielsen (1936) [see He a(1)] performed measurements at 293°K for $E/p > 3 \times 10^{-2}$. The gas purity was better than 99.8%.

(2) Bowe (1960) [see He a(4)] performed measurements at room temperature for $E/p > 6 \times 10^{-2}$ (this range corresponds to a field distortion due to space charge below 1%), using the induced current method in a 2.54-cm-long drift chamber. As in the case of helium, this author used a calcium purifier, operating at 450°C, to improve the purity of the reagent-grade tank neon, but no effect on w has been observed. Bowe places an estimate of 5% on the overall error of his w determinations.

(3) Pack and Phelps (1961) [see He a(3)] performed measurements at 77°K and 300 ± 3°K for $E/p > 4 \times 10^{-4}$. These authors state that they were unable to take data below this E/p value, because the loss of electrons to the grid wires at the maximum available pressure (660 torr at 300°K) was too large in their apparatus. The impurity content was like that for helium.

The results of (3), which are the most significant because of accuracy, extent of the E/p range, and their midway position, are plotted in Fig. 4.3.1, together with those of (1) and (2). It is apparent that even at the lowest E/p values the slope is always far from the 45° value, which characterizes the condition of isothermicity. Compared to the data of (3), the results of (1) are slightly higher, whereas those of (2) are lower, particularly for $E/p > 0.3$, where the disagreement becomes significantly large. The results of (2) are also well represented by the relation:

$$w = 8.9 \times 10^5 \left(\frac{E}{p_0}\right)^{0.432} \quad (\text{cm sec}^{-1})$$

Here we have disregarded the large mobilities observed by English and Hanna (1953) [see Xe a(1)], which seem attributable to impurities.

b and c. Diffusion to mobility ratio and magnetic drift velocity (LMF case)
The only data available for these quantities are a set of old measurements by Bailey (1924) in neon contaminated by 1% of helium; here they will not be considered for the derivation of collision parameters.

d. Free electron diffusion
(1) Gardener, Kisdnasamy, Rössle, and Wolfendale (1957) and Coxell and Woldendale (1960) performed measurements of the flashing efficiency-time

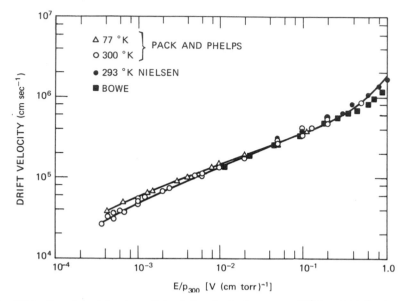

Fig. 4.3.1. Experimental values of the drift velocity in neon at 77°K and 300°K by Pack and Phelps, and previous measurements by Nielsen and Bowe [after Pack and Phelps (1961)].

delay characteristics of flash tubes, filled wth commercial-grade neon. From these data Lloyd (1960) derived the diffusion coefficient for electrons at room temperature.

(2) Cavalleri and Loria (1964) and Cavalleri, Gatti, and Interlenghi (1965) measured the diffusion coefficient for electrons at 304°K, using the light chamber technique.

(3) Nelson and Davis (1969) [see He $e(2)$] performed measurements at 300 ± 2°K in neon with a minimum purity of 99.999%. The authors found poor agreement among the individual Dp_0 values, even at the highest pressures they considered for the computation of their average value (30 and 40 torr only).

The results of (1), (2), and (3) are, respectively, $Dp_0 = (13.7 \pm 0.8) \times 10^5$, $Dp_0 = (22.3 \pm 1.4) \times 10^5$, and $Dp_0 = (18 \pm 3) \times 10^5$ cm^2 sec^{-1} torr.

e. Microwave conductivity ratio $(\omega^2 \gg \nu_m{}^2)$

(1) Fundingsland, Faire, and Penico (1954) [see He $f(2)$] measured the conductivity ratio in late isothermal afterglows of microwave discharges produced in room-temperature neon.

(2) Gilardini and Brown (1957) performed measurements from room temperature up to 16,000°K by applying a microwave heating field to a late-afterglow isothermal plasma, contained in a cubic quartz bottle centered in

a parallelepiped S-band cavity [see He $f(1)$]. Theory shows that at the chosen pressures (from 13 to 5 torr, the lowest for the highest electron temperatures) the electron energy becomes uniform (but in general the distribution is no longer Maxwellian) inside the bottle, and the conductivity ratio becomes pressure independent. Data have been taken both in pure neon (uniform electron density because of dominant recombination losses) and in neon contaminated with 0.01–0.1% argon (electron density distribution as per diffusion losses). The data of Gilardini and Brown supersede preliminary measurements of Phelps, Fundingsland, and Brown (1951), obtained in the same laboratory using a similar resonant cavity, but less sensitive measuring techniques [see Kr $b(1)$]. The spread of the individual data is within 2%.

(3) Chen (1964) [see He $f(4)$] measured the conductivity ratio in neon discharge late isothermal afterglows over the gas temperature range 200–600°K.

(4) Bhattacharya, Verdeyen, Adler, and Goldstein (1967) [see He $f(5)$] observed, in the case of neon at pressures below 100 torr, that electron temperatures during the reactor ionizing pulse are higher than room temperature and depend on the gas pressure, because the electron energy relaxation times are 1 order of magnitude larger than in helium. The conductivity ratio and the radiation temperature have been measured at different gas pressures; the results, plotted in Fig. 10 of their paper for the energy range 300–2800°K, can be represented within the experimental errors by a straight-line approximation.

(5) Hoffmann and Skarsgard (1969) [see He $f(6)$] measured the conductivity ratio over the energy range 450–10,000°K.

The results for isothermal plasmas or uniform electron heating, all satisfying the conditions for a Maxwellian velocity distribution of electrons, are given in Table 4.3.1. The results of (2), which were obtained under nonuniform heating conditions, are plotted in Fig. 4.3.2. In the above we have disregarded the results of Volkov, Zinov'ev, and Malyuta (1968), which were obtained from free-space propagation measurements at millimeter wavelengths across a rather dense afterglow plasma ($0.2\omega < \omega_p < \omega$); in fact, these conditions do not meet the requirements discussed in Section 3.11 for accurate determinations of the collision parameters.

Observing the data of Table 4.3.1, we note that within the experimental errors the one-temperature value of (1) agrees with the corresponding values of (2) and (4); therefore, (1) will not be considered further for an independent determination of the collision cross sections.

f. Thermal diffusivity. The value of $\mathfrak{D}p_0$, extrapolated to zero density from Nygaard's (1967a) measurements (see He h) is 2.6×10^6 cm^2 sec^{-1} torr. The author performed the experiment at $p = 1$ and 2 torr, and verified

Table 4.3.1 Experimental Values of the Microwave Conductivity Ratio in Neon

The expression representing Hoffmann's and Skarsgard's (5) results has been determined so as to provide within reading errors a good fit to the curve published in their paper.

Authors	Temperature Range (°K)	$10^{-7} \omega \rho$ (sec . torr)$^{-1}$
(1)	300	2.85
(2)	300	3.03
(3)	200–600	$2.94 \left(\dfrac{T_e}{300}\right) + 0.54 \left(\dfrac{T_e}{300}\right)^{0.5}$
(4)	300–2800	$2.77 \left(\dfrac{T_e}{300}\right)$
(5)	450–10,000	$2.9 \left[1 - 0.11 \left(\dfrac{T_e}{300}\right)^{0.5}\right] \left(\dfrac{T_e}{300}\right)$

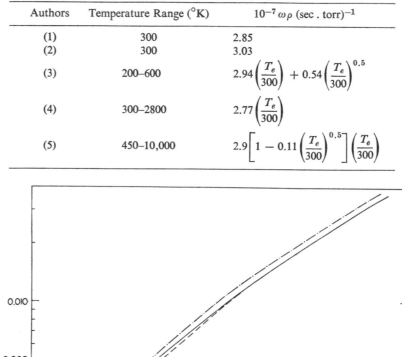

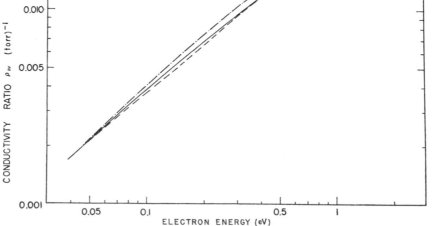

Fig. 4.3.2. Experimental curves of the average conductivity ratio measured in neon by Gilardini and Brown [e(2)], as a function of the electron energy at the cavity center in an equivalent uniform plasma, for neon contaminated with argon (solid line) and for pure neon (short-dash line), and as a function of the average electron energy in the cavity (long-and-short-dash line) [after Gilardini and Brown (1957)].

that diffusion cooling effects were noticeable at $p = 0.5$ torr and below. To depopulate metastable states, which are usually present in a rare-gas afterglow and can be a significant source of electron heating, Nygaard has also made some measurements of $\mathfrak{D}p_0$ in neon contaminated with 0.1% argon; no difference from the pure neon case has been found.

Determination of Collision Parameters

The most significant determinations of the collision cross section Q_m, derived from analyses of the experimental data reported above, are given in Tables 4.3.2 and 4.3.3 and are plotted in Fig. 4.3.3.

The drift velocity data of Nielsen and of Pack and Phelps [see $a(1)$ and $a(3)$] have not been considered in Table 4.3.2, since Frost and Phelps (1964), who tried to analyze them, found an apparent 30% discrepancy in the cross sections needed to fit the 77 and 300°K data at low E/p. For the reduction of the room-temperature single data $d(1)$, $d(2)$, $d(3)$ and f, it has been assumed that $h = 1.8$, in accordance with the slope at 0.039 eV of the $\omega\rho$ curve of Fig. 4.3.2.

The most reliable results are those derived from conductivity ratio measurements performed over wide electron energy ranges by Gilardini and Brown (curve 5 in Fig. 4.3.3) and by Hoffmann and Skarsgard (curve 8); considering the spread of the experimental data, preference is given to curve 5, which is also closer to the results of most other authors. Curve a represents O'Malley's (1963) calculated results, obtained by applying the atomic effective range theory formulas (1.5.1)–(1.5.5) to Q_t and s_3 data of Ramsauer and Kollath (1929, 1932) ($A = 0.24$, $B = 0.317$, $A_1 = 1.66$).

4.4 ARGON

Experimental Data [dc parameters for $E/p < 1$ $(D/\mu \simeq 7$ eV)]

a. Drift velocity
(1) Nielsen (1936) [see He $a(1)$] performed measurements at 293°K for $E/p > 6 \times 10^{-2}$. The gas purity was better than 99.8%.
(2) Herreng (1942, 1943) determined drift velocities at 295°K for $E/p > 3 \times 10^{-3}$ from the pulse duration of the collector induced current, after ionization from a narrow beam of pulsed X-rays, parallel to the electrodes and at variable distances from the collector. Tank argon was used in these experiments.
(3) Unpublished data of Errett (1951) [see He $a(2)$] were reported by Frost and Phelps (1964) for $E/p > 0.1$ and $T_g = 300$°K.
(4) Colli and Facchini (1952) determined drift velocities at room temperature for $E/p > 1 \times 10^{-2}$ from the pulse duration of the collector induced

Table 4.3.2 Summary of Q_m Determinations in Neon

Case No.	Source	Transport Parameters	Range of Swarm Mean Energies	Type of Analysis	Q_m (Å²)
1	Bowe (1960)	$a(2)$	0.38–8.0	C [eqs. (2.6.11), (2.3.47), (2.3.46'), $h = 1.314$]	$1.7u^{0.157}$
2	⋯	$d(1)$	0.039	D [eq. (2.6.23'), $h = 1.8$]	0.52 ± 0.03
3	⋯	$d(2)$	0.039	D [eq. (2.6.23'), $h = 1.8$]	0.32 ± 0.02
4	⋯	$d(3)$	0.039	D [eq. (2.6.23'), $h = 1.8$]	0.39 ± 0.07
5	Gilardini and Brown (1957)	$e(2)$	0.039–2.0	C [eq. (2.6.49) modified]	See Table 4.3.3(a)
6	Chen (1964)	$e(3)$	0.026–0.078	C [eq. (2.6.49)]	$0.107 + 2.17u^{0.5}$
7	⋯	$e(4)$	0.039–0.36	C [eq. (2.6.41), $h = 2$]	$2.04u^{0.5}$
8	Hoffmann and Skarsgard (1969)	$e(5)$	0.06–1.3	A [eqs. (2.3.31), (2.6.40)]	See Table 4.3.3(b)
9	⋯	f	0.039	D [eq. (2.6.74), $h = 1.8$]	0.48

Table 4.3.3 The Momentum Transfer Cross Section $Q_m(u)$, Determined by Gilardini and Brown (a) and by Hoffmann and Skarsgard (b), for Electrons in Neon

u (eV)	Q_m (Å²) (a)	(b)	u (eV)	Q_m (Å²) (a)	(b)
0.020	0.348		0.25	1.03	0.870
0.025	0.383		0.30	1.09	0.920
0.030	0.414		0.40	1.20	1.01
0.040	0.469	0.460	0.50	1.29	1.09
0.050	0.516	0.500	0.60	1.37	1.14
0.060	0.558	0.530	0.80	1.50	1.24
0.080	0.631	0.590	1.0	1.59	1.30
0.10	0.695	0.635	1.2	1.65	1.34
0.12	0.752	0.675	1.5	1.69	1.40
0.15	0.827	0.725	2.0		1.46
0.20	0.940	0.800			

current, after a single alpha-particle ionization along a track parallel to the electrodes ($d = 2.6$ cm). They used tank argon (purity 98%), treated for about 3 hours at 450°C in a calcium-magnesium filled, convection current purifier. The estimated error from the scatter of the individual data is ±5%.

(5) Kirschner and Toffolo (1952) determined drift velocities at room

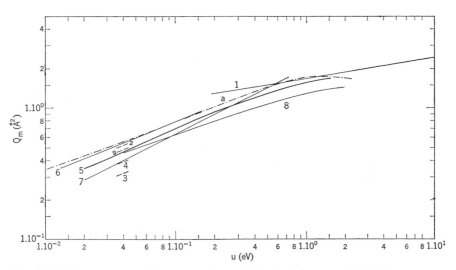

Fig. 4.3.3. Momentum transfer cross sections of electrons in neon, derived from swarm experiments (numbers as in Table 4.3.2) and deduced from single scattering experiments (curve *a* shows O'Malley's results).

temperature for $E/p > 0.15$ from the waveform of the collector induced current, after a long-lasting X-ray flash filling the whole chamber ($d \simeq 1.5$ cm). The tank argon (stated purity 99.9%) used in these experiments was carefully purified by convection over hot calcium. The authors estimate that their determinations are accurate to within 5%.

(6) Bowe (1960) [see Ne $a(2)$] performed measurements at room temperature for $E/p > 6 \times 10^{-2}$. The effectiveness of removing impurities by circulation of the original tank gas (stated purity 99.99%) over hot calcium turnings is shown in Fig. 4.4.1.

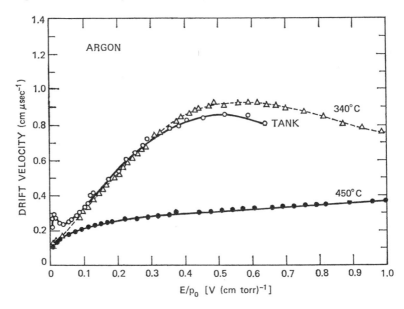

Fig. 4.4.1. Drift velocity in argon, measured by Bowe at $p_0 = 680$ torr immediately after filling the chamber (open circles) and with the purifier operated at 340°C (triangles) and at 450°C (points) [after Bowe (1960)].

(7) Nagy, Nagy, and Dési (1960) determined drift velocities at room temperature over the range $1.5 \times 10^{-2} < E/p < 0.95$ from the duration of the pulses of the collector induced current, after ionization of the gas from fission fragments (see Section 3.2; $d = 5$ cm). The authors used tank argon (purity 99.7%), treated in a hot calcium purifier.

(8) Pack and Phelps (1961) [see He $a(3)$] performed measurements at 77°K and at 300 ± 3°K over the range $2 \times 10^{-4} < E/p < 0.5$. The impurity content was the same as for helium.

(9) Levine and Uman (1964) determined drift velocities at room temperature for $E/p > 1.5 \times 10^{-2}$ from the ramp duration of the integrated

induced current at the collector of a drift chamber, after a flash of single alpha-particle ionization along a track parallel to the electrodes at 4.52 cm from the collector. No attempt at improving the purity of the argon (stated as 99.99%) by purification techniques was made by these authors, but they took great care not to contaminate the argon during gas handling in the filling system. Individual results show significant scatter, often in excess of 10%.

(10) Wagner, Davis, and Hurst (1967) [see He $a(6)$] performed measurements at room temperature over the range $2 \times 10^{-2} < E/p < 1$; these supersede previous results of Bortner, Hurst, and Stone (1957), who used an earlier version of the same method (two proportional counters at different distances from a collimated beam of ionizing alpha-particles, instead of flashes of photoelectrons emitted by a cathode and detected by an electron multiplier at the collector) and found higher mobilities, probably due to impurities, in argon purified only by fractional distillation from cold traps. The impurity content of the argon used by Wagner et al. was the same as for helium.

(11) Grünberg (1968) [see He $a(7)$] performed measurements at room temperature in argon of high purity (99.99%) for $E/p > 4 \times 10^{-2}$ and pressures as high as 41 atm.

(12) Allen and Prew (1970) performed measurements at 295°K for $E/p > 0.3$, using the induced current method and very high pressures, from 3 to 100 atm. The gas was from standard cylinders, but properly dried (impurity content $< \sim 40$ ppm); estimated measuring accuracy is $\pm 5\%$.

The results of (8), which seem to be the most significant, since they are accurate, cover the widest E/p range, and fall in a position midway among all available results, are plotted in Fig. 4.4.2 [those of (1) and (6) are also included]. All other data, except those of (9) below $E/p = 0.2$, lie within $\pm 10\%$ of the values of (8). The results of (9), which agree with (8) at $E/p = 1$, become progressively higher than those of (8) at lower E/p values; below $E/p = 0.2$ the discrepancy becomes so large that we suggest the rejection of these data, also considering their large scatter. The results of (6) are also well represented by the relation:

$$w = 3.7 \times 10^5 \left(\frac{E}{p_0}\right)^{0.25} \quad (\text{cm sec}^{-1})$$

We have completely disregarded here all large mobilities observed by Hudson (1944), by Klema and Allen (1950), and by English and Hanna (1953) [see Xe $a(1)$], which are now recognized by all authors as due to impurities. A critical, enlightening discussion of the significance of these measurements and of the related gas-purity problems can be found in Loeb (1955).

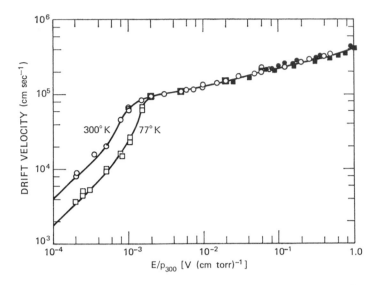

Fig. 4.4.2. ● Nielsen's [$a(1)$], ■ Bowe's [$a(6)$], and ○□ Pack's and Phelps's [$a(8)$] experimental values of the drift velocity in argon. For $E/p < 2 \times 10^{-4}$ electrons are in thermal equilibrium with the gas; the corresponding μp_0 values are 1.64 and 4.0×10^7 cm^2 V^{-1} sec^{-1} torr, respectively, at 77 and 300°K [after Pack and Phelps (1961)].

The experimental results of (11) and (12) show that up to 100 atm ($N \simeq 3 \times 10^{21}$ cm^{-3}) no pressure effect exceeds the accuracy limits; these experiments refer to conditions under which electron energies are larger than 0.1 eV, so that the corresponding atom densities at which $l_m = \lambda_e$ happen to be always larger than $\sim 2 \times 10^{22}$ cm^{-3} (see Fig. 4.4.9 for Q_m values). Hence the failure to observe pressure effects in argon, as compared to helium, may be due to the higher values of these critical densities.

Argon has a large Ramsauer minimum at ~ 0.3 eV, so that electrons experience far fewer collisions with molecules at these energies than at lower and higher energies; above $E/p \simeq 3 \times 10^{-3}$, the dominating influence of this minimum in the computation of the drift velocity integral (2.4.20) is such that this velocity becomes rather insensitive to any choice of the values of the electron collision cross section above 0.7 eV. This makes it difficult to derive the $Q_m(u)$ curve from measurements of the transport coefficients. This problem can be solved, however, by adding to argon a gas like hydrogen, which has a large Q_m cross section at the energies of the Ramsauer minimum of argon.

Since this approach has actually been used for the derivation of the $Q_m(u)$ curve in argon, as we shall discuss later, we collect here also any pertinent

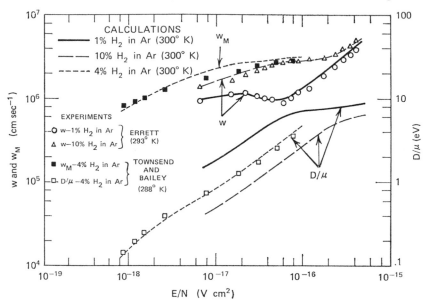

Fig. 4.4.3. Drift velocity (w), characteristic energy (D/μ), and magnetic drift velocity (w_M) in mixtures of 1, 4, and 10% H_2 in argon. The points show experimental data; the curves are computed using for argon the Q_m curve 1 of Fig. 4.4.9 and for H_2 the Q_m curve 2 of Fig. 4.7.6 [after Engelhardt and Phelps (1964)].

information regarding measurements in hydrogen-argon mixtures. Engelhardt and Phelps (1964) report unpublished data by:

(i) Errett (1951) [see (3) above], for mixtures containing 1 and 10% of hydrogen, $E/p > 0.2$, and $T_g = 293°K$ [see Fig. 4.4.3].

(ii) Pack, for mixtures containing 1.5% of hydrogen, $3 \times 10^{-4} < E/p < 0.5$, and $T_g = 77°K$ and $300°K$ [see Fig. 4.4.4].

b. Diffusion to mobility ratio

(1) Townsend and Bailey (1922) performed measurements at room temperature for $E/p > 0.1$, using the diffusion chambers described under He $b(1)$. The authors used commercial argon with 10% nitrogen, which was removed by Rayleigh's method (first it was made to recombine with oxygen in the presence of a solution of caustic potash, under the action of a discharge; then the gas was passed over a hot copper foil and dried by phosphor pentoxide). These authors also performed measurements in mixtures of 4% hydrogen in argon.

(2) Warren and Parker (1962) [see He $b(2)$] performed measurements at

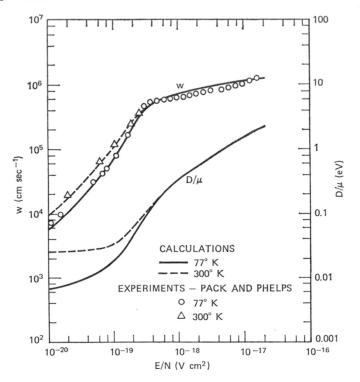

Fig. 4.4.4. Drift velocity (w) and characteristic energy (D/μ) in a mixture of 1.5% H_2 in argon. The points show experimental data; the curves are computed as in the case of Fig. 4.4.3 [after Engelhardt and Phelps (1964)].

77°K and 87°K over the range $1.5 \times 10^{-4} < E/p < 4 \times 10^{-2}$. These authors preferred, over the E/p range where D/μ is large compared to kT_g/e, to perform measurements at 87°K, the boiling point of argon, because of the better precision afforded by the higher gas densities thus achievable in the diffusion chamber.

(3) Wagner, Davis, and Hurst (1967) [see He $b(4)$] determined the quantity D_L/μ at room temperature over the range $1.5 \times 10^{-2} < E/p < 1$. For the impurity content of the gas see $a(10)$.

All the results are given in Fig. 4.4.5. They can be considered representative of argon at 77°K, even if some of them were taken at higher temperatures; in fact, all D/μ measured at these temperatures are large compared to kT_g/e and therefore coincide with the values one should obtain at 77°K. The results of Townsend and Bailey for the hydrogen-argon mixtures are given in Fig. 4.4.3.

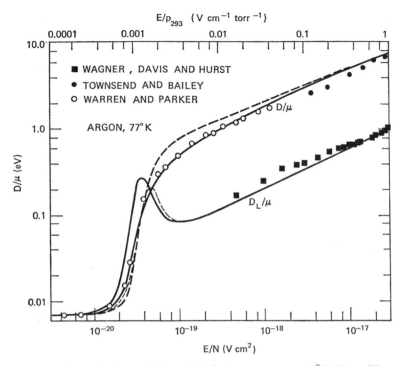

Fig. 4.4.5. Experimental values of D/μ and D_L/μ in pure argon at 77°K. The solid curves are computed using the Q_m curve 1 of Fig. 4.4.9, whereas the broken curve is calculated assuming the Q_m curve a [after Lowke and Parker (1969)].

c. Magnetic drift velocity (LMF case). The only experimental measurements are those of Townsend and Bailey (1922), who used the tube and conditions discussed under $b(1)$ above; the corresponding magnetic deflection coefficient Ψ is plotted in Fig. 4.4.6. The magnetic drift velocities in hydrogen-argon mixtures are given in Fig. 4.4.3.

d. Free electron diffusion

(1) Cavalleri and Loria (1964) [see Ne $d(2)$] performed measurements of the diffusion coefficient for electrons at room temperature.

(2) Nelson and Davis (1969) [see He $e(2)$] performed measurements at 300 ± 2°K in argon with a minimum purity of 99.995 %.

The results of (1) and (2) are, respectively, $Dp_0 = (90 \pm 8) \times 10^4$ and $Dp_0 = (80 \pm 2) \times 10^4$ cm² sec⁻¹ torr.

e. Direct current conductivity. All the experiments, in which the dc conductivity was measured in shock-heated argon, are listed and reviewed in Section 4.17. Actually the results of the measurements performed at low

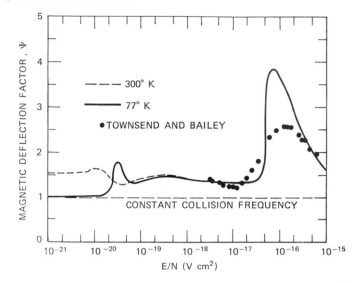

Fig. 4.4.6. Magnetic deflection coefficient Ψ in argon. The points show experimental data obtained at room temperature; the curves are computed using the Q_m curve 1 of Fig. 4.4.9 [after Frost and Phelps (1964)].

electron densities, namely at low Mach numbers and pressures, could be used in principle for the purpose of the determination of the electron-neutral collision cross sections, while the results of the measurements performed at high electron densities should provide good tests for the validity of the conductivity expressions given by the Spitzer-Härm theory or by the later improvements of this theory. No author, however, has been able to derive from the experimental data significant values of the electron-neutral collision cross section in argon; in fact, at low Mach numbers and pressures: (1) conductivities are small, and therefore large measurement errors are possible; (2) equilibrium ionization is seldom reached within the hot region, and then the electron density cannot be properly estimated; (3) impurities may contribute significantly, but by an unpredictable amount, to the overall ionization. Hence dc conductivity measurements in shock-heated argon can be useful only for the study of electron-ion and electron-electron collisions; for this reason we have postponed to the appropriate Section 4.17 a review of all the pertinent experiments. Since the same considerations apply to the similar cases of krypton and xenon, the experiments performed in these gases will also be discussed in Section 4.17 only.

f. Microwave conductivity ratio. No experiment performed in argon afterglow plasmas has given valuable results for the derivation of the electron collision parameters. Preliminary results $[\omega\rho = 3.0 \times 10^7 \ (\text{sec} \cdot \text{torr})^{-1}$ at

300°K], obtained by Phelps, Fundingsland, and Brown (1951) [see Kr $b(1)$], have provided collision cross-section values that are inconsistent with other available determinations; this may signify that the electrons were not in thermal equilibrium with the gas. Fundingsland, Faire, and Penico (1954) [see He $f(2)$] performed measurements in argon, but did not report any definite value of the conductivity ratio in this gas. The results of Volkov, Zinov'ev, and Malyuta (1968) are disregarded for the same reasons as were given in the case of neon.

A few attempts at measuring the conductivity ratio under thermal equilibrium conditions in the high-temperature ionized layer of a shock tube are described in the literature. The data of Brandewie and Williams (1964) and Krylov and Tusnov (1966) were obtained using shock tubes as waveguides for a signal of frequency $\omega < \omega_p$; these data are disregarded for the reasons discussed in Section 3.11C. Makios (1966) used his Lecher wire interferometer for measurements in the plasma of electromagnetically produced shock waves in T-tubes; the results are of little value, however, since most of the actual plasma parameters are unknown. Thus the only data worth consideration are those which Daiber and Waldron (1966) obtained using the shock tube as a waveguide, under standard low-perturbation conditions $\omega_p^2 \ll \omega^2$ (see description in Section 3.11A, and particularly Fig. 3.11.4). These data are shown in Fig. 4.4.7 as apparent collision cross sections, namely, as $\omega\rho$

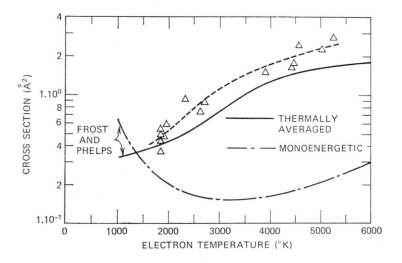

Fig. 4.4.7. Daiber's and Waldron's experimental values of the microwave conductivity ratio in argon shock waves, plotted as apparent collision cross sections as a function of the electron temperature. The solid line shows the curve computed using the Q_m curve 1 of Fig. 4.4.9 [after Daiber and Waldron (1966)].

values divided by the standard density 3.54×10^{16} cm^{-3}, and by the corresponding mean electron speed $[(8/3)\sqrt{2kT_e/\pi m}]$. The authors have verified by appropriate computations that the time to thermal equilibrium for the gas (30 μsec) and the electron energy relaxation times (8–80 μsec for initial shock tube pressures of 100–15 torr) are both sufficiently shorter than the test time (400 μsec). Although this is not the case for the ionization equilibrium time, which is much longer, this fact does not affect in any way the significance of the conductivity ratio measurements, adequate time resolution being provided.

g. Cyclotron resonance

(1) Fehsenfeld (1963) [see He i(2)] obtained at 300°K:

$$\Delta\omega_I = (4.8 \pm 0.5)10^7 p \quad (\text{sec}^{-1})$$

(2) Tice and Kivelson (1967) [see He i(3)] measured the resonance line shape in argon, but the collision cross sections they derived from these data

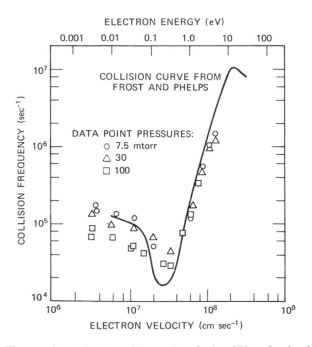

Fig. 4.4.8. The experimental values of Bruce, Crawford and Harp for the electron collision frequency as a function of the electron velocity, derived from pulse response observations in argon magnetoplasmas at cyclotron resonance. The solid line shows the results of computations based on the Q_m curve 1 of Fig. 4.4.9 [after Bruce, Crawford, and Harp (1968)].

are too small in comparison to the typical values we present later in this Section (see, e.g., Fig. 4.4.9); a possible explanation is that the electrons are not in thermal equilibrium with the gas. These data will therefore be disregarded hereafter.

(3) Bruce, Crawford, and Harp (1968) applied to argon their method, which is described in Sections 2.6D and 3.14. The collision data derived from pulse response observations in 99.99% pure argon at three different pressures (7.5, 30, and 100 mtorr) are shown in Fig. 4.4.8; the effect of the microwave field nonuniformity in the waveguide was properly taken into account by the authors in the derivation of the above results. Below 3×10^7 cm sec^{-1}, however, the data as derived should be rejected, since the induced velocity does not satisfy the condition of being sufficiently larger than the thermal values ($\simeq 1 \times 10^7$ cm sec^{-1}) to neglect the effect of the agitational motion.

Determination of Collision Parameters

Only one derivation of the collision cross section Q_m, that of Frost and Phelps (1964) [see also Engelhardt and Phelps (1964)], is worth consideration; these authors performed a numerical analysis of type A [eqs. (2.3.25), (2.4.20), (2.6.1)] in order to find the best fit to the average curves of the drift velocity at 77°K and 300°K [based on the data of $a(1)$, $a(3)$, $a(6)$, and $a(8)$] and of the characteristic energy at 77°K (this is the D/μ curve of Fig. 4.4.5). The resulting $Q_m(u)$ curve is given in Table 4.4.1 and is plotted in

Table 4.4.1 The Momentum Transfer Cross Section $Q_m(u)$, Determined by Frost and Phelps for Electrons in Argon

u (eV)	Q_m (Å²)	u (eV)	Q_m (Å²)
0.0	8.05	0.50	0.283
0.01	6.10	0.65	0.470
0.02	3.74	0.8	0.68
0.03	2.80	1.0	1.05
0.04	2.29	1.5	1.74
0.05	1.84	2.0	2.48
0.07	1.14	3.0	4.07
0.09	0.56	4.0	5.8
0.11	0.342	6.0	8.7
0.14	0.235	8.0	11.7
0.17	0.196	10.0	13.8
0.20	0.177	12.0	14.5
0.25	0.156	15.0	13.2
0.32	0.151	20.0	10.4
0.40	0.182		

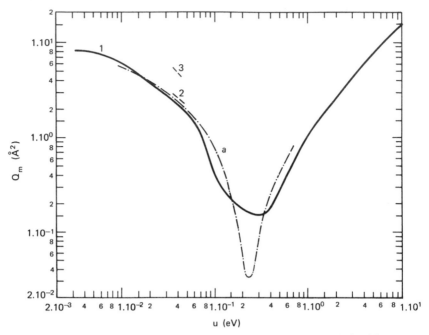

Fig. 4.4.9. Momentum transfer cross sections of electrons in argon, derived from swarm experiments [curve 1 represents the values of Table 4.4.1 of Frost and Phelps, whereas points 2 and 3 are appropriate to Nelson's and Davis's [d(2)] and Fehsenfeld's [g(1)] experimental results] and deduced from single scattering experiments (curve *a* shows Golden's results).

Fig. 4.4.9 (curve 1). Frost and Phelps state that the determination of the Q_m curve for $u > 0.7$ eV was very difficult, because of the pronounced Ramsauer minimum near 0.3 eV, and that for this reason they preferred to derive the $Q_m(u)$ curve for $u > 0.7$ eV from Errett's measurements of mobility in a mixture of 10% H_2 and 90% Ar. For these computations the hydrogen cross sections of Engelhardt and Phelps, which will be discussed in Section 4.7, were used. As shown in Figs. 4.4.3 and 4.4.4, calculations indicate that the assumed cross sections are reasonably consistent with all the results of mixture experiments.

Frost's and Phelp's Q_m curve provides satisfactory agreement also with the D_L/μ values of Wagner, Davis, and Hurst, plotted in Fig. 4.4.5 [according to the computations performed by Lowke and Parker (1969)]. Order of magnitude agreement is also obtained with the less accurate measurements of the magnetic drift velocity performed by Townsend and Bailey [see Fig. 4.4.6, where in the region above $E/N = 3 \times 10^{-17}$ V cm² the curve has been calculated taking into account also inelastic collisions for energies larger than 11.5 eV, using the inelastic cross section by Maier-Leibnitz (1935)], of the microwave conductivity ratio performed in shock tubes by Daiber and Waldron (see Fig. 4.4.7) and of the microwave pulse response in magnetic fields performed by Bruce, Crawford, and Harp [see Fig. 4.4.8, $v \geqslant 3 \times 10^7$ cm sec^{-1} only]. Cavalleri's and Loria's value $d(1)$ of the free electron diffusion is in excellent agreement with the value of 89×10^4 cm^2 sec^{-1} torr, which follows from application of the Einstein relationship to the Frost and Phelps computed value of the drift velocity at room temperature. The Q_m values of Frost and Phelps should, however, be increased by a factor of 10% (point 2 of Fig. 4.4.9) to obtain a diffusion value in agreement with the results of $d(2)$.

Fehsenfeld (1963) determined Q_m from the $g(1)$ value of $\Delta\omega_I$, using eq. (2.6.97) with $h = -1$; the result (point 3 of Fig. 4.4.9) is:

$$Q_m(u = 0.039 \text{ eV}) = 4.9 \pm 0.5 \text{ Å}^2$$

On Fig. 4.4.9 a curve (a) of Golden (1966), computed using the atomic effective range theory formulas (1.5.1), (1.5.3), and (1.5.5) and the experimental Q_t measurements of Golden and Bandel (1965) ($A = -1.65$, $B = 1.11$, $A_1 = 11.6$), is also shown. Golden claims that in a gas like argon, where there is a strong energy dependence of the cross section, a much better measure of Q_m is obtained from these precise measurements of Q_t than from the analysis of the swarm-experiment data.

4.5 KRYPTON

Experimental Data [dc parameters for $E/p < 1(D/\mu \simeq 10 \text{ eV})$]

a. Drift velocity

(1) Bowe (1960) [see Ne a(2)] performed measurements at room temperature for $E/p > 0.1$. As in the case of helium, this author used a calcium purifier, operating at 450°C, to improve the purity of the reagent-grade krypton, but no effect on w was observed.

(2) Pack, Voshall, and Phelps (1962) performed measurements at 195°K, 300°K, and 368°K over the range $2.6 \times 10^{-4} < E/p < 1$, using the double-shutter tube of Pack and Phelps (1961) and adopting the drift distances 5.08 and 10.16 cm. The impurity content of the reagent-grade krypton was less than 0.01 mole %.

All the results are given in Fig. 4.5.1; those of (2) are regarded as the most significant because of accuracy and extent of the E/p range. The results of (1) for $E/p > 0.3$ are well represented by the relation:

$$w = 2.3 \times 10^5 \left(\frac{E}{p_0}\right)^{0.25} \quad (\text{cm sec}^{-1})$$

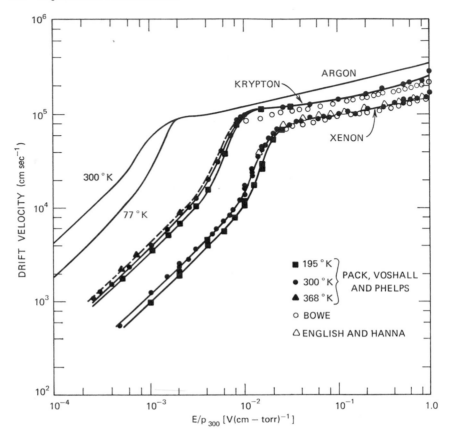

Fig. 4.5.1. △ English's and Hanna's [Xe $a(1)$], ○ Bowe's [Kr $a(1)$, Xe $a(2)$], and ■ ● ▲ Pack's, Voshall's, and Phelps's [Kr $a(2)$, Xe $a(3)$] experimental values of the drift velocities in krypton and xenon. For $E/p < 2 \times 10^{-3}$ in krypton and for $E/p < 5 \times 10^{-3}$ in xenon, electrons are in thermal equilibrium with the gas; the corresponding μp_0 values are for krypton 3.25, 3.73, and 3.98×10^6 cm^2 V^{-1} sec^{-1} torr, respectively, at 195, 300, and 368°K, and for xenon 0.89 and 1.09×10^6 cm^2 V^{-1} sec^{-1} torr, respectively, at 195 and 300°K. The drift velocity curves are similar to the curve for argon, which is shown for comparison [after Pack, Voshall, and Phelps (1962)].

We have disregarded large mobilities observed by English and Hanna (1953) [see Xe $a(1)$], which seem attributable to impurities.

b. Microwave conductivity ratio $(\omega^2 \gg v_m{}^2)$

(1) Phelps, Fundingsland, and Brown (1951) performed measurements in an S-band resonant cavity, using the reflection method to determine the characteristics of the late-afterglow isothermal plasma, contained in the

cavity and produced by a microwave discharge in room-temperature krypton; in particular they determined $\Delta(1/Q_L)$ from measurements of the power ratio at resonance (see Section 3.11B).

(2) Chen (1964) [see He $f(4)$] performed measurements in helium-krypton mixtures over the temperature range 200–600°K. The results for pure krypton are obtained by subtracting the contributions to $\omega\rho$ due to electron-ion [eq. (2.6.43)] and electron-helium collisions.

(3) Hoffmann and Skarsgard (1969) [see He $f(6)$] performed measurements over the temperature range 300–10,000°K.

The results are given in Table 4.5.1. In the above we have disregarded the results of Volkov, Zinov'ev, and Malyuta (1968), as in the case of neon, and

Table 4.5.1 Experimental Values of the Microwave Conductivity Ratio in Krypton

Authors	Temperature Range (°K)	$10^{-8}\,\omega\rho\;(\text{sec}\cdot\text{torr})^{-1}$
(1)	300	7.75
(2)	200–600	$33.3\left(\dfrac{T_e}{300}\right)^{0.5} - 37.8\left(\dfrac{T_e}{300}\right) + 12.3\left(\dfrac{T_e}{300}\right)^{1.5}$
(3)	300–10,000	See Fig. 4.5.2

the inconsistent results obtained by Bethke and Ruess (1964) in shock tubes (see Section 3.11C).

Observing the data of Table 4.5.1, we note that, within the experimental errors, the one-temperature value of (1) agrees with the corresponding value of (2); therefore (1) will not be further considered for an independent determination of the collision cross sections.

Determination of Collision Parameters

The most significant determinations of the collision cross section Q_m, derived from analyses of the experimental data reported above, are given in Tables 4.5.2 and 4.5.3 and are plotted in Fig. 4.5.3.

The greatest confidence is placed here on curve 2 of Frost and Phelps, which is derived from drift velocity data over a wide energy range and appears to be the closest curve to a hypothetical average, drawn through all the reported data. Curve a represents O'Malley's (1963) calculated results, obtained by applying the atomic effective range theory formulas (1.5.1)–(1.5.5) to Q_t and s_9 data of Ramsauer and Kollath (1929) ($A = -3.7$, $B = 1.84$, $A_1 = 12.8$).

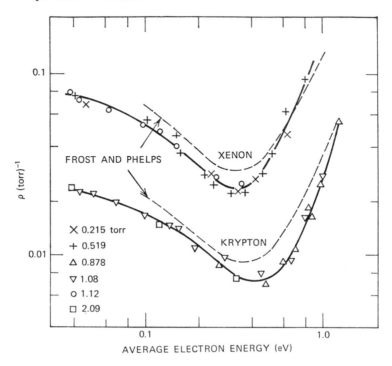

Fig. 4.5.2. Hoffmann's and Skarsgard's experimental values of the microwave conductivity ratios in krypton and xenon. The solid curves are computed using the cross-section values of Table 4.5.3, determined by the same authors [column (*b*)] for the best fit to the measured points; the dashed curves are computed using the values of column (*a*) of the same table [after Hoffmann and Skarsgard (1969)].

Calculated values of D/μ and of D_L/μ, based on the Q_m curve 2, are given in the paper by Lowke and Parker (1969).

4.6 XENON

Experimental Data [dc parameters for $E/p < 1(D/\mu \simeq 6\,\text{eV})$]

a. Drift velocity

(1) English and Hanna (1953) determined drift velocities at room temperature in pure xenon over the range $5 \times 10^{-2} < E/p < 0.5$, using their modification of the collector induced current method discussed in Section 3.2B.

(2) Bowe (1960) [see Ne *a*(2)] performed measurements at room temperature for $E/p > 0.1$. As in the case of helium, this author used a calcium

Table 4.5.2 Summary of Q_m Determinations in Krypton

Case No.	Source	Transport Parameters	Range of Swarm Mean Energies	Type of Analysis	Q_m (Å^2)
1	Bowe (1960)	a(1)	1.6–3.0	D [eqs. (2.6.11), (2.3.47), (2.3.46'), $h = 3$]	$4.0u$
2	Frost and Phelps (1964)	a(2) [195 and 300°K only]	0.026–4	A [eqs. (2.3.25), 2.4.20)]	See Table 4.5.3(a)
3	Chen (1964)	b(2)	0.026–0.078	C [eq. (2.6.49)]	$65.6 - 279u^{0.5} + 314u$
4	Hoffmann and Skarsgard (1969)	b(3)	0.039–1.3	A [eqs. (2.3.31), (2.6.40)]	See Table 4.5.3(b)

Table 4.5.3 The Momentum Transfer Cross Sections $Q_m(u)$, Determined by Frost and Phelps (a) and by Hoffmann and Skarsgard (b), for Electrons in Krypton

	Q_m (Å²)			Q_m (Å²)	
u (eV)	a	b	u (eV)	a	b
0.0	30.7		0.40	0.620	0.920
0.01	26.0		0.50	0.523	0.610
0.02	19.7		0.60	0.493	0.400
0.03	16.0		0.80	0.526	0.205
0.04	13.5	10.4	1.0	0.617	0.215
0.05	11.4	9.00	1.3	0.867	0.335
0.06	10.0	8.00	1.6	1.32	0.590
0.08	8.16	6.30	2.0	2.07	1.09
0.10	6.80	5.30	3.0	4.84	
0.13	5.33	4.20	5.0	10.2	
0.16	3.97	3.45	7.0	15.0	
0.20	2.47	2.65	10.0	19.3	
0.25	1.48	1.95	12.0	22.0	
0.30	0.983	1.50	20.0	18.0	

purifier, operating at 450°C, to improve the purity of the reagent-grade xenon, but no effect on w was observed.

(3) Pack, Voshall, and Phelps (1962) [see Kr a(2)] performed measurements at 195°K and 300°K over the range $5 \times 10^{-4} < E/p < 1$. The impurity content was the same as for krypton.

All the results are given in Fig. 4.5.1; those of (3) are regarded as the most significant because of accuracy and extent of the E/p range. The results of (2) for $E/p > 0.2$ are well represented by the relation:

$$w = 1.6 \times 10^5 \, (E/p_0)^{0.25} \quad (\text{cm sec}^{-1})$$

The results of (1) are slightly larger than those of the other authors; this can be due to impurities, since the mobilities of English and Hanna for many other gases have also been found to be too large for this reason.

b. Microwave conductivity ratio $(\omega^2 \gg \nu_m^2)$

(1) Phelps, Fundingsland, and Brown (1951) [see Kr b(1)] performed measurements in late isothermal afterglows of discharges produced in room-temperature xenon.

(2) Chen (1964) [see He f(4) and Kr b(2)] performed measurements in helium-xenon mixtures over the temperature range 200–600°K.

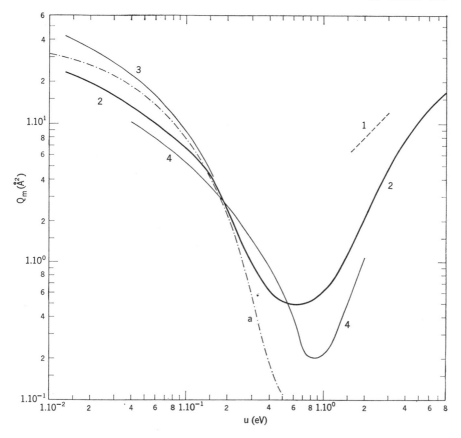

Fig. 4.5.3. Momentum transfer cross sections of electrons in krypton, derived from swarm experiments (numbers as in Table 4.5.2) and deduced from single scattering experiments (curve *a* shows O'Malley's results).

(3) Hoffmann and Skarsgard (1969) [see He $f(6)$] performed measurements over the temperature range 300–6200°K.

(4) Coffey (1970) performed 4.9-GHz propagation measurements for the determination of the conductivity ratio in late isothermal afterglows of pulsed dc xenon discharges; for this purpose the author used a slotted-line bridge. In addition to measuring ω_p at 300°K, Coffey performed measurements of the conductivity ratio for electron temperatures from 1500°K to 8000°K, electron heating having been produced by a 4.6-GHz microwave signal of adequate power. At the same time and in each case, radiation temperatures were observed (here $T_r = T_e$, the distributions being Maxwellian).

Table 4.6.1 Experimental Values of the Microwave Conductivity Ratio in Xenon

Authors	Temperature Range (°K)	$10^{-9}\,\omega\rho$ (sec · torr)$^{-1}$
(1)	300	2.6
(2)	200–600	$9.68\left(\dfrac{T_e}{300}\right)^{0.5} - 11.24\left(\dfrac{T_e}{300}\right) + 3.70\left(\dfrac{T_e}{300}\right)^{1.5}$
(3)	300–6200	See Fig. 4.5.2
(4)	$\begin{cases} 300 \\ 1500\text{–}8000 \end{cases}$	$\begin{aligned} &1.88 \\ &128T_e^{-0.5} - 4.85 + 5.77 \times 10^{-2}T_e^{0.5} \end{aligned}$

The impurity content of the gas used was of the order of 100 ppm. The scatter of the individual $\omega\rho$ measurements is often larger than 10%.

The results are given in Table 4.6.1. As in the case of krypton and for the same reasons, the measurements of Volkov, Zinov'ev, and Malyuta (1968) and of Bethke and Ruess (1964) are not considered here.

c. Thermal diffusivity. Nygaard (1967b) (see He *h*) performed measurements in xenon, but unfortunately, whereas the $\mathfrak{D}p_0$ results at the lowest electron densities fall in the correct range, a physically appropriate curve cannot be drawn through the data plotted as a function of the electron density; therefore, these results will not be further considered for the derivation of the electron collision parameters.

d. Cyclotron resonance

(1) Narasinga Rao, Verdeyen, and Goldstein (1961) applied to an afterglow xenon plasma their method, based on the determination of the σ_{Tr} versus ν_m maximum, as described in Sections 2.6D and 3.14A. However, these authors do not report any experimental result, but state only that the method yielded collision frequency values in good agreement with published ones.

(2) Coffey (1970) performed measurements of the line half-width $\Delta\omega_{0.5}$ and of the displacement from resonance $\delta\omega_s$, which corresponds to the σ_{Tr} versus ν_m maximum. For the purpose of these measurements Coffey used a C-band frequency-modulated pulse, as discussed in Section 3.14A, sent through the late isothermal afterglow of a pulsed dc xenon discharge; the plasma waveguide housing was positioned in the center of a solenoid with a longitudinal field variation of 1.5%. A transient microwave radiometer was included for the determination of the radiation temperatures. The results

for a 300°K plasma are:

$$\Delta\omega_{0.5} = (1.73 \pm 0.03) \times 10^9 p_0 \quad (\text{sec}^{-1})$$
$$\delta\omega_s = (1.8 \pm 0.3) \times 10^9 p_0 \quad (\text{sec}^{-1})$$

Coffey has also determined $\Delta\omega_{0.5}$ values in the presence of microwave heating; his method appears open to criticism, however, being based on formulas appropriate to the resonance line shapes for the constant ν_m case only. These data will not be further discussed here.

Determination of Collision Parameters

The most significant determinations of the collision cross section Q_m, derived from analyses of the experimental data reported above, are given in Tables 4.6.2 and 4.6.3 and are plotted in Fig. 4.6.1. For the reduction of the room-temperature $\omega\rho$ value $b(1)$, it has been assumed that $h = -0.5$, in accordance with the slope of the Q_m curve 2 at 0.039 eV.

The greatest confidence is placed here on curve 2 of Frost and Phelps, which is derived from drift velocity data over a wide energy range and appears to be the closest curve to a hypothetical average, drawn through all the reported data. Curve a represents O'Malley's (1963) calculated results, obtained by applying the atomic effective range theory formulas (1.5.1)–(1.5.5) to Q_t and the s_3 data of Ramsauer and Kollath (1929) ($A = -6.5$, $B = 6.10$, $A_1 = 23.2$).

Calculated values of D/μ and of D_L/μ, based on the Q_m curve 2, are given in the paper of Lowke and Parker (1969).

4.7 HYDROGEN

Experimental Data [dc parameters for $E/p < 10(D/\mu \simeq 1 \text{ eV})$]

a. Drift velocity

(1) Bradbury and Nielsen (1936) performed measurements at 293°K for $E/p > 3 \times 10^{-2}$, using their double-shutter method ($d = 5.93$ cm); the estimated accuracy from the scatter of the experimental points is of the order of 2–3%. Commercial tank gas, purified by passing over hot copper gauze and through a series of liquid-air traps, was used.

(2) Pack and Phelps (1961) [see He $a(3)$] performed measurements at 77, 195, 300 $\pm$ 3, and 375°K over the range $2.5 \times 10^{-4} < E/p < 10$. The most important impurity contents of the reagent-grade hydrogen were 0.02 and 0.01 mole % of nitrogen and hydrocarbons, respectively.

(3) Lowke (1963) performed measurements at 77.6°K over the range $1 \times 10^{-3} < E/p < 3$ and at 293°K for $E/p > 4 \times 10^{-3}$, using a double-shutter tube with a drift distance of ~6 cm; the estimated accuracy of the

Table 4.6.2 Summary of Q_m Determinations in Xenon

Case No.	Source	Transport Parameters	Range of Swarm Mean Energies	Type of Analysis	Q_m (Å²)
1	Bowe (1960)	a(2)	1–2.4	D [eqs. (2.6.11), (2.3.47), (2.3.46′), $h = 3$]	$7.4u$
2	Frost and Phelps (1964)	a(3)	0.026–3	A [eqs. (2.3.25), (2.4.20)]	See Table 4.6.3(a)
3	· · ·	b(1)	0.039	D [eq. (2.6.41), $h = -0.5$]	66.8
4	Chen (1964)	b(2)	0.026–0.078	C [eq. (2.6.49)]	$191 - 830u^{0.5} + 940u$
5	Hoffmann and Skarsgard (1969)	b(3)	0.039–0.8	A [eqs. (2.3.31), (2.6.40)]	See Table 4.6.3(b)
6	Coffey (1970)	b(4)	0.20–1.0	C [eq. (2.6.49)]	$19.7 - 21.4u^{-0.5} + 7.53u^{-1}$
7	Coffey (1970)	b(4), a(2)	0.039	D [eqs. (2.6.41), (2.6.97), (2.6.105), $h = -0.7$]	49.0

Table 4.6.3 The Momentum Transfer Cross Sections $Q_m(u)$, Determined by Frost and Phelps (a) and by Hoffmann and Skarsgard (b), for Electrons in Xenon

u (eV)	Q_m (Å^2) (a)	(b)	u (eV)	Q_m (Å^2) (a)	(b)
0.0	176		0.50	1.38	0.840
0.01	116		0.60	1.28	0.610
0.02	80		0.80	1.61	0.670
0.03	61.3		1.0	2.47	1.30
0.04	48.0	35.5	1.3	3.90	2.95
0.05	39.5	32.0	1.6	5.60	5.50
0.06	33.5	28.0	2.0	8.25	9.50
0.08	25.6	22.0	3.0	17.0	
0.10	20.4	17.8	4.0	24.8	
0.13	15.1	13.2	5.0	30.8	
0.16	12.0	10.0	6.5	32.0	
0.20	8.40	7.50	8.0	33.7	
0.25	5.35	5.25	10.0	32.0	
0.30	3.15	3.70	16.0	25.2	
0.40	1.75	1.70	20.0	20.5	

measurements is $\pm 1\%$ at 293°K and $\pm 2\%$ at 77.6°K. This author gave particular attention to the effect of impurities and verified, in accordance with theory, that impurities of more than 1% of nitrogen are required to change the drift velocity results in hydrogen by 1%; this result indicates that nitrogen impurities, the ones most likely to be present, but always in amounts well below 1%, will not have any significance for the measurements. Furthermore, Lowke used a silver-palladium alloy osmosis tube [Crompton and Elford (1962)] to admit the gas; according to Young (1963) this method can produce hydrogen with an impurity level of a few parts in 10^{10}.

(4) Hurst and Parks (1966) used a drift chamber equipped with a Geiger-Müller counter, as described in Section 3.4E. Measurements were performed at room temperature over the range $4 \times 10^{-2} < E/p < 1$.

(5) Prasad and Smeaton (1967) used a double-shutter drift tube, similar to the one designed by Pack, Phelps, and Frost (1960), with a fixed drift distance of ~ 7 cm and very short gating pulses. Measurements were performed at 293°K for $E/p > 2.5$ in hydrogen with a 99.98% purity. The estimated accuracy of the w measurements is $\pm 1\%$.

(6) Wagner, Davis, and Hurst (1967) [see He a(6)] performed measurements at room temperature over the range $4 \times 10^{-2} < E/p < 2.5$. The impurity content of hydrogen was less than 0.001 mole %.

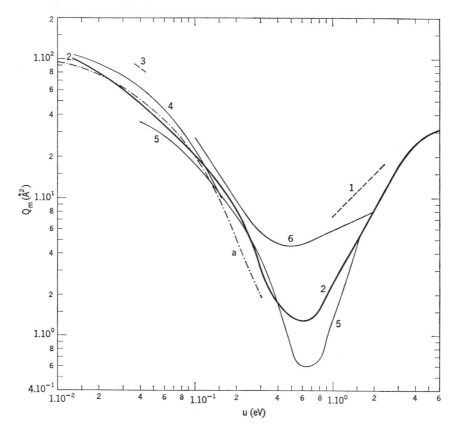

Fig. 4.6.1. Momentum transfer cross sections of electrons in xenon, derived from swarm experiments (numbers as in Table 4.6.2) and deduced from single scattering experiments (curve a shows O'Malley's results). Coffey's result (case 7) is not represented since it falls exactly on curve 2.

(7) Grünberg (1967, 1968) [see He a(7)] performed measurements at room temperature and at pressures from 1 to 41 atm, for $E/p > 2 \times 10^{-3}$ in hydrogen purified by passage through a palladium filter.

(8) Robertson (1971) performed measurements up to $E/p = 8.6$ in normal hydrogen and in pure parahydrogen at 76.8°K for $E/p > 2.6 \times 10^{-3}$ and in normal hydrogen at 293°K for $E/p > 0.33$; a small bakable double-shutter tube ($d = 5$ cm typically), appropriate for low-temperature measurements in parahydrogen (see Section 3.4C), was used. The purity of the normal hydrogen was as in (3) above. Parahydrogen, whose fractional content in normal hydrogen is only 25%, was obtained with a content not less than 98% using

a closed conversion system operating at liquid-helium temperatures in the presence of a paramagnetic catalyst (activated iron oxide); the hydrogen to be converted was admitted in the system through a silver-palladium alloy osmosis tube. Measurements for $E/p > 0.3$ are corrected for diffusion according to relation (4.2.1); the estimated accuracy is better than 1 %. In parahydrogen the data are in excellent agreement (within 0.5 %) with previous results similarly obtained in the same laboratory by Crompton and McIntosh (1968).

All the results for normal hydrogen are in good agreement within the combined experimental errors, except for high-pressure deviations to be discussed later. The results of (3) and (8), which the writer regards as the most significant because of accuracy, consistency, and gas temperature range, are summarized, together with the results of (8) for pure parahydrogen, in Table 4.7.1. At room temperature (4) and (7) find velocities slightly above those of the table at low E/p values, and slightly below at large E/p values; (6), on the other hand, finds velocities always slightly above those of the table. The linear dependence of w on E/p is found for $E/p < 1 \times 10^{-3}$; then the most significant μp_0 data for isothermal electrons are those of (2), which are reported for future discussion in Table 4.7.5.

A significant pressure variation of the drift velocity at fixed E/p and T_g values was first disclosed by the high-pressure results of (7) at room temperature; this effect has been shown [Crompton and Robertson (1971)] to be present also in the 77°K measurements of (8) at pressures below 1 atm. Since at these pressures molecular densities are of the order of 10^{18}–10^{19} cm^{-3}, the effect appears also when $l_m \gg \lambda_e$, so that the explanation must be different from that for the case of helium and must be related in some way to the molecular properties of hydrogen. A possible interpretation of this type will be discussed at the end of the section, after having derived from the swarm data the electron scattering characteristics of hydrogen molecules. Here we note only that the pressure-dependent drift velocity can be represented [Frommhold (1968)] by an equation of the form:

$$w = \frac{w_0}{1 + \alpha N} \qquad (4.7.1)$$

where w_0 is the zero-pressure limit of the drift velocity. The 77°K drift velocities of Table 4.7.1 are extrapolated w_0 values in all cases in which the pressure effect was noticeable ($E/p < {\sim}0.15$). The α values in normal hydrogen and in parahydrogen at 77°K, which result from the data of (8), are the most accurate and valuable for interpreting the effect; they are plotted in Fig. 4.7.1 as a function of the corresponding D/μ values.

Table 4.7.1 Experimental Values of the Drift Velocity and of the Magnetic Deflection Coefficient in Hydrogen

Drift velocity values in normal hydrogen for $E/N \geqslant 1 \times 10^{-17}$ V cm² are from Robertson [a(8)], whereas values below this E/N are from Lowke [a(3)].

$10^{17}E/N$ (V cm²)	E/p_{293} (V cm⁻¹ torr⁻¹)	Normal Hydrogen $10^{-5}w$ (cm sec⁻¹) [$T_g = 293°K$; a(3) and a(8)]	Normal Hydrogen $10^{-5}w$ (cm sec⁻¹) [$T_g = 77°K$; a(8)]	Parahydrogen $10^{-5}w$ (cm sec⁻¹) [$T_g = 77°K$; a(8)]	Normal Hydrogen Ψ [$T_g = 293°K$; c(3)]
9.10×10^{-3}	3.0×10^{-3}	1.88×10^{-1}	3.05×10^{-1}	3.06×10^{-1}	
1.214×10^{-2}	4.0	2.83	3.87	3.90	
1.82	6.0	3.74	5.30	5.39	
2.43	8.0	4.63	6.52	6.68	
3.03	1.0×10^{-2}	6.76	7.63	7.84	
4.55	1.5	8.78	1.00×10^{0}	1.05×10^{0}	1.21
6.07	2.0	1.26×10^{0}	1.22	1.31	1.18
9.10	3.0	1.59	1.62	1.78	1.17
1.214×10^{-1}	4.0	2.18	1.98	2.21	1.16
1.82	6.0	2.70	2.61	2.97	1.16
2.43	8.0	3.15	3.16	3.61	1.15
3.03	1.0×10^{-1}	4.09	3.63	4.14	1.15
4.55	1.5	4.85	4.62	5.18	1.16
6.07	2.0	5.97	5.36	5.92	1.15
9.10	3.0	6.23	6.46	6.95	1.16
1.0×10^{0}	3.30		6.71	7.15	1.16
2.0	6.60	8.37	8.70	8.93	1.16
4.0	1.32×10^{0}	1.15×10^{1}	1.16×10^{1}	1.17×10^{1}	1.13
6.0	1.98	1.41	1.42	1.43	1.12
8.0	2.64	1.65	1.66	1.67	1.11
1.0×10^{1}	3.30	1.87	1.88	1.89	1.10
1.4	4.62	2.27	2.28	2.28	1.08
2.0	6.60	2.81	2.81	2.81	1.07
2.6	8.58	3.30	3.30	3.31	1.06

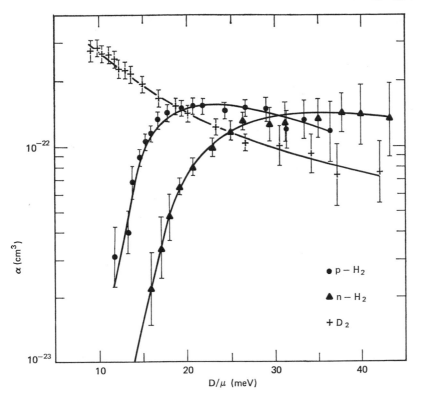

α (cm^3)

D/μ (meV)

• p – H$_2$

▲ n – H$_2$

+ D$_2$

Fig. 4.7.1. The slopes α of the experimental w_0/w versus N curves [see eq. (4.7.1)] as a function of D/μ for hydrogen, parahydrogen, and deuterium at 77°K [after Crompton and Robertson (1971)].

b. Diffusion to mobility ratio

(1) Townsend and Bailey (1921) performed measurements at room temperature for $E/p > 0.2$, using the 4-cm diffusion chamber described under He b(1). Consistent values (within 5% of the original ones) were later obtained when the measurements were repeated with both 2- and 4-cm diffusion chambers [Townsend and Bailey (1922), Brose (1925)].

(2) Crompton and Sutton (1952) performed measurements at room temperature for $E/p > 5 \times 10^{-2}$. The quantity that they measured directly is (see Section 3.5) the ratio of the ratios $R(b)/R(c)$ in a long ($d = 2.0$ cm, $b/d = 0.25$ or 0.5, $c/d = 1.3$) and in a short ($d = 1.0$ cm, $b/d = 0.5$, $c/d = 1.5$) diffusion tube; Huxley's relationship (3.5.10) was used to derive D/μ from the measured current ratios. Measurements obtained with the two types of apparatus agree to within 3%. Hydrogen was admitted into the system through the walls of a heated palladium tube.

(3) Hall (1955a) used the techniques of Crompton and Sutton to perform room-temperature measurements of D/μ, and of $D_T/\mu_{\parallel}$ at various magnetic fields, over the range $2 \times 10^{-1} < E/p < 1$, using strip and annular electrodes tubes ($d = 2$ cm, $b/d = 0.25$). Hall (1955c) also attempted for the first time to perform the experiment at low temperatures (81°K), extending measurements down to $E/p \simeq 2 \times 10^{-2}$.

(4) Cochran and Forester (1962) also determined, using eq. (3.5.10), D/μ values at room temperature over the range $0.2 < E/p < 5$ from measurements of the ratio $R(b)/R(c)$ in a diffusion tube ($d = 3.0$ cm, $b/d = 0.1, 0.2$, 0.3, or 0.5, $c/d = 1.5$). In a purifier unit oxygen impurities in the hydrogen were catalytically combined with the hydrogen itself to form water; the gas was then passed through a drying tube and liquid-air traps before being admitted into the system.

(5) Warren and Parker (1962) [see He $b(2)$] performed measurements at 77°K over the range $2 \times 10^{-4} < E/p < 3$.

(6) Crompton, Elford, and McIntosh (1968) performed measurements at 77.3°K over the range $6.6 \times 10^{-4} < E/p < 4$ and at 293°K over the range $6 \times 10^{-3} < E/p < 2$, choosing the experimental conditions, equipment, and techniques that are capable of providing the most accurate results according to the discussion of Sections 3.5, 3.6, and 3.7. Thus hydrogen was obtained by diffusion through a silver-palladium alloy osmosis tube [see $a(3)$ above]; a diffusion chamber similar to the one of Fig. 3.7.1, but with the collecting electrode properly modified for operation at 77°K, was used; contact potential differences were accurately compensated for by the application of an appropriate voltage; and space-charge effects were corrected by linearly extrapolating data to zero current. Since these measurements are the most accurate (the authors place on them a conservative error limit of $\pm 2\%$), they supersede all other results previously obtained by the same Canberra group [Crompton and Jory (1962), Crompton and Elford (1963)]. These earlier results actually show excellent agreement with the latest values; in particular, those of Crompton and Jory at 293°K complement them up to $E/p = 5$.

(7) Naidu and Prasad (1968) performed measurements at room temperature for $E/p > 7$ in a diffusion tube ($d = 7$ cm), whose collector plate was divided into a central disk ($b = 0.5$ cm, $b/d \simeq 0.07$) and three concentric annuli (radii 1.0, 1.5, and 2.0 cm, respectively); D/μ values were derived according to eq. (3.5.10) from the current ratios to adjacent pairs of these elements, a proper correction being introduced to take into account the large size of the cathode hole source (radius 0.25 cm). The overall accuracy of the data is estimated to be better than $\pm 2\%$. The stated purity of hydrogen was 99.98%.

(8) Crompton and McIntosh (1968) performed measurements in pure

parahydrogen, prepared with the techniques described in $a(8)$ above, at 77°K over the range $6.6 \times 10^{-4} < E/p < 2$. They used the same experimental techniques adopted by (6), but replaced the thermionic electron source with a radioactive source (as explained in Section 3.7).

(9) Wagner, Davis, and Hurst (1967) [see He $b(4)$] determined the quantity D_L/μ at room temperature over the range $4 \times 10^{-2} < E/p < 2$. For the impurity content of the gas, see $a(6)$ above. Earlier, similar measurements, due to Hurst and Parks (1966) [see also $a(4)$], agree well with the more recent results.

The most accurate measurements in normal hydrogen at 77°K and at room temperature are those of (6), which are summarized in Table 4.7.2, together with those of (7) for the points $E/p = 7$ and 10. The results of (1), and of (2) for $E/p > 1$, are in good agreement with these data within the combined experimental errors. Below $E/p = 1$ the results of (2) and (3) are in very good agreement between themselves, but are systematically a few percentages lower than those of the table; it is possible that the data of (2) and (3) are inaccurate because the relatively large b/d ratios of the tubes employed should have required low field strengths, and no correction for the effects of contact potential differences was in use at that time. Values of D/μ much larger than those of all other authors were found by (4); no satisfactory explanation of this discrepancy has yet been given. Whereas the low-temperature experiment of (3) failed to show any significant change from the room-temperature data, the results of (5), the only other available data for 77°K, agree with those of Table 4.7.2, but only to within 15–20%.

The interpolated results of (8) in pure parahydrogen are also given in the Table 4.7.2. The D_L/μ room-temperature results of (9) are plotted in Fig. 4.7.2.

c. Magnetic drift velocity (LMF case)

(1) Townsend and Bailey (1921) performed measurements using the original Townsend magnetic deflection method and the tube and conditions discussed under $b(1)$ above.

(2) Hall (1955a) used the modified Townsend method and the same tube as in the $b(3)$ experiment to perform room-temperature measurements over the range $5 \times 10^{-2} < E/p < 4$.

(3) Creaser (1967) adopted Jory's tube (Fig. 3.9.6) and the most advanced techniques for measurements at 293°K over the range $2 \times 10^{-2} < E/p < 9$; two flux densities, 40 and 60 G, were used. Hydrogen was admitted to the apparatus through a heated palladium osmosis tube and a liquid-nitrogen trap. The author estimates a maximum error of 2–3%.

Table 4.7.2 Experimental Values of the Characteristic Energy in Hydrogen

Values in normal hydrogen at $T_g = 293°K$ are from Crompton, Elford, and McIntosh [b(6)] for $E/p_{293} \leqslant 2$, from Crompton and Jory [b(6)] for $3 \leqslant E/p_{293} \leqslant 5$, and from Naidu and Prasad [b(7)] for $E/p_{293} \geqslant 7$.

$10^{17}E/N$ (V cm^2)	E/p_{293} (V cm^{-1} torr^{-1})	Normal Hydrogen D/μ (eV) [$T_g = 293°K$; b(6) and b(7)]	Normal Hydrogen D/μ (eV) [$T_g = 77°K$; b(6)]	Parahydrogen D/μ (eV) [$T_g = 77°K$; b(8)]
2.0×10^{-3}	6.60×10^{-4}		6.76×10^{-3}	6.83×10^{-3}
3.0	9.90		6.85	6.95
4.0	1.32×10^{-3}		6.96	7.05
6.0	1.98		7.23	7.30
8.0	2.64		7.55	7.61
1.0×10^{-2}	3.30		7.90	7.95
1.4	4.62		8.67	8.66
1.82	6.0	2.58×10^{-2}	9.50	9.48
2.43	8.0	2.62	1.07×10^{-2}	1.05×10^{-2}
3.03	1.0×10^{-2}	2.65	1.18	1.15
4.55	1.5	2.75	1.44	1.36
6.07	2.0	2.85	1.66	1.51
9.10	3.0	3.08	2.01	1.75
1.214×10^{-1}	4.0	3.30	2.30	1.95
1.82	6.0	3.75	2.80	2.32
2.43	8.0	4.18	3.24	2.69
3.03	1.0×10^{-1}	4.59	3.66	3.05
4.55	1.5	5.62	4.70	3.98
6.07	2.0	6.68	5.77	5.01
9.10	3.0	8.88	7.93	7.18
1.214×10^{0}	4.0	1.11×10^{-1}	1.02×10^{-1}	9.45
1.82	6.0	1.55	1.47	1.40×10^{-1}
2.43	8.0	1.96	1.88	1.84
3.03	1.0×10^{0}	2.33	2.26	2.22
4.55	1.5	3.10	3.04	3.02
6.07	2.0	3.74	3.69	3.68
9.10	3.0	4.87	4.77	
1.214×10^{1}	4.0	5.78	5.69	
1.517	5.0	6.62		
2.125	7.0	7.8		
3.03	1.0×10^{1}	1.02×10^{0}		

The above w_M results can be divided by the drift velocity data of Table 4.7.1 to obtain the magnetic deflection coefficient Ψ. This quantity, for the data of (3), is also given in Table 4.7.1; the results of (1) intersect those of (3) at $E/p \simeq 2$; above this value they are slightly smaller and below are slightly larger, the poorest agreement (5%) taking place at the lowest E/p (= 0.25). The results of (2) are significantly lower than those of the table;

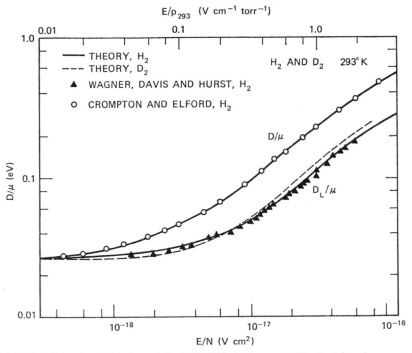

Fig. 4.7.2. Experimental values of D_L/μ [b(9)] in hydrogen at 293 K; D/μ values measured at 293°K by Crompton and Elford[b(6)] are shown for comparison. The curves for hydrogen and deuterium are computed using the Engelhardt and Phelps electron collision cross sections [after Lowke and Parker (1969)].

probably these data must be rejected, on the ground of the criticism already made concerning Hall's D/μ results and of the fact that she did not correct for all secondary effects, which were properly considered, however, by Creaser (see Section 3.9A).

d. Magnetic drift velocity (SMF case). Bernstein (1962) (see He *d*) performed measurements at room temperature in hydrogen from standard tanks (0.19% N_2, 0.04% O_2) without using any particular purification process. The results are plotted in Fig. 4.7.3 as $e/m\mu_M p_0$ (corresponding to $\omega\rho$ in the microwave case) versus E/ω_b.

e. Free electron diffusion. Nelson and Davis (1969) [see He *e*(2)] obtained $Dp_0 = (121 \pm 2) \times 10^3$ cm² sec⁻¹ torr at 300 ± 2°K in hydrogen with a minimum purity of 99.999%.

f. Diffusion in crossed fields. Bernstein (1962) performed measurements of the $D_{||}/\mu_T$ ratio as a function of E/ω_b, using the setup shown in Fig. 3.9.7.

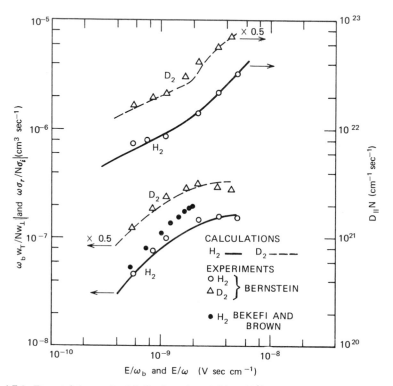

Fig. 4.7.3. Bernstein's results (d, f) plotted as $3.54 \times 10^{16} D_{\parallel}p_0$ ($= D_{\parallel}N$) and $e/(3.54 \times 10^{16})m\mu_M p_0$ ($= \omega_b w_T/Nw_{\perp}$) versus E/ω_b for hydrogen and deuterium; Bekefi's and Brown's (g) $\omega\rho/3.54 \times 10^{16}$ data ($= -\omega\sigma_r/N\sigma_i$) are also given as a function of E/ω for hydrogen. The curves are computed using the Engelhardt and Phelps electron collision cross sections [after Engelhardt and Phelps (1963)].

Combining these values with the μ_T data that he had derived from the corresponding measurements of magnetic drift velocities previously reported (see d above), Bernstein computed $D_{\parallel}$. In Fig. 4.7.3 representative $D_{\parallel}p_0$ values are plotted as a function of E/ω_b.

g. Microwave conductivity ratio. We shall consider only Bekefi's and Brown's (1958) extensive measurements of this parameter in late-afterglow hydrogen plasmas. Their data supersede earlier results obtained using less sensitive techniques at low pressures ($\omega^2 \gg \nu_m{}^2$) in less pure hydrogen by Phelps, Fundingsland, and Brown (1951), [see Kr $b(1)$; preliminary value, $\omega\rho = 6.6 \times 10^8$ (sec · torr)$^{-1}$ at 300°K] and by Varnerin (1951) (transient slotted-line measurements of the standing-wave-pattern variations due to an

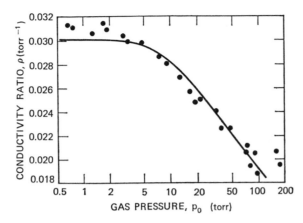

Fig. 4.7.4. Bekefi's and Brown's conductivity ratio results (g) as a function of the gas pressure; the solid line is computed assuming $Q_m = 32.7 \, u^{0.3}$ (Å²) [after Bekefi and Brown (1958)].

afterglow plasma filling a $\lambda_g/4$ long section of a short-circuited S-band waveguide).

Bekefi and Brown used very pure hydrogen, obtained by decomposition of uranium hydride [Rose (1956)], to fill a cubic quartz bottle centered in a parallelepiped S-band cavity [see He $f(1)$]. They performed three sets of experiments by measuring the conductivity ratio as a function of, respectively, (1) the gas pressure at 300°K, (2) the gas temperature from 300°K to 600°K at low pressures ($\nu_m^2 \ll \omega^2$), and (3) the rms value of the microwave heating field (averaged over the plasma volume), applied to an isothermal (300°K) low pressure ($\nu_m^2 \ll \omega^2$) plasma. The results of (1) and (3) are plotted in Figs. 4.7.4 and 4.7.5, respectively; the results of (2) can be represented by the expression:

$$\omega\rho = 5.30 \times 10^8 (T_g/300)^{0.75} \quad (\text{sec} \cdot \text{torr})^{-1}$$

For comparison purposes some of the results of (3) are also shown in Fig. 4.7.3.

h. Cyclotron resonance. Tice and Kivelson (1967) [see He $i(3)$] performed line-shape measurements in hydrogen-nitrogen mixtures at room temperature.

Determination of Collision Parameters

(1) Bekefi and Brown (1958) applied:
(a) an analysis of type C [eq. (2.6.41) and the curves of Fig. 2.6.3, or eq. (2.6.44)] to the room-temperature ρ versus p_0 data of Fig. 4.7.4, thus

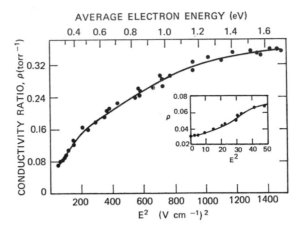

Fig. 4.7.5. Bekefi's and Brown's conductivity ratio results (g) as a function of the rms value of the microwave heating field [after Bekefi and Brown (1958)].

obtaining over a narrow range around thermal energies:

$$Q_m = 23.7u^{0.3} \text{ Å}^2 \quad (u \text{ in eV})$$

(b) an analysis of type C [eq. (2.6.49)] to the above reported ρ versus T_g relationship (for the temperature range 300–600°K), thus obtaining for thermal electron energies:

$$Q_m = 20.5u^{0.25} \text{ Å}^2 \quad (u \text{ in eV})$$

(c) an analysis of type C [systems of eqs. (2.6.13) and (2.6.20); (2.6.41) and (2.6.58), (2.6.19)] to the ρ versus E^2 data of Fig. 4.7.5, together with Crompton and Sutton's D/μ versus E/p data [see b(2)].

The Bekefi and Brown curve of Q_m versus u, which is shown in Fig. 4.7.6, represents an average of (a) and (b) results below 0.35 eV (estimated accuracy ±3%), and represents the (c) results above 0.70 eV (estimated accuracy ±7%); in between the curve is extrapolated.

(2) Engelhardt and Phelps (1963) [see also Frost and Phelps (1962)] per-formed a numerical analysis of type A in order to find the best fit to the average curves of the drift velocity at 77°K [based on the 77°K data of a(2) and of the room-temperature data of a(1) for $E/p > 1$] and of the charac-teristic energy at 77°K [based on the 77°K data of b(5) and on the room-temperature data of b(1) for $E/p > 1$ and of b(2) for $E/p > 4$].†

† According to Tables 4.7.1 and 4.7.2, the transport parameters for $E/p > 1$ become independent of the gas temperature within the various authors' experimental errors, so that measurements performed at room temperature can be attributed to 77°K also.

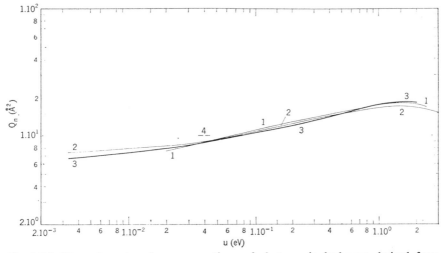

Fig. 4.7.6. Momentum transfer cross sections of electrons in hydrogen derived from swarm experiments [(1) Bekefi and Brown, (2) Engelhardt and Phelps, (3) Crompton, Gibson, and McIntosh, (4) Tice and Kivelson].

Below $D/\mu \simeq 0.08$ eV only rotational transitions are of importance [the threshold for the lowest vibrational transition $v(0 \to 1)$ is 0.52 eV]; in normal hydrogen, cooled at 77°K without modifying the room-temperature para/ ortho ratio, the populations of the rotational levels are as follows: $\alpha_0 = 0.2487$, $\alpha_1 = 0.7500$, $\alpha_2 = 0.0013$, so that only two cross sections for inelastic collisions of the first kind are of importance: Q_{02} and Q_{13}. Engelhardt and Phelps assumed that, except for a common multiplying factor M_R, q_{02}, and q_{13} are given by Gerjuoy's and Stein's expression (1.5.18), modified by the polarization factor of Dalgarno and Moffett; $\mathcal{Q} = 0.473$ is taken on the basis of the measurements of Harrick and Ramsey (1952). The collision cross section $Q_m(u)$ and the factor M_R are determined by successive adjustments to initial estimates until good agreement is secured between the experimental values of w and D/μ and the values computed using (2.4.20) and (2.6.1), where f^0 is obtained by solving numerically eq. (2.3.51) through the linear equation system (2.3.55). Actually the experimental and theoretical values can be compared more conveniently by replacing w and D/μ with the derived quantities v_m^* and v_u^* [see eqs. (2.6.6) and (2.6.7)], since v_m^* discrepancies are minimized by adjusting $Q_m(u)$, whereas v_u^* discrepancies are minimized by adjusting inelastic cross sections (see the discussion in Section 2.6A). The resulting final Q_m cross section is given in Table 4.7.3 and plotted in Fig. 4.7.6; the resulting M_R factor for the rotational cross sections is approximately 1.5.

Table 4.7.3 The Momentum Transfer Cross Sections $Q_m(u)$, Determined by Engelhardt and Phelps (a) and by Crompton, Gibson, and McIntosh (b), for Electrons in Hydrogen

u (eV)	Q_m (Å²) (a)	Q_m (Å²) (b)	u (eV)	Q_m (Å²) (a)	Q_m (Å²) (b)
0	7.20	6.4	0.40	14.2	13.9
0.01	7.90	7.3	0.50	14.8	14.7
0.02	8.35	8.0	0.60	15.2	15.6
0.03	8.60	8.5	0.80	16.0	16.7
0.04	8.90	8.96	1.0	16.6	17.4
0.05	9.30	9.28	1.3	16.9	18.1
0.06	9.55	9.56	1.6	17.0	18.3
0.08	10.1	10.1	2.0	16.7	18.0
0.10	10.6	10.5	3.0	15.3	
0.13	11.3	11.0	4.0	14.0	
0.16	11.9	11.5	6.0	11.7	
0.20	12.5	12.0	8.0	10.2	
0.25	13.0	12.5	10.0	9.10	
0.30	13.5	13.0			

Above $D/\mu \simeq 0.08$ eV vibrational transitions must also be considered. Engelhardt and Phelps have investigated the region up to $D/\mu \simeq 1$ eV, assuming that only the cross section for excitation of the $v = 1$ level was significant. Their analysis is based on taking for Q_m, Q_{02}, and Q_{13} the results of the lower energy analysis, on assuming as the initial estimate for the vibration excitation cross section Q_v Ramien's (1931) experimental results, and on determining the Q_v curve that provides the best fit with w and D/μ experimental values. In these calculations the term for the collisions of the second kind in eq. (2.3.51) is neglected, and the equation is solved by the method of backward prolongation. The resulting Q_v curve is plotted in Fig. 4.7.7.

Engelhardt and Phelps have also used their cross sections to compute $e/m\mu_M p_0$ (or $\omega\rho$) and $D_{||}p_0$ versus E/ω_b (or E/ω); the results, plotted in Fig. 4.7.3, show good agreement with the SMF data of Bernstein (d and f), but not with the microwave data of Bekefi and Brown (g). With the same cross sections Ψ has been computed; the results, plotted in Fig. 4.7.8, although showing fair agreement with the data of Townsend [$c(1)$], are larger than those of Creaser [$c(3)$], more than the overall experimental scatter of the data would permit.

The work of Engelhardt and Phelps was extended by Lowke and Parker (1969), who used these authors' cross sections (except for $\mathcal{Q}$, which was taken to be 0.56) for computing D/μ and D_L/μ values in room-temperature (293°K)

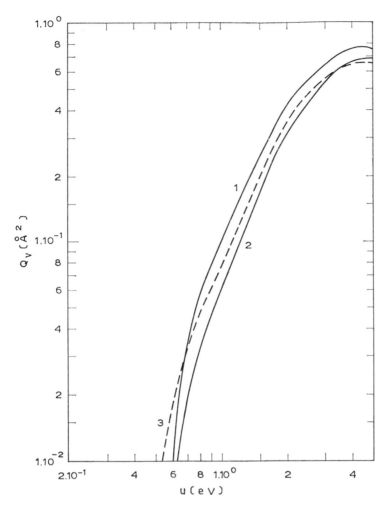

Fig. 4.7.7. Vibrational excitation cross sections, derived from swarm experiments, for electrons in hydrogen [(1) Engelhardt and Phelps, (2) Crompton, Gibson, and McIntosh] and for electrons in deuterium [(3) Engelhardt and Phelps].

hydrogen. Since at this temperature the populations of the rotational levels are as follows: $\alpha_0 = 0.135$, $\alpha_1 = 0.670$, $\alpha_2 = 0.112$, $\alpha_3 = 0.079$, $\alpha_4 = 0.003$, at least four levels must be considered. Lowke and Parker considered these levels individually for D/μ below 0.08 eV and preferred to adopt the continuous approximation (1.5.22), modified by the polarization factor of Dalgarno and Moffett, for D/μ above 0.08 eV. Comparison with experimental results is shown in Fig. 4.7.2.

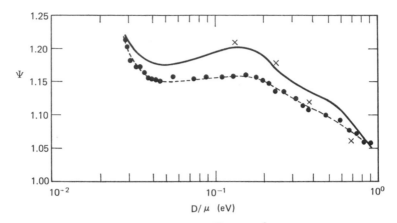

Fig. 4.7.8. Magnetic deflection coefficient Ψ' at $300°K$, computed by Engelhardt and Phelps using their electron collision cross sections (solid curve). The results are compared with experimental data of Townsend [c(1)] (crosses) and of Creaser [c(3)] (dashed curve) [after Creaser (1967)].

(3) An analysis similar to the above of Engelhardt and Phelps was carried out by Crompton, Gibson, and McIntosh (1969) for parahydrogen at 77°K. This is a particularly simple situation, since the populations of the rotational levels are $\alpha_0 = 0.995$, $\alpha_1 = 0$, $\alpha_2 = 0.005$, so that only the cross section Q_{02} need be considered. These authors used, as initial estimates for their analysis, the theoretical values of q_{02} of Lane and Geltman (1967) and the Q_v values of Engelhardt and Phelps to derive the Q_m curve that provides the best fit to the experimental data of w and D/μ; the agreement was then found to improve if the q_{02} theoretical values of Henry and Lane (1969) were adopted. Adjustments to all these cross sections were then made in subsequent numerical tests and iterations until the best agreement (within 1%) with experimental data was secured over the entire E/p range.

Subsequently Gibson (1970) determined the q_{13} cross section by an extension of the procedure to the $a(3)$, $a(8)$, and $b(6)$ data for normal hydrogen at 77°K, using the already determined Q_m, q_{02}, and Q_v cross sections. Furthermore, the fact that, according to Table 4.7.1, for $E/p > 2$ drift velocities in parahydrogen and normal hydrogen at 77°K and in normal hydrogen at 293°K agree to within $\pm 1\%$, with even closer correspondence at the highest values of E/p, indicate [Crompton, Gibson, and Robertson (1970)] that the vibrational cross section, to which drift velocity is sensitive over this E/p range, is practically independent of the initial rotational state, as theoretically found by Henry (1970).

The resulting cross sections are plotted in Figs. 4.7.6 and 4.7.7 and are given in Tables 4.7.3 and 4.7.4. The estimated accuracy of Q_m is $\pm 5\%$; also, the error limit to be placed on rotational cross sections is $\pm 5\%$ below 0.3 eV,

Table 4.7.4 The Rotational Excitation Cross Sections $q_{02}(u)$ and $q_{13}(u)$, Determined by Crompton, Gibson, and McIntosh and by Gibson, for Electrons in Hydrogen

Threshold values are 0.0439 and 0.0727 eV, respectively.

u (eV)	$10^2 q_{02}$ (Å^2)	$10^2 q_{13}$ (Å^2)	u (eV)	$10^2 q_{02}$ (Å^2)	$10^2 q_{13}$ (Å^2)
0.047	1.85		0.13	8.9	4.1
0.050	2.7		0.15	9.9	4.7
0.055	3.5		0.20	12.0	6.0
0.060	4.2		0.25	13.7	7.4
0.065	4.8		0.30	16.0	8.8
0.070	5.3		0.35	18.5	10.2
0.08	6.0	1.7	0.40	21.0	11.8
0.09	6.8	2.5	0.45	23.6	13.3
0.10	7.4	2.95	0.50	26.3	14.9
0.11	7.9	3.35			

whereas above this energy it deteriorates rapidly. The cross sections of these authors are regarded as the most accurate and reliable set presently available for hydrogen, since they were obtained from the most precise experimental data, taking advantage, particularly for high-energy extrapolations, of all recent theoretical treatments, and using sequentially the transport data in parahydrogen and in normal hydrogen to determine the cross sections with a minimum of mutual interactions.

When the Q_v cross section is compared with the corresponding values measured with electron beam methods, substantial differences are observed. All the authors find near the threshold a linear rise of the vibrational $v(0 \rightarrow 1)$ cross section as a function of the electron energy, but Ehrhardt, Langhans, Linder, and Taylor (1968), using an electrostatic analyzer, measure a slope which is about twice that of the cross section obtained from the swarm analysis of Crompton, Gibson, and McIntosh, and Burrow and Schulz (1969), using Schulz's (1958) trapped-electron method, measure a slope about four times that of the swarm analysis; a convincing explanation of these discrepancies has not yet been given [Crompton, Gibson, and Robertson (1970)].

(4) The collision cross sections Q_m derived by Engelhardt and Phelps and by Crompton, Gibson, and McIntosh provide the best fit to the electron mobility data at 77°K for $E/p > \sim 5 \times 10^{-3}$. The same Q_m curves can be adopted to compute μ and D values for isothermal electrons at various gas temperatures; the results that we have thus obtained by means of numerical computations are given in Table 4.7.5, where they are compared with the experimental results of $a(2)$ and of e. The data of $a(2)$ were chosen because they provide the largest T_g coverage at the lowest E/p values. Satisfactory agreement with the values derived from the cross section of Crompton et al.

Table 4.7.5 Mobility and Diffusion for Thermal Electrons in Hydrogen and Deuterium, Computed Using Cross Sections $Q_m(u)$ of Engelhardt and Phelps (α) and of Crompton, Gibson, and McIntosh (β), and Compared with the Experimental Values of Pack and Phelps [a(2)] and of Nelson and Davis (e) for Hydrogen, and of Pack, Voshall, and Phelps [a(1)] for Deuterium

T_g(°K)	μp_0 (cm² V⁻¹ sec⁻¹ torr)				Dp_0 (cm² sec⁻¹ torr)		
	Comp. α	Comp. β	Exp. H₂ [a(2)]	Exp. D₂ [a(1)]	Comp. α	Comp. β	Exp. H₂ (e)
77	9.54×10^6	10.1×10^6	9.64×10^6	9.66×10^6	6.33×10^4	6.73×10^4	
195	5.60	5.77	5.82		9.41	9.69	
300	4.29	4.37	4.52	4.38	1.11×10^5	1.13×10^5	1.21×10^5
373	3.72	3.79	3.79		1.20	1.22	

is found at all temperatures except at 77°K; however, at this temperature the experimental μ value derived by $a(2)$ from their measurements at the lowest E/p ratios can be slightly too low, electron energies at $E/p \simeq 1 \times 10^{-3}$ being still $\sim 3\%$ above the thermal value (see Table 4.7.2).

(5) The cyclotron resonance measurements of Tice and Kivelson (h) yield the velocity-independent value $Q_m = 10 \pm 1$ Å^2 at about 0.039 eV.

Several possible theoretical explanations of the pressure dependence of the drift velocities at high densities have been discussed by Frommhold (1968). He concludes that electron trapping by some low-energy short-living resonance states is likely to take place and suggests that here rotational resonances, as described in a theoretical paper by Kouri (1966), can be the main mechanism for electron trapping. This explanation is in accordance with relation (4.7.1) [the mean rate of electron trapping is properly proportional to pressure, since this rate multiplied by the lifetime of the trapped state is equal to $(w_0 - w)/w = \alpha N$], and the features of the α versus D/μ curves shown in Fig. 4.7.1 provide, when associated with the positions and widths of the rotational levels, additional evidence in its support. In fact, Crompton and Robertson (1971) have been able to obtain theoretical curves that fit rather well the normal hydrogen and parahydrogen curves of Fig. 4.7.1, assuming two δ-function resonance-capture cross sections, located at energies of 3.7×10^{-2} eV [associated with the $J(0 \rightarrow 2)$ transition] and of 6.0×10^{-2} eV [associated with the $J(1 \rightarrow 3)$ transition], respectively. These energy values seem plausible if one recalls that the normal $J(0 \rightarrow 2)$ transition requires a threshold energy of 4.4×10^{-2} eV, and that the $J(1 \rightarrow 3)$ transition requires a threshold energy of 7.3×10^{-2} eV [see eq. (1.5.16)]. Theoretically, the existence of these resonances can be proved or disproved, depending on the assumed form of the interaction potential [Kouri (1968), Henry and Lane (1969), Kouri, Sams, and Frommhold (1969a)]; on the other hand, no other interpretation than rotational resonances can explain easily the observed features. The formation of trapped states of finite lifetimes would also imply electron aftercurrents at the collector, due to the late arrival of the negative ions; since such currents have not yet been observed, Kouri, Sams, and Frommhold (1969b) conclude that the lifetime of the lower resonance state in hydrogen cannot exceed 4×10^{-9} sec.

4.8 DEUTERIUM

Experimental Data [dc parameters for $E/p < 7$ $(D/\mu \simeq 1$ eV$)$]

a. Drift velocity

(1) Pack, Voshall, and Phelps (1962) [see Kr $a(2)$] performed measurements at 77°K and at 300°K for $E/p > 7.5 \times 10^{-4}$. The impurities in the deuterium are believed to have been less than 0.5%.

(2) McIntosh (1966) used a drift tube similar to that of Fig. 3.4.2 for measurements at 293°K; the author places an error limit of $\pm 1\%$ for $6 \times 10^{-3} < E/p < 1.2$ and of $\pm 2\%$ for $1.2 < E/p < 5$. Deuterium was admitted to the system through a heated palladium osmosis tube.

(3) Robertson (1971) [see H_2 $a(8)$] performed measurements at 76.8°K over the range $2.6 \times 10^{-3} < E/p < 3.3$ in very pure deuterium [see (2) above]. As was done for hydrogen, data have been corrected for diffusion effects, and accuracy is estimated to be better than 1%. The data are in excellent agreement (within 0.5%) with previous results similarly obtained in the same laboratory by Crompton, Elford, and McIntosh (1968).

The most accurate results are those of (2) and (3); they are summarized in Table 4.8.1. Below $E/p \simeq 0.15$ the 77°K data are extrapolated zero-pressure values, since density effects in accordance with eq. (4.7.1) have been observed for $p > 100$ torr; the corresponding α values are shown in Fig. 4.7.1. Compared to the data of Table 4.8.1, the results of (1) are generally higher by as much as 7%, except at 293°K and low E/p, where good agreement is found; μp_0 values of (1) for isothermal electrons ($E/p < 1 \times 10^{-3}$) are reported in Table 4.7.5.

b. Diffusion to mobility ratio

(1) Hall (1955b) performed measurements at 288°K for $E/p > 0.1$ with a diffusion chamber equal to the long one of Crompton and Sutton (1952) [see H_2 $b(2)$ and $b(3)$]. Deuterium was obtained by dropping heavy water (99.75% pure) slowly onto molten sodium in an evacuated vessel; all traces of heavy-water vapor were then removed by keeping the gas over liquid air for several hours.

(2) Warren and Parker (1962) [see He $b(2)$] performed measurements at 77°K over the range $2 \times 10^{-4} < E/p < 3$ in deuterium containing 2% of HD and traces of other gases.

(3) McIntosh (1966) used the diffusion tube of Fig. 3.7.1 for measurements at 293°K over the range $6 \times 10^{-3} < E/p < 2$ in very pure deuterium [see $a(2)$ above]. The author places an error limit of $\pm 1\%$ on his D/μ data.

(4) Crompton, Elford, and McIntosh (1968) [see H_2 $b(6)$] performed measurements at 77.3°K over the range $6.6 \times 10^{-4} < E/p < 1$. For the purity of the gas, see $a(2)$ above.

The very accurate results of (3) and (4) are summarized in Table 4.8.1. Compared to (3), the results of (1) agree well for $E/p > 0.6$ but are higher below, probably because here too Hall did not correct for the effects of contact potential differences. Compared to (4), the results of (2) agree fairly well over the entire range.

Table 4.8.1 Experimental Values of the Drift Velocity, of the Characteristic Energy, and of the Magnetic Deflection Coefficient in Deuterium

$10^{17}E/N$ (V cm²)	E/p_{293} (V cm⁻¹ torr⁻¹)	$10^{-5}w$ (cm sec⁻¹) [$T_g = 293°$K; a(2)]	$10^{-5}w$ (cm sec⁻¹) [$T_g = 77°$K; a(3)]	D/μ (eV) [$T_g = 293°$K; b(3)]	D/μ (eV) [$T_g = 77°$K; b(4)]	Ψ [$T_g = 293°$K; c(2)]
2.0×10^{-3}	6.60×10^{-4}				6.80×10^{-3}	
3.0	9.90				6.85	
4.0	1.32×10^{-3}				6.92	
6.0	1.98				7.12	
8.0	2.64		2.80×10^{-1}		7.33	
1.0×10^{-2}	3.30		3.41		7.56	
1.4	4.62		4.57		8.01	
1.82	6.0	2.81×10^{-1}	5.71	2.57×10^{-2}	8.48	1.21
2.43	8.0	3.72	7.26	2.60	8.99	1.18
3.03	1.0×10^{-2}	4.62	8.71	2.63	9.65	1.17
4.55	1.5	6.78	1.20×10^{0}	2.71	1.09×10^{-2}	1.16
6.07	2.0	8.82	1.50	2.80	1.21	1.15
9.10	3.0	1.26×10^{0}	2.01	3.02	1.45	1.15
1.214×10^{-1}	4.0	1.59	2.43	3.25	1.69	1.16
1.82	6.0	2.17	3.09	3.76	2.19	1.15
2.43	8.0	2.64	3.59	4.29	2.72	1.16
3.03	1.0×10^{-1}	3.05	3.97	4.83	3.26	1.15
4.55	1.5	3.81	4.65	6.28	4.76	1.16
6.07	2.0	4.37	5.09	7.84	6.35	1.16
9.10	3.0	5.18	5.71	1.10×10^{-1}	9.68	1.16
1.214×10^{0}	4.0	5.82	6.21	1.40	1.29×10^{-1}	1.16
1.82	6.0	6.95	7.20	1.93	1.85	1.15
2.43	8.0	7.98	8.17	2.37	2.32	1.15
3.03	1.0×10^{0}	8.93	9.09	2.77	2.72	1.14
4.55	1.5	1.10×10^{1}	1.11×10^{1}	3.65	3.61	1.13
6.07	2.0	1.28	1.29	4.41	4.38	1.12
9.10	3.0	1.59	1.60		5.74	1.10
1.214×10^{1}	4.0	1.86			7.00	1.09
1.517	5.0	2.11				1.08
2.125	7.0					1.06

c. Magnetic drift velocity (LMF case)

(1) Hall (1955b) [see H$_2$ c(2)] used the modified Townsend method for measurements at 288°K over the 0.1 < E/p < 5 range in deuterium, obtained as described under b(1) above.

(2) Creaser (1967) [see H$_2$ c(3)] performed measurements at 293°K in very pure deuterium [see a(2)] over the range 0.02 < E/p < 8.

The magnetic deflection coefficient Ψ can be obtained by dividing the magnetic drift velocities by the w data of Table 4.8.1; Ψ thus computed as a function of E/p for the data of (2) is also given in Table 4.8.1. Hall's data are lower, by as much as 10%, probably for the same reasons already discussed for hydrogen.

d. Magnetic drift velocity (SMF case). Bernstein (1962) (see He d) performed measurements at room temperature in deuterium from standard tanks; some of his data are shown in Fig. 4.7.3.

e. Diffusion in crossed fields. Relevant $D_{\parallel}p_0$ values obtained by Bernstein (1962) (see H$_2$ f) are also shown in Fig. 4.7.3.

f. Cyclotron resonance. Tice and Kivelson (1967) [see He i(3)] performed line-shape measurements in deuterium-nitrogen mixtures at room temperature, obtaining the same results as in hydrogen-nitrogen mixtures.

Determination of Collision Parameters

The only derivation of collision parameters for deuterium based on a large set of data is due to Engelhardt and Phelps (1963), who followed the same approach that they also adopted for hydrogen; the pertinent average curves, which are analyzed for $T_g = 77°$K only, are based on the results of a(1), b(1), and b(2). As in the case of hydrogen, these authors considered only Q_{02} and Q_{13} among the collisions of the first kind of importance at 77°K. Actually, since in deuterium cooled at this temperature without changing the room temperature para/ortho ratio the populations of the rotational levels are $\alpha_0 = 0.565$, $\alpha_1 = 0.330$, and $\alpha_2 = 0.101$, the resulting approximation for deuterium is less satisfactory than that for hydrogen.

Best-fit results have been obtained by Engelhardt and Phelps, using the Q_m cross section of hydrogen; the rotational cross section (1.5.18) of Gerjouy and Stein ($\mathcal{Q} = 0.473$), modified by the polarization factor of Dalgarno and Moffett and multiplied by the M_R factor 1.5; and an appropriate Q_v cross section [see Fig. 4.7.7]. Not only the calculated values of w and D/μ, but also those of $e/m\mu_M p_0$ and $D_{\parallel}p_0$, agree well with the experimental results, as shown in Fig. 4.7.3. In deuterium the slope near threshold of the Q_v cross section shown in Fig. 4.7.7 is 54% of that in hydrogen; this value compares well with the figure of 47% measured by Burrow and Schulz (1969), using the trapped-electron method.

Values of D_L/μ at 293°K, calculated using the cross sections of Engelhardt and Phelps (except for $\mathscr{D}$, which has been taken to be 0.574), but including seven levels of rotational excitation, are shown in Fig. 4.7.2. No experimental data exist for comparison.

In their analysis Engelhardt and Phelps adopted the same Q_m cross section in deuterium as in hydrogen; the mobility data of Table 4.7.5 for isothermal electrons, which depend on the Q_m cross sections only, justify this assumption. On the contrary, Gibson (1970), analyzing the most recent w measurements, presents some evidence that there is probably a small difference between the two cross sections.

The pressure dependence of the drift velocity measurements at high densities can be interpreted in terms of Frommhold's theory, discussed for the case of hydrogen. Moreover, the deuterium data plotted in Fig. 4.7.1, which, compared to those for hydrogen, show large α values for low D/μ energies, actually confirm this theory, since in deuterium the threshold energy for the $J(0 \to 2)$ transition has half the value of the corresponding energy in hydrogen.

4.9 NITROGEN

Experimental Data [dc parameters for $E/p < 12$ $(D/\mu \simeq 1.3$ eV)]

a. Drift velocity

(1) Nielsen (1936) [see He $a(1)$] performed measurements at 293°K for $E/p > 4 \times 10^{-2}$; the estimated accuracy from the scatter of the experimental points is of the order of 2%. Commercial tank gas, purified by passing over hot copper gauze and through a series of liquid-air traps, was used.

(2) Unpublished data of Errett (1951) [see He $a(2)$] were reported by Engelhardt, Phelps, and Risk (1964) for the range $0.3 < E/p < 1$ and $T_g = 300$°K.

(3) Colli and Facchini (1952) [see Ar $a(4)$] performed measurements at room temperature over the range $3 \times 10^{-2} < E/p < 2.4$, both in tank and in purified nitrogen.

(4) Unpublished data obtained at room temperature by Crompton and Hall, using an electrical shutter apparatus, were reported by Huxley (1959a) for the range $2 \times 10^{-2} < E/p < 8 \times 10^{-2}$.

(5) Bowe (1960) [see Ne $a(2)$] performed measurements at room temperature over the range $5 \times 10^{-2} < E/p < 2$; purification of the original tank nitrogen (stated purity 99.996%) was achieved by forced circulation of the gas through the calcium purifier, operated at 350°C.

(6) Nagy, Nagy, and Dési (1960) [see Ar $a(7)$] performed measurements at room temperature over the range $6 \times 10^{-2} < E/p < 0.87$, using nitrogen with a stated purity of 99.9%.

(7) Comunetti and Huber (1960), using their method and apparatus, as described in Sections 3.2, 3.3, and 3.4F, performed measurements in nitrogen (stated purity 99.8%) at 303°K over the range $0.75 < E/p < 1.9$. The estimated accuracy of the w measurements is 1.5%.

(8) Pack and Phelps (1961) [see He $a(3)$] performed measurements at 77, 195, 300 ± 3, and 373°K over the range $1 \times 10^{-4} < E/p < 10$. The impurity content was the same as for helium.

(9) Lowke (1963) [see H_2 $a(3)$] performed measurements at 77.6°K over the range $1 \times 10^{-3} < E/p < 2$ and at 293°K for $E/p > 4 \times 10^{-3}$. Nitrogen was prepared by heating sodium azide and admitted to the apparatus through several liquid-air traps. Lowke verified that, in accordance with theory, impurities of more than 0.5% of hydrogen are required to change the drift velocity results in nitrogen by 1%; therefore hydrogen impurities, whose content was found to be less than 0.01%, have no significance for these measurements.

(10) Levine and Uman (1964) [see Ar $a(9)$] performed measurements at room temperature over the range $5 \times 10^{-2} < E/p < 1.7$. Individual results show significant scatter, often in excess of 10%.

(11) Fisher-Treuenfeld (1965), using his method and apparatus, as described in Sections 3.2, 3.3, and 3.4B, performed measurements at room temperature for $E/p > 5 \times 10^{-2}$. The estimated error limit of the w measurements is 5%.

(12) Prasad and Smeaton (1967) [see H_2 $a(5)$] performed measurements at 293°K for $E/p > 6$ in nitrogen with 99.98% purity.

(13) Wagner, Davis, and Hurst (1967) [see He $a(6)$] performed measurements at room temperature over the range $1 \times 10^{-2} < E/p < 2$; the impurity content of nitrogen was less than 0.003 mole %. These data supersede previous results obtained at Oak Ridge, using earlier versions of the same basic method [Bortner, Hurst, and Stone (1957), Hurst, Stockdale, and O'Kelly (1963), Christophorou, Hurst, and Hadjiantoniou (1966)]. Christophorou and Christodoulides (1969) [see H_2O $a(4)$] state that they recently obtained data within 1–3% of those of Wagner, Davis, and Hurst and of Lowke [see (9) above].

(14) Grünberg (1967, 1968) [see He $a(7)$] performed measurements at room temperature for $E/p > 3 \times 10^{-2}$ in high-purity nitrogen (99.999%) and over the pressure range from 1 to 40 atm.

(15) Allen and Prew (1970) [see Ar $a(12)$] performed measurements at 295°K over the range $0.3 < E/p < 3.7$, using pressures from 3 to 70 atm. The accuracy is ± 4%, and the gas purity similar to that in the case of argon (impurity content $< \sim 20$ ppm).

With the exception of the results at very high molecule densities, where a marked pressure dependence is observed, all other results agree well within the combined experimental errors. The results of (9), which are the most significant if we consider both accuracy and extent of the E/p range, are summarized in Table 4.9.1; at 77°K the drift velocity shows a minimum at $E/p \simeq 1 \times 10^{-1}$. At room temperature (3) and (4), as well as (10) when $E/p < 1$, find velocities generally higher than those of (9), whereas the results of (5) are up to 6% lower; (14) also finds velocities slightly below those of (9) for $E/p > 0.2$, whereas at the lowest E/p larger velocities are observed. The results of (5) over the range $6 \times 10^{-2} < E/p < 0.3$ are well represented by the relation:

$$w = 5.4 \times 10^5 (E/p_0)^{0.25} \quad (\text{cm sec}^{-1})$$

The linear dependence of w on E/p is found for $E/p < 2 \times 10^{-3}$; then the most significant μp_0 data for isothermal electrons are those of (8), which are reported for future discussion in Table 4.9.3. Measurements by Klema and Allen (1950) have not been included in the above list, because of the poor results they obtained in argon, a sensitive test to the impurity content, and because of the inferior techniques these authors used in their experiments.

Pressure-dependent results were obtained at the highest densities by (9) and (14), whereas for the less accurate data of (15) the effect remains within the error limits. The results can be interpreted by means of (4.7.1); the order of magnitude of the resulting α values is 10^{-22}–10^{-23} cm^3, as in the case of hydrogen.

We mention also, as an item of interest in the discussion of the collision parameters, that Pack and Phelps (1966) measured drift velocities of nearly isothermal electrons in nitrogen-carbon dioxide mixtures at 300 and 529°K, using a single-grid drift chamber.

b. Diffusion to mobility ratio

(1) Townsend and Bailey (1921) [see H$_2$ b(1)] performed measurements at room temperature for $E/p > 0.42$.

(2) Crompton and Sutton (1952) [see H$_2$ b(2)] performed measurements at room temperature for $E/p > 5 \times 10^{-2}$. Commercially pure nitrogen (stated purity 99.5%) was admitted into the diffusion chamber after being passed over copper turnings heated to 400°C and through two liquid-air traps containing glass beads.

(3) Hall (1955c) [see H$_2$ b(3)] attempted for the first time low-temperature (83°K) measurements also in nitrogen, down to E/p as low as $\sim 2 \times 10^{-2}$. Further data obtained at room temperature by Crompton and Hall (un-published) were reported by Huxley (1959a) [see also Huxley and Crompton (1962)] for the range $2 \times 10^{-2} < E/p < 9 \times 10^{-2}$; the estimated maximum

Table 4.9.1 Experimental Values of the Drift Velocity, of the Characteristic Energy, and of the Magnetic Deflection Coefficient in Nitrogen

D/μ values at 293°K are from Crompton and Elford [b(6)] for $E/p_{293} \leqslant 5$ and from Naidu and Prasad [b(9)] for $E/p_{293} \geqslant 7$; D/μ values at 77°K were read on Fig. 8 of Warren's and Parker's paper and are therefore affected by reading errors.

$10^{17} E/N$ (V cm²)	E/p_{293} (V cm⁻¹ torr⁻¹)	$10^{-5} w$ (cm sec⁻¹) [$T_g = 293°K$; a(9)]	$10^{-5} w$ (cm sec⁻¹) [$T_g = 77°K$; a(9)]	D/μ (eV) [$T_g = 293°K$; b(6) and b(9)]	D/μ (eV) [$T_g = 77°K$; b(5)]	Ψ [$T_g = 293°K$; c(2)]
1.214×10^{-2}	4.0×10^{-3}	4.55×10^{-1}	1.41×10^{0}		7.6×10^{-3}	
1.82	6.0	6.72	1.93		8.3	
2.43	8.0	8.77	2.35		9.0	
3.03	1.0×10^{-2}	1.07×10^{0}	2.69		9.8	
4.55	1.5	1.49	3.22		1.30×10^{-2}	
6.07	2.0	1.82	3.48		1.55	
9.10	3.0	2.28	3.68		2.40	1.55
1.214×10^{-1}	4.0	2.55	3.67	4.72×10^{-2}	3.20	1.52
1.82	6.0	2.81	3.53	6.56	5.1	1.48
2.43	8.0	2.96	3.42	8.48	7.0	1.45
3.03	1.0×10^{-1}	3.09	3.39	1.04×10^{-1}	9.2	1.38
4.55	1.5	3.43	3.58	1.45	1.35×10^{-1}	1.34
6.07	2.0	3.76	3.85	1.83	1.75	1.28
9.10	3.0	4.28	4.33	2.56	2.65	1.23
1.214×10^{0}	4.0	4.76	4.80	3.27	3.25	1.16
1.82	6.0	5.68	5.73	4.49		1.11
2.43	8.0	6.66	6.73	5.40		1.09
3.03	1.0×10^{0}	7.72	7.77	6.08		1.06
4.55	1.5	1.02×10^{1}	1.03×10^{1}	7.25		1.05
6.07	2.0	1.27	1.28	8.00		1.05
9.10	3.0	1.71		9.09		1.07
1.214×10^{1}	4.0	2.11		9.85		1.08
1.517	5.0	2.50		1.04×10^{0}		1.11
2.125	7.0	3.23		1.13		
3.03	1.0×10^{1}	4.20		1.25		

error is 2–3%. Nitrogen was obtained by thermal decomposition of sodium azide.

(4) Cochran and Forester (1962) [see H_2 $b(4)$] performed measurements at room temperature over the range $0.2 < E/p < 5$; commercial tank nitrogen (stated purity 99.99%) was admitted into the diffusion chamber through liquid-air traps.

(5) Warren and Parker (1962) [see He $b(2)$] performed measurements at 77°K over the range $2 \times 10^{-4} < E/p < 3$.

(6) Crompton and Elford (1963) performed measurements at 293°K over the range $6 \times 10^{-3} < E/p < 5$, using a diffusion chamber of construction very similar to the one of Fig. 3.7.1 ($d = 10$ cm, $b/d = 0.05$) and the most appropriate techniques (compensating voltages, correction of space-charge effects). The accuracy of the data is estimated to be of the order of 1%. The oxygen impurity content of the reagent-grade nitrogen was less than 5 ppm; consistent results have also been obtained using nitrogen prepared by thermal decomposition of sodium azide.

(7) Jory (1965) performed measurements at 293°K over the range $4 \times 10^{-2} < E/p < 8$, using the apparatus shown in Fig. 3.9.6. The impurity content of the reagent-grade nitrogen was less than 25 ppm, of which 15 ppm consisted of monatomic gases.

(8) Unpublished data obtained at room temperature by Ellis (1966) were reported by Naidu and Prasad (1968) for $E/p > 7$.

(9) Naidu and Prasad (1968) [see H_2 $b(7)$] performed measurements at room temperature for $E/p > 7$ in 99.98% pure nitrogen.

(10) Wagner, Davis, and Hurst (1967) [see He $b(4)$] determined the quantity D_L/μ at room temperature over the range $1 \times 10^{-2} < E/p < 2$. For the impurity content of the gas, see $a(13)$.

The most reliable and accurate room-temperature results for D/μ are those of (6) for $E/p \leqslant 5$ and of (9) for $E/p \geqslant 7$; these data are also summarized in Table 4.9.1. Below $E/p = 5$ the results of (3) and of (7) agree very well with those of (6), whereas the results of (2) and (4) are in good agreement among themselves but become significantly larger than those of (6) as E/p decreases (up to 40% at $E/p = 0.05$), probably because of the impurity content of the gas. Above $E/p = 7$ the results of (2) agree well with those of (9), whereas those of (8) are consistently lower. Over the entire E/p range of interest to us the results of (1) are significantly and consistently lower than all other measurements.

The results of (5) for D/μ at 77°K [here, too, the preceding low-temperature experiment of (3) had failed to show any significant change from the room-temperature data] are also given in Table 4.9.1. The results of (10) for D_L/μ are shown in Fig. 4.9.1.

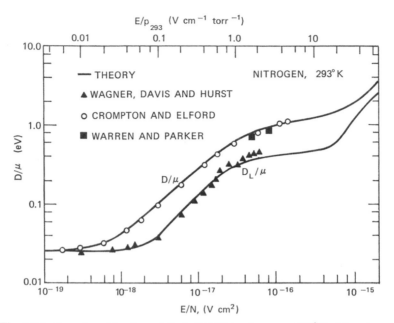

Fig. 4.9.1. Experimental values of D_L/μ [$b(10)$] in nitrogen at 293°K; D/μ values measured by Warren and Parker [$b(5)$] and by Crompton and Elford [$b(6)$] are shown for comparison. The solid curves are computed using the Engelhardt, Phelps, and Risk electron collision cross sections [after Lowke and Parker (1969)].

c. Magnetic drift velocity (LMF case)

(1) Townsend and Bailey (1921) [see H_2 $c(1)$] performed w_M measurements, using the same tube and conditions as in $b(1)$ above.

(2) Jory (1965) used the apparatus shown in Fig. 3.9.6 for measurements at 293°K over the range $4 \times 10^{-2} < E/p < 8$ [see $b(7)$ above]; several flux densities (20, 40, 60, 80, 100, and 120 G) were used. The estimated maximum error is 2–3%.

The results of (1) and (2) can be divided by the drift velocity data of Table 4.9.1 to obtain the magnetic deflection coefficient Ψ. This quantity, for the data of (2), is also given in Table 4.9.1; the data of (1) agree well within the combined experimental errors.

d. Free electron diffusion. Nelson and Davis (1969) [see He $e(2)$] obtained $Dp_0 = (268 \pm 8) \times 10^3$ cm^2 sec^{-1} torr at 300 ± 2°K from measurements performed at 10 torr in nitrogen with a minimum purity of 99.997%.

e. Microwave conductivity ratio. Radiation temperature measurements in the afterglow of a pulsed dc discharge have shown a relaxation of electron energies from the active discharge values to the isothermal regime that is much slower than expected [Formato and Gilardini (1959), Noon and Holt (1966), Noon, Blaszuk, and Holt (1968)]. Therefore all antecedent measurements of the microwave conductivity ratio that were performed in afterglows without verifying the isothermicity condition [as in the case of Anderson and Goldstein (1956a)] are questionable and will not be considered here, with only the following two exceptions.

(1) Phelps, Fundingsland, and Brown (1951) [see Kr $b(1)$], performing measurements very late in the afterglow of pulsed microwave discharges, appear to have obtained consistent results for thermal electrons at gas temperatures from 303°K to 511°K and at low pressure, such that $\nu_m{}^2 \ll \omega^2$.

(2) Van Lint (1964) measured at $\sim$300°K with an X-band waveguide interferometer the conductivity ratio of the decaying plasma generated by a pulse of high-energy electrons [see also O_2 $c(1)$]. The $\omega\rho$ value for thermal electrons is identified with the asymptotic value during the late afterglow.

The results of (1) can be represented by the relation

$$\omega\rho = 2.7 \times 10^8(T_g/300) \quad (\text{sec} \cdot \text{torr})^{-1}$$

whereas the results of (2), which extend also over the region $|\sigma_r/\sigma_i| \sim 1$, provide the low-pressure $(\nu_m{}^2 \ll \omega^2)$ asymptotic value $\omega\rho = 2 \times 10^8$ $(\text{sec} \cdot \text{torr})^{-1}$ for $T_g \simeq 300$°K.

f. Radiation temperature under microwave heating conditions. Formato and Gilardini (1962) performed X-band measurements of the radiation temperatures in the afterglow of a pulsed dc nitrogen discharge, both in the absence and in the presence of microwave heating signals, as discussed at the end of Section 3.12. They also measured the corresponding conductivity ratios, so that by applying eqs. (2.6.61), (3.12.13), and (3.12.15) (since local thermal equilibrium conditions prevail) a velocity-independent G-factor can be obtained. In this way Formato and Gilardini found $G = (3.95 \pm 0.25) \times 10^{-4}$ at radiation temperatures between 3000 and 5500°K and no significant dependence of this value on temperature.

g. Energy relaxation time. Mentzoni and Row (1963) determined this parameter from energy decay measurements, as discussed in Section 3.13B, in the isothermal afterglow of a mildly driven pulsed dc discharge in nitrogen; the gas was also heated to high temperatures, up to about 900°K. The authors verified that no systematic change in the energy relaxation time takes place as a function of the postdischarge time (over the range 0.2–10 msec). If measurements at pressures below 2 torr, where considerable reading errors were present, are disregarded, the other results can be represented within the experimental errors by the relation:

$$p_0\tau_r = 2.4(T_g/300)^{1/2} \quad (\mu\text{sec} \cdot \text{torr})$$

h. Cyclotron resonance. Tice and Kivelson (1967) performed measurements of the resonance line shape in active nitrogen at room temperature, using a 9.5-GHz resonant cavity and the arrangement discussed in Section 3.14A [these results supersede earlier ones by Bayes, Kivelson, and Wong (1962)]. The experimental line derivative, obtained at a pressure of 6.80 torr, is shown in Fig. 4.9.2.

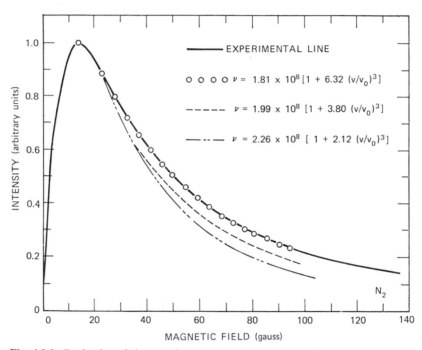

Fig. 4.9.2. Derivative of the experimental cyclotron resonance line, measured at 6.80-torr pressure in nitrogen (solid line), and calculated line shapes for three different $Q_m(u)$ cross sections, all having the form $Q_m = A(u^{-0.5} + Bu)$ [after Tice and Kivelson (1967)].

Determination of Collision Parameters

Engelhardt, Phelps, and Risk (1964) performed a numerical analysis of type A, the technique being virtually unchanged from that of Engelhardt and Phelps (1963) for hydrogen. These authors considered the average curves for $T_g = 77°K$ of the drift velocity [based on the 77°K data of $a(8)$ and $a(9)$ and on the room-temperature data of $a(1)$ for $E/p > 0.4$ and of $a(2)$ for $E/p > 0.3$] and of the characteristic energy [based on the 77°K data of $b(5)$

Table 4.9.2 The Momentum Transfer Cross Section $Q_m(u)$, Determined by Engelhardt, Phelps, and Risk for Electrons in Nitrogen

u (eV)	Q_m (Å^2)	u (eV)	Q_m (Å^2)	u (eV)	Q_m (Å^2)
0.0	1.00	0.0400	3.86	1.60	12.90
0.0004	1.20	0.0484	4.24	1.80	16.95
0.0009	1.33	0.0651	4.90	2.00	24.01
0.0016	1.43	0.0786	5.33	2.20	28.76
0.0025	1.56	0.1030	6.04	2.60	29.88
0.0036	1.69	0.1156	6.31	2.80	28.01
0.0049	1.80	0.1502	7.12	3.00	21.63
0.0064	1.94	0.226	8.22	3.30	17.19
0.0081	2.05	0.332	9.34	3.60	14.66
0.0103	2.20	0.445	9.95	4.00	12.62
0.0144	2.49	1.00	9.98	4.50	11.52
0.0221	2.94	1.20	10.51	6.00	10.30
0.0324	3.50	1.40	11.45	10.0	9.51

and on the room-temperature data of $b(1)$ for $E/p > 0.5$, of $b(4)$ for $E/p > 0.2$, and of $b(6)$ for $E/p > 0.15$].

By successive adjustments to initial estimates a good agreement between computed and experimental data is secured using the collision cross section $Q_m(u)$ values of Table 4.9.2 and:

(a) for $D/\mu < 0.02$ eV, Gerjuoy's and Stein's rotational excitation cross sections (1.5.18) and (1.5.19), with $\mathcal{Q} = 1.04$ and no polarization correction;

(b) over the range $0.02 < D/\mu < 0.046$, an arbitrary set of Gerjuoy's and Stein's rotational excitation cross sections, with thresholds increased by a factor of 2 and magnitudes appropriately decreased;

(c) over the range $0.046 < D/\mu < 0.08$, the rotational excitation cross sections as represented by the continuous approximation (1.5.22);

(d) over the range $0.08 < D/\mu < 1.3$, the rotational excitation cross sections as in (c), the $v(0 \rightarrow 1)$ vibrational excitation cross section of Fig. 4.9.3 for energies from the threshold value 0.29 eV up to 1.7 eV, and Schulz's cross sections of Fig. 4.9.4, normalized to $\Sigma Q_v = 5.5$ Å^2 at 2.2 eV, for excitation of the $v = 1$ to $v = 8$ vibrational levels from electrons with energies above 1.7 eV.

The necessity of using different expressions for rotational excitation cross sections follows from the fact that in nitrogen there are at least 20 rotational states that are significantly populated even at 77°K, since the rotational constant of nitrogen is about 30 times smaller than the constant of hydrogen.

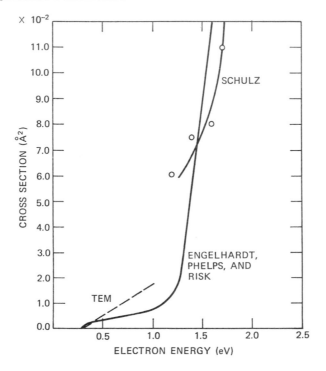

Fig. 4.9.3. Cross section for the $v(0 \to 1)$ vibrational excitation of nitrogen, determined from the analysis of swarm data by Engelhardt, Phelps, and Risk. Also shown are the results of beam experiments by Schulz, and near threshold by Burrow and Schulz (TEM; trapped-electron method) [after Burrow and Schulz (1969)].

More recently Phelps (1968) has observed that equally good agreement with experiment is obtained by using the Sampson and Mjolsness (1965) correction (see Section 1.5) to Gerjuoy's and Stein's cross sections and adopting the value $\mathscr{Q} = 1.10$, recommended by Strogryn and Strogryn (1966). Poor agreement, on the contrary, results from the use of the theories of either Dalgarno and Moffett (1963) or Geltman and Takayanagi (1966).

Engelhardt, Phelps, and Risk also computed Ψ as a function of E/p; their results, plotted in Fig. 4.9.5 together with the data of $c(1)$ and $c(2)$, show a fair agreement with experiments, although the discrepancies at low E/p and in the vicinity of $E/p = 2$ are outside experimental errors. Using the same cross sections (a) for $D/\mu < 0.066$ and (c) and (d) for $D/\mu > 0.066$, Lowke and Parker (1969) computed D/μ and D_L/μ in nitrogen at 293°K; the theoretical and experimental values, shown in Fig. 4.9.1, agree particularly well at low E/p, whereas above $E/p = 0.5$ the theoretical values of D_L/μ appear to be too small.

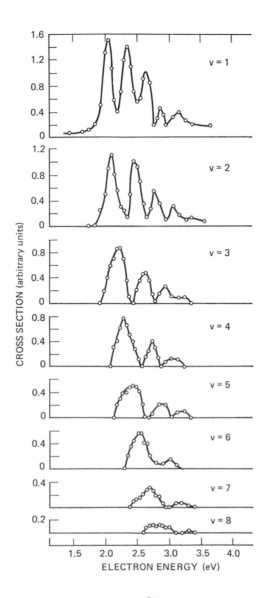

Fig. 4.9.4. Cross sections (approximately Å² units) for excitation of the first eight vibrational levels of nitrogen, measured using beam techniques [after Schulz (1964)].

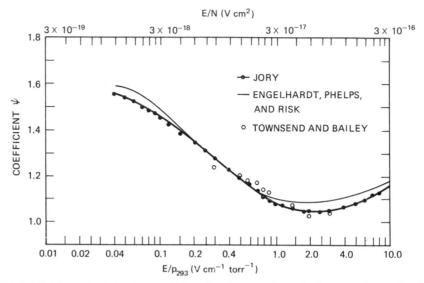

Fig. 4.9.5. Magnetic deflection coefficient in nitrogen: theoretical curve of Engelhardt, Phelps, and Risk compared to the experimental data of Townsend and Bailey [c(1)] and of Jory [c(2)] [after Jory (1965)].

The collision cross section of Table 4.9.2 is such as to provide the best fit to the electron drift velocities at 77°K for $E/p > {\sim}4 \times 10^{-4}$. As in the case of hydrogen, we have also computed μ and D values for isothermal electrons at various gas temperatures, using the same Q_m curve; the results are given in Table 4.9.3. Agreement with the experimental results may be considered quite satisfactory. According to the computations of Pack and Phelps (1966), good agreement can also be predicted for the case of their drift velocity measurements of isothermal electrons in nitrogen-carbon dioxide mixtures.

Table 4.9.3 Mobility and Diffusion for Thermal Electrons in Nitrogen, Computed Using the Cross Section $Q_m(u)$ of Engelhardt, Phelps, and Risk, and Compared with the Experimental Values of Pack and Phelps [a(8)] and of Nelson and Davis (d)

$T_g(°K)$	μp_0 (cm^2 V^{-1} sec^{-1} torr) Comp.	Exp. [a(8)]	Dp_0 (cm^2 sec^{-1} torr) Comp.	Exp. (d)
77	3.48×10^7	3.68×10^7	2.31×10^5	
195	1.57	1.61	2.64	
300	1.06	1.09	2.75	2.68×10^5
373	0.87	0.89	2.79	

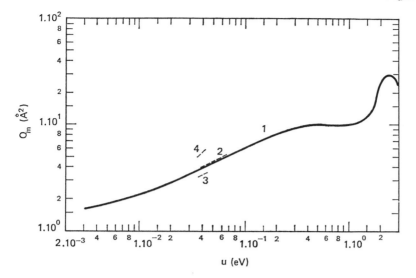

Fig. 4.9.6. Momentum transfer cross sections of electrons in nitrogen, derived from swarm experiments [(1) Engelhardt, Phelps, and Risk, (2) Phelps, Fundingsland, and Brown, (3) Van Lint, (4) Tice and Kivelson].

The cross-section Q_m of Table 4.9.2 is plotted in Fig. 4.9.6 (curve 1), together with the following cross-section curves (u in eV), derived from microwave data:

2. $Q_m = 20u^{0.5}$ Å² from $e(1)$ data ($0.039 < \bar{u} < 0.066$);
3. $Q_m = 16.5u^{0.5}$ Å² from $e(2)$ data ($\bar{u} = 0.039$);
4. $Q_m = 0.140(u^{-0.5} + 828u)$ Å² from h data ($\bar{u} = 0.039$) (see Fig. 4.9.2, where two other, different cross sections are depicted for comparison purposes).

In Fig. 4.9.3 comparable results from beam experiments are also shown: the low-energy portion of the analyzer data of Schulz (1964), and the cross-section slope at threshold measured by Burrow and Schulz (1969), using the trapped-electron method.

Gerjuoy's and Stein's rotational excitation cross sections with $\mathcal{Q} = 0.98$ were adopted by Mentzoni and Row (1963) for computing electron energy relaxation times; thus they used eq. (2.6.70), when the continuous approximation (1.5.22) could be accepted, or in general a more rigorous expression based on the evaluation of eq. (2.5.10), where ν_u is given by (1.5.20) and the distribution is regarded as Maxwellian. Having assumed $\langle \nu_m \rangle / p_0 = \omega \rho = 4.2 \times 10^9 \bar{u}$ ($\bar{u}$ in eV), on the basis of experimental verification, Mentzoni and Row found satisfactory agreement between experimental and theoretical results.

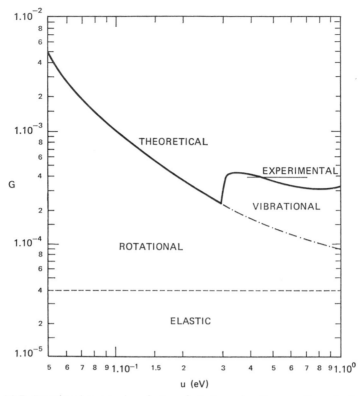

Fig. 4.9.7. Fractional energy loss factor of electrons in nitrogen: theoretical curve computed using the Engelhardt, Phelps, and Risk cross sections, compared with the experimental result of Formato and Gilardini (f).

Using the cross sections and approximations adopted by Lowke and Parker for the computation of D/μ, we have evaluated the $G(v)$ parameter in room-temperature nitrogen; the results are shown in Fig. 4.9.7. In the same figure the result of Formato and Gilardini (see f above) is also shown.

For an interpretation of the pressure dependence of the drift velocity at high molecular densities the reader is referred to the discussion of hydrogen. Here too Frommhold's theory seems to provide the most appropriate explanation of the observed features.

4.10   OXYGEN

Experimental Data [dc parameters for $E/p < 3$ ($D/\mu \simeq 1$ eV)]

a. Drift velocity

(1) Nielsen and Bradbury (1937) [see He $a(1)$] performed measurements at 293°K for $E/p > 0.5$; the estimated accuracy from the scatter of the

experimental points is of the order of 3%. Commercial tank oxygen, purified by passing through a series of liquid-air traps, was used.

(2) Doehring (1952) determined drift velocities at $\sim$293°K for $E/p > 0.6$, using a double-shutter tube ($d = 4.0$ cm), properly designed for measurements in attaching gases. The author used oxygen dried, purified, and finally distilled at liquid-air temperatures.

(3) Pack and Phelps (1966) performed measurements at 300°K and 525°K in research-grade-purity oxygen for $E/p > 4 \times 10^{-2}$ in double-grid [see He a(3)] and in single-grid drift chambers. The transit time data have been corrected for the combined effects of electron diffusion and attachment as discussed in Section 3.3. Furthermore these authors performed measurements in mixtures of O_2 and CO_2, since CO_2 molecules, which have a large vibrational excitation cross section at low electron energies, keep the electron energy close to the isothermal value also for moderately large E/p. Experimental data, taken at 300°K and 529°K for ratios of oxygen to carbon dioxide densities of up to 14, show that Blanc's law (2.6.32) is here satisfied within the error limits.

All the results are given in Fig. 4.10.1; if we consider the relatively large scatter of each author's experimental data, the various results agree fairly well. Since the drift velocities of thermal electrons in CO_2 are known (see Section 4.12), Blanc's law applied to the results of (3) in mixtures yields for isothermal electrons in oxygen:

$$\mu p_0 = 1.15 \times 10^7 (T_g/300)^{-0.5} \quad (\text{cm}^2 \text{ V}^{-1} \text{ sec}^{-1} \text{ torr})$$

b. Diffusion to mobility ratio

(1) Unpublished data of D/μ, and B/p values for which $D_T/\mu_{||} = kT_g/e$ [see eq. (3.9.13)], obtained at room temperature for $E/p > 0.25$ by Healey and Kirkpatrick using Bailey's method for attaching gases, were reported by Healey and Reed (1941).

(2) Rees (1965) adopted the Townsend-Huxley method and a tube similar to the one of Fig. 3.7.2 to perform measurements of D/μ at 293°K for $E/p > 0.4$ in reagent-grade-purity oxygen (impurity content: 190 ppm of nitrogen, 190 ppm of argon). The author estimates the results to be accurate to within ±2%.

The most accurate and reliable D/μ data are those of (2), which are given in Table 4.10.1. The D/μ results of (1) disagree badly by as much as 60%; both the D/μ and B/p data of (1) will therefore be disregarded in any future consideration.

c. Microwave conductivity ratio

(1) Van Lint (1959) measured the conductivity ratio at $\sim$300°K, using the waveguide interferometer method to determine the characteristics at X-band

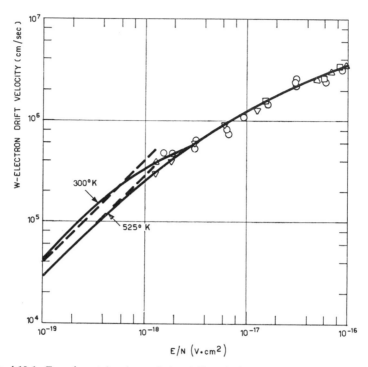

Fig. 4.10.1. Experimental values of the drift velocity in oxygen, and theoretical curve computed using the Hake-Phelps electron collision parameters. Symbols designate the following: □ Nielsen and Bradbury [293°K, $a(1)$], △ Doehring [293°K, $a(2)$], ○ Pack and Phelps [300°K, $a(3)$], ▽ Pack and Phelps [525°K, $a(3)$]; dashed lines represent drift velocities for isothermal electrons determined by Pack and Phelps in O_2—CO_2 mixtures [after Pack and Phelps (1966)].

Table 4.10.1 Experimental Values of the Characteristic Energy Measured by Rees [$b(2)$] in Oxygen at $T_g = 293°K$

$10^{17}E/N$ (V cm^2)	E/p_{293} (V cm^{-1} torr^{-1})	D/μ (eV)
1.214×10^0	4.0×10^{-1}	1.88×10^{-1}
1.517	5.0	2.07
1.82	6.0	2.25
2.43	8.0	2.58
3.03	1.0×10^0	2.93
4.55	1.5	4.29
6.07	2.0	6.01
9.10	3.0	9.77

of the decaying plasma generated by the traversal through the gas of a short pulse (0.2–5 μsec) of 20-MeV electrons emitted from a linear accelerator. An aluminum scatterer was placed at the accelerator exit window, made of a thin titanium foil, to spread the beam uniformly over the volume of test gas. The author used 99.9% pure research-grade oxygen; individual results show a very large scatter.

(2) Carruthers (1962) measured the conductivity ratio at room temperature, using the experimental arrangement discussed in Section 3.11B and based on the use of a 50-MHz resonant circuit containing an electron swarm generated by X-rays.

(3) Mentzoni (1965) performed, over the gas temperature range 300–900°K and using a transient X-band interferometer bridge, measurements of the conductivity ratio in the late isothermal afterglow of pulsed dc oxygen discharges; for this purpose the cylindrical quartz tube containing the plasma was gold coated on the outer surface, so as to function as a waveguide supporting a "quasi" TE_{11} mode. The isothermicity condition was verified by means of radiometer measurements.

The results of (1) and (3) satisfy the condition $\nu_m^2 \ll \omega^2$ and are, respectively, $\omega\rho = 1.35 \times 10^8 \pm 10\%$ (sec $\cdot$ torr)$^{-1}$ at 300°K and $\omega\rho = 1.0 \times 10^8$ $(T_g/300)$(sec $\cdot$ torr)$^{-1}$. The results of (2) cover the region where $|\sigma_r/\sigma_i| \sim 1$; they were taken in the range 1–10 torr and can be represented, within the experimental scatter, by the simple relationship (p in torr):

$$\log |\sigma_r/\sigma_i| = -0.35 + 0.63 \log p$$

Whereas all experiments cited previously refer to temperatures below 1000°K, where the gas is completely molecular, experiments above 3000°K provide information regarding the atomic species that becomes dominant at these temperatures, when pressures are of the order of a few torr. These conditions are attained in shock tubes; the only significant work performed in this area is due to Daiber and Waldron (1966) (see Ar f), who obtained, from experiments in 5% O_2 + 95% Ar mixtures and in the 3300–3800°K temperature range, a value of the apparent oxygen-atom collision cross section, defined as in case of argon, of 12 ± 2 Å^2.

d. Radiation temperature under microwave heating conditions. Formato and Gilardini (1962) performed measurements as in the case of nitrogen, but using a negative glow discharge between parallel wires. They found $G = (4.8 \pm 1.0) \times 10^{-3}$ at radiation temperatures between 850 and 1900°K, and the data seem to indicate that G decreases when the temperature increases. Gilardini (1964) extended the measurements in the presence of a transverse magnetic field, so as to attain higher electron energies, and found $G = (3.9 \pm 0.6) \times 10^{-3}$, independently of temperature over the range 3000–12,000°K.

e. Energy relaxation time.   Mentzoni and Row (1965), using the method and setup reported for nitrogen, obtained data that can be represented, together with earlier results obtained in the same laboratory and within the experimental errors, by the relation:

$$p_0 \tau_r = 0.44(T_g/300)^{1/2} \quad (\mu\text{sec} \cdot \text{torr})$$

f. Cyclotron resonance

(1) Fehsenfeld (1963) [see He $i(2)$] obtained at 300°K:

$$\Delta\omega_I = (2.8 \pm 0.3) \times 10^7 p \quad (\text{sec}^{-1})$$

The author estimates that atomic oxygen formed by the discharge may alter the measured $\Delta\omega_I$ value by as much as 25%.

(2) Veatch, Verdeyen, and Cahn (1966) performed measurements of the $\delta\omega_s$ displacement from resonance and of the resonance line width and shape. No values have been published, however, for these quantities.

Determination of Collision Parameters

Hake and Phelps (1967) performed a numerical analysis of type A, very similar to the one of Engelhardt and Phelps (1963) for hydrogen. For this purpose they considered the average w and D/μ curves through the room-temperature data discussed under a and b above. A good agreement between computed and experimental data is secured by using:

(a) the collision cross section $Q_m(u)$ values of Table 4.10.2 (below $u = 0.06$ eV the constant Q_m value is chosen to fit the room-temperature mobility obtained from measurements in O_2-CO_2 mixtures);

(b) the rotational excitation cross sections of Gerjuoy and Stein, as given by the continuous approximation (1.5.22) with $\mathcal{Q} = 1.8$;

(c) above 0.195 eV vibrational excitation cross sections consisting of a set of narrow spikes, delayed in energy relative to the excitation thresholds [to account for the small cross sections measured near thresholds by Schulz and Dowell (1962)], spaced uniformly by 0.16 eV, and having appropriate strengths as required for a good fit with the transport coefficient data.

The cross section Q_m of Table 4.10.2 is plotted in Fig. 4.10.2 (curve 1), together with other cross sections derived from the microwave data. The results of $c(3)$ and $f(2)$ indicate a dependence $\nu_m \propto v^2$, and the appropriate equations of Section 2.6 yield, respectively, $Q_m = 7.4u^{0.5}$ (Å^2) (curve 3) and

Table 4.10.2 The Momentum Transfer Cross Section $Q_m(u)$, Determined by Hake and Phelps for Electrons in Oxygen

u (eV)	Q_m (Å²)	u (eV)	Q_m (Å²)	u (eV)	Q_m (Å²)
0	3.0	0.30	4.6	2.0	6.8
0.06	3.0	0.40	4.7	2.5	6.1
0.08	3.4	0.50	5.1	3.0	5.7
0.10	4.4	0.60	5.5	4.0	5.5
0.12	4.8	0.80	6.8	5.0	5.6
0.15	5.0	1.0	7.6	6.0	6.0
0.20	4.9	1.2	7.9	8.0	7.2
0.25	4.7	1.5	7.7	10.0	8.0

$Q_m = 7.8u^{0.5}$ (Å²) (point 4), two consistent results. The results of $c(1)$ and $f(1)$, on the other hand, do not provide any indication concerning the energy dependence of Q_m; if we assume $h = 1$ in accordance with the mobility results, we obtain, respectively, $Q_m = 2.7$ Å² and $Q_m = 1.4$ Å², whereas if we assume $h = 2$ in accordance with the previous microwave data, we obtain $Q_m = 2.0$ Å² and $Q_m = 6.4$ Å², always at $u = 0.039$ eV. The alternative cross sections, derived from the results of $c(1)$, are shown in Fig. 4.10.2 (points 2), whereas those derived from the results of $f(1)$ are rejected because the spread is too large. Hake and Phelps have verified that the results of $c(2)$ are reasonably consistent with their Q_m curve; these authors also remark that one way of making all microwave results consistent with the w data would be to assume a resonance peak in the Q_m curve at about 0.15 eV.

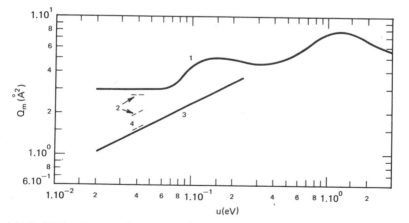

Fig. 4.10.2. Momentum transfer cross sections of electrons in oxygen, derived from swarm experiments [(1) Hake and Phelps, (2) Van Lint, (3) Mentzoni, (4) Veatch, Verdeyen, and Cahn].

If Gerjuoy's and Stein's rotational excitation cross sections are considered, eq. (2.6.70) can be made consistent with the experimental results of e if we neglect the first term, which is justified here since elastic losses turn out to be much smaller than inelastic ones also at room temperature. In this case agreement with the experimental $p_0\tau_r$ values is obtained, assuming $\mathcal{Q} = 2.8$. Using the corresponding eq. (1.5.23) and attributing the measured $G = 4.8 \times 10^{-3}$ value [see d above] to the energy of 0.1 eV, we find that radiation temperature measurements yield $\mathcal{Q} = 3.0$ if Hake's and Phelp's Q_m value is taken, or $\mathcal{Q} = 2.65$ if Mentzoni's (curve 3) value is chosen. Good agreement is thus found between the microwave determinations of the quadrupole moment, whereas a smaller value was required to fit the experimental drift velocities; spectroscopic techniques yield still lower values [see Appendix A, Table A.3].

Vibrational excitation collision cross sections adopted by Hake and Phelps represent only a first attempt to obtain a set of data consistent with known experiments. Therefore we shall not discuss them further here; the interested reader is referred to the original paper and to the recent electron-beam work of Spence and Schulz (1970), who confirm that the dominant feature of the vibrational cross section is represented by the existence of spikes but modify the spacing to about 0.11 eV.

Theoretical predictions of D_L/μ at 293°K based on the cross sections of Hake and Phelps are given by Lowke and Parker (1969).

Finally, we note that all the foregoing data refer to molecular oxygen. The only significant information regarding atomic oxygen is that provided by Daiber's and Waldron's ρ measurements, namely, $Q_m = 12 \pm 2$ Å² in the hypothesis of a velocity-independent collision cross section.

4.11 CARBON MONOXIDE

Experimental Data [dc parameters for $E/p < 7 \ (D/\mu \simeq 0.7 \text{ eV})$]

a. Drift velocity

(1) Pack, Voshall, and Phelps (1962) [see Kr a(2)] performed measurements at 77, 195, and 300°K for $E/p > 1 \times 10^{-3}$. The impurity content of the gas was 2.75×10^{-2} mole % of H_2 and 3×10^{-2} mole % of CO_2. However, at 300°K and below 10 torr, the authors found electron losses by attachment, probably due to compounds formed by the reaction of CO with the metallic parts of the vacuum system.

(2) Wagner, Davis, and Hurst (1967) [see He a(6)] performed measurements at room temperature over the range $5 \times 10^{-2} < E/p < 4$ in carbon monoxide with a purity of 99.5 mole %.

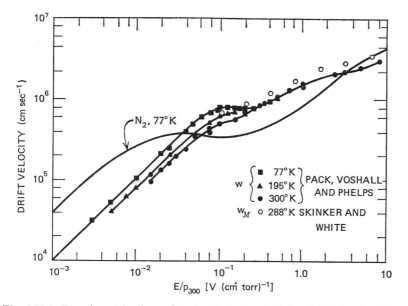

Fig. 4.11.1. Experimental values of Pack, Voshall and Phelps [$a(1)$] for the drift velocity at 77, 195, and 300°K and of Skinker and White (c) for the magnetic drift velocity at ~288°K in carbon monoxide. For $E/p < 3 \times 10^{-2}$, electrons are in thermal equilibrium with the gas; the corresponding μp_0 values are given in Table 4.11.2. The Pack and Phelps [N_2 $a(8)$] drift velocities for nitrogen are shown for comparison [after Pack, Voshall, and Phelps (1962)].

The results of (1) are shown in Fig. 4.11.1. Worth of mention in this figure is the minimum which characterizes the 77°K curve; the same feature was found at 77°K in nitrogen. The results of (2) agree with those of (1) within the scatter of both sets of data.

b. Diffusion to mobility ratio

(1) Skinker and White (1923) performed measurements at room temperature for $E/p > 0.1$, using the original Townsend method and the 4-cm diffusion chamber described under He $b(1)$.

(2) Warren and Parker (1962) [see He $b(2)$] performed measurements at 77°K over the range $1 \times 10^{-3} < E/p < 6$.

(3) Wagner, Davis, and Hurst (1967) [see He $b(4)$] determined the quantity D_L/μ at room temperature over the range $5 \times 10^{-2} < E/p < 4$. For gas purity, see $a(2)$.

The results of (2) and (3) are given in Fig. 4.11.2. The results of (1) agree well within the combined experimental errors with those of (2), E/p being large enough to overcome gas temperature differences.

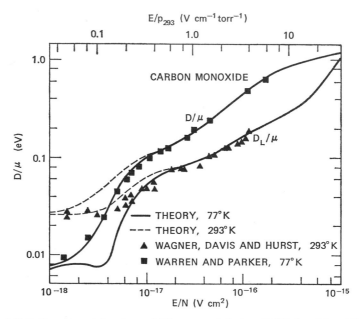

Fig. 4.11.2. Experimental values of Warren and Parker [$b(2)$] for D/μ at 77°K and of Wagner, Davis, and Hurst [$b(3)$] for D_L/μ at ~293°K in carbon monoxide. The curves are computed using the Hake-Phelps electron collision parameters [after Lowke and Parker (1969)].

c. Magnetic drift velocity (LMF case). The only experimental measurements are those of Skinker and White (1923). The results, which were obtained using the original Townsend method and the same tube and conditions as in $b(1)$ above, are plotted in Fig. 4.11.1 and yield Ψ values in the range 1.2–1.3.

d. Free electron diffusion. Nelson and Davis (1969) [see He $e(2)$] obtained $Dp_0 = 1.62 \times 10^5$ cm^2 sec^{-1} torr at 300 ± 2°K in chemically pure carbon monoxide (purity 99.5%).

e. Energy relaxation time. Mentzoni and Donohoe (1968) measured, using the energy decay method discussed in Section 3.13B, the value of the energy relaxation time pertaining to the isothermal afterglow of a pulsed dc discharge in carbon monoxide (main impurity content: helium, 268 ppm). The result at $T_g = 300$°K is

$$p_0\tau_r = 113 \pm 11 \quad (\text{nsec} \cdot \text{torr})$$

f. Cyclotron resonance. Tice and Kivelson (1967) [see He $i(3)$] performed line-shape measurements in CO-N$_2$ mixtures at room temperature.

Determination of Collision Parameters

Hake and Phelps (1967) performed a numerical analysis of type A, very similar to the one for oxygen. For this purpose they considered the average curves for $T_g = 77°K$ of the drift velocity [based on the results of $a(1)$] and of the characteristic energy [based on the results of $b(2)$ and on the room-temperature data of $b(1)$ for $E/p > 0.5$].

Table 4.11.1 The Momentum Transfer Cross Section $Q_m(u)$, Determined by Hake and Phelps for Electrons in Carbon Monoxide

u (eV)	Q_m (Å²)	u (eV)	Q_m (Å²)	u (eV)	Q_m (Å²)
0	60	0.040	5.2	1.3	31
0.0001	50	0.070	6.1	1.5	35
0.0010	40	0.10	7.3	1.7	30
0.0020	25	0.20	10.0	1.9	20
0.0040	14.0	0.40	13.5	2.1	15.0
0.0070	9.8	0.70	15.5	2.4	13.0
0.010	7.8	1.0	17.0	3.0	12.0
0.020	5.9	1.2	23	4.0	11.7

A good agreement between computed and experimental data is secured by using the collision cross section $Q_m(u)$ values of Table 4.11.1 and:

(a) for $D/\mu < 0.02$ eV, Takayanagi's cross sections (1.5.27) and (1.5.28) with $\mathcal{M} = 4.6 \times 10^{-2}$ au for the excitation of 16 rotational levels;

(b) over the range $0.02 < D/\mu < 0.06$, the rotational excitation cross sections given by an appropriate continuous approximation, only slightly better than that of formulas (1.5.30) and (1.5.31) (good results are obtained also below $D/\mu = 0.02$ when use of the continuous approximation is extended into that region);

(c) over the range $0.1 < D/\mu < 0.6$, the rotational excitation cross sections as in (b), and above the threshold energy 0.266 eV the $v(0 \rightarrow 1)$ vibrational excitation cross section shown in Fig. 4.11.3.

Hake and Phelps have also computed Ψ as a function of E/p; their theoretical results (Ψ in the range 1.3–1.5 for $0.2 < E/p < 5$) are significantly larger than the experimental ones. Using the same cross sections, Lowke and Parker (1969) computed D_L/μ at 77°K and at 293°K, and w and D/μ at 293°K. Here too there is considerable discrepancy (see Fig. 4.11.2) between calculated and experimental characteristic energy values, particularly from $E/p = 0.1$ to $E/p = 0.5$; also, the calculated drift velocities at 293°K differ from the experimental values by up to 7% in this E/p region. The above

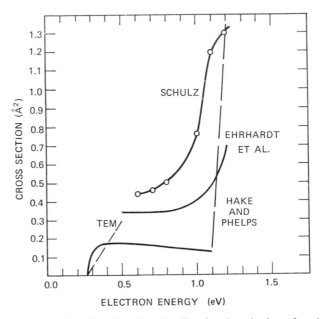

Fig. 4.11.3. Cross section for the $v(0 \to 1)$ vibrational excitation of carbon monoxide, determined from the analysis of swarm data by Hake and Phelps. Also shown are the results of beam experiments by Schulz, by Ehrhardt, Langhans, Linder, and Taylor, and near threshold by Burrow and Schulz (TEM: trapped-electron method) [after Burrow and Schulz (1969)].

results suggest that a better set of cross sections will have to be found if a more satisfactory agreement with experiments is desired.

As in the case of hydrogen, we have also computed μ and D values for isothermal electrons at various gas temperatures, using the Q_m values of Table 4.11.1; the results are given in Table 4.11.2. Agreement with the experimental data is not as good as for hydrogen and nitrogen.

Table 4.11.2 Mobility and Diffusion for Thermal Electrons in Carbon Dioxide, Computed Using the Cross Section $Q_m(u)$ of Hake and Phelps, and Compared with the Experimental Values of Pack, Voshall, and Phelps [$a(1)$] and Nelson and Davis (d)

$T_g(^\circ K)$	μp_0 (cm^2 V^{-1} sec^{-1} torr)		Dp_0 (cm^2 sec^{-1} torr)	
	Comp.	Exp. [$a(1)$]	Comp.	Exp. (d)
77	9.98×10^6	9.38×10^6	6.62×10^4	
195	7.87	7.23	1.32×10^5	
300	6.34	5.88	1.64	1.62×10^5

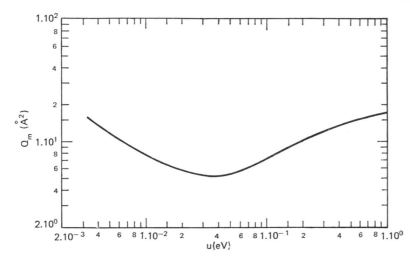

Fig. 4.11.4. Momentum transfer cross sections of electrons in carbon monoxide, derived by Hake and Phelps from swarm experiments.

It can be easily verified that the cross section Q_m of Table 4.11.1, which is also plotted in Fig. 4.11.4, can be regarded as approximately the sum of a dipole term, which can be calculated by means of Altshuler's formula (1.5.32), and a nitrogen-like cross section (CO and N_2 are isoelectronic and exhibit similar low-energy electronic states). The dipole moment value $\mathcal{M} = 4.6 \times 10^{-2}$ ($= 0.177$ debye) found by Hake and Phelps agrees rather well with the Stark shift value of 4.4×10^{-2} measured by Burrus (1958); the fact that in the cross section only the contribution of the dipole moment was considered, that of the quadrupole moment being neglected, is the probable cause of this small difference in $\mathcal{M}$ values [Singh (1970)]. The cyclotron resonance results of Tice and Kivelson (f) indicate that Q_m at thermal energies is given dominantly by Altshuler's term, namely, that $h \simeq -1$, and that $Q_m = 1.9 \pm 0.2$ Å² at about 0.039 eV; this value and the velocity dependence law differ so markedly from Hake's and Phelp's Q_m curve that we reject these results on the hypothesis that the measured line shapes were probably affected by some secondary effect (e.g., reaction of CO with active nitrogen could have occurred).

In Fig. 4.11.3 the low-energy portions of the analyzer data of Schulz (1964) and of Ehrhardt, Langhans, Linder, and Taylor (1968), and the cross-section slope at threshold, measured by Burrow and Schulz (1969) using the trapped-electron method, are also shown for comparison purposes. The values of the vibrational excitation cross section of Hake and Phelps below 1.1 eV are very close to those obtained using Takayanagi's (1966) formula for the dipole interaction in the Born approximation, and assuming the $v(0 \rightarrow 1)$ transition probability given by Penner (1959). As discussed at the end of Section 1.5, however, the most recent theoretical developments indicate that a polarization interaction term of comparable size must be added to the dipole one. Above 1.1 eV the data of Hake and Phelps are only approximate, and have been chosen, on the basis of the existence of a very large resonance observed by Schulz, for the sole purpose of insuring a reasonable behavior of the calculated distribution functions at energies above 1 eV.

The rotational excitation cross sections of Gerjuoy and Stein for the quadrupole interaction ($\mathcal{Q} = 1.3$) and of Takayanagi for the dipole interaction ($\mathcal{M} = 4.6 \times 10^{-2}$) were used by Mentzoni and Donohoe (1968) for computing electron energy relaxation times in Maxwellian plasmas; they found, for small deviations from the isothermal $T_g = 300°K$ condition, $p_0 \tau_r = 338$ (nsec $\cdot$ torr). This value is about three times larger than the experimental one; no significant explanation of this discrepancy has yet been found.

4.12 CARBON DIOXIDE

Experimental Data [dc parameters for $E/p < 10$ $(D/\mu \simeq 1$ eV)]

a. Drift velocity

(1) Unpublished data of Errett (1951) [see He $a(2)$] are reported by Hake and Phelps (1967) for $E/p > 0.5$ and $T_g = 293°K$.

(2) English and Hanna (1953) [see Xe $a(1)$] performed room-temperature measurements in 99.956% pure carbon dioxide over the range $1 \times 10^{-2} < E/p < 1$.

(3) Pack, Voshall, and Phelps (1962) [see Kr $a(2)$] performed measurements at 195 (boiling point), 300, and 410°K for $E/p > 1 \times 10^{-2}$. The impurity content of the gas was 5×10^{-3} mole % of H_2.

(4) Levine and Uman (1964) [see Ar $a(9)$] performed measurements at room temperature over the range $0.3 < E/p < 2$. Individual results show significant scatter, as in argon and nitrogen.

(5) Elford (1966) used a drift tube similar to that of Fig. 3.4.2 for measurements at 293°K; the author places an error limit of ± 0.5% for $0.1 < E/p < 3$ and of ± 1% for $4 < E/p < 7$.

(6) Wagner, Davis, and Hurst (1967) [see He $a(6)$] performed measurements at room temperature over the range $8 \times 10^{-2} < E/p < 2$ in carbon dioxide with an impurity content of 0.2 mole %. These data supersede previous results, obtained at Oak Ridge using earlier versions of the same basic method [Bortner, Hurst, and Stone (1957), Hurst, Stockdale, and O'Kelly

(1963), Hurst, O'Kelly, Wagner, and Stockdale (1963), Christophorou, Hurst, and Hadjiantoniou (1966)].

(7) Lehning (1968) adopted the induced current method to perform room-temperature measurements over the range $0.1 < E/p < 10$ at very high pressures ($10 < p < 33$ atm); the accuracy was 2–3%.

(8) Allen and Prew (1970) [see Ar $a(12)$] performed measurements at 295°K over the range $3.3 \times 10^{-2} < E/p < 3.3 \times 10^{-1}$, using pressures from 2.3 to 17 atm and analytical-grade carbon dioxide (air content ~20 ppm).

The results of the various authors agree well within the combined experimental errors. Those of (5), which are the most significant if we consider both accuracy and extent of the E/p range, are summarized in Table 4.12.1. Below $E/p = 1$ drift velocities correspond to a constant μp_0 value, which is equal to 5.0×10^5 cm^2 V^{-1} sec^{-1} torr, and, according to the results of (3), they are practically independent of temperature from 195 to 410°K. Therefore in CO_2, on the basis of w measurements alone, it is not possible to specify the E/p value below which electrons are certainly isothermal; such a value ($E/p \simeq 0.5$) is given instead from D/μ measurements.

The results of (7) and (8) indicate that at pressures above 10 atm the drift velocity decreases with increasing pressure, but the data do not satisfy relationship (4.7.1). According to Lehning, the effect becomes larger when E/p decreases, and remains constant below $E/p = 1$, thus indicating that it is mainly a function of the average electron energy or of D/μ.

b. Diffusion to mobility ratio

(1) Skinker (1922) performed measurements at ~288°K for $E/p > 0.2$, using the original Townsend method, his approximate formulas, and the 4-cm diffusion chamber described under He $b(1)$. No significant presence of negative ions was observed after all the diffusing electrons had been deflected out of the central strip by means of a transverse magnetic field.

(2) Unpublished data of D/μ, and B/p values for which $D_T/\mu_{||} = kT_g/e$ [see eq. (3.9.13)], obtained at room temperature for $E/p > 2$ by Rudd using Bailey's method for attaching gases, were reported by Healey and Reed (1941).

(3) Cochran and Forester (1962) [see H_2 $b(4)$] performed D/μ measurements at room temperature over the range $0.8 < E/p < 5$. The diffusion chamber was the same as that used for nonattaching gases, but no inconsistency of results was reported by the authors as being due to the presence of negative ions, so that to have considered as negligible the influence of attachment appears to have been justified. Commercial tank carbon dioxide (stated purity 99.95%) was used after being further purified by distillation from liquid-air traps.

Table 4.12.1 Experimental Values of the Drift Velocity and of the Characteristic Energy in Carbon Dioxide

D/μ values at 195°K were read on Fig. 12 of Warren and Parker's paper and are therefore affected by reading errors.

$10^{17}E/N$ (V cm²)	E/p_{293} (V cm⁻¹ torr⁻¹)	$10^{-5}w$ (cm sec⁻¹) [$T_g = 293°K$; a(5)]	D/μ (eV) [$T_g = 293°K$; b(5)]	D/μ (eV) [$T_g = 195°K$; b(4)]
3.03×10^{-1}	1.0×10^{-1}	5.41×10^{-1}	2.55×10^{-2}	1.75×10^{-2}
4.55	1.5	8.11	2.56	1.80
6.07	2.0	1.08×10^{0}	2.58	1.85
9.10	3.0	1.62	2.61	1.95
1.214×10^{0}	4.0	2.16	2.65	2.05
1.82	6.0	3.24	2.75	2.15
2.43	8.0	4.32	2.86	2.30
3.03	1.0×10^{0}	5.42	3.01	2.50
4.55	1.5	8.24	3.38	2.90
6.07	2.0	1.13×10^{1}	3.86	3.45
9.10	3.0	1.82	5.68	5.4
1.214×10^{1}	4.0	2.73	9.47	9.2
1.82	6.0	5.46	3.38×10^{-1}	3.35×10^{-1}
2.43	8.0		6.97	
3.03	1.0×10^{1}		1.01×10^{0}	

(4) Warren and Parker (1962) [see He $b(2)$] performed D/μ measurements at 195°K and 300°K over the range $5 \times 10^{-3} < E/p < 1.2$. The considerations on attachment mentioned in (3) apply here also.

(5) Rees (1964) adopted the Townsend-Huxley method and a tube similar to the one of Fig. 3.7.2 in order to perform measurements of D/μ at 293°K for $E/p > 0.1$. Both commercial carbon dioxide (purified by distillation from liquid-air traps and dried by phosphorus pentoxide) and reagent-grade carbon dioxide were used. The author remarks that, in the E/p range we consider, he did not observe any effect attributable to attachment; in fact, consistent results were always obtained, even when using the tube according to the original Townsend method. The results are estimated to be accurate within $\pm 1\%$.

(6) Wagner, Davis, and Hurst (1967) [see He $b(4)$] determined the quantity D_L/μ at room temperature over the range $8 \times 10^{-2} < E/p < 1$. For the impurity content of the gas, see $a(6)$. These data supersede previous measurements, performed using a similar technique, but exhibiting a large scatter [Hurst, O'Kelly, Wagner, and Stockdale (1963)].

The most reliable and accurate room-temperature results of D/μ are those of (5), which are summarized in Table 4.12.1. The results of (2), except the value at $E/p = 2$, and of (4) agree well with (5) within the combined experimental errors. The results of (1) and of (3) disagree badly and are rejected. The D/μ results of (4) at 195°K are also given in Table 4.12.1. Below $E/p = 1$ the values of D/μ are nearly thermal; the results of (6) indicate, however, that D_L/μ remains consistently about 25% higher over the same range. Surprisingly, the measured values do not appear to tend to the thermal level.

c. *Magnetic drift velocity (LMF case).* Skinker's (1922) w_M measurements are rejected on the basis of his poor D/μ results [see b above], obtained with the same tube and conditions.

d. *Free electron diffusion.* Nelson and Davis (1969) [see He $e(2)$] obtained $Dp_0 = 1.37 \times 10^4$ cm² sec⁻¹ torr at 300 ± 2°K from measurements performed in 99.99% pure carbon dioxide.

e. *Cyclotron resonance*
(1) Fehsenfeld (1963) [see He $i(2)$] obtained at 300°K:

$$\Delta\omega_I = (1.41 \pm 0.14) \times 10^9 p \quad (\text{sec}^{-1})$$

(2) Tice and Kivelson (1967) [see He $i(3)$] performed line-shape and $\Delta\omega_I$ measurements in CO_2-N_2 mixtures at room temperature [these results supersede previous ones by Bayes, Kivelson, and Wong (1962)].

Determination of Collision Parameters

Hake and Phelps (1967) performed a numerical analysis of type A, very similar to the one for oxygen. For this purpose they considered the drift velocity data $a(1)$, $a(3)$, and $a(5)$ and D/μ data $b(1)$, $b(2)$, $b(4)$, and $b(5)$. A good agreement between computed and experimental data is secured, neglecting rotational excitation losses compared to vibrational ones (the lowest threshold of these states is, in fact, very low: 0.083 eV), and using:

(a) the collision cross section $Q_m(u)$ values of Table 4.12.2 [below $u = 0.1$

Table 4.12.2   The Momentum Transfer Cross Section $Q_m(u)$, Determined by Hake and Phelps for Electrons in Carbon Dioxide

u (eV)	Q_m (Å²)	u (eV)	Q_m (Å²)	u (eV)	Q_m (Å²)
0	600	0.10	52	2.2	5.3
0.0010	540	0.15	42	2.8	8.5
0.0020	380	0.20	34	3.6	15.8
0.0040	270	0.30	18.0	4.0	17.1
0.0070	200	0.42	9.7	4.5	17.0
0.010	170	0.60	5.7	5.2	13.4
0.020	120	0.85	4.4	6.4	10.5
0.040	85	1.0	4.1	8.0	11.7
0.070	64	1.5	4.1	10.0	12.9

eV a constant ν_m value ($\nu_m/p_0 = 3.5 \times 10^9$ sec^{-1} torr^{-1}) is taken consistently with the temperature independence of the thermal mobility];

(b) vibrational excitation cross sections consisting of a set of four resonances (at 0.08, 0.3, 0.6, and 0.9 eV), the first two of which are assumed to have high-energy tails, with cross sections given by the formula of Takayanagi (1966) for the dipole interaction in the Born approximation.

The constant $\nu_m = 3.5 \times 10^9 p_0$ (sec^{-1}) value yields $Dp_0 = 1.3 \times 10^4$ cm² sec^{-1} torr at 300°K in good agreement with the results of d. For a constant collision frequency, $e(1)$ predicts a much lower value: $\nu_m = 2.45 \times 10^9 p_0$ (sec^{-1}) and $e(2)$ a larger one $\nu_m = 4.5 \times 10^9 p_0$ (sec^{-1}). The cross section values of Table 4.12.2 and the ν_m values from $e(1)$ and $e(2)$ experiments are plotted in Fig. 4.12.1 as curve 1 and as points 2 and 3, respectively.

The existence of vibrational resonances at 0.3, 0.6, and 0.9 eV, and of the high-energy tail for the first one, has been assumed on the basis of observations by Schulz. However, the validity of Takayanagi's formula for the energy dependence of the vibrational excitation cross sections in CO_2 has been recently questioned [Claydon, Segal, and Taylor (1970)], so that its use by Hake and Phelps is now open to criticism. For this reason, and since these

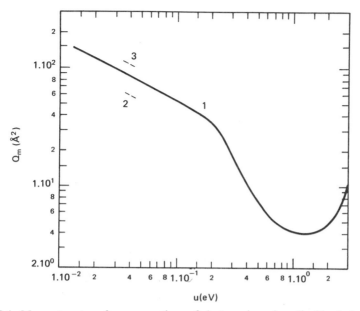

Fig. 4.12.1. Momentum transfer cross sections of electrons in carbon dioxide, derived from swarm experiments [(1) Hake and Phelps, (2) Fehsenfeld, (3) Tice and Kivelson].

cross sections are under reevaluation, taking into account the results of recent electron-beam measurements [private communication by Phelps; preliminary results in Phelps (1970)], no detailed information concerning the vibrational excitation cross sections of Hake and Phelps is reported here.

Theoretical predictions of D_L/μ at 293°K, based on the Hake-Phelps cross sections, are given by Lowke and Parker (1969).

Here, too, the attachment of thermal electrons to molecules to form unstable, short-lived ions is regarded as the most probable cause of the pressure dependence of drift velocity at very high pressures (see Section 4.7). Since $1/w$ is not linearly related to pressure [eq. (4.7.1)], Lehning interprets this fact as due to a pressure dependence of the ion lifetime before detachment; this happens if the neighboring CO_2 molecules stabilize the temporarily attached electron to a certain extent.

4.13 WATER VAPOR

Experimental Data [dc parameters for $E/p < 20$ $(D/\mu \simeq 1.6 \text{ eV})$]

a. Drift velocity

(1) Pack, Voshall, and Phelps (1962) [see Kr $a(2)$] performed measurements at 300°K and 443°K for $E/p > 2 \times 10^{-2}$, using water vapor prepared

by multiple distillation. These authors found no evidence of attachment in pure water vapor.

(2) Lowke and Rees (1963) performed measurements at 293°K over the range $0.35 < E/p < 15$, using a double-shutter tube; the estimated accuracy is better than 2%. The water vapor was obtained from distilled water, de-ionized, and outgassed. Measurements in nitrogen-water vapor mixtures at $E/p = 0.2$ yielded a mobility value that, for a high water content, equals within the experimental error the isothermal value of mobility in pure water vapor.

(3) Ryzko (1966) performed measurements at room temperature for $E/p > 8$ in the drift chamber described in Section 3.4A; the author used pure distilled water, properly outgassed. He estimates that the drift velocity values are subject to an error of less than 5%.

(4) Christophorou and Christodoulides (1969) performed accurate measurements at room temperature over the range $1 < E/p < 12$, in a drift chamber such as the one shown in Fig. 3.4.5; triple-distilled water (purity 99.99%) was used. These measurements were subsequently extended for low E/p (<3) from 298°K up to $\sim$440°K by Christophorou and Pittman (1970).

Below $E/p = 6$ the drift velocity depends linearly on E/p (isothermal electrons); the corresponding μp_0 values (in units of $cm^2 \ V^{-1} \ sec^{-1}$ torr) are satisfactorily consistent among the various authors and increase from 6.6 [(2) and (4)]–6.75 $\times$ 10^4 [(1)] at room temperature up to 8.5 $\times$ 10^4 at $\sim$440°K. Above $E/p = 6$, the results that we regard as the most reliable ones at room temperature are those of (4) up to $E/p = 12$ and those of (3) above this value; a corresponding plot is shown in Fig. 4.13.1. The results of (1) agree well with the above ones up to $E/p \simeq 18$; the results of (2) over the common E/p range are lower, with a maximum discrepancy of 7%, which just exceeds the combined experimental errors.

Data on the electron collision cross sections in water vapor have also been obtained from drift velocity measurements of isothermal electrons in room-temperature ethylene containing small percentages of H_2O (max. 8%), for E/p values below 0.1. In this region the results satisfy the relationship:

$$\frac{1}{\mu p} = \frac{1}{(\mu p)_1}\left(1 + \frac{p_2}{p_1} S\right) \tag{4.13.1}$$

where index 1 refers to ethylene, index 2 to water vapor, and S is approximately 134 according to measurements by:

(5) Hurst, Stockdale, and O'Kelly (1963), who determined drift velocities from the pulse duration of the collector induced current following a single alpha-particle ionization along a track parallel to the electrodes;

(6) Hurst and Parks (1966) [see H_2 $a(4)$].

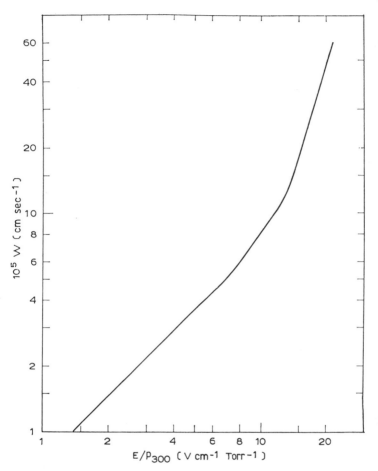

Fig. 4.13.1. Drift velocity in water vapor at room temperature, according to the experimental results of Christophorou and Christodoulides [a(4)] for $E/p < 12$ and of Ryżko [a(3)] for $E/p > 12$.

Hurst, Stockdale, and O'Kelly found that the same S value holds also in the case of ethylene-heavy water (D_2O) mixtures.

b. Diffusion to mobility ratio. The only data available below $E/p = 20$ are those of Bailey and Duncanson (1930), who measured D/μ and $D_T/\mu_{\|}$. However, above $E/p = 20$ these data are well below later, more accurate determinations performed by Crompton, Rees, and Jory (1965); this fact casts considerable doubts on the correctness of the older results, which are therefore omitted from our consideration. It seems also that other, more

Table 4.13.1 Summary of Q_m Determinations in Water Vapor

Since data are for the thermal energy range only, cross sections are expressed more significantly in terms of the electron energy measured in units of the average energy in a 300°K plasma ($\bar{u} = 0.0388$ eV); the electron energy in this scale is denoted as u_r ($= u/0.0388$).

Case No.	Source	Transport Parameters	Range of Swarm Mean Energies	Type of Analysis	Q_m (Å^2)
1	Pack, Voshall, and Phelps (1962)	$a(1)$	0.039–0.057	D [eq. (2.6.24)]	$940u_r^{-1}/(1 + 0.18u_r^{0.5})$
2	Christophorou and Christodoulides (1969)	$a(2), a(4)$	0.039	D [eq. (2.6.23'), $h = -1$]	790
3	Christophorou and Pittman (1970)	$a(4)$	0.039–0.057	D [eq. (2.6.23')]	$810u_r^{-1.085}$
4	Christophorou, Hurst, and Hendrick (1966)	$a(5), a(6)$	0.039	D [eq. (2.6.32'), $\beta = \frac{4}{7}$]	680
5	Takeda and Dougal (1960)	c	0.039	D [eq. (2.6.41), $h = -1$]	400 ± 80
6	Tice and Kivelson (1967)	d	0.039	D [eq. (2.6.95), $h = -1$]	840 ± 120

412

recent attempts to measure D/μ in water vapor have failed to provide consistent results; since D_L/D may attain in this gas values as large as 6 [Lowke and Parker (1969)], the use of formula (3.5.10) instead of (3.5.9) can be quite erroneous in small chambers, and this may be a possible explanation of the failures.

c. Microwave conductivity ratio ($\omega^2 \gg \nu_m^2$). Takeda and Dougal (1960) obtained $\omega\rho = (1.5 \pm 0.3) \times 10^{10}$ (sec $\cdot$ torr)$^{-1}$ from transmission measurements through a late-afterglow plasma contained in an X-band waveguide; isothermal conditions at 300°K are assumed. These authors also performed measurements of the conductivity ratio in the presence of microwave heating, but we shall disregard these results, since the derivation of the corresponding electron temperatures, based on the D/μ data of Bailey and Duncanson, is open to question.

d. Cyclotron resonance. Tice and Kivelson (1967) [see He i(3)] performed line-shape and $\Delta\omega_I$ measurements in H_2O-N_2 mixtures at room temperature [their results supersede earlier ones by Bayes, Kivelson, and Wong (1962)].

Determination of Collision Parameters

No analysis of type A is available for water vapor. Hence the only data that can be handled easily are the transport parameter values for isothermal electrons; the Q_m values so derived are given in Table 4.13.1.

Except for the results of a(1) and a(4), all values are for 300°K only. The choice $h = -1$ is based on accepting the functional dependence of Q_m on v given by Altshuler's formula (1.5.32). Equation (4.13.1) reduces to (2.6.32'), since experiments were conducted under conditions for which $p_1 \simeq p$; Christophorou, Hurst, and Hendrick assume that in ethylene Q_m is constant and has the value 4.37 Å^2 (see Section 4.15). According to the mentioned results of a(5), the collision cross section of H_2O holds also for D_2O.

Except for the value obtained by Takeda and Dougal, all other Q_m determinations are consistent and provide values about twice those predicted by Altshuler's formula (1.5.32) with the appropriate $\mathscr{M}$ value of 0.73 ($= 1.85$ debye). Crawford (1968) has shown that about half the difference can be removed if close-coupling scattering calculations with an appropriate interaction potential are performed.

4.14 ALKALI-METAL VAPORS

Experimental Data

A major source of uncertainty in most of the experiments that will now be reported is faulty knowledge of the vapor pressure. This pressure is usually determined from the value of the temperature of the liquid-phase reservoir, but because of a combination of diffusion and wall-coating effects it takes a long time (a few hours) before the pressure reaches equilibrium. This can become a significant source of error, particularly when the system is subsequently sealed off from the liquid reservoir.

a. Drift velocity

(1) Chanin and Steen (1964) performed measurements in pure cesium (99.99%) at temperatures between 566°K and 725°K for $0.8 < E/p < 20$, using the induced current technique with a variable drift distance (up to $d = 5$ cm) and a pulsed discharge electron source. The reproducibility of the data was approximately 50%, but significant space-charge distortion could have altered the mobility measurements [Ward (1966)]. The results are plotted in Fig. 4.14.1.

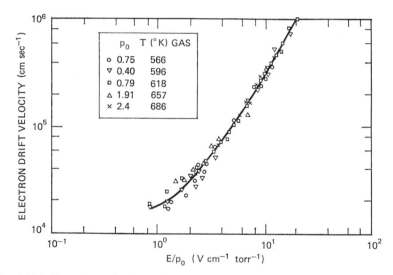

Fig. 4.14.1. Experimental values of the drift velocity in cesium [after Chanin and Steen (1964)].

(2) Nolan and Phelps (1965), on the other hand, measured drift velocities in argon contaminated by small amounts of cesium (concentration ratios between 10^{-8} and 10^{-5}) at temperatures between 520°K and 580°K and for $1 \times 10^{-2} < E/p < 0.15$. Under these conditions the contribution of cesium to the total collision cross section for the mixture is negligible, but the change in the distribution function from the corresponding pure argon case is noticeable, since high-energy electrons excite the first electronic states of cesium (threshold energies: 1.386 eV for the $6^2P_{1/2}$ state and 1.454 eV for the

$6^2P_{3/2}$ state) and are then transferred to the low-energy region, where they experience far fewer collisions because of the Ramsauer minimum of Q_m in argon. Drift velocities larger than those in pure argon are thus observed for concentration ratios $> \sim 10^{-6}$ and for $E/p > 2 \times 10^{-2}$. The method used by Nolan and Phelps was thoroughly described in Section 3.4G.

   b. *Direct current conductivity.* Measurements in hot diode cells filled with pure cesium vapor have been performed by:

   (1) Mirlin, Pikus, and Yur'ev (1962) ($1300 < T_g < 1700°K$);
   (2) Roehling (1963) ($1100 < T_g < 1900°K$);
   (3) Jermokhin et al. (1969) ($1400 < T_g < 2100°K$).

   A sample of data taken from the results of Jermokhin et al. is shown in Fig. 4.14.2; the data of (2) also agree with the curves of the figure within the large scatter of this type of data.

   (4) Harris (1963) performed conductivity measurements in cesium-seeded inert gases (He, Ne, Ar, N_2), using the hot diode cell discussed in Section 3.8B. The operating temperatures ranged from 1500 to 2000°K, whereas the total pressure was 1 atm and the cesium vapor pressures (p_c) ranged from 0.1 to 10 torr. For each inert gas at a fixed temperature Harris found that σ_{dc} versus p_c shows a maximum (σ_{max}), and that the values of these maxima for the four gases at $T_g = 1750°K$ are given by (σ in S m^{-1}, p_c in torr):

$$\sigma_{max} = 7.7(p_{c\ max})^{-1/2}$$

These results can easily be interpreted with reference to eqs. (2.4.20), (2.6.31), and (3.8.7). In fact, if n_0 is the electron density predicted by Saha's equation for the cesium at a reference pressure p_0, and if the electron collision frequencies in the inert gas (ν_g) and in cesium (ν_c) are energy independent, we have:

$$\sigma_{dc} = \frac{e^2 n_0}{m p_0^{1/2}} \frac{p_c^{1/2}}{\nu_g + (\nu_c/p_c)p_c} \qquad (4.14.1)$$

This expression has a maximum when p_c equals $\nu_g/(\nu_c/p_c)$, that is, when the collision frequencies of the two gases are equal. The maxima are:

$$\sigma_{max} = \frac{e^2 n_0}{2 m p_0^{1/2}} \left(\frac{p_c}{\nu_c}\right)(p_{c\ max})^{-1/2} \qquad (4.14.2)$$

in accordance with the experimental results. Furthermore, Harris found that, within the experimental errors (estimated to be about 40%), the changes of conductivity with the cell temperature and with the seeding gas (potassium) [Harris (1964)] are due only to electron density changes, as predicted by Saha's equation (3.8.7).

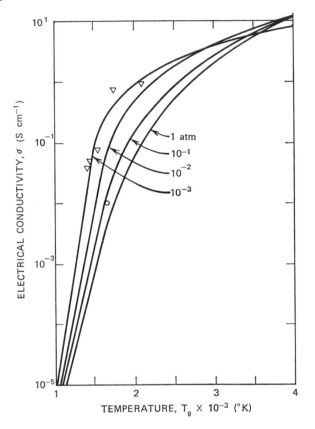

Fig. 4.14.2. The results of Jermokhin et al. for dc electrical conductivity measurements in cesium, plotted as a function of the gas temperature and for different pressures (∇ : 10^{-3} atm, $\bigcirc$: 10^{-1} atm); the solid curves were computed by the same authors using curve 3 of Fig. 4.14.5 [after Jermokhin et al. (1969)].

(5) Sakao and Sato (1969), using a four-electrode arrangement, performed hot cell conductivity measurements in potassium-seeded argon (0.013 and 0.13 % mole fraction of potassium in argon at 1 atm, $1670 < T_g < 1870°K$). The results are shown in Fig. 4.14.3.

c. Direct current conductivity in magnetic fields. Mullaney and Dibelius (1961) used the method and techniques discussed in Sections 2.6C and 3.9D for measurements in cesium. In accordance with eq. (2.6.93), at $T_g = 1125°K$ these authors found (B in gauss):

$$\sigma_{dc}/\sigma_{dc\ T} = 1 + 6.25 \times 10^{-8}B^2$$

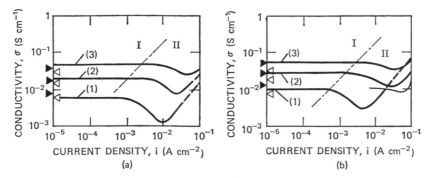

Fig. 4.14.3. Sakao's and Sato's experimental values of the dc electrical conductivity in potassium-seeded argon [(a) 0.013 % and (b) 0.13 % mole fraction of K_2 in Ar] as a function of the current density in the hot cell for different gas temperatures [(1) 1400°C, (2) 1500°C, (3) 1600°C]; in region I electrons can be considered isothermal. Closed triangles at the ordinate indicate theoretical values computed with the cross sections assumed by Sakao and Sato for the best fit to the measured conductivities; open triangles are theoretical values computed using cross sections proposed earlier by Frost (1961) [after Sakao and Sato (1969)].

d. Microwave conductivity ratio $(\omega^2 \gg \nu_m{}^2)$

(1) Chen and Raether (1962) performed conductivity ratio determinations in the temperature range 450–550°K, by means of transmission and reflection measurements in the late isothermal afterglow of pulsed dc discharges, established in pure cesium and in cesium-helium mixtures. The afterglow was housed in a cylindrical Pyrex tube, inserted coaxially in an X-band square waveguide. The vapor was obtained through vacuum distillation of 99.9 % pure cesium.

(2) Balfour and Harris (1966) performed X-band cavity measurements, using the reflection method to study the characteristics of the isothermal afterglow of a pulsed dc discharge in cesium. Two tubes were adopted: a straight one of small diameter (6 mm), inserted through the end plates of the cavity, and another forming a cylindrical bottle within the cavity. The discharge tube and cavity were maintained at 680°K.

(3) Balfour, Hart, and Haynes (1966) performed waveguide transmission measurements, using a 35-GHz interferometer. A waveguide section was heated to 1600°K, so that thermal ionization of cesium vapor was established. The measured electron densities were found to be consistent with Saha's equation.

The results are given in Table 4.14.1. The data of (3) are converted into equivalent values for the limit $\omega_p{}^2 \ll \omega^2$, since this condition does not hold in the experiment.

Table 4.14.1   Experimental Values of the Microwave Conductivity Ratio in Cesium

Authors	Temperature Range ($^\circ$K)	$10^{-10} \omega \rho$ (sec $\cdot$ torr)$^{-1}$
(1)	450–550	$\left[1.55 \times 10^{-3} \left(\dfrac{p_{He}}{p_{Cs}} \right) + 0.595 \right] T_e^{1/2} - 28.2 + 472 T_e^{-1/2}$
(2)	680	3.35
(3)	1600	7.0

e. Cyclotron resonance

(1) Meyerand and Flavin (1964), using the setup shown in Fig. 3.14.3, performed measurements of cyclotron resonance curves at X-band in cesium over the vapor temperature range 570–1160°K.

(2) Ingraham (1966), using the setup shown in Fig. 3.14.2, performed C-band measurements of the line width $\Delta\omega_{0.5}$ in cesium over the temperature range 550–3100°K.

Meyerand and Flavin do not report original experimental data; they give only derived collision cross sections, which we shall consider later. Ingraham's experimental results are shown in Fig. 4.14.4.

Determination of Collision Parameters

The various determinations in cesium are described in Table 4.14.2 and are shown in Fig. 4.14.5. Since in the case of thermal cesium plasmas it has been customary for most authors to refer the cross-section values derived from the high-gas-temperature results to the corresponding energies $\hat{u}$ (rather than to $\bar{u}$), here too all data obtained from type D analyses are given and plotted following this practice. Roehling's (1963) determinations of Q_m have not been included, since this author finds a dependence $Q_m \propto v^4$, which does not permit the use of the correct formula (2.6.23′) $[\gamma_p{}^q \rightarrow \infty]$.

The only known analysis of type A is due to Postma (1969a) [case 1], who found the best fit for the experimental data of $a(1)$. However, unlike the situation for the other gases discussed so far, electronic excitation levels, which in cesium are very low, have here to be included in the analysis. For this purpose Postma used the excitation cross sections of the $6^2P_{3/2}$ and $6^2P_{1/2}$ resonant states measured by Zapesochnyi (1967) and lumped them together into a single transition with an excitation energy of 1.42 eV and a cross section equal to the sum of the individual ones.

Nolan and Phelps (1965) used their w measurements in argon-cesium mixtures [see $a(2)$] for a similar analysis, but for the purpose of determining

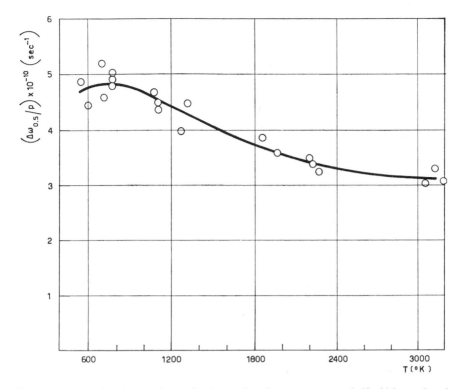

Fig. 4.14.4. Ingraham's experimental values of cyclotron resonance half-widths, reduced to 1 torr pressure, as a function of cesium temperature [after Ingraham (1966)].

the above-mentioned excitation transitions. In their case this was possible, since cesium concentrations were very low and the cross section Q_m could be identified with that of pure argon, already known (see Fig. 4.4.9, curve 1). Thus Nolan and Phelps found good results, assuming a linear cross section of slope 71 $\text{Å}^2 \text{ eV}^{-1}$, from a single excitation threshold at 1.386 eV up to about 1.8 eV.

Since Postma's results are derived from a type A analysis, they should be regarded as the most reliable ones. However, doubts have been cast on them, since drift velocities measured in cesium arc columns at slightly higher average electron energies (0.25–0.5 eV) are inconsistent with the Q_m curve derived by Postma, but lend support to a different Q_m curve, having the same Ramsauer-like shape but the minimum located at a much higher energy (~0.35 eV) [Nigham (1967), Postma (1969b), Andersen and Nigham (1970)].

Very few data are available for potassium. Harris (1964) [see b(4)] found

Table 4.14.2 Summary of Q_m Determinations in Cesium

Case No.	Source	Transport Parameters	Range of Swarm Mean Energies	Type of Analysis	Q_m (Å^2)
1	Postma (1969a)	$a(1)$	$\sim 0.1 - \sim 0.25$	A [eqs. (2.3.51), (2.4.20)]	See Fig. 4.14.5
2	Mirlin, Pikus, and Yur'ev (1962)	$b(1)$	0.17–0.22	D [eq. (2.6.23'), $h = 0$]	~ 200 at $u = 0.13$
3	Jermokhin et al. (1969)	$b(3)$	0.18–0.27	B [eqs. (2.3.31), (2.4.20)]	See Fig. 4.14.5
4	Harris (1963)	$b(4)$	0.23	D [eq. (4.1.42)]	~ 300 at $u = 0.15$
5	Mullaney and Dibelius (1961)	c	0.145	D [eq. (2.6.93)]	36 at $u = 0.1$
6	Chen and Raether (1962)	$d(1)$	0.058–0.071	D [eq. (2.6.49)]	$2050 - 1360u^{-0.5} + 282u^{-1}$
7	Balfour and Harris (1966)	$d(2)$	0.088	D [eq. (2.6.41), $h = 1$]	440
8	Balfour, Hart, and Haynes (1966)	$d(3)$	0.21	D [eq. (2.6.41), $h = 1$]	600
9	Meyerand and Flavin (1964)	$e(1)$	0.074–0.15	D [see p. 120]	See Fig. 4.14.5
10	Ingraham (1966)	$e(2)$	0.071–0.80	D [eq. (2.6.97), $h = 0$]	See Fig. 4.14.5

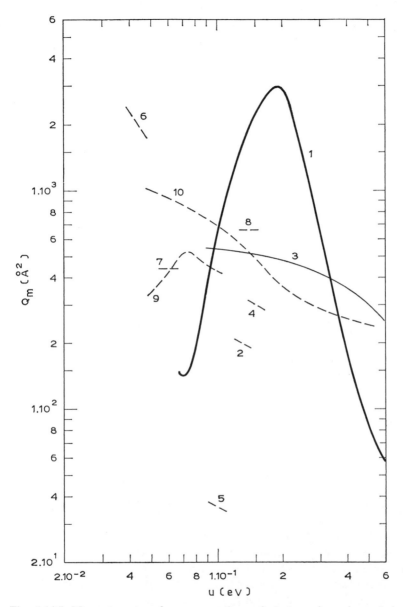

Fig. 4.14.5. Momentum transfer cross sections of electrons in cesium, derived from swarm experiments (numbers as in Table 4.14.2).

that the same Q_m value that he had derived from cesium experiments predicts conductivity values in agreement with the experimental results in potassium. Sakao and Sato (1969) [see $b(5)$], using eq. (2.6.23') and $h = 1$, obtained $Q_m = 230$ Å^2 at about 0.15 eV, but for the derivation they assumed velocity-independent Q_m values for argon (0.23 Å^2) and for positive ions (4×10^4 Å^2) too.

4.15   OTHER GASES

Experimental Data

a. Drift velocity

(1) Nielsen and Bradbury (1937) [see He $a(1)$] performed measurements at 293°K in N_2O for $E/p > 0.3$ and in NH_3 for $E/p > 0.2$. These authors used nitrous oxide obtained by fractional distillation, at reduced pressures and liquid-air temperatures, of commercial cylinder gas of high purity, and ammonia obtained by liquefaction over metallic sodium and subsequent fractional distillation, at reduced pressures and liquid-air and solid CO_2 temperatures, of commercial cylinders of anhydrous ammonia gas.

(2) English and Hanna (1953) [see Xe $a(1)$] performed measurements at room temperature in 99% pure methane over the range $7 \times 10^{-2} < E/p < 0.63$.

(3) McCutchen (1958) determined drift velocities in triply distilled mercury vapor at room temperature, from the pulse duration of the collector induced current, after ionization by collimated alpha-particles parallel to the electrodes.

(4) Pack, Voshall, and Phelps (1962) [see Kr $a(2)$] performed measurements in N_2O at 195 and 300°K for $E/p > 4 \times 10^{-2}$ and in NH_3 at 195, 300, and 381°K for $E/p > 1 \times 10^{-2}$. The gas samples were obtained from liquid nitrous oxide (98% pure) and liquid ammonia (99.99% pure), the ammonia gas being changed frequently to avoid dissociation effects.

(5) At Oak Ridge National Laboratory a large number of gases, mostly hydrocarbons, have been investigated, using the methods and tubes described in Sections 3.2B and 3.4E. In many gases measurements have been extended to E/p values as low as necessary in order to find the linear w versus E/p relation, which is appropriate to thermal equilibrium between electrons and gas molecules. The best data appear to be the following: for methane, those of Wagner, Davis, and Hurst (1967) [see He $a(6)$], obtained at room temperature in a gas of minimum purity of 99.0 mole %; for ethylene, those of Christophorou, Hurst, and Hendrick (1966) (confirmed by a few subsequent measurements), obtained with the drift tube of Fig. 3.4.5 at nine different temperatures ranging from 298°K to 473°K in a gas of 99.5% stated

purity, and further purified by fractional distillation; for propane, butane, pentane, and benzene, those of Christophorou, Hurst, and Hadjiantoniou (1966), obtained at 298°K with a drift tube similar to that of Fig. 3.4.5; for thirty more complex organic polar molecules, those obtained at 298°K by Christophorou and Christodoulides (1969) [see H_2O $a(4)$] and extended in a few cases to the range 298–450°K by Christophorou and Pittman (1970), both experiments being performed with a drift tube similar to that of Fig. 3.4.5 in gases of tested purity greater than 99%, and in some cases as high as 99.99%. Furthermore, a large number of mixtures of polar and nonpolar gases in ethylene have been analyzed under isothermal conditions by Hurst, Stockdale, and O'Kelly (1963) and by Christophorou and his coworkers, as mentioned above. The necessity of studying some gases in mixtures with a carrier such as ethylene is due to the fact that these gases at room temperature have high electron attachment cross sections, so that at usual pressures no electron would be left to trigger the electron detector on the collecting electrode.

(6) Cottrell and Walker (1965), Cottrell, Pollock, and Walker (1968), and Pollock (1968) used double-shutter drift tubes ($d = 8.04$ cm between shutters) for measurements at room temperature in methane, acetylene, silane, and in their deuterated forms, in arsine, phosphine, ethane, propane, ethylene, and in five more complex organic polar gases; minimum E/p values range from ~ 0.1 to 2.5, so that in a few cases the authors attained the region where the linear dependence of w on E/p begins.

(7) Hamilton and Stockdale (1966), following the work of Oak Ridge National Laboratory, also used a drift tube similar to that of Fig. 3.4.5 to perform measurements in pure, triply distilled ethylene at 298°K and 353°K for $1 \times 10^{-2} < E/p < 0.2$, and in mixtures of ten polar gases with ethylene at 298°K (except at 353°K for 2-aminoethanol only) for $E/p < 0.1$ (isothermal range).

(8) Bowman and Gordon (1967) used a double-shutter drift tube, similar to that of Pack and Phelps (1961) [see He $a(3)$] with drift distances equal to 5.08 and 10.16 cm, for measurements in research-grade methane, ethane, ethylene, propene, and 1-butene, and in acetylene of 99.5% minimum purity. Data were taken at various temperatures (at least two for each gas) in the interval from 225 to 370°K; the ratio E/p ranged from a maximum of about 10 to values low enough to satisfy the condition of electron thermal equilibrium. The w reproducibility is about 8% for acetylene, and within 6% for all other gases.

(9) At the Institut für Angewandte Physik of Hamburg University, Huber (1968, 1969) and Lehning (1969) adopted the induced current method to perform measurements at 295°K up to very high pressures (accuracy 2–3%). Huber investigated propane over the range $1.3 \times 10^{-2} < E/p < 13.2$ and

found that w decreases with increasing pressure ($p > 3$ atm), and that this decrease is largest for isothermal electrons ($E/p < 0.2$). In ethane for $1 \times 10^{-2} < E/p < 10$ Huber found that the pressure dependence ($p > 6$ atm) has a resonance-like behavior over the range $3 \times 10^{-2} < E/p < 3.3$ with a maximum at $E/p = 0.2$, which corresponds to an electron energy of about 0.08 eV. In methane Lehning, who investigated the range $2 \times 10^{-2} < E/p < 5$, found an opposite behavior: when $E/p < 0.6$, w increases with increasing pressure ($p > 10$ atm), and relationship (4.7.1) is satisfied, but α is negative.

b. Diffusion to mobility ratio

(1) Bannon and Brose (1928) and Brose and Keyston (1935) performed D/μ measurements at room temperature in ethylene ($E/p > 1.26$) and in methane ($E/p > 5.7 \times 10^{-2}$), respectively, using the original Townsend method and very similar equipment. They also verified that in the swarms drifting from the cathode to the collector there is no appreciable fraction of negative ions.

(2) At Sydney, using Bailey's methods and equipment as described in Sections 3.6E and 3.9B, a few determinations of D/μ and $D_T/\mu_{\parallel}$ at room temperature in different gases showing significant electron attachment were performed in the 1920s and 1930s. These investigations were extensively reviewed by Healey and Reed (1941). The most valuable data are the D/μ results in pentane for $E/p > 1.25$ by McGee and Jaeger (1928), and the D/μ and $D_T/\mu_{\parallel}$ results in ammonia for $E/p > 4$ and in hydrogen chloride for $E/p > 10$ by Bailey and Duncanson (1930), in nitrous oxide for $E/p > 1$ by Bailey and Rudd (1932), in nitric oxide for $E/p > 0.5$ by Bailey and Somerville (1934), and in bromine for $E/p > 4$ by Bailey, Makinson, and Somerville (1937).

(3) Cochran and Forester (1962) [see H_2 b(4)] performed D/μ measurements at room temperature over the range $0.2 < E/p < 5$ in methane, ethylene, and cyclopropane. The gases, whose stated purities were, respectively, 99.0%, 99.5%, and 99.5%, were further purified by distillation at liquid-air temperatures.

(4) Cottrell and Walker (1967) performed measurements at room temperature in a diffusion tube ($d \simeq 5.0$ cm), whose collector plate was divided into a central disk A_1 ($b = 0.248$ cm, $b/d \simeq 0.05$) and three concentric annuli A_2, A_3, and A_4 [outer radii: (A_2) 0.526 cm, (A_3) 1.026 cm]; following Warren's and Parker's approach, by means of calibration curves D/μ is derived from the ratios of the currents of A_1, $A_1 + A_2$, $A_1 + A_2 + A_3$ to the total current. The gases that have been investigated are methane, silane, and their deuterated forms, and ethane, propane, ethylene, and methanol; minimum E/p values range from ~0.1 to 1.5.

(5) Wagner, Davis, and Hurst (1967) [see He b(4)] determined the quantity D_L/μ at room temperature in methane $(1 \times 10^{-2} < E/p < 0.4)$ and in ethylene $(1.5 \times 10^{-2} < E/p < 1)$. The minimum purity of the gases was 99.0 mole %.

c. *Magnetic drift velocity (LMF case)*. Using the original Townsend magnetic deflection method and similar equipment, Bannon and Brose (1928), McGee and Jaeger (1928), and Brose and Keyston (1935) performed w_M measurements at room temperature in ethylene $(E/p > 1.25)$, in pentane $(E/p > 0.6)$, and in methane $(E/p > 5.7 \times 10^{-2})$, respectively.

d. *Free electron diffusion.* Nelson and Davis [see He e(2)] performed measurements at room temperature in methane (minimum purity 99.95%) and in ethylene (minimum purity 99.9%).

e. *Microwave conductivity ratio* $(\omega^2 \gg \nu_m^2)$. Mentzoni and Donohoe (1966) used a waveguide interferometer bridge and a transient transmission radiometer to perform ρ and T_r measurements at 10 GHz in the afterglow of a pulsed dc discharge in nitric oxide. Since the authors dealt with the early afterglow only, the electrons did not attain thermal equilibrium with the gas $(T_g = 293°K)$, and the measurements cover the thermal decay interval from 12,000 to about 1200°K.

f. *Energy relaxation time.* Narasinga Rao and Taylor (1967) performed measurements of τ_r in the afterglow of a nitric oxide pulsed discharge, located in a C-band waveguide, using the energy decay method, modified by the introduction of a magnetic field for the purpose of cyclotron heating of electrons (see Section 3.14B).

g. *Cyclotron resonance.* Tice and Kivelson (1967) [see He i(3)] performed line-shape and $\Delta\omega_I$ measurements at room temperature in mixtures of nitrogen and methane, deuteromethane, carbon fluoride, nitrous oxide, ammonia, sulfur dioxide, and hydrogen cyanide. The last four gases had also been investigated previously with the same method by Bayes, Kivelson, and Wong (1962), who included in their analysis as well mixtures of nitrogen with trifluoromethane and with nitromethane. The results of Bayes et al. will be disregarded, however, since the collision cross sections that can be derived from their results are consistently much larger than the usually accepted values of this parameter in the same gases.

Determination of Collision Parameters

(1) Pollock (1968) performed numerical analyses of type A for methane and silane in order to find the best fit to the available w and D/μ data. Since in these two gases dipole and quadrupole moments are absent, rotational

Table 4.15.1 Summary of Q_m Determinations in Some Nonpolar Gases

The use of u_r is the same as in Table 4.13.1.

Case No.	Gas	Formula	Transport Parameter	Energy Range (eV)	Pertinent Equation	Q_m (Å²)
1	Methane	CH_4	$a(8)$	0.030–0.048	(2.6.24)	$100u_r^{-1.5}/(8.21 + 5.26u_r^{0.5})$
			g	0.039	(2.6.95)	$(8.9 \pm 0.9)u_r^{-1}$
2	Deutero-methane	CD_4	g	0.039	(2.6.95)	Same as methane
3	Carbon fluoride	CF_4	g	0.039	(2.6.95)	$(2.1 \pm 0.2)u_r^{-1}$
4	Ethane	C_2H_6	$a(8)$	0.029–0.048	(2.6.24)	$100u_r^{-1.5}/(5.64 + 7.54u_r^{0.5})$
5	Ethylene	C_2H_4	$a(5)$	0.039–0.062	(2.6.24)	4.37 ± 0.26
			$a(8)$	0.030–0.048	(2.6.24)	$\begin{cases}100u_r^{-1.5}/(5.83 + 4.02u_r^{0.5}), & u_r < 0.8 \\ 12.2u_r^{-1}, & u_r > 0.8\end{cases}$
6	Acetylene	C_2H_2	$a(8)$	0.039–0.048	(2.6.24)	$100u_r^{-1}/(1.40 + 1.13u_r^{0.5})$

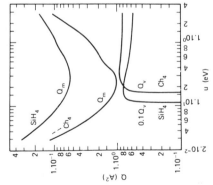

Fig. 4.15.1. Collision cross sections for momentum transfer and for vibrational excitation in methane and in silane. The solid curves are from Pollock's (1968) numerical analyses. Two other Q_m determinations in methane are reported in Table 4.15.1: the room-temperature value from g is shown in the figure; the Q_m expression from $a(8)$ practically coincides with Pollock's curve over the pertinent range.

excitation of molecules can be neglected. Using the Q_m and Q_v cross sections of Fig. 4.15.1 Pollock obtained transport coefficients which agree closely with his own drift velocity measurements [a(6)] and with the characteristic energy measurements of b(3) and b(4) over the range $0.4 < E/p < 3$ in CH_4 and $2 < E/p < 13$ in SiH_4. Worth of notation are for each gas the existence of a Ramsauer minimum in the Q_m cross section, justified by the similarity of the electronic structure to that of the rare gases, and the close correspondence between the positions of the Q_v maximum and of the Q_m minimum.

(2) No other analysis of type A is available for the gases under consideration in this section. Therefore the only data that can be handled easily are the transport parameter values for isothermal electrons; also, the microwave data e will be considered for the derivation of the collision cross sections, since the assumption of a Maxwellian distribution function is reasonably well justified by the authors. For convenience of analysis polar and nonpolar molecules will be discussed separately.

No velocity-dependence law of common validity holds for nonpolar molecules; therefore significant cross-section data can be derived only from experiments whose results cover a range of conditions, such as temperature and magnetic field, appropriate for evaluating the v dependence of Q_m as well as its value. Table 4.15.1 contains all of the pertinent data from these determinations. From the above it is apparent that the collision cross sections are obtained from the transport parameters for thermal electrons, so that their expressions hold over a limited range of electron energies around the thermal values only. It seems more significant and practical, therefore, to write these Q_m expressions in terms of the electron energy, measured not in electron volts, but in units of the average energy $\bar{u}$ of a 300°K plasma (0.0388 eV); we designate as u_r the electron energy defined and measured accordingly.

In the case of polar molecules the v^{-2} dependence of Q_m predicted by Altshuler's formula (1.5.32) seems generally verified for thermal electrons and for molecules whose electric dipole moments are larger than 0.4 debye; in fact, the dipole term of scattering becomes dominant in these cases ($Q_m \propto \mathcal{M}^2$). Therefore, under the assumption $h = -1$, collision cross sections have been obtained also from all experiments in which a value of mobility at one temperature only was measured. All available data of collision cross sections for thermal electrons are given in Table 4.15.2, where the various gases are listed in the order of increasing dipole moments. The Q_m determinations, which have been based on eq. (2.6.32′), are derived from mobility measurements in mixtures with ethylene; the experimental data satisfy relationship (4.13.1) and are handled as discussed in Section 4.13 for the case of water vapor-ethylene mixtures.

Table 4.15.2 Summary of Q_m Determinations in Polar Gases for Thermal Electrons

Naming of organic compounds follows the nomenclature rules either of the International Union of Chemistry or of the *Handbook of Chemistry and Physics* (1970) (see "Table of Physical Constants of Organic Compounds," compiled by S. Patai and J. Zabicky); synonyms of common usage are also given within parentheses. When available, dipole moment values are taken from the *Handbook of Chemistry and Physics* (1970) (see "Table of Selected Values of Electric Dipole Moments for Molecules in the Gas Phase," compiled by R. D. Nelson, Jr., D. R. Lide, Jr., and A. M. Maryott; for the other gases (Nos. 4, 18, 40, 42, 45, and 46) values are based on the compilations by McClellan (1963) and by Landolt and Börnstein (1951). The use of u_r is the same as in Table 4.13.1.

Case No.	Gas	Formula	Dipole Moment (debye)	Transport Parameter	Energy Range (eV)	Pertinent Equation	Q_m (Å²)
1	Nitrous oxide	N_2O	0.167	$a(4)$	0.025–0.039	(2.6.23')	$31u_r^{-0.5}$
				g	0.039	(2.6.95)	$(16.4 \pm 1.6)u_r^{-1}$
2	Arsine	AsH_3	0.20	$a(6)$	0.039	(2.6.23')	230
3	1-Butene	C_4H_8	0.34	$a(8)$	0.039–0.048	(2.6.23')	$51.5u_r^{-1.5}$
4	1,3-Dimethylbenzene (*m*-xylene)	C_8H_{10}	0.35	$a(5)$	0.039	(2.6.32')	82
5	Propene	C_3H_6	0.366	$a(8)$	0.039–0.048	(2.6.24)	$100u_r^{-0.5}/(2.01 + 1.94u_r^{0.5})$
6	Chlorotrifluoromethane	$CClF_3$	0.50	$a(5)$	0.039	(2.6.23')	59.8
7	Phosphine	PH_3	0.58	$a(6)$	0.039	(2.6.23')	170
8	Trimethylamine	C_3H_9N	0.612	$a(5)$	0.039	(2.6.23')	93.4
9	1,2-Dimethylbenzene (*o*-xylene)	C_8H_{10}	0.62	$a(5)$	0.039	(2.6.32')	158
10	Diethylamine	$C_4H_{11}N$	0.92	$a(5)$	0.039	(2.6.23')	144
				$a(7)$	0.039	(2.6.32')	133 ± 10
11	Hydrogen sulfide	H_2S	0.97	$a(5)$	0.039	(2.6.32')	331

No.	Name	Formula					
12	Dimethylamine	C_2H_7N	1.03	$a(5)$	0.039–0.058	(2.6.23')	$162u_r^{-0.895}$
13	Ethoxyethane (ethyl ether)	$C_4H_{10}O$	1.15	$a(7)$	0.039	(2.6.23')	142 ± 9
14	1-Aminopropane (propylamine)	C_3H_9N	1.17	$a(5)$	0.039	(2.6.23')	272
15	Aminoethane (ethylamine)	C_2H_7N	1.22	$a(5)$	0.039	(2.6.23')	253
16	Methoxymethane (methyl ether)	C_2H_6O	1.30	$a(5)$	0.039	(2.6.32')	224
				$a(5)$	0.039	(2.6.23')	272
				$a(6)$	0.039	(2.6.23')	150
17	Chlorodifluoromethane	$CHClF_2$	1.42	$a(5)$	0.039	(2.6.23')	422
18	2-Bromotoluene	C_7H_7Br	1.45	$a(5)$	0.039	(2.6.32')	398
19	Ammonia	NH_3	1.47	$a(4)$	0.025–0.050	(2.6.24)	$1000u_r^{-1}/(1.71 + 0.414u_r^{0.5})$
				g	0.039	(2.6.95)	$(418 \pm 40)u_r^{-1}$
20	Dichloromethane (methylene chloride)	CH_2Cl_2	1.60	$a(5)$	0.039	(2.6.23')	603
21	Sulfur dioxide	SO_2	1.63	g	0.039	(2.6.95)	$(448 \pm 30)u_r^{-1}$
22	1-Butanol (butyl alcohol)	$C_4H_{10}O$	1.66	$a(5)$	0.039–0.058	(2.6.23')	$468u_r^{-0.977}$
23	2-Propanol (isopropyl alcohol)	C_3H_8O	1.66	$a(7)$	0.039	(2.6.32')	355 ± 16
24	1-Propanol (propyl alcohol)	C_3H_8O	1.68	$a(5)$	0.039–0.058	(2.6.23')	$445u_r^{-0.943}$
				$a(7)$	0.039	(2.6.32')	378

Table 4.15.2. (continued)

Case No.	Gas	Formula	Dipole Moment (debye)	Transport Parameter	Energy Range (eV)	Pertinent Equation	Q_m (Å²)
25	Chlorobenzene	C_6H_5Cl	1.69	$a(5)$	0.039	(2.6.32')	520
				$a(5)$	0.039–0.058	(2.6.23')	$647u_r^{-1.187}$
				$a(7)$	0.039	(2.6.32')	342 ± 25
26	Ethanol (ethyl alcohol)	C_2H_6O	1.69	$a(5)$	0.039–0.058	(2.6.23')	$435u_r^{-1.012}$
				$a(6)$	0.039	(2.6.23')	220
				$a(7)$	0.039	(2.6.32')	378 ± 46
27	Bromobenzene	C_6H_5Br	1.70	$a(5)$	0.039	(2.6.32')	555
28	Methanol (methyl alcohol)	CH_4O	1.70	$a(5)$	0.039	(2.6.32')	388
				$a(5)$	0.039–0.058	(2.6.23')	$443u_r^{-1.056}$
				$a(6)$	0.039	(2.6.23')	385
29	1,3-Dichlorobenzene	$C_6H_4Cl_2$	1.72	$a(5)$	0.039	(2.6.32')	437
30	Methyl methanoate (methyl formate)	$C_2H_4O_2$	1.77	$a(5)$	0.039	(2.6.23')	458
31	Chloromethane (methyl chloride)	CH_3Cl	1.87	$a(5)$	0.039–0.058	(2.6.23')	$673u_r^{-1.157}$
				$a(6)$	0.039	(2.6.23')	560
32	1,2-Diaminoethane (ethylenediamine)	$C_2H_8N_2$	1.99	$a(5)$	0.039	(2.6.23')	582
				$a(7)$	0.039	(2.6.32')	471 ± 12
33	Chloroethane (ethyl chloride)	C_2H_5Cl	2.05	$a(5)$	0.039	(2.6.23')	631
34	1-Chloropropane (propyl chloride)	C_3H_7Cl	2.05	$a(5)$	0.039	(2.6.23')	653
35	1-Chlorobutane (butyl chloride)	C_4H_9Cl	2.05	$a(5)$	0.039–0.058	(2.6.23')	$704u_r^{-1.116}$

36	1-Bromobutane (butyl bromide)	C_4H_9Br	2.08	$a(5)$	0.039	(2.6.23')	678
37	1,2-Ethanediol (ethylene glycol)	$C_2H_6O_2$	2.28	$a(5)$	0.039	(2.6.32)	756
38	1,2-Dichlorobenzene	$C_6H_4Cl_2$	2.50	$a(5)$	0.039	(2.6.32')	872
39	Propanal (propionaldehyde)	C_3H_6O	2.52	$a(5)$	0.039–0.058	(2.6.23')	$892u_r^{-1.135}$
40	2-Aminoethanol	C_2H_7NO	2.59	$a(7)$	0.046	(2.6.32')	600 ± 120
41	Butanal (butyraldehyde)	C_4H_8O	2.72	$a(5)$	0.039	(2.6.23')	854
42	2-Butanone	C_4H_8O	2.77	$a(5)$	0.039	(2.6.23')	966
				$a(5)$	0.039	(2.6.32')	958
43	2-Propanone (acetone)	C_3H_6O	2.88	$a(5)$	0.039	(2.6.32')	1000
				$a(5)$	0.039–0.058	(2.6.23')	$1040u_r^{-1.142}$
				$a(6)$	0.039	(2.6.23')	680
				$a(7)$	0.039	(2.6.32')	1100 ± 100
44	Hydrogen cyanide	HCN	2.98	g	0.039	(2.6.95)	$(1430 \pm 70)u_r^{-1}$
45	2,4-Pentanedione (acetylacetone)	$C_5H_8O_2$	3.03	$a(5)$	0.039	(2.6.23')	986
46	Cyclopentanone	C_5H_8O	3.30	$a(5)$	0.039–0.058	(2.6.23')	$1084u_r^{-1.057}$
47	Nitromethane	CH_3NO_2	3.46	$a(5)$	0.039	(2.6.32')	1054
48	1-Nitropropane	$C_3H_7NO_2$	3.66	$a(5)$	0.039	(2.6.32')	1396
49	Propenenitrile (acrylonitrile)	C_3H_3N	3.87	$a(5)$	0.039–0.058	(2.6.23')	$1361u_r^{-0.973}$
50	Ethanenitrile (acetonitrile)	C_2H_3N	3.92	$a(5)$	0.039	(2.6.23')	1453
51	Propanenitrile (propionitrile)	C_3H_5N	4.02	$a(5)$	0.039	(2.6.23')	1456
52	Butanenitrile (butyronitrile)	C_4H_7N	4.07	$a(5)$	0.039–0.058	(2.6.23')	$1531u_r^{-1.092}$

Altshuler's formula (1.5.32) predicts $Q_m \propto \mathcal{M}^2$; this dependence law, when $\mathcal{M} > 0.4$ debye, appears to be substantially verified by the experimental results given in Table 4.15.2. However, for a more significant analysis of the Q_m versus $\mathcal{M}$ relationship, it is convenient to remove, as far as possible, any effect of systematic errors, and this can be partially accomplished by considering the data of one author only. Thus the largest and most recent set of data given in Table 4.15.2, which is derived by applying eq. (2.6.23′) to the Oak Ridge measurements [a(5)] (water is also included according to Table 4.13.1), shows (Fig. 4.15.2) that Altshuler's predictions are in better agreement with experiment than the more sophisticated calculations of Mittleman and von Holdt [see Section 1.5]; for $\mathcal{M} > 0.6$ debye the agreement is usually within 50%. Furthermore, the experimental Q_m values,

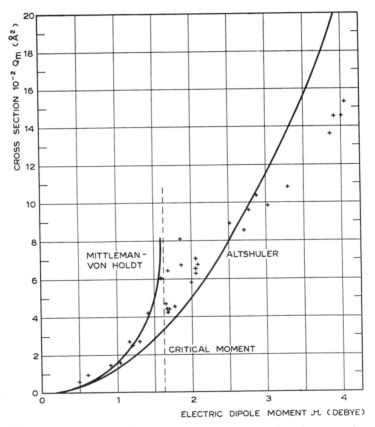

Fig. 4.15.2. Momentum transfer cross sections of electrons in polar gases, determined by applying eq. (2.6.23′) to the Oak Ridge measurements [a(5)] (see Table 4.15.2), and plotted as a function of the dipole moment $\mathcal{M}$, for $\mathcal{M} > 0.4$ only. The theoretical predictions of Altshuler and of Mittleman and von Holdt are also shown.

higher than those predicted by Altshuler for all $\mathcal{M} < {\sim}2.6$ debye, and smaller when $\mathcal{M} > {\sim}2.6$ debye, suggest a resonance-type contribution superimposed on the smooth variation of Q_m with $\mathcal{M}$; thus these results support qualitatively the calculations of Takayanagi and Itikawa (1968) and of Itikawa (1969) on the existence of a potential resonance and on the approximate location of its inversion point (see Section 1.5). There is also experimental evidence that the Born approximation, which does not include the effect of the electron binding when $\mathcal{M} > 1.625$ debye, is inadequate for analyzing the data in detail at and around this critical $\mathcal{M}$ value.

In nitric oxide ($\mathcal{M} = 0.153$ debye) the microwave results of e and f can be elaborated. The ρ measurements for the energy range $0.15 < \bar{u} < 1.5$ eV provide $Q_m = 2.61u^{-0.1}$ (Å²)[eqs. (2.6.41), (2.3.31)]; the τ_r measurements indicate a large G value, in the range 0.1–0.2, relatively independent of the electron energy.

Here, too, Frommhold's theory (see Section 4.7) seems the most appropriate one for the interpretation of the observed pressure dependence of the drift velocity at high densities in propane and ethane. On the other hand, failure to satisfy the condition $l_m \gg \lambda_e$ (see Section 4.2) is the only reason that may be invoked to explain the negative α values measured by Lehning in methane.

4.16 AIR AND OTHER MIXTURES

A. Experimental Data for Dry Air and N_2–O_2^i Mixtures [dc parameters for $E/p < 10$ $(D/\mu \simeq 1.2$ eV)]

a. Drift velocity

(1) Nielsen and Bradbury (1937) [see He a(1)] performed measurements at 293°K for $E/p > 0.2$ in samples of dry air purified from condensable components by passage through liquid-air traps.

(2) Hessenauer (1967) performed measurements at 293°K for $E/p > 0.2$, using Fisher-Treuenfeld's method discussed in Sections 3.2B and 3.3; proper consideration was given by this author to the presence of attachment (see Fig. 3.3.6). Hessenauer investigated both dry ambient air, purified by passage through liquid-air traps or through silica gel and further dehumidified by the presence of P_2O_5, and 9 : 1 nitrogen-oxygen mixtures. The author estimates that drift velocities are accurate within $\pm 8\%$ ($E/p > 1$).

The results of (2) for air, which agree with those of (1) within the combined experimental errors, are summarized in Table 4.16.1.

b. Diffusion to mobility ratio

(1) Bailey (1925) used his original method, appropriate to electronegative gases, as described in Section 3.6E, to perform measurements in dry air at room temperature over the range $0.5 < E/p < 2.5$.

Table 4.16.1   Experimental Values of the Drift Velocity and of the Characteristic Energy in Dry Air

Drift velocity data are taken from a smooth curve drawn through Hessenauer's [a(2)] results.

$10^{17}E/N$ (V cm^2)	E/p_{288} (V cm^{-1} torr^{-1})	$10^{-5}w$ (cm sec^{-1}) [$T_g = 293°$K, $a(2)$]	D/μ (eV) [$T_g = 288°$K, $b(2)$]
2.98×10^{-1}	1.0×10^{-1}		9.9×10^{-2}
5.97	2.0	6.0	1.46×10^{-1}
8.95	3.0	7.3	1.79
1.193×10^0	4.0	8.2	2.06
1.79	6.0	10.0	2.61
2.385	8.0	11.4	3.23
2.98	1.0×10^0	12.5	3.87
4.47	1.5	15.4	5.39
5.97	2.0	17.8	6.58
8.95	3.0	22.3	8.19
1.193×10^1	4.0	26.6	9.28
1.49	5.0	30.6	1.00×10^0
2.98	1.0×10^1	52	1.18

(2) Crompton, Huxley, and Sutton (1953) extended to air the earlier work performed at the same laboratory in hydrogen and nitrogen [see H_2 b(2)]; the current to the central disk, which collects the negative ions, was disregarded and only the ratios of the currents to the annuli were considered. Measurements were performed at 288°K for $E/p > 0.1$.

(3) Rees and Jory (1964) adopted the Townsend-Huxley method and a tube similar to the one of Fig. 3.7.2 to perform measurements at 293°K for $E/p > 1.5$ in dry ambient air, stored over P_2O_5 and purified by passage through three liquid-air traps. The authors estimate that their results are accurate to better than $+2\%$, -1%.

(4) Raja Rao and Govinda Raju (1971) also adopted the Townsend-Huxley method to perform room-temperature measurements in dry air for $E/p > 1$. They used a diffusion chamber [$d = 1, 2, 4$, and 8 cm; $b = 0.5$ (central disk), 1.0, 1.5, 3.2, and 4.25 cm (annuli)] with a glow discharge electron source.

The results of (2), which agree well with (3) and (4) within the combined experimental errors, are summarized in Table 4.16.1; the results of (1) also agree fairly well with the others. No consideration is given to the work of Huxley and Zaazou (1949), who adopted Townsend's method but used eq. (3.5.10) without adequate consideration of the presence of negative ions.

c. Magnetic drift velocity (LMF case).   The results obtained by Huxley and Zaazou (1949), using their modification of Townsend's magnetic deflection method, discussed in Section 3.9A, are of scant significance, since no consideration was given to the presence of negative ions.

d. Direct-current conductivity.   This conductivity was measured in pressure-driven shock tubes by Lamb and Lin (1957), Lin, Neal, and Fyfe (1962), and Morsell (1967), using induction probe techniques. Since air at high temperatures is a mixture of many different atomic and molecular species (N_2, O_2, N, O, NO, Ar), among which a large number of dissociation, exchange, and ionization reactions take place, nonequilibrium conditions of concentrations, ionization, and temperature prevail in the region behind the shock front, were both dc and microwave conductivity measurements are usually performed. This fact has been verified both experimentally and theoretically [see, in particular, Lin and Teare (1963) and Tevelow (1967)]. Because of the uncertainty of the actual conditions to which conductivity measurements have thus to be related, the experimental work in shock tubes seems at present still inadequate as a source of reliable data for electron collision parameters in atmospheric gases. Similar remarks apply to the measurements performed by Lin and Kivel (1959) in a pressure-driven shock tube filled with a mixture of 95% O_2 + 5% N_2.

e. Microwave conductivity ratio.   All measurements of this type have been performed in shock-heated air [Daiber and Glick (1961), Tevelow (1967)], and therefore the considerations mentioned under *d* above apply here too. Thus the results of Daiber and Glick cannot be properly used because of the lack of reliable information on the axial electron density distribution in the shock (see Section 3.11C, case *b*), and Tevelow's results show a large spread and important deviations from the expected equilibrium values (see Fig. 3.11.14).

Theoretical Evaluations for Dry Air

Using their cross sections for electrons in O_2 (see Section 4.10) and those of Engelhardt, Phelps, and Risk for electrons in N_2 (see Section 4.9), Hake and Phelps (1967) computed v_m^* and v_u^* [eqs. (2.6.6) and (2.6.7)] for room-temperature air, free from argon and condensable components. The calculated points, plotted in Fig. 4.16.1, agree to within 20% with the smooth curves based on the results of *a*(1), *b*(2), and *b*(3). The 0.9% Ar present in air in these experiments is expected to contribute less than 1% to the above collision frequencies.

Microwave conductivities for high-temperature air in thermodynamic equilibrium have been computed and reported by various authors, using different sets of electron collision cross sections for the air constituents. The

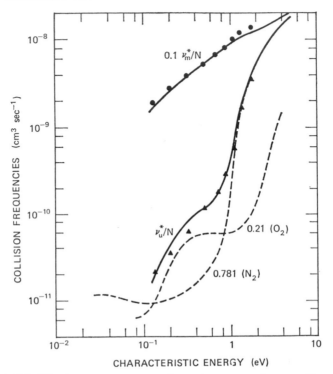

Fig. 4.16.1. Effective momentum transfer and energy exchange collision frequencies of electrons in dry air. The solid curves are derived from the experimental data of Nielsen and Bradbury [a(1)], Crompton, Huxley, and Sutton [b(2)], and Rees and Jory [b(3)]; the points are calculated using the Hake and Phelps cross sections for O_2 and the Engelhardt, Phelps, and Risk cross sections for N_2. The dashed curves show the ν_u^*/N values for pure O_2 and N_2 multiplied by their respective fractional concentrations in standard air [after Hake and Phelps (1967)].

most significant determinations are those of Shkarofsky, Bachynski, and Johnston (1961) and of Hochstim (1965); however, no accurate experimental verification is possible at present for the reasons discussed under d and e above.

Experimental Data for Other Mixtures

For the benefit of the interested reader we list here a bibliography of the available transport parameter data for all mixtures that have not been mentioned so far because they neither have been considered for the determination of collision parameters of constituent gases according to the methods discussed in Section 2.6, nor can be checked against properly derived theoretical values. Combustion gases, either seeded with easily ionized metal vapors or not, are excluded from this review.

a. Drift velocity. Klema and Allen (1950) [Ar–N_2, Ar–O_2]; Errett (1951) [Ar–N_2, Ar–CO_2]; Colli and Facchini (1952) [Ar–N_2]; Kirschner and Toffolo (1952) [Ar–N_2]; English and Hanna (1953) [Ne–CO_2, Ar–N_2, Ar–CO_2, Ar–CH_4, Kr–CO_2, Kr–CH_4, Xe–CO_2]; Colli and de Leonardis (1953) [Ar–C_2H_5OH]; Bortner, Hurst, and Stone (1957) [Ar–N_2, Ar–CH_4, Ar–air, O_2–CH_4, CO_2–CH_4]; McCutchen (1958) [CO_2–Hg]; Nagy, Nagy, and Dési (1960) [Ar–N_2, Ar–CO_2, Ar–CH_4]; Comunetti and Huber (1960) [Ar–N_2, Ar–CO_2]; Hurst, O'Kelly, and Bortner (1961) [Ar–H_2O]; Lowke and Rees (1963) [N_2–H_2O]; Levine and Uman (1964) [Ar–N_2, Ar–CO_2]; Fischer (1966) [He–Ne–C_2H_5OH]; Christophorou, Compton, and Dickson (1968) [N_2–HCl, N_2–HBr, N_2–HI].

b. Diffusion to mobility ratio (D/μ *and* $D_T/\mu_\parallel$). Bailey and Healey (1935) [He–Cl_2, CO_2–Cl_2]; Bailey, Makinson, and Somerville (1937) [He–Br_2, CO_2–Br_2]; Healey (1938) [He–I_2, CO_2–I_2].

4.17 POSITIVE IONS

The aim of this section is to determine whether the theoretical expression (2.6.25) of Spitzer and Härm for elastic collisions between electrons and positive ions provides transport parameter data in agreement with experiments.

a. Direct-current conductivity. In shock-heated rare gases, at initial pressures of a few torr, electron-ion and electron-electron collisions dominate over electron-atom collisions if equilibrium temperatures are greater than $\sim$10,000°K. The electrical conductivity predicted by Spitzer and Härm for a plasma dominated by charged particle interactions is given by eq. (2.6.30), multiplied by the numerical factor 0.5816 and modified according to Fig. 2.6.2 for low $\mathscr{L}$ values. At temperatures below 10,000°K both electron-atom collisions and charged particle interactions need to be considered, and the electrical conductivity must be computed with the procedures discussed in Section 2.6A. In particular, on the basis of the Chapman-Enskog method Devoto and his coworkers have computed the conductivity for helium [Devoto and Li (1968)], argon [Devoto (1967a)], krypton, and xenon [Devoto (1969)], using Frost's and Phelps's (1964) electron-atom collision cross sections (see Sections 4.2, 4.4, 4.5, and 4.6) and a shielded Coulomb potential for charged particle interactions.

No clear-cut verification of the theoretical results of Spitzer and Härm and of Devoto is obtained from reported measurements. A short list of the experiments is now given; in general, authors have assumed that at the conductivity maximum thermodynamic equilibrium has been reached, so that electron density can be computed from Saha's equation (3.8.7).

(1) Lin, Resler, and Kantrowitz (1955) performed measurements in argon using their induction probe, and obtained results that in the temperature range 10,000–14,000°K agree well both with Spitzer's and Härm's original values and with Devoto's calculations, which yield very similar data. The results are, on the other hand, well below theoretical expectations in the range 6000–10,000°K, but the authors believe that this discrepancy arises because of departure from thermal equilibrium conditions.

(2) Pain and Smy (1960) performed similar measurements and obtained results that are, in general, considerably higher and lower than the theoretical ones below and above 11,000°K, respectively.

(3) Lau (1964) and Lau and Mills (1965) measured over the 5500–8000°K range the ac, low-frequency conductivity in travelling and reflected argon shocks with a Savic-Boult probe [see Section 3.8D]; under thermal equilibrium conditions the results, for which an accuracy of $\pm 15\%$ is claimed, agree well with Devoto's theoretical data.

(4) Johnsen and Rehder (1966) performed measurements in krypton over the temperature range 9000–11,000°K, using an induction probe; the results are about 30% higher than Devoto's theoretical expectations, which in turn are higher than the Spitzer-Härm values. It has been suggested [Zauderer (1970)] that in these experiments the shock velocity, determined from optical measurements, can be lower than the true value by as much as 15%; if temperatures are correspondingly corrected, a much better agreement with Devoto's data is obtained.

(5) Zauderer (1970) performed measurements in xenon over the temperature range 5500–9000°K, using both an induction probe and a Savic-Boult probe; over the entire range the results are significantly lower than Devoto's data.

(6) Enomoto, Goda, and Hashiguchi (1970) performed measurements in xenon over the temperature range 6000–10,000°K, using an induction probe; here too results are lower than Devoto's data for temperatures above 8000°K, but a reasonable agreement is obtained for temperatures between 7000 and 8000°K.

(7) Vasil'eva et al. (1970) determined from the Q changes of a solenoid (see Section 3.8D) the dc conductivities in argon, krypton and xenon at temperatures above 4500°K. Most of the experimental data show a large spread and are in bad agreement with the theoretical predictions; only at the highest temperatures (e.g. above 7000°K in xenon) the difference between the measured and the theoretical values becomes less than the estimated experimental error of 20%.

Low current conductivities in a Q-machine (see Fig. 3.8.5) were measured by Rynn (1964); in a potassium plasma at temperatures from 2390 to 3040°K this author found an average conductivity equal to 1.2 times the Spitzer-Härm value, with an rms error of about 10%. However, at $\mathscr{L} = 14$, appropriate to Rynn's experiment, Fig. 2.6.2 indicates that the conductivity must be slightly higher than the Spitzer-Härm value, which reduces the discrepancy significantly.

On the contrary Krasnikov, Kulik, and Norman (1971) report that conductivity measurements recently performed by Russian workers in high-temperature cells ($T_g > 1800°K$) filled with cesium indicate that the ratio of the measured conductivities (positive ion contribution only) to the Spitzer-Härm value is less than 1 (for $\mathscr{L} < \sim 10$) and decreases with decreasing $\mathscr{L}$.

b. Microwave conductivity ratio.   Measurements in isothermal plasmas should provide results in accordance with expression (2.6.43), the electron density being defined according to (3.11.26). The first data were obtained in helium afterglows by Anderson and Goldstein (1955), who measured sets of ΔA and $\Delta \phi$ values [see He $f(3)$] and the ΔA time variations upon introduction of a disturbing wave. The helium gas contribution was subtracted from the measured $\omega \rho_{av}$ values, and eq. (2.6.52) was used for the analysis of the $\rho \propto \Delta A$ time variations. Thus, for a 300°K plasma and electron peak densities from 3×10^{10} to 4×10^{11} electrons cm^{-3}, with an assumed diffusion-controlled distribution, these authors found the best agreement with experiments for $A' = 1.8 \times 10^{-6}$ m^3 sec^{-1} °K$^{3/2}$ and $B = 1.4 \times 10^{13}$ (m °K)$^{-3}$.

About 10 years later Chen (1964) [see He $f(4)$ and Ne $e(3)$] performed ΔA and $\Delta \phi$ measurements, thus obtaining $\omega \rho_{av}$ as a function of $\bar{n}/p$, in neon afterglows covering the temperature range 200–600°K. A best fit to the experimental data according to (2.6.43) and assuming a diffusion-controlled n distribution yields $A' = 1.8 \times 10^{-6}$ m^3 sec^{-1} °K$^{3/2}$ and $B = 4.0 \times 10^{14}$ (m °K)$^{-3}$ for electron peak densities in the range 6.2–9.6×10^{10} electrons cm^{-3}. These same values provide fair agreement also with the results that Mentzoni (1965) [see O$_2$ $c(3)$] obtained in oxygen at $p = 1.5$ and 3.0 torr during the early nonisothermal decay times of an afterglow.

All of the above experiments yield results in good accordance with (2.6.43): the A' values are in excellent agreement with theory, whereas the B values differ significantly; however, the B coefficient appears in the logarithmic term, so that it plays a secondary role only, and for the same reason its determination is subject to much larger errors. A conductivity ratio that is a few times larger than the theoretical value was measured by Takeda (1961), who determined the X-band microwave conductivity of a plasma post in a waveguide during the afterglow of a pulsed argon discharge (see Section 3.11C, case e).

The same author performed conductivity measurements in neon and argon shock-generated plasmas, using a similar method [Takeda and Roux (1961)] and X-band and K-band microwave reflection probes, placed at a shock-tube end [Takeda, Tsukishima, and Funahashi (1966)] (see Section 3.11C, case c). In these cases, however, the measured conductivities were used for determining the electron temperatures on the basis of (2.6.43); the resulting temperatures agree fairly well with the theoretically predicted values, except at low Mach numbers.

c. Energy relaxation time. In rare gases under dominating electron-ion collisions the energy relaxation time is predicted from theory [eqs. (2.6.69) and (2.6.43)] to be:

$$\tau_r = \frac{M}{2m} \frac{T_e^{3/2}}{A'\mathscr{L}n} \tag{4.17.1}$$

Dougal and Goldstein (1958) measured this time in decaying plasmas of helium, neon, and rare-gas mixtures at different gas temperatures: 77, 225, and 300°K (T_e is believed to be equal to T_g). The techniques used by these authors were discussed in Section 3.13B. The results, which were obtained

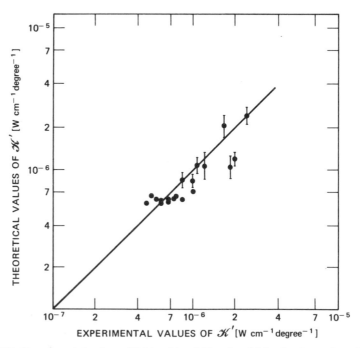

Fig. 4.17.1. Experimental values of Sekiguchi and Herndon (d) for the thermal conductivity $\mathscr{K}'$ compared with the theoretical values [eqs. (2.6.76) and (3.13.19) and the Spitzer-Härm coefficients for electron–electron collisions][after Sekiguchi and Herndon (1958)].

for $1 \times 10^{11} < n < 6 \times 10^{11}$ electrons cm^{-3} and assumed uniform distributions, are not consistent with formula (4.17.1) and show that τ_r varies from values about one third of the theoretical one at 300°K to values approximately twice the theoretical one at 77°K. Unexpected nonisothermal conditions and the presence of appreciable quantities of molecular ions and of metastable excited atoms have been invoked as explanations for the disagreement. On the other hand, the observed dependence of τ_r on M and on n was as predicted by (4.17.1).

d. Thermal diffusivity. Sekiguchi and Herndon (1958) performed in helium and neon afterglows measurements of τ_r and $\mathcal{D}'$ with the methods and techniques discussed in Section 3.13, and of the corresponding electron density values with the impedance post method discussed in Section 3.11C, case *e*. Measured τ_r data are used to determine electron temperatures on the basis of eq. (4.17.1) and of $A'\mathscr{L}$ values consistent with the 300°K experimental results of Dougal and Goldstein, discussed in *c* above.

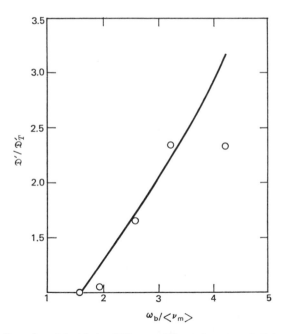

Fig. 4.17.2. Experimental values of Rostas, Bhattacharya, and Cahn (open circles) for the ratio $\mathcal{D}'/\mathcal{D}'_T$, plotted as a function of $\omega_b/\langle \nu_m \rangle$ and normalized to unity at $\omega_b/\langle \nu_m \rangle = 1.56$; measurements refer to a cylindrical tube filled by a neon afterglow plasma under dominating electron-ion collisions conditions. The solid line is the theoretical curve computed using the formulas of Chapter 2 [after Rostas et al. (1963)].

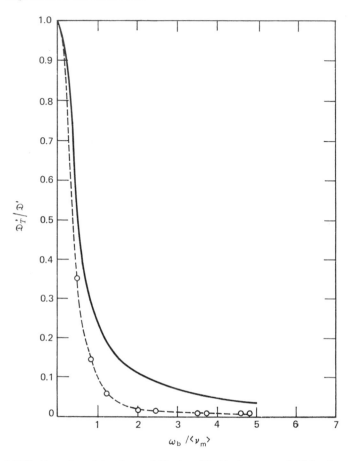

Fig. 4.17.3. Experimental values of Rostas, Bhattacharya, and Cahn (open circles) for the ratio $\mathfrak{D}_T'/\mathfrak{D}'$ plotted as a function of $\omega_b/\langle\nu_m\rangle$; measurements refer to a rectangular tube filled by a neon afterglow plasma under dominating electron-ion collisions conditions. The solid line is the theoretical curve computed using the formulas of Chapter 2 [after Rostas et al. (1963)].

The experimental $\mathscr{K}'$ data, found by the steady-state method at small temperature gradients, are compared in Fig. 4.17.1 with the theoretical values [see eqs. (2.6.76) and (3.13.19) and the Spitzer-Härm numerical coefficients for electron-electron collision effects], estimated using the measured density values and the electron temperatures derived from τ_r data; agreement is quite satisfactory. The experimental $\mathscr{K}'$ data, found by the transient method, are functions of the microwave heating power but approach consistently the values obtained by the steady-state method as the heating power decreases.

The results of Sekiguchi and Herndon supersede earlier, less accurate data obtained in xenon and neon afterglows by Goldstein and Sekiguchi (1958). The thermal diffusivity was also measured in helium and neon afterglows by Nygaard (1967a) [see He f], who obtained for $n > 1 \times 10^{11}$ electrons cm^{-3} results in good agreement with those of Sekiguchi and Herndon.

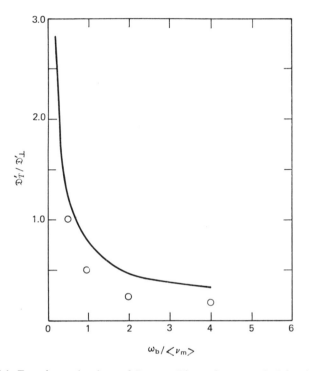

Fig. 4.17.4. Experimental values of Rostas, Bhattacharya, and Cahn (open circles) for the ratio $\mathfrak{D}'_T/\mathfrak{D}'_\perp$ plotted as a function of $\omega_b/\langle \nu_m \rangle$; measurements refer to a rectangular tube filled by a neon afterglow plasma under dominating electron-ion collisions conditions. The solid line is the theoretical curve computed using the formulas of Chapter 2 [after Rostas et al. (1963)].

Thermal diffusivities in good agreement with the theoretical predictions of Spitzer and Härm were also obtained by Rynn (1964) in potassium quiescent plasmas generated in a Q-machine. Adopting the approach considered in Section 3.13C, this author found experimentally ε values whose average ratio to the theoretical Spitzer-Härm value is 0.98 with a standard deviation of 0.065.

   e. *Thermal diffusivity in magnetic fields.*   Rostas, Bhattacharya, and Cahn (1963), using the methods and techniques extensively discussed in Section

3.14B, determined the ratios $\mathcal{D}_T'/\mathcal{D}$ and $\mathcal{D}_T'/\mathcal{D}_\perp'$ versus $\omega_b/\langle \nu_m \rangle$ in neon afterglows, under dominating electron-ion collision conditions. The results are shown in Figs. 4.17.2–4.17.4; electron densities are of the order of 10^{11}–10^{12} electrons cm^{-3}, and magnetic fields are up to 1200 Oe. On the same figures the theoretical predictions, computed using the formulas reported in Section 2.6D, are given by the solid curves.

Fundamental Constants and Useful Physical Data

Electronic charge	$e = 1.602 \times 10^{-19}$ C
Electronic mass	$m = 9.109 \times 10^{-31}$ kg
Electronic charge/mass ratio	$e/m = 1.759 \times 10^{11}$ C kg^{-1}
Atomic mass unit	$M_0 = 1.660 \times 10^{-27}$ kg
Vacuum permittivity	$\varepsilon_0 = 8.854 \times 10^{-12}$ F m^{-1}
Vacuum permeability	$\mu_0 = 1.257 \times 10^{-6}$ H m^{-1}
Velocity of light in vacuum	$c = (\varepsilon_0 \mu_0)^{-\frac{1}{2}} = 2.9979 \times 10^8$ m sec^{-1}
Planck's constant	$h = 2\pi\hbar = 6.625 \times 10^{-34}$ J sec
Atomic unit of length	$a_0 = \varepsilon_0 h^2/\pi m e^2 = 0.5292$ Å
Atomic unit of cross section	$\pi a_0^2 = 0.8806$ Å^2
Atomic unit of dipole moment	$ea_0 = 8.478 \times 10^{-28}$ C m
	$= 2.542$ debye (or 10^{-18} esu)
Atomic unit of quadrupole moment	$ea_0^2 = 4.486 \times 10^{-36}$ C m^2
	$= 1.345 \times 10^{-26}$ esu
Boltzmann constant	$k = 1.3805 \times 10^{-23}$ J °K^{-1}
	$= 8.617 \times 10^{-5}$ eV °K^{-1}
Unit of energy	1 eV $= 1.602 \times 10^{-19}$ J
	1 rydberg $= 2.180 \times 10^{-18}$ J
	$= 13.605$ eV
Temperature such that kT is 1 eV	$T = 1.1605 \times 10^4$ °K
Mean energy of a particle:	
At 0°C	$\frac{3}{2}kT = 5.657 \times 10^{-21}$ J
	$= 3.531 \times 10^{-2}$ eV
At 300°K	$= 6.212 \times 10^{-21}$ J
	$= 3.878 \times 10^{-2}$ eV
Velocity of an electron with an energy:	
Of 1 eV	$v = 5.931 \times 10^5$ m sec^{-1}
Equal to the mean value at 0°C	$= 1.115 \times 10^5$ m sec^{-1}
Equal to the mean value at 300°K	$= 1.168 \times 10^5$ m sec^{-1}

de Broglie wavelength of an electron
with an energy:

Of 1 eV $\qquad\qquad\qquad\qquad\qquad\lambda_e = 12.26$ Å

Equal to the mean value at 0°C $\qquad= 65.2$ Å

Equal to the mean value at 300°K $\qquad= 62.3$ Å

Number density of an ideal gas:

At 760 torr and 0°C

(Loschmidt's number) $\qquad\qquad N = 2.687 \times 10^{19}$ cm^{-3}

At 1 torr and 0°C $\qquad\qquad\qquad= 3.536 \times 10^{16}$ cm^{-3}

At 1 torr and 300°K $\qquad\qquad\quad= 3.219 \times 10^{16}$ cm^{-3}

Collision parameters corresponding to
a cross section of 1 Å^2 in an
ideal gas at 1 torr and 0°C:

Mean free path $\qquad\qquad\qquad\quad l = 0.2828$ cm

Collision probability $\qquad\qquad\quad\mathscr{P} = 3.536$ cm^{-1}

Collision frequency of the mean
energy electron $\qquad\qquad\qquad\quad \nu = 3.94 \times 10^7$ sec^{-1}

Plasma frequency at 10^{10}
electrons cm^{-3} $\qquad\qquad\qquad f_p = 897.9$ MHz

Debye electron shielding length in a
300°K plasma with a density of
10^{10} electrons cm^{-3} $\qquad\qquad \lambda_D = 12.0$ μm

Electron cyclotron frequency at
1 G $\qquad\qquad\qquad\qquad\qquad\quad f_c = 2.80$ MHz

Table A.1 **Weights of Some Atoms and Corresponding Energy Loss Factors of Electrons in Elastic Collisions**

Atom:	He	Ne	Ar	Kr	Xe	K	Cs
Atomic weight (in atomic units):	4.003	20.18	39.94	83.8	131.3	39.10	132.9
Energy loss factor ($2m/M \times 10^4$):	2.74	0.544	0.275	0.131	0.0836	0.281	0.0826

Table A.2 **Weights of Some Molecules and Corresponding Energy Loss Factors of Electrons in Elastic Collisions**

Molecule:	H_2	D_2	N_2	O_2	CO	CO_2	H_2O	NO
Molecular weight (in atomic units):	2.016	4.03	28.02	32.00	28.01	44.01	18.02	30.01
Energy loss factor ($2m/M \times 10^4$)	5.44	2.72	0.392	0.343	0.392	0.249	0.609	0.366

Table A.3   Rotational Constants [Herzberg (1945, 1950)] and Quadrupole Moments [Strogryn and Strogryn (1966)] of Some Molecules

Molecule:	$^1H_2$	$^2D_2$	$^{14}N_2$	$^{16}O_2$	$^{12}C^{16}O$	$^{12}C^{16}O_2$	$^{14}N^{16}O$
Rotational constant ($B_0 \times 10^4$ eV):	73.5	37.1	2.48	1.78	2.38	0.485	2.10
Quadrupole moment ($\mathcal{Q}$ in atomic units):	+0.492	+0.483	−1.13	−0.29	−1.85	−3.2	−1.3

Application of Dingle Functions to the Evaluation of the Microwave Conductivity

Dingle, Arndt, and Roy (1957) introduced and tabulated the integrals of importance in the theory of electrical conductivity in semiconductors. The following ones, presently known as Dingle functions, are conveniently used for the evaluation of the microwave conductivity ratio of an electron swarm:

$$\mathfrak{A}_p(x) = \frac{1}{\Gamma(p+1)} \int_0^\infty \frac{w^p \exp(-w)}{w+x} \, dw \tag{B.1}$$

$$\mathfrak{C}_p(x) = \frac{1}{\Gamma(p+1)} \int_0^\infty \frac{w^p \exp(-w)}{w^2+x^2} \, dw \tag{B.2}$$

$$\mathfrak{E}_p(x) = \frac{1}{\Gamma(p+1)} \int_0^\infty \frac{w^p \exp(-w)}{1+xw^3} \, dw \tag{B.3}$$

In fact, the integral that appears in the conductivity ratio expressions:

$$D_i^h(x) = \frac{1}{\Gamma(i+1)} \int_0^\infty \frac{y^i \exp(-y)}{y^h + x^2} \, dy \tag{2.6.45}$$

for the integer values $h = -3, -2, -1, 1, 2, 3$ is given by:

$$D_i^{-3}(x) = (i+1)(i+2)(i+3)\mathfrak{E}_{i+3}(x^2) \tag{B.4}$$

$$D_i^{-2}(x) = \frac{(i+1)(i+2)}{x^2} \mathfrak{C}_{i+2}\left(\frac{1}{x}\right) \tag{B.5}$$

$$D_i^{-1}(x) = \frac{i+1}{x^2} \mathfrak{A}_{i+1}\left(\frac{1}{x^2}\right) \tag{B.6}$$

$$D_i^1(x) = \mathfrak{A}_i(x^2) \tag{B.7}$$

$$D_i^2(x) = \mathfrak{C}_i(x) \tag{B.8}$$

$$D_i^3(x) = \frac{1}{x^2} \mathfrak{E}_i\left(\frac{1}{x^2}\right) \tag{B.9}$$

The following recurrence relations have been established by Dingle and his coworkers:

$$(p + 1)\mathfrak{A}_{p+1} + x\mathfrak{A}_p = 1 \tag{B.10}$$

$$(p + 1)(p + 2)\mathfrak{C}_{p+2} + x^2\mathfrak{C}_p = 1 \tag{B.11}$$

$$(p + 1)(p + 2)(p + 3)x\mathfrak{C}_{p+3} + \mathfrak{C}_p = 1 \tag{B.12}$$

Application of these relations to eqs. (B.4), (B.5), and (B.6) yields:

$$D_i^{-3}(x) = \frac{1}{x^2}[1 - \mathfrak{C}_i(x^2)] \tag{B.13}$$

$$D_i^{-2}(x) = \frac{1}{x^2}\left[1 - \frac{1}{x^2}\mathfrak{C}_i\left(\frac{1}{x}\right)\right] \tag{B.14}$$

$$D_i^{-1}(x) = \frac{1}{x^2}\left[1 - \frac{1}{x^2}\mathfrak{A}_i\left(\frac{1}{x^2}\right)\right] \tag{B.15}$$

Dingle and his coworkers have tabulated their functions for x values from 0 up to 20 and for integer and semi-integer p values from -0.5 to 4.0 for $\mathfrak{A}_p$, from -0.5 to 5.0 for $\mathfrak{C}_p$, and from 0 to 6.5 for $\mathfrak{C}_p$. Recurrence relations (B.10), (B.11), and (B.12) can be applied to extend these tabulations to other integer and semi-integer p values, but the authors caution that accuracy is preserved only if higher orders are obtained from lower orders when x is small, and lower orders are obtained from higher orders when x is large. Brown and Hindley (1964) have extended the tabulation of $\mathfrak{A}_p$ and $\mathfrak{C}_p$ up to $x = 10^4$.

The following differentiation formulas, valid for $p > -1$, can be of interest when conductivity maxima or minima are investigated:

$$\mathfrak{A}_p' = \frac{1}{x}[(p + x)\mathfrak{A}_p - 1] = \frac{1}{x}[p\mathfrak{A}_p - (p + 1)\mathfrak{A}_{p+1}] \tag{B.16}$$

$$\mathfrak{C}_p' = \frac{1}{x}[(p - 1)\mathfrak{C}_p - (p + 1)\mathfrak{C}_{p+1}] \tag{B.17}$$

$$\mathfrak{C}_p' = \frac{p + 1}{3x}(\mathfrak{C}_{p+1} - \mathfrak{C}_p) \tag{B.18}$$

Dingle functions $\mathfrak{A}_p$, $\mathfrak{C}_p$, and $\mathfrak{C}_p$ are not truly independent of each other. In fact, if we consider them as complex functions of the complex argument x, we should find that along the real x axis $\mathfrak{C}_p$ is simply related to the $\mathfrak{A}_p$ values along the imaginary x axis, whereas $\mathfrak{C}_p$ can also be expressed as a simple combination of $\mathfrak{A}_p$ values over the real axis and the 60° lines. Although the appropriate formulas are given by Dingle and his coworkers, they are presently of little value since no tabulation of the $\mathfrak{A}_p$ function over the entire complex plane is yet available.

Bibliography

R. A. Abram and A. Herzenberg (1969), *Chem. Phys. Letters* **3**, 187.

N. L. Allen and B. A. Prew (1970), *J. Phys. B* **3**, 1113.

W. P. Allis (1956), Motion of Ions and Electrons, in *Handbuch der Physik*, Vol. XXI, p. 383 (Springer-Verlag, Berlin).

W. P. Allis, S. J. Buchsbaum, and A. Bers (1963), *Waves in Anisotropic Plasmas* (M.I.T. Press, Cambridge, Mass.).

W. P. Allis and D. J. Rose (1954), *Phys. Rev.* **93**, 84.

D. Alpert, C. G. Matland, and A. O. McCoubrey (1951), *Rev. Sci. Instr.* **22**, 370.

S. Altshuler (1957), *Phys. Rev.* **107**, 114.

R. J. Andersen and W. L. Nigham (1970), *J. Appl. Phys.* **41**, 3750.

J. M. Anderson (1960), *J. Appl. Phys.* **31**, 511.

J. M. Anderson (1961), *Rev. Sci. Instr.* **32**, 975.

J. M. Anderson and L. Goldstein (1955), *Phys. Rev.* **100**, 1037.

J. M. Anderson and L. Goldstein (1956a), *Phys. Rev.* **102**, 388.

J. M. Anderson and L. Goldstein (1956b), *Phys. Rev.* **102**, 933.

R. W. B. Ardill and W. D. Davison (1968), *Proc. Roy. Soc.* **A304**, 465.

J. E. Bailey, R. E. B. Makinson, and J. M. Somerville (1937), *Phil. Mag.* **24**, 177.

V. A. Bailey (1924), *Phil. Mag.* **47**, 379.

V. A. Bailey (1925), *Phil. Mag.* **50**, 825.

V. A. Bailey (1930), *Phil. Mag.* **9**, 560, 625.

V. A. Bailey and W. E. Duncanson (1930), *Phil. Mag.* **10**, 145.

V. A. Bailey and R. H. Healey (1935), *Phil. Mag.* **19**, 725.

V. A. Bailey and J. D. McGee (1928), *Phil. Mag.* **6**, 1073.

V. A. Bailey and J. B. Rudd (1932), *Phil. Mag.* **14**, 1033.

V. A. Bailey and J. M. Somerville (1934), *Phil. Mag.* **17**, 1169.

P. M. Bakshi, R. E. Haskell, and R. J. Papa (1967), Air Force Cambridge Res. Lab., Rept. 67-0527 (Bedford, Mass.).

D. Balfour and J. H. Harris (1966), *Proceedings of the Seventh International Conference on Phenomena in Ionized Gases, Belgrade*, Vol. I, p. 32 (Gradevinska Knjiga Publishing House, Belgrade).

D. Balfour, D. A. Hart, and J. A. Haynes (1966), *Proceedings of the Seventh International Conference on Phenomena in Ionized Gases, Belgrade*, Vol. II, p. 235 (Gradevinska Knjiga Publishing House, Belgrade).

J. Bannon and H. L. Brose (1928), *Phil. Mag.* **6**, 817.

J. N. Bardsley, A. Herzenberg, and F. Mandl (1966), *Proc. Phys. Soc.* **89**, 321.

D. R. Bates and A. Dalgarno (1962), in *Atomic and Molecular Processes* (Ed. D. R. Bates), Chapter 7 (Academic Press, New York).

K. D. Bayes, D. Kivelson, and S. C. Wong (1962), *J. Chem. Phys.* **37**, 1217.

G. Bekefi (1966), *Radiation Processes in Plasmas* (John Wiley & Sons, New York).

G. Bekefi and S. C. Brown (1958), *Phys. Rev.* **112**, 159.

G. Bekefi and S. C. Brown (1961), *J. Appl. Phys.* **32**, 25.

G. Bekefi, J. L. Hirshfield, and S. C. Brown (1961), *Phys. Fluids* **4**, 173.

I. B. Bernstein (1969), in *Advances in Plasma Physics* (Eds. A. Simon and W. B. Thomson), Vol. 3, p. 127 (Interscience Publishers, New York).

M. J. Bernstein (1962), *Phys. Rev.* **127**, 335.

G. W. Bethke and A. D. Ruess (1964), *Phys. Fluids* **7**, 1446.

A. K. Bhattacharya, J. T. Verdeyen, F. T. Adler, and L. Goldstein (1967), *J. Appl. Phys.* **38**, 527.

M. A. Biondi (1954), *Phys. Rev.* **93**, 1136.

M. A. Biondi (1963), in *Advances in Electronics and Electron Physics* (Ed. L. Marton), Vol. 18, p. 67 (Academic Press, New York).

M. A. Biondi (1968), in *Atomic and Electron Physics* (Ed. L. Marton), Vol. 7, Part B, p. 78 (Academic Press, New York).

J. C. Blackburn and F. L. Tevelow (1965), Harry Diamond Lab., Rept. TR-1275 (Washington, D.C.).

E. Blue and J. E. Stanko (1969), *J. Appl. Phys.* **40**, 4061.

R. J. Blum (1965), *Proc. IEEE* **53**, 523.

F. Boeschoten (1964), *Plasma Phys.* **6**, 339.

D. Bohm, E. H. S. Burhop, and H. S. W. Massey (1949), in *The Characteristics of Electrical Discharges in Magnetic Fields* (Eds. A. Guthrie and R. K. Wakerling), NNES Div. 1, Vol. 5 (McGraw-Hill Book Co., New York).

D. Bohm and E. P. Gross (1949), *Phys. Rev.* **75**, 1851, 1864.

T. E. Bortner, G. S. Hurst, and W. G. Stone (1957), *Rev. Sci. Instr.* **28**, 103.

C. Bottcher (1971), *Chem. Phys. Letters* **9**, 57.

J. C. Bowe (1960), *Phys. Rev.* **117**, 1411, 1416.

J. C. Bowe and R. K. Langs (1966), *Bull. Am. Phys. Soc.*, Ser. II, **11**, 232.

C. R. Bowman and D. E. Gordon (1967), *J. Chem. Phys.* **46**, 1878.

N. E. Bradbury and R. A. Nielsen (1936), *Phys. Rev.* **49**, 388.

G. L. Braglia (1970a), *Lett. Nuovo Cimento* **3**, 295.

G. L. Braglia (1970b), *Nuovo Cimento* **70B**, 169.

G. L. Braglia and L. Ferrari (1970), *Nuovo Cimento* **67B**, 167.

G. L. Braglia and L. Ferrari (1971), *Nuovo Cimento* **2B**, 254; **4B**, 245, 262.

J. Brambring (1964), *Z. Physik* **179**, 532, 539.

R. A. Brandewie and E. M. Williams (1964), *J. Appl. Phys.* **35,** 2299.

J. M. Breare and A. von Engel (1963), *Proceedings of the Sixth International Conference on Ionization Phenomena in Gases, Paris* (Eds. P. Hubert and E. Crémieu-Alcan), Vol. I, p. 317.

J. M. Breare and A. von Engel (1964), *Proc. Roy. Soc.* **A282,** 390.

E. L. Breig and C. C. Lin (1965), *J. Chem. Phys.* **43,** 3839.

H. L. Brose (1925), *Phil. Mag.* **50,** 536.

H. L. Brose and J. L. Keyston (1935), *Phil. Mag.* **20,** 902.

R. M. Brown and N. K. Hindley (1964), *J. Appl. Phys.* **35,** 3418.

S. C. Brown (1959), *Basic Data of Plasma Physics* (Technology Press of M.I.T. and John Wiley & Sons, New York).

S. C. Brown and D. J. Rose (1952), *J. Appl. Phys.* **23,** 711, 719, 1028.

R. L. Bruce, F. W. Crawford, and R. S. Harp (1968), *J. Appl. Phys.* **39,** 2088.

F. V. Bunkin (1957), *Zh. Eksperim. i Teor. Fiz.* **32,** 811 [English transl.: *Soviet Phys. JETP* **5,** 665].

D. S. Burch and L. G. H. Huxley (1967), *Australian J. Phys.* **20,** 625.

P. G. Burke (1969), in *Atomic Physics* (*Proceedings of the First International Conference on Atomic Physics*), p. 265 (Plenum Press, New York).

P. G. Burke and K. Smith (1962), *Rev. Mod. Phys.* **34,** 458.

P. D. Burrow and G. J. Schulz (1969), *Phys. Rev.* **187,** 97.

C. A. Burrus (1958), *J. Chem. Phys.* **28,** 427.

J. A. Carruthers (1962), *Can. J. Phys.* **40,** 1528.

G. Cavalleri (1969a), *Phys. Rev.* **179,** 186.

G. Cavalleri (1969b), *Phys. Rev. Letters* **23,** 907.

G. Cavalleri, E. Gatti, and A. M. Interlenghi (1965), *Nuovo Cimento* **40B,** 450.

G. Cavalleri, E. Gatti, and P. Principi (1964), *Nuovo Cimento* **31,** 302, 318.

G. Cavalleri and G. Loria (1964), Rendiconti della LXV Riunione Annuale dell'AEI, Palermo, Paper 92.

G. Cavalleri and G. Sesta (1968), *Phys. Rev.* **170,** 286.

G. Cavalleri and G. Sesta (1969), *Phys. Rev.* **177,** 434.

R. J. Chaffin and J. B. Beyer (1968), *IEEE Trans. Microwave Theory Tech.* **MTT-16,** 37.

E. S. Chang (1970), *Phys. Rev. A* **2,** 1403.

L. M. Chanin and R. D. Steen (1964), *Phys. Rev.* **136,** A138.

S. Chapman and T. G. Cowling (1952), *The Mathematical Theory of Non-uniform Gases* (University Press, Cambridge).

C. L. Chen (1963), *Phys. Rev.* **131,** 2550.

C. L. Chen (1964), *Phys. Rev.* **135,** A627.

C. L. Chen, C. C. Leiby, and L. Goldstein (1961), *Phys. Rev.* **121,** 1391.

C. L. Chen and M. Raether (1962), *Phys. Rev.* **128,** 2679.

F. F. Chen (1965), in *Plasma Diagnostic Techniques* (Eds. R. H. Huddlestone and S. L. Leonard), Chapter 4 (Academic Press, New York).

L. G. Christophorou and A. A. Christodoulides (1969), *J. Phys. B* **2,** 71.

L. G. Christophorou, R. N. Compton, and N. W. Dickson (1968), *J. Chem. Phys.* **48,** 1949.

L. G. Christophorou, G. S. Hurst, and A. Hadjiantoniou (1966), *J. Chem. Phys.* **44**, 3506.

L. G. Christophorou, G. S. Hurst, and W. G. Hendrick (1966), *J. Chem. Phys.* **45**, 1081.

L. G. Christophorou and D. Pittman (1970), *J. Phys. B* **3**, 1252.

C. R. Claydon, G. A. Segal, and H. S. Taylor (1970), *J. Chem. Phys.* **52**, 3387.

L. W. Cochran and D. W. Forester (1962), *Phys. Rev.* **126**, 1785.

D. M. Coffey (1965), Air Force Cambridge Res. Lab., Rept. 65-324 (Bedford, Mass.).

D. M. Coffey (1970), Air Force Cambridge Res. Lab., Rept. 70-0107 (Bedford, Mass.).

I. M. Cohen and M. D. Kruskal (1965), *Phys. Fluids* **8**, 920.

L. Colli and M. T. de Leonardis (1953), *J. Appl. Phys.* **24**, 255.

L. Colli and U. Facchini (1952), *Rev. Sci. Instr.* **23**, 39.

R. E. Collin (1960), *Field Theory of Guided Waves* (McGraw-Hill Book Co., New York).

A. Comunetti and P. Huber (1960), *Helv. Phys. Acta* **33**, 911.

M.-J. Cottereau, J. Guilly, and P. Valentin (1970), *Compt. Rend.* **B270**, 729.

M.-J. Cottereau and P. Valentin (1967), *Compt. Rend.* **B264**, 130.

M.-J. Cottereau and P. Valentin (1968), *Compt. Rend.* **B267**, 440.

T. L. Cottrell, W. J. Pollock, and I. C. Walker (1968), *Trans. Faraday Soc.* **64**, 2260.

T. L. Cottrell and I. C. Walker (1965), *Trans. Faraday Soc.* **61**, 1585.

T. L. Cottrell and I. C. Walker (1967), *Trans. Faraday Soc.* **63**, 549.

H. Coxell and A. W. Wolfendale (1960), *Proc. Phys. Soc.* **75**, 378.

O. H. Crawford (1967), *J. Chem. Phys.* **47**, 1100.

O. H. Crawford (1968), *Chem. Phys. Letters* **2**, 461.

R. P. Creaser (1967), *Australian J. Phys.* **20**, 547.

R. W. Crompton (1967), *J. Appl. Phys.* **38**, 4093.

R. W. Crompton (1969), in *Advances in Electronics and Electron Physics* (Ed. L. Marton), Vol. 27, p. 1 (Academic Press, New York).

R. W. Crompton and M. T. Elford (1957), *J. Sci. Instr.* **34**, 405.

R. W. Crompton and M. T. Elford (1962), *J. Sci. Instr.* **39**, 480.

R. W. Crompton and M. T. Elford (1963), *Proceedings of the Sixth International Conference on Ionization Phenomena in Gases, Paris* (Eds. P. Hubert and E. Crémieu-Alcan), Vol. I, p. 337.

R. W. Crompton, M. T. Elford, and J. Gascoigne (1965), *Australian J. Phys.* **18**, 409.

R. W. Crompton, M. T. Elford, and R. L. Jory (1967), *Australian J. Phys.* **20**, 369.

R. W. Crompton, M. T. Elford, and A. I. McIntosh (1968), *Australian J. Phys.* **21**, 43.

R. W. Crompton, M. T. Elford, and A. G. Robertson (1970), *Australian J. Phys.* **23**, 667.

R. W. Crompton, D. K. Gibson, and A. I. McIntosh (1969), *Australian J. Phys.* **22**, 715.

R. W. Crompton, D. K. Gibson, and A. G. Robertson (1970), *Phys. Rev. A* **2**, 1386.

R. W. Crompton, L. G. H. Huxley, and D. J. Sutton (1953), *Proc. Roy. Soc.* **A218**, 507.

R. W. Crompton and R. L. Jory (1962), *Australian J. Phys.* **15**, 451.

R. W. Crompton and R. L. Jory (1965), *Proceedings of the Fourth International Conference on the Physics of Electronic and Atomic Collisions, Quebec*, p. 118 (Science Bookcrafters, Hastings-on-Hudson, N.Y.).

R. W. Crompton, B. S. Liley, A. I. McIntosh, and C. A. Hurst (1966), *Proceedings of the Seventh International Conference on Phenomena in Ionized Gases, Belgrade*, Vol. I, p. 86 (Gradevinska Knjiga Publishing House, Belgrade).

R. W. Crompton and A. I. McIntosh (1968), *Australian J. Phys.* **21**, 637.

R. W. Crompton, J. A. Rees, and R. L. Jory (1965), *Australian J. Phys.* **18**, 541.

R. W. Crompton and A. G. Robertson (1971), *Australian J. Phys.* **24**, 543.

R. W. Crompton and D. J. Sutton (1952), *Proc. Roy. Soc.* **A215**, 467.

H. M. Cronson (1966), *Phys. Fluids* **9**, 581.

J. W. Daiber and H. S. Glick (1961), *Proceedings of the Symposium on Electromagnetics and Fluid Dynamics of Gaseous Plasma*, p. 323 (Polytechnic Institute of Brooklyn, Brooklyn, N.Y.).

J. W. Daiber and H. F. Waldron (1966), *Phys. Rev.* **151**, 51.

A. Dalgarno and R. J. Moffett (1963), *Proc. Natl. Acad. Sci. India* **A33**, 511.

N. D'Angelo and N. Rynn (1960), *Rev. Sci. Instr.* **31**, 1326.

N. D'Angelo and N. Rynn (1961), *Phys. Fluids* **4**, 1303.

B. Davydov (1935), *Phys. Z. Sowjet.* **8**, 59.

J. L. Delcroix (1965), *Plasma Physics*, Vol. 1 (John Wiley & Sons, New York).

J. L. Delcroix (1968), *Plasma Physics*, Vol. 2 (John Wiley & Sons, New York).

J. F. Delpech and J. C. Gauthier (1969), Ninth International Conference on Phenomena in Ionized Gases, Bucharest, Paper 4.2.3.10.

J. F. Delpech and J. C. Gauthier (1971), *Rev. Sci. Instr.* **42**, 958.

R. S. Devoto (1966), *Phys. Fluids* **9**, 1230.

R. S. Devoto (1967a), *Phys. Fluids* **10**, 354.

R. S. Devoto (1967b), *Phys. Fluids* **10**, 2105.

R. S. Devoto (1969), *AIAA J.* **7**, 199.

R. S. Devoto and C. P. Li (1968), *J. Plasma Phys.* **2**, 17.

R. H. Dicke (1946), *Rev. Sci. Instr.* **17**, 268.

R. B. Dingle, D. Arndt, and S. K. Roy (1957), *Appl. Sci. Res.* **B6**, 144, 155, 245.

A. Doehring (1952), *Z. Naturforsch.* **7a**, 253.

K. V. Donskoi, Yu. A. Dunaev, and A. I. Prokof'ev (1962), *Zh. Tekh. Fiz.* **32**, 1095 [English transl.: *Soviet Phys.-Tech. Phys.* **7**, 805 (1963)].

A. A. Dougal and L. Goldstein (1958), *Phys. Rev.* **109**, 615.

H. Dreicer (1960), *Phys. Rev.* **117**, 343.

G. F. Drukarev (1965), *The Theory of Electron-Atom Collisions* (English transl.) (Academic Press, London).

M. J. Druyvesteyn and F. M. Penning (1940), *Rev. Mod. Phys.* **12**, 87.

R. A. Duncan (1957), *Australian J. Phys.* **10**, 54.

S. Dushman (1962), *Scientific Foundations of Vacuum Technique* (Ed. J. M. Lafferty) (John Wiley & Sons, New York).

H. Ehrhardt, L. Langhans, F. Linder, and H. S. Taylor (1968), *Phys. Rev.* **173**, 222.

M. T. Elford (1966), *Australian J. Phys.* **19**, 629.

A. G. Engelhardt and A. V. Phelps (1963), *Phys. Rev.* **131**, 2115.

A. G. Engelhardt and A. V. Phelps (1964), *Phys. Rev.* **133**, A375.

A. G. Engelhardt, A. V. Phelps, and C. G. Risk (1964), *Phys. Rev.* **135**, A1566.

W. N. English and G. C. Hanna (1953), *Can. J. Phys.* **31**, 768.

Y. Enomoto, N. Goda, and S. Hashiguchi (1970), *J. Phys. Soc. Japan* **29**, 1400.

D. Errett (1951), doctoral thesis, Purdue University (Lafayette, Ind.).

456 Bibliography

U. Fano (1970), *Comments At. Mol. Phys.* **2**, 47.

A. Farkas (1935), *Orthohydrogen, Parahydrogen and Heavy Hydrogen* (University Press, Cambridge).

F. C. Fehsenfeld (1963), *J. Chem. Phys.* **39**, 1653.

H. Fields, G. Bekefi, and S. C. Brown (1963), *Phys. Rev.* **129**, 506.

J. Fischer (1966), *Proceedings of the Seventh International Conference on Phenomena in Ionized Gases, Belgrade*, Vol. I, p. 68 (Gradevinska Knjiga Publishing House, Belgrade).

W. F. v. Fischer-Treuenfeld (1965), *Z. Phys.* **185**, 336.

D. Formato and A. Gilardini (1960), *Proceedings of the Fourth International Conference on Ionization Phenomena in Gases, Uppsala*, Vol. I, p. 99 (North-Holland Publishing Co., Amsterdam).

D. Formato and A. Gilardini (1962), *Proceedings of the Fifth International Conference on Ionization Phenomena in Gases, Munich*, p. 660 (North-Holland Publishing Co., Amsterdam).

R. J. Freiberg and L. A. Weaver (1968), *Phys. Rev.* **170**, 335.

L. Frommhold (1968), *Phys. Rev.* **172**, 118.

L. S. Frost (1961), *J. Appl. Phys.* **32**, 2029.

L. S. Frost and A. V. Phelps (1962), *Phys. Rev.* **127**, 1621.

L. S. Frost and A. V. Phelps (1964), *Phys. Rev.* **136**, A1538.

A. E. Fuhs (1965), *Instrumentation for High Speed Plasma Flow* (Gordon & Breach Science Publishers, New York).

A. Funahashi and S. Takeda (1968), *J. Appl. Phys.* **39**, 2117.

O. T. Fundingsland, A. C. Faire, and A. J. Penico (1954), in *Rocket Exploration of the Upper Atmosphere* (Eds. R. L. F. Boyd and M. J. Seaton), p. 339 (Interscience Publishers, New York).

A. A. Ganichev, V. E. Golant, A. P. Zhilinskii, B. Z. Khotimskii, and V. N. Shilin (1964), *Zh. Tekh. Fiz.* **34**, 77 [English transl.: *Soviet Phys.-Tech. Phys.* **9**, 58].

M. Gardener, S. Kisdnasamy, E. Rössle, and A. W. Wolfendale (1957), *Proc. Phys. Soc.* **A70**, 687.

W. R. Garrett (1971), *Phys. Rev. A* **3**, 961.

K. H. Geissler (1968), *Plasma Phys.* **10**, 127.

K. H. Geissler (1970), *Phys. Fluids* **13**, 935.

S. Geltman and K. Takayanagi (1966), *Phys. Rev.* **143**, 25.

E. Gerjuoy and S. Stein (1955), *Phys. Rev.* **97**, 1671.

D. K. Gibson (1970), *Australian J. Phys.* **23**, 683.

A. Gilardini (1957), *Proceedings of the Third International Conference on Ionization Phenomena in Gases, Venetia*, p. 374 (Società Italiana di Fisica, Milan).

A. Gilardini (1963), Selenia S.p.A. Res. Lab., Tech. Final Rept. Contr. AF 61(052)-39 (Rome).

A. Gilardini (1964), in *Fisica del Plasma*, Convegno indetto dalla Società Lombarda di Fisica, Milano, p. 189 (Ed. C.N.R., Rome).

A. L. Gilardini and S. C. Brown (1957), *Phys. Rev.* **105**, 25, 31.

H. S. Glick and T. G. Jones (1969), *AIAA J.* **7**, 4.

V. E. Golant (1960), *Zh. Tekh. Fiz.* **30**, 1265 [English transl.: *Soviet Phys.-Tech. Phys.* **5**, 1197].

P. D. Goldan and L. Goldstein (1965), *Phys. Rev.* **138**, A39.

D. E. Golden (1966), *Phys. Rev.* **151**, 48.

D. E. Golden and H. W. Bandel (1965), *Phys. Rev.* **138**, A14.

L. Goldstein, J. M. Anderson, and G. L. Clark (1953), *Phys. Rev.* **90**, 151, 486.

L. Goldstein, M. A. Lampert, and R. H. Geiger (1952), *Elec. Commun.* **29**, 243.

L. Goldstein and T. Sekiguchi (1958), *Phys. Rev.* **109**, 625.

L. Gould and S. C. Brown (1953), *J. Appl. Phys.* **24**, 1053.

L. Gould and S. C. Brown (1954), *Phys. Rev.* **95**, 897.

R. Grünberg (1967), *Z. Physik* **204**, 12.

R. Grünberg (1968), *Z. Naturforsch.* **23a**, 1994.

R. Grünberg (1969), *Z. Naturforsch.* **24a**, 1838.

R. C. Gunton and T. M. Shaw (1965), *Phys. Rev.* **140**, A748.

A. V. Gurevich (1956), *Zh. Eksperim. i Teor. Fiz.* **30**, 1112 [English transl.: *Soviet Phys. JETP* **3**, 895].

A. Guthrie (1963), *Vacuum Technology* (John Wiley & Sons, New York).

R. D. Hake, Jr., and A. V. Phelps (1967), *Phys. Rev.* **158**, 70.

B. I. H. Hall (1955a), *Proc. Phys. Soc.* **B68**, 334.

B. I. H. Hall (1955b), *Australian J. Phys.* **8**, 468.

B. I. H. Hall (1955c), *Australian J. Phys.* **8**, 551.

N. Hamilton and J. A. D. Stockdale (1966), *Australian J. Phys.* **19**, 813.

Handbook of Chemistry and Physics (1970), 50th Edition (Ed. R. C. Weast) (The Chemical Rubber Co., Cleveland, Ohio).

S. Hara (1969), *J. Phys. Soc. Japan* **27**, 1592.

R. S. Harp and R. M. Moser (1967), *Rev. Sci. Instr.* **38**, 1795.

N. J. Harrick and N. F. Ramsey (1952), *Phys. Rev.* **88**, 228.

L. P. Harris (1963), *J. Appl. Phys.* **34**, 2958.

L. P. Harris (1964), *J. Appl. Phys.* **35**, 1993.

R. E. Haskell, R. J. Papa, and P. M. Bakshi (1967), Air Force Cambridge Res. Lab., Rept. 67-0528 (Bedford, Mass.).

J. B. Hasted (1964), *Physics of Atomic Collisions* (Butterworths & Co., London).

M. A. Heald and C. B. Wharton (1965), *Plasma Diagnostics with Microwaves* (John Wiley & Sons, New York).

R. H. Healey (1938), *Phil. Mag.* **26**, 940.

R. H. Healey and J. W. Reed (1941), *The Behaviour of Slow Electrons in Gases* (Amalgamated Wireless, Ltd., Sydney).

R. J. W. Henry (1970), *Phys. Rev. A* **2**, 1349.

R. J. W. Henry and N. F. Lane (1969), *Phys Rev.* **183**, 221.

H. Hermansdorfer (1968), in *Plasma Diagnostics* (Ed. W. Lochte-Holtgreven), Chapter 8 (North-Holland Publishing Co., Amsterdam).

P. Herreng (1942), *Compt. Rend.* **215**, 79.

P. Herreng (1943), *Compt. Rend.* **217**, 75.

G. Herzberg (1945), *Infrared and Raman Spectra* (Van Nostrand Co., Princeton, N.J.).

G. Herzberg (1950), *Molecular Spectra and Molecular Structure.* I: *Spectra of Diatomic Molecules* (Van Nostrand Co., Princeton, N.J.).

H. Hessenauer (1967), *Z. Physik* **204,** 142.

J. L. Hirshfield and S. C. Brown (1958), *J. Appl. Phys.* **29,** 1749.

A. R. Hochstim (1965), Convair Res. Paper P-124 IDA.

P. S. Hoeper, W. Franzen, and R. Gupta (1968), *Phys. Rev.* **168,** 50.

C. R. Hoffmann and H. M. Skarsgard (1969), *Phys. Rev.* **178,** 168.

L. B. Holmes and H. D. Weymann (1969), *Phys. Fluids* **12,** 1200.

T. Holstein (1946), *Phys. Rev.* **70,** 367.

J. A. Hornbeck (1951), *Phys. Rev.* **83,** 374.

B. Huber (1968), *Z. Naturforsch.* **23a,** 1228.

B. Huber (1969), *Z. Naturforsch.* **24a,** 528.

D. E. Hudson (1944), U.S. At. Energy Comm. Rept. MDDC 564.

C. A. Hurst and B. S. Liley (1965), *Australian J. Phys.* **18,** 521.

G. S. Hurst, L. B. O'Kelly, and T. E. Bortner (1961), *Phys. Rev.* **123,** 1715.

G. S. Hurst, L. B. O'Kelly, E. B. Wagner, and J. A. Stockdale (1963), *J. Chem. Phys.* **39,** 1341.

G. S. Hurst and J. E. Parks (1966), *J. Chem. Phys.* **45,** 282.

G. S. Hurst, J. A. Stockdale, and L. B. O'Kelly (1963), *J. Chem. Phys.* **38,** 2572.

L. G. H. Huxley (1937), *Phil. Mag.* **23,** 210.

L. G. H. Huxley (1940), *Phil. Mag.* **30,** 396.

L. G. H. Huxley (1959a), *J. Atmospheric Terrest. Phys.* **16,** 46.

L. G. H. Huxley (1959b), *Australian J. Phys.* **12,** 171.

L. G. H. Huxley and R. W. Crompton (1955), *Proc. Phys. Soc.* **B68,** 381.

L. G. H. Huxley and R. W. Crompton (1962), in *Atomic and Molecular Processes* (Ed. D. R. Bates), Chapter 10 (Academic Press, New York).

L. G. H. Huxley, R. W. Crompton, and M. T. Elford (1966), *Brit. J. Appl. Phys.* **17,** 1237.

L. G. H. Huxley and A. A. Zaazou (1949), *Proc. Roy. Soc.* **A196,** 402.

R. C. Hwa (1958), *Phys. Rev.* **110,** 307.

J. C. Ingraham (1966), *Proceedings of the Seventh International Conference on Phenomena in Ionized Gases, Belgrade,* Vol. I, p. 57 (Gradevinska Knijga Publishing House, Belgrade).

J. C. Ingraham and S. C. Brown (1965), *Phys. Rev.* **138,** A1015.

J. C. Ingraham and J. J. McCarthy (1962), M.I.T. Res. Lab. Electron., Q.P.R. 64, p. 76 (Cambridge, Mass.).

Y. Itikawa (1963), *J. Phys. Soc. Japan* **18,** 1499.

Y. Itikawa (1969), *J. Phys. Soc. Japan* **27,** 444.

Y. Itikawa (1971), *J. Phys. Soc. Japan* **30,** 835; **31,** 1532.

Y. Itikawa and K. Takayanagi (1969a), *J. Phys. Soc. Japan* **26,** 1254.

Y. Itikawa and K. Takayanagi (1969b), *J. Phys. Soc. Japan* **27,** 1293.

N. V. Jermokhin, B. M. Kovaliov, P. P. Kulik, V. A. Riabii, and I. S. Sokolovskii (1969), Ninth International Conference on Phenomena in Ionized Gases, Bucharest, Paper 4.2.3.3.

R. Johnsen and L. Rehder (1966), *Proceedings of the Seventh International Conference on Phenomena in Ionized Gases, Belgrade,* Vol. II, p. 746 (Gradevinska Knijga Publishing House, Belgrade).

T. W. Johnston (1966), *J. Math. Phys.* **7**, 1453.

R. L. Jory (1965), *Australian J. Phys.* **18**, 237.

A. J. Kelly (1966), *J. Chem. Phys.* **45**, 1723, 1733.

D. C. Kelly, H. Margenau, and S. C. Brown (1957), *Phys. Rev.* **108**, 1367.

E. H. Kennard (1938), *Kinetic Theory of Gases* (McGraw-Hill Book Co., New York).

C. Kenty (1928), *Phys. Rev.* **32**, 624.

T. Kihara and O. Aono (1963), *J. Phys. Soc. Japan* **18**, 837.

E. J. M. Kirschner and D. S. Toffolo (1952), *J. Appl. Phys.* **23**, 594.

E. D. Klema and J. S. Allen (1950), *Phys. Rev.* **77**, 661.

D. J. Kouri (1966), *J. Chem. Phys.* **45**, 154.

D. J. Kouri (1968), *J. Chem. Phys.* **49**, 5205.

D. J. Kouri, W. N. Sams, and L. Frommhold (1969a), *Sixth International Conference on the Physics of Electronic and Atomic Collisions, Abstracts of Papers*, p. 153 (M.I.T. Press, Cambridge, Mass.).

D. J. Kouri, W. N. Sams, and L. Frommhold (1969b), *Phys. Rev.* **184**, 252.

Yu. G. Krasnikov, P. P. Kulik, and G. E. Norman (1971), Tenth International Conference on Phenomena in Ionized Gases, Oxford, Invited Paper.

J. H. Krenz (1965), *Phys. Fluids* **8**, 1871.

C. H. Kruger, M. Mitchner, and U. Daybelge (1968), *AIAA J.* **6**, 1712.

V. V. Krylov and Yu. I. Tusnov (1966), *Teplofiz. Vysokikh Temp.* **4**, 768 [English transl.: *High Temp.* **4**, 718].

V. V. Krylov and Yu. I. Tusnov (1967), *Teplofiz. Vysokikh Temp.* **5**, 1 [English transl.: *High Temp.* **5**, 1].

R. W. La Bahn and J. Callaway (1966), *Phys. Rev.* **147**, 28.

P. Laborie, J.-M. Rocard, and J. A. Rees (1968), *Tables de sections efficaces électroniques et coefficients macroscopiques* (Dunod, Paris).

L. Lamb and S. C. Lin (1957), *J. Appl. Phys.* **28**, 754.

H. H. Landolt and R. Börnstein (1951), *Zahlenwerte und Funktionen*, 6th Edition, Vol. 1: *Atom und Molekularphysik*, Part 3 (Springer-Verlag, Berlin).

R. Landshoff (1949), *Phys. Rev.* **76**, 904.

R. Landshoff (1951), *Phys. Rev.* **82**, 442.

N. F. Lane and S. Geltman (1967), *Phys. Rev.* **160**, 53.

J. Lau (1964), *Can. J. Phys.* **42**, 1548.

J. Lau and E. Mills (1965), *Can. J. Phys.* **43**, 1334.

P. A. Lawson and J. Lucas (1965a), *Proc. Phys. Soc.* **85**, 177.

P. A. Lawson and J. Lucas (1965b), *Brit. J. Appl. Phys.* **16**, 1813.

J. H. Leck (1964), *Pressure Measurements in Vacuum Systems* (Chapman and Hall, London).

W. Legler (1970), *Phys. Letters* **31A**, 129.

H. Lehning (1968), *Phys. Letters* **28A**, 103.

H. Lehning (1969), *Phys. Letters* **29A**, 719.

J. L. Levine and T. M. Sanders, Jr. (1967), *Phys. Rev.* **154**, 138.

N. E. Levine and M. A. Uman (1964), *J. Appl. Phys.* **35**, 2618.

C. P. Li and R. S. Devoto (1968), *Phys. Fluids* **11**, 448.

R. L. Liboff (1959), *Phys. Fluids* **2**, 40.

B. S. Liley (1967), *Australian J. Phys.* **20**, 527.

S. C. Lin and B. Kivel (1959), *Phys. Rev.* **114**, 1026.

S. C. Lin, R. A. Neal, and W. I. Fyfe (1962), *Phys. Fluids* **5**, 1633.

S. C. Lin, E. L. Resler, and A. Kantrowitz (1955), *J. Appl. Phys.* **26**, 95.

S. C. Lin and J. D. Teare (1963), *Phys. Fluids* **6**, 355.

J. L. Lloyd (1960), *Proc. Phys. Soc.* **75**, 387.

L. B. Loeb (1922), *Phys. Rev.* **19**, 24; **20**, 397.

L. B. Loeb (1924), *Phys. Rev.* **23**, 157.

L. B. Loeb (1955), *Basic Processes of Gaseous Electronics* (University of California Press, Berkeley and Los Angeles).

C. L. Longmire (1963), *Elementary Plasma Physics* (Interscience Publishers, New York).

J. J. Lowke (1962), *Australian J. Phys.* **15**, 39.

J. J. Lowke (1963), *Australian J. Phys.* **16**, 115.

J. J. Lowke (1971), Tenth International Conference on Phenomena in Ionized Gases, Oxford, Paper 1.1.1.5.

J. J. Lowke and J. H. Parker, Jr. (1969), *Phys. Rev.* **181**, 302.

J. J. Lowke and J. A. Rees (1963), *Australian J. Phys.* **16**, 447.

B. H. Mahan (1960), *J. Chem. Phys.* **33**, 959.

H. Maier-Leibnitz (1935), *Z. Physik* **95**, 499.

W. Makios (1965), in *Fundamental Studies of Ions and Plasmas*, Vol. II, p. 423 (*AGARD Conference Proceedings*, No. 8, Ed. H. D. Wilsted) (Technical Editing and Reproduction, London); *Proceedings of the Seventh International Conference on Phenomena in Ionized Gases, Belgrade*, Vol. II, p. 750 (Gradevinska Knijga Publishing House, Belgrade).

W. Makios (1966), *Z. Naturforsch.* **21a**, 2040.

W. Makios (1967), *Rev. Sci. Instr.* **38**, 352.

N. Marcuvitz (1951), *Waveguide Handbook* (McGraw-Hill Book Co., New York).

H. Margenau (1946), *Phys. Rev.* **69**, 508.

H. Margenau and L. M. Hartmann (1948), *Phys. Rev.* **73**, 309.

H. S. W. Massey (1956), Theory of Atomic Collisions, in *Handbuch der Physik*, Vol. XXXVI, p. 232 (Springer-Verlag, Berlin).

H. S. W. Massey and E. H. S. Burhop (1952), *Electronic and Ionic Impact Phenomena* (Clarendon Press, Oxford).

C. H. Mayer (1956), *IRE Trans. Microwave Theory Tech.* **MTT-4**, 24.

M. A. Mazing and N. A. Vrublevskaya (1966), *Zh. Eksperim. i Teor. Fiz.* **50**, 343 [English transl.: *Soviet Phys. JETP* **23**, 228].

A. L. McClellan (1963), *Tables of Experimental Dipole Moments* (W. H. Freeman and Co., San Francisco).

C. W. McCutchen (1958), *Phys. Rev.* **112**, 1848.

E. W. McDaniel (1964), *Collision Phenomena in Ionized Gases* (John Wiley & Sons, New York).

J. D. McGee and J. C. Jaeger (1928), *Phil. Mag.* **6**, 1107.

A. I. McIntosh (1966), *Australian J. Phys.* **19**, 805.

M. H. Mentzoni and J. Donohoe (1966), *Can. J. Phys.* **44**, 693.

M. H. Mentzoni and J. Donohoe (1968), *Can. J. Phys.* **46**, 1323.

M. H. Mentzoni and K. V. Narasinga Rao (1965), *Phys. Rev. Letters* **14**, 779.

M. H. Mentzoni and R. V. Row (1963), *Phys. Rev.* **130**, 2312.

R. G. Meyerand, Jr., and R. K. Flavin (1964), *Proceedings of the Third International Conference on the Physics of Electronic and Atomic Collisions, London*, p. 59 (North-Holland Publishing Co., Amsterdam).

D. N. Mirlin, G. E. Pikus, and V. G. Yur'ev (1962), *Zh. Tekh. Fiz.* **32**, 766 [English transl.: *Soviet Phys.-Tech. Phys.* **7**, 559].

A. C. G. Mitchell and M. W. Zemansky (1934), *Resonance Radiation and Excited Atoms* (University Press, Cambridge).

M. H. Mittleman and V. P. Myerscough (1966), *Phys. Letters* **23**, 545.

M. H. Mittleman and R. E. von Holdt (1965), *Phys. Rev.* **140**, A726.

B. L. Moiseiwitsch (1962), in *Atomic and Molecular Processes* (Ed. D. R. Bates), Chapter 9 (Academic Press, New York).

P. M. Morse, W. P. Allis, and E. S. Lamar (1935), *Phys. Rev.* **48**, 412.

A. L. Morsell (1967), *Phys. Fluids* **10**, 2171.

G. J. Mullaney and N. R. Dibelius (1961), *ARS J.* **31**, 1575.

G. J. Mullaney, P. H. Kydd, and N. R. Dibelius (1961), *J. Appl. Phys.* **32**, 668.

H. M. Musal, Jr. (1969), *Proc. IEEE* **57**, 98.

T. Nagy, L. Nagy, and S. Dési (1960), *Nucl. Instr. Methods* **8**, 327.

M. S. Naidu and A. N. Prasad (1968), *J. Phys. D* **1**, 763.

K. V. Narasinga Rao and R. L. Taylor (1967), *Bull. Am. Phys. Soc.*, Ser. II, **12**, 810

K. V. Narasinga Rao, J. T. Verdeyen, and L. Goldstein (1961), *Proc. IRE* **49**, 1877.

D. R. Nelson and F. J. Davis (1969), *J. Chem. Phys.* **51**, 2322.

R. A. Nielsen (1936), *Phys. Rev.* **50**, 950.

R. A. Nielsen and N. E. Bradbury (1937), *Phys. Rev.* **51**, 69.

W. L. Nigham (1967), *Phys. Fluids* **10**, 1085.

J. F. Nolan and A. V. Phelps (1965), *Phys. Rev.* **140**, A792.

J. H. Noon, P. R. Blaszuk, and E. H. Holt (1968), *J. Appl. Phys.* **39**, 9.

J. H. Noon and E. H. Holt (1966), *Phys. Rev.* **150**, 121.

C. E. Normand (1930), *Phys. Rev.* **35**, 1217.

K. J. Nygaard (1967a), *Phys. Rev.* **157**, 138.

K. J. Nygaard (1967b), *Phys. Letters* **25A**, 567.

T. F. O'Malley (1963), *Phys. Rev.* **130**, 1020.

T. F. O'Malley, L. Rosenberg, and L. Spruch (1962), *Phys. Rev.* **125**, 1300.

J. L. Pack and A. V. Phelps (1961), *Phys. Rev.* **121**, 798.

J. L. Pack and A. V. Phelps (1966), *J. Chem. Phys.* **44**, 1870; **45**, 4316.

J. L. Pack, R. E. Voshall, and A. V. Phelps (1962), *Phys. Rev.* **127**, 2084.

H. J. Pain and P. R. Smy (1960), *J. Fluid Mech.* **9**, 390.

J. H. Parker, Jr. (1963), *Phys. Rev.* **132**, 2096.

J. H. Parker, Jr., and J. J. Lowke (1969), *Phys. Rev.* **181**, 290.

S. L. Paveri-Fontana (1970), *Lett. Nuovo Cimento* **4**, 1259.

S. S. Penner (1959), *Quantitative Molecular Spectroscopy and Gas Emissivities* (Addison-Wesley Publishing Co., Reading, Mass.).

K. B. Persson (1961), *J. Appl. Phys.* **32**, 2631.

A. V. Phelps (1960), *J. Appl. Phys.* **31**, 1723.

A. V. Phelps (1968), *Rev. Mod. Phys.* **40**, 399.

A. V. Phelps (1969), *Can. J. Phys.* **47**, 1783.

A. V. Phelps (1970), *Bull. Am. Phys. Soc.*, Ser. II, **15**, 423.

A. V. Phelps, O. T. Fundingsland, and S. C. Brown (1951), *Phys. Rev.* **84**, 559.

A. V. Phelps, J. L. Pack, and L. S. Frost (1960), *Phys. Rev.* **117**, 470.

M. Pirani and J. Yarwood (1961), *Principles of Vacuum Engineering* (Reinhold Publishing Corp., New York).

G. H. Plantinga (1961), *Philips Res. Rept.* **16**, 462.

W. J. Pollock (1968), *Trans. Faraday Soc.* **64**, 2919.

A. J. Postma (1969a), *Physica* **43**, 465.

A. J. Postma (1969b), *Physica* **44**, 38.

A. N. Prasad and J. D. Craggs (1962), in *Atomic and Molecular Processes* (Ed. D. R. Bates), Chapter 6 (Academic Press, New York).

A. N. Prasad and G. P. Smeaton (1967), *Brit. J. Appl. Phys.* **18**, 371.

E. M. Purcell (1947), in *Technique of Microwave Measurements* (Ed. C. G. Montgomery), Chapter 8 (McGraw-Hill Book Co., New York).

C. Raja Rao and G. R. Govinda Raju (1971), *J. Phys. D* **4**, 769.

H. Ramien (1931), *Z. Physik* **70**, 353.

C. Ramsauer and R. Kollath (1929), *Ann. Physik* **3**, 536.

C. Ramsauer and R. Kollath (1932), *Ann. Physik* **12**, 529, 837.

P. A. Redhead, J. B. Hobson, and E. V. Kornelsen (1968), *The Physical Basis of Ultrahigh Vacuum* (Chapman and Hall, London).

J. A. Rees (1964), *Australian J. Phys.* **17**, 462.

J. A. Rees (1965), *Australian J. Phys.* **18**, 41.

J. A. Rees and R. L. Jory (1964), *Australian J. Phys.* **17**, 307.

R. W. Roberts and T. A. Vanderslice (1963), *Ultrahigh Vacuum and Its Applications* (Prentice-Hall, Englewood Cliffs, N.J.).

A. G. Robertson (1971), *Australian J. Phys.* **24**, 445.

B. B. Robinson and I. B. Bernstein (1962), *Ann. Phys.* **18**, 110.

D. Roehling (1963), *Advan. Energy Conversion* **3**, 69.

D. J. Rose (1956), *Phys. Rev.* **104**, 273.

F. Rostas, A. K. Bhattacharya, and J. H. Cahn (1963), *Phys. Rev.* **129**, 495.

N. Rynn (1964), *Phys. Fluids* **7**, 284.

H. Ryżko (1966), *Arkiv Fysik* **32**, 1.

F. Sakao and H. Sato (1969), *Phys. Fluids* **12**, 2063.

D. H. Sampson (1965), *Phys. Rev.* **137**, A4.

D. H. Sampson and R. C. Mjolsness (1965), *Phys. Rev.* **140**, A1466.

P. Savic and G. T. Boult (1962), *J. Sci. Instr.* **39**, 258.

H. Schlumbohm (1965), *Z. Physik* **182**, 306, 317.

H. J. Schmitt, G. Meltz, and P. J. Freyheit (1965), *Phys. Rev.* **139,** A1432.

L. Schott (1968), in *Plasma Diagnostics* (Ed. W. Lochte-Holtgreven), Chapter 11 (North-Holland Publishing Co., Amsterdam).

G. J. Schulz (1958), *Phys. Rev.* **112,** 150.

G. J. Schulz (1964), *Phys. Rev.* **135,** A988.

G. J. Schulz and J. T. Dowell (1962), *Phys. Rev.* **128,** 174.

S. Schweitzer and M. Mitchner (1966), *AIAA J.* **4,** 1012.

T. Sekiguchi and R. G. Herndon (1958), *Phys. Rev.* **112,** 1.

B. Sherman (1960), *J. Math. Analysis Application* **1,** 342.

M. Shimizu (1963), *J. Phys. Soc. Japan* **18,** 811.

I. P. Shkarofsky (1961), *Can. J. Phys.* **39,** 1619.

I. P. Shkarofsky (1963), *Can. J. Phys.* **41,** 1753, 1776.

I. P. Shkarofsky (1968), *Plasma Phys.* **10,** 169.

I. P. Shkarofsky, M. P. Bachynski, and T. W. Johnston (1961), in *Electromagnetic Effects of Re-entry,* p. 24 (Pergamon Press, London).

I. P. Shkarofsky, T. W. Johnston, and M. P. Bachynski (1966), *The Particle Kinetics of Plasmas* (Addison-Wesley Publishing Co., Reading, Mass.).

A. Simon (1955), *Phys. Rev.* **98,** 317.

Y. Singh (1970), *J. Phys. B* **3,** 1222.

M. F. Skinker (1922), *Phil. Mag.* **44,** 994.

M. F. Skinker and J. V. White (1923), *Phil. Mag.* **46,** 630.

H. R. Skullerud (1969), *J. Phys. B* **2,** 696.

M. S. Sodha and P. K. Kaw (1969), in *Advances in Electronics and Electron Physics* (Ed. L. Marton), Vol. 27, p. 187 (Academic Press, New York).

D. Spence and G. J. Schulz (1970), *Phys. Rev. A* **2,** 1802.

L. Spitzer, Jr. (1952), *Astrophys. J.* **16,** 299.

L. Spitzer, Jr. (1956), *Physics of Fully Ionized Gases* (Interscience Publishers, New York).

L. Spitzer, Jr., and R. Härm (1953), *Phys. Rev.* **89,** 977.

R. C. Stabler (1963), *Phys. Rev.* **131,** 679.

H. A. Steinhertz (1963), *Handbook of High Vacuum Engineering* (Reinhold Publishing Corp., New York).

A. Stevenson (1952), *Rev. Sci. Instr.* **23,** 93.

M. W. P. Strandberg, M. Tinkham, I. H. Solt, Jr., and C. F. Davis, Jr. (1956), *Rev. Sci. Instr.* **27,** 596.

J. A. Stratton (1941), *Electromagnetic Theory* (McGraw-Hill Book Co., New York).

D. E. Strogryn and A. P. Strogryn (1966), *Mol. Phys.* **11,** 371.

P. D. Strum (1958), *Proc. IRE* **46,** 93.

E. J. Stubbe (1968), *Proc. IEEE* **56,** 1483.

G. W. Sutton and A. Sherman (1965), *Engineering Magnetohydrodynamics* (McGraw-Hill Book Co., New York).

W. C. Taft, K. C. Stotz, and E. H. Holt (1963), *IEEE Trans. Instr. Meas.* **IM-12,** 90.

K. Takayanagi (1965), *J. Phys. Soc. Japan* **20,** 562.

K. Takayanagi (1966), *J. Phys. Soc. Japan* **21,** 507.

K. Takayanagi and Y. Itikawa (1968), *J. Phys. Soc. Japan* **24**, 160.

K. Takayanagi and Y. Itikawa (1970), in *Advances in Atomic and Molecular Physics* (Ed. R. Bates and I. Estermar), Vol. VI, p. 105 (Academic Press, New York).

S. Takeda (1961), *J. Phys. Soc. Japan* **16**, 1267.

S. Takeda and A. A. Dougal (1960), *J. Appl. Phys.* **31**, 412.

S. Takeda and E. H. Holt (1959), *Rev. Sci. Instr.* **30**, 722.

S. Takeda and M. Roux (1961), *J. Phys. Soc. Japan* **16**, 95.

S. Takeda, T. Tsukishima, and A. Funahashi (1966), *Proceedings of the Seventh International Conference on Phenomena in Ionized Gases, Belgrade*, Vol. II, p. 767 (Gradevinska Knijga Publishing House, Belgrade).

J. D. Teare (1963), in *Ionization in High-Temperature Gases* (Ed. K. E. Shuler), p. 217 (Academic Press, New York).

F. L. Tevelow (1967), *J. Appl. Phys.* **38**, 1765.

F. L. Tevelow (1972), Harry Diamond Labs., Rept. HDL-TR-1604 (Washington, D.C.).

F. L. Tevelow and F. J. Tischer (1971), Harry Diamond Labs., Rept. HDL-TR-1544 (Washington, D.C.).

R. Tice and D. Kivelson (1967), *J. Chem. Phys.* **46**, 4743, 4748.

J. S. Townsend (1915), *Electricity in Gases* (Clarendon Press, Oxford).

J. S. Townsend (1948), *Electrons in Gases* (Hutchison Scientific and Technical Publishers, London).

J. S. Townsend and V. A. Bailey (1921), *Phil. Mag.* **42**, 873.

J. S. Townsend and V. A. Bailey (1922), *Phil. Mag.* **43**, 593; **44**, 1033.

J. S. Townsend and V. A. Bailey (1923), *Phil. Mag.* **46**, 657.

J. S. Townsend and H. T. Tizard (1913), *Proc. Roy. Soc.* **A88**, 336.

D. G. Truhlar and J. K. Rice (1970), *J. Chem. Phys.* **52**, 4480.

T. Tsukishima and S. Takeda (1962), *J. Appl. Phys.* **33**, 3290.

J. E. Turner and K. Fox (1966), *Phys. Letters* **23**, 547.

P. E. Vandenplas (1968), *Electron Waves and Resonances in Bounded Plasmas* (John Wiley & Sons, London).

V. A. J. Van Lint (1959), General Atomic Division of General Dynamics Corp., Rept. TR 59-43 (San Diego, Calif.).

V. A. J. Van Lint (1964), *IEEE Trans. Nucl. Sci.* **NS-11**, 266.

L. J. Varnerin, Jr. (1951), *Phys. Rev.* **84**, 563.

R. N. Varney and L. H. Fischer (1968), in *Atomic and Electron Physics* (Ed. L. Marton), Vol. 7, Part B, p. 29 (Academic Press, New York).

R. V. Vasil'eva, K. V. Donskoi, B. M. Dobrynin, J. Kh. Manerov, G. K. Tumakaev, and V. A. Shingarkina (1970), *Zh. Tekh. Fiz.* **40**, 605 [English transl.: *Soviet Phys.-Tech. Phys.* **15**, 467].

G. E. Veatch, J. T. Verdeyen, and J. H. Cahn (1966), *Bull. Am. Phys. Soc.*, Ser. II, **11**, 496.

J. R. Viegas (1971), *Phys. Fluids* **14**, 541.

J. R. Viegas and C. H. Kruger (1969), *Phys. Fluids* **12**, 2050.

Yu. M. Volkov, O. A. Zinov'ev, and D. D. Malyuta (1968), *Teplofiz. Vysokikh Temp.* **6**, 209 [English transl.: *High Temp.* **6**, 207].

E. B. Wagner, F. J. Davis, and G. S. Hurst (1967), *J. Chem. Phys.* **47**, 3138.

H. B. Wahlin (1923), *Phys. Rev.* **21**, 517.

H. B. Wahlin (1924), *Phys. Rev.* **23**, 169.

H. B. Wahlin (1926), *Phys. Rev.* **27**, 588.

H. B. Wahlin (1931), *Phys. Rev.* **37**, 260.

A. L. Ward (1966), *Proceedings of the Seventh International Conference on Phenomena in Ionized Gases, Belgrade*, Vol. I, p. 65 (Gradevinska Knjiga Publishing House, Belgrade).

R. W. Warren and J. H. Parker, Jr. (1962), *Phys. Rev.* **128**, 2661.

L. A. Weaver and R. J. Freiberg (1968), *J. Appl. Phys.* **39**, 1550, 4283.

H. D. Weymann (1969), *Phys. Fluids* **12**, 1193.

J. H. Whealton and S. B. Woo (1969), *J. Appl. Phys.* **40**, 3060.

L. Wijnberg (1966), *J. Chem. Phys.* **44**, 3864.

B. L. Wright (1966), M.I.T. Res. Lab. Electron., Q.P.R. 80, p. 99 (Cambridge, Mass.).

B. L. Wright and G. Bekefi (1971), *Phys. Fluids* **14**, 1764, 1773.

S. Yano, S. Matsunaga, T. Hiramoto, and H. Shirakata (1964), Symposium international sur la production MHD d'énergie électrique, Paris, Session 26, Paper 26.

J. R. Young (1963), *Rev. Sci. Instr.* **34**, 891.

R. A. Young (1970), *Phys. Rev. A* **2**, 1983.

I. P. Zapesochnyi (1967), *Teplofiz. Vysokikh Temp.* **5**, 7 [English transl.: *High Temp.* **5**, 6].

B. Zauderer (1970), *AIAA J.* **8**, 645.

Author Index

Subject Index

Except when otherwise noted, collision and transport parameters refer always to electrons.

Absorption coefficient, 272, 273, 279, 281, 282

Acetylene, deuterated form, 423
  drift velocity, 423
  momentum transfer collision cross section data, 426

Adiabatic method of cross section calculation, 43

Afterglow, electron energy decay, 234, 235, 244, 324, 384, 385
  measurements, of conductivity ratio, 233-235, 244, 247, 249, 254, 259-265, 300, 322, 324, 330, 331, 342, 348, 352, 353, 366, 367, 384, 385, 395, 413, 417, 425, 439
  of cyclotron resonance, 301-306, 325, 354
  of electron diffusion, 222, 225, 228, 229, 321
  of radiation temperature, 279, 283-285, 384, 385, 425
  of thermal coefficients, 288-297, 307, 309, 325, 333, 385, 400, 425, 440-444
  recombination light, 291-293, 295-297, 307, 309, 325

Air and nitrogen-oxygen mixtures, attachment frequency determination, 157, 158
  diffusion to mobility ratio, 433, 434
  drift velocity, 433, 434

effective collision frequencies, 435, 436
magnetic drift velocity, 435
shock-heated, electrical conductivities, 435

Alpha-particles, 132, 135, 136, 138, 143, 163, 164, 167-170, 194, 195, 314, 335, 337, 363, 410, 422

Ambipolar diffusion, 221, 222
  in a magnetic field, 232, 233

Ammonia, cyclotron resonance, in active nitrogen mixtures, 425
  diffusion to mobility ratio, 424
  dipole moment, 429
  drift velocity, 422
  momentum transfer collision cross section data, 429

Anode, secondary emission by photons at, 183, 184. *See also* Electron collector

Argon, conductivity ratio, 342-344, 347, 439
  in oxygen mixtures, 395
  cyclotron resonance, 344, 345, 347
  in active nitrogen mixtures, 344
  dc conductivity, 99, 341, 342, 437, 438
  in cesium seeded argon, 204, 415
  in potassium seeded argon, 204, 415-417, 422
  determination of collision parameters, 345-347
  diffusion coefficient, 341, 347